Methods in Cell Biology

Correlative Light and Electron Microscopy II

Volume 124

Series Editors

Leslie Wilson
Department of Molecular, Cellular and Developmental Biology
University of California
Santa Barbara, California

Phong Tran
University of Pennsylvania
Philadelphia, USA &
Institut Curie, Paris, France

Methods in Cell Biology

Correlative Light and Electron Microscopy II

Volume 124

Edited by

Thomas Müller-Reichert

Experimental Center,
Medical Faculty Carl Gustav Carus,
Technische Universität Dresden,
Dresden, Germany

Paul Verkade

Wolfson Bioimaging Facility,
Schools of Biochemistry and Physiology & Pharmacology,
University of Bristol, Bristol, United Kingdom

ELSEVIER

AMSTERDAM • BOSTON • HEIDELBERG • LONDON
NEW YORK • OXFORD • PARIS • SAN DIEGO
SAN FRANCISCO • SINGAPORE • SYDNEY • TOKYO
Academic Press is an imprint of Elsevier

Academic Press is an imprint of Elsevier
32 Jamestown Road, London NW1 7BY, UK
525 B Street, Suite 1800, San Diego, CA 92101-4495, USA
225 Wyman Street, Waltham, MA 02451, USA
The Boulevard, Langford Lane, Kidlington, Oxford OX5 1GB, UK

First edition 2014

Notices
Knowledge and best practice in this field are constantly changing. As new research and experience broaden our understanding, changes in research methods, professional practices, or medical treatment may become necessary.

Practitioners and researchers must always rely on their own experience and knowledge in evaluating and using any information, methods, compounds, or experiments described herein. In using such information or methods they should be mindful of their own safety and the safety of others, including parties for whom they have a professional responsibility.

To the fullest extent of the law, neither the Publisher nor the authors, contributors, or editors, assume any liability for any injury and/or damage to persons or property as a matter of products liability, negligence or otherwise, or from any use or operation of any methods, products, instructions, or ideas contained in the material herein.

ISBN: 978-0-12-801075-4
ISSN: 0091-679X

For information on all Academic Press publications
visit our website at store.elsevier.com

Contents

Contributors

Alin Achim
Visual Information Laboratory, University of Bristol, Bristol, United Kingdom

Claude Antony
European Molecular Biology Laboratory, Meyerhofstrasse 1, Heidelberg, Germany; Institut de Génétique et Biologie Moléculaire et Cellulaire, 1 rue Laurent Fries, BP 10142, and Department of Structural Biology and Genomics, Institut de Génétique et de Biologie Moléculaire et Cellulaire, INSERM, U964, CNRS UMR7104, Illkirch, France

Vera Bandmann
Biomineralization, INM-Leibniz Institute for New Materials, Saarbrücken, Germany

Sebastian Banhart
Junior Research Group Sexually Transmitted Bacterial Pathogens (NG 5), Robert Koch Institute, Berlin, Germany

Norbert Bannert
Centre for HIV and other Retroviruses (FG 18), Robert Koch Institute, Berlin, Germany

Derron Bishop
Department of Physiology, Indiana University School of Medicine-Muncie, Muncie, Indiana, USA

Erik Bos
Department of Molecular Cell Biology, Section Electron Microscopy, Leiden University Medical Center, Leiden, The Netherlands

Filip Braet
School of Medical Sciences (Discipline of Anatomy and Histology)—The Bosch Institute; Charles Perkins Centre, and Australian Centre for Microscopy & Microanalysis, The University of Sydney, Sydney, NSW, Australia

Robert J. Bryson-Richardson
School of Biological Sciences, Faculty of sciences, Monash University, Melbourne, Victoria, Australia

Raffaella Carzaniga
Electron Microscopy Unit, London Research Institute, Cancer Research UK, London, United Kingdom

Katherine Celler
Molecular Biotechnology, Institute of Biology Leiden, Leiden University, PO Box 9505, Leiden, The Netherlands

Delfine Cheng
School of Medical Sciences (Discipline of Anatomy and Histology)—The Bosch Institute, The University of Sydney, Sydney, NSW, Australia

Lucy M. Collinson
Electron Microscopy Unit, London Research Institute, Cancer Research UK, London, United Kingdom

Niels de Jonge
Innovative Electron Microscopy, INM-Leibniz Institute for New Materials, Saarbrücken, Germany

Cédric Delevoye
Centre de Recherche, Institut Curie, and Structure and Membrane Compartments, CNRS UMR144, Paris, France

Kyle C. Dent
Diamond Light Source, Harwell Science and Innovation Campus, Didcot, and Oxford Particle Imaging Centre, Division of Structural Biology, Wellcome Trust Centre for Human Genetics, University of Oxford, Oxford, United Kingdom

Marie-Charlotte Domart
Electron Microscopy Unit, London Research Institute, Cancer Research UK, London, United Kingdom

Elizabeth Duke
Diamond Light Source, Harwell Science and Innovation Campus, Didcot, Oxfordshire, United Kingdom

Jeroen C. Eikenboom
Department of Thrombosis and Hemostasis, Einthoven Laboratory for Experimental Vascular Medicine, Leiden University Medical Center, Leiden, The Netherlands

Frank G.A. Faas
Department of Molecular Cell Biology, Section Electron Microscopy, Leiden University Medical Center, Leiden, The Netherlands

Isabel Fernández de Castro
Cell Structure Laboratory, Centro Nacional de Biotecnología, Consejo Superior de Investigaciones Científicas (CNB-CSIC), Cantoblanco, Madrid, Spain

Pascal Fua
Computer Vision Laboratory, EPFL, Lausanne, Switzerland

Kimberley H. Gibson
Max Planck Institute of Molecular Cell Biology and Genetics, Dresden, Germany

Kay Grünewald
Oxford Particle Imaging Centre, Division of Structural Biology, Wellcome Trust Centre for Human Genetics, University of Oxford, Oxford, United Kingdom

Christoph Hagen
Oxford Particle Imaging Centre, Division of Structural Biology, Wellcome Trust Centre for Human Genetics, University of Oxford, Oxford, United Kingdom

Xavier Heiligenstein
European Molecular Biology Laboratory, Meyerhofstrasse 1, Heidelberg, Germany; Institut Curie, Centre de Recherche, and Structure and Membrane Compartments, CNRS UMR144, Paris, France

Jeffrey Henriquez
School of Medical Sciences (Discipline of Anatomy and Histology)—The Bosch Institute, The University of Sydney, Sydney, NSW, Australia

Lorna Hodgson
School of Biochemistry, University of Bristol, Bristol, United Kingdom

Anthony Holtmaat
Department of Neuroscience, University Medical Center (CMU), University of Geneva, Geneva, Switzerland

Jacob P. Hoogenboom
Department of Imaging Physics, Faculty of Applied Sciences, Delft University of Technology, Delft, The Netherlands

William A. Horton
Research Department, Shriners Hospital for Children, Portland, Oregon, USA

Ilse Hurbain
Centre de Recherche, Institut Curie; Structure and Membrane Compartments, CNRS UMR144, and Cell and Tissue Imaging Facility (PICT-IBiSA), CNRS UMR144, Paris, France

Minh Huynh
Australian Centre for Microscopy & Microanalysis, The University of Sydney, Sydney, NSW, Australia

Anne Jorstad
Computer Vision Laboratory, EPFL, Lausanne, Switzerland

Douglas R. Keene
Research Department, Shriners Hospital for Children, Portland, Oregon, USA

Martin Kerschensteiner
Research Unit Therapy Development, Institute of Clinical Neuroimmunology, Ludwig-Maximilian University, and Munich Center for Systems Neurology (SyNergy), Nordallee, Munich, Germany

Graham W. Knott
BioEM Facility, Centre of Electron Microscopy, EPFL, Lausanne, Switzerland

Roman I. Koning
Department of Molecular Cell Biology, Section Electron Microscopy, Leiden University Medical Center, Leiden, The Netherlands

Abraham J. Koster
Department of Molecular Cell Biology, Section Electron Microscopy, Leiden University Medical Center, Leiden, The Netherlands

Marko Lampe
European Molecular Biology Laboratory, and Department of Translational Pulmonology, Translational Lung Research Center (TLRC), University of Heidelberg, Heidelberg, Germany

Michael Laue
Advanced Light and Electron Microscopy (ZBS 4), Robert Koch Institute, Berlin, Germany

Nalan Liv
Department of Imaging Physics, Faculty of Applied Sciences, Delft University of Technology, Delft, The Netherlands

Bohumil Maco
BioEM Facility, Centre of Electron Microscopy, and Computer Vision Laboratory, EPFL, Lausanne, Switzerland

Kazimierz Madela
Advanced Light and Electron Microscopy (ZBS 4), Robert Koch Institute, Berlin, Germany

Judith Mantell
Wolfson Bioimaging Facility, University of Bristol, Bristol, United Kingdom

Jana Meissner
Max Planck Institute of Molecular Cell Biology and Genetics, Dresden, Germany

Thomas Misgeld
Munich Center for Systems Neurology (SyNergy), and Institute of Neuronal Cell Biology, Technische Universität München, Nordallee, Munich, Germany

Marjon J. Mourik
Department of Molecular Cell Biology, Section Electron Microscopy, Leiden University Medical Center, Leiden, The Netherlands

David Nam
Visual Information Laboratory, University of Bristol, Bristol, United Kingdom

Ivana Nikic
Research Unit Therapy Development, Institute of Clinical Neuroimmunology, Ludwig-Maximilian University, Munich, Nordallee, Germany

Viola M.J. Oorschot
Monash Micro Imaging, Monash University, Melbourne, Victoria, Australia

Benjamin S. Padman
Department of Biochemistry and Molecular Biology, and Monash Micro Imaging, Monash University, Melbourne, Victoria, Australia

Diana B. Peckys
Innovative Electron Microscopy, INM-Leibniz Institute for New Materials, Saarbrücken, Germany

Christopher J. Peddie
Electron Microscopy Unit, London Research Institute, Cancer Research UK, London, United Kingdom

Ina Peristere
European Molecular Biology Laboratory, Meyerhofstrasse 1, Heidelberg, Germany

Janett Piesker
Advanced Light and Electron Microscopy (ZBS 4), Robert Koch Institute, Berlin, Germany

Georg Ramm
Department of Biochemistry and Molecular Biology, and Monash Micro Imaging, Monash University, Melbourne, Victoria, Australia

Graca Raposo
Centre de Recherche, Institut Curie; Structure and Membrane Compartments, and Cell and Tissue Imaging Facility (PICT-IBiSA), CNRS UMR144, Paris, France

Cristina Risco
Cell Structure Laboratory, Centro Nacional de Biotecnología, Consejo Superior de Investigaciones Científicas (CNB-CSIC), Cantoblanco, Madrid, Spain

Lynn Y. Sakai
Research Department, Shriners Hospital for Children, Portland, Oregon, USA

Jean Salamero
Centre de Recherche, Institut Curie; Cell and Tissue Imaging Facility (PICT-IBiSA), and Spatio-Temporal Modeling Imaging and Cellular Dynamics, CNRS UMR144, Paris, France

Laura Sanz-Sánchez
Cell Structure Laboratory, Centro Nacional de Biotecnología, Consejo Superior de Investigaciones Científicas (CNB-CSIC), Cantoblanco, Madrid, Spain

Gerry Shami
School of Medical Sciences (Discipline of Anatomy and Histology)—The Bosch Institute, The University of Sydney, Sydney, NSW, Australia

Nicholas R. Smith
Department of Cell and Developmental Biology, Oregon Health Sciences University, Portland, Oregon, USA

Tamar E. Sztal
School of Biological Sciences, Faculty of sciences, Monash University, Melbourne, Victoria, Australia

Erin M. Tranfield
Instituto Gulbenkian de Ciência, Rua da Quinta Grande 6, Oeiras, Portugal, and European Molecular Biology Laboratory, Meyerhofstrasse 1, Heidelberg, Germany

Sara L. Tufa
Research Department, Shriners Hospital for Children, Portland, Oregon, USA

Jack A. Valentijn
Department of Molecular Cell Biology, Section Electron Microscopy, Leiden University Medical Center, Leiden, The Netherlands

Karine M. Valentijn
Department of Molecular Cell Biology, Section Electron Microscopy, Leiden University Medical Center, Leiden, The Netherlands

Gilles P. van Wezel
Molecular Biotechnology, Institute of Biology Leiden, Leiden University, PO Box 9505, Leiden, The Netherlands

Jean-Marc Verbavatz
Max Planck Institute of Molecular Cell Biology and Genetics, Dresden, Germany

Paul Verkade
Wolfson Bioimaging Facility, Schools of Biochemistry and Physiology & Pharmacology, University of Bristol, Bristol, United Kingdom

Daniela Vorkel
Max Planck Institute of Molecular Cell Biology and Genetics, Dresden, Germany

Joost Willemse
Molecular Biotechnology, Institute of Biology Leiden, Leiden University, PO Box 9505, Leiden, The Netherlands

Melissa H. Wong
Department of Cell and Developmental Biology, Oregon Health Sciences University, Portland, Oregon, USA

Anja Zimmermann
Centre for HIV and other Retroviruses (FG 18), Robert Koch Institute, Berlin, Germany

Preface

I do not search, I find.

(Pablo Picasso)

Within a rather short period of time, we present a second *MCB* volume on Correlative Light and Electron Microscopy (CLEM). Immediately after the completion of the first volume, we realized that numerous additional new as well as established approaches—not mentioned in the first volume—needed to be presented so that interested colleagues could follow these detailed and well-established protocols. It was also by popular demand that we decided to look for techniques and approaches that were not published in the first volume. We are aware of the fact that also the second book on CLEM techniques is by no means exhaustive. However, we hope that this second volume—as well as the first one—will serve as an inspiring source for further research involving the combination of imaging modalities.

Since the publication of the first book, the motivation to apply correlative approaches continues to be strong. Motivations for CLEM include locating and imaging certain "spots" in cells and tissues, collecting dynamic as well as structural information about the same specimen, and/or simply increasing sample size and throughput. Certainly, we all wish to "find" without extensive "searching." Whatever the motivations are for each individual researcher, it is always the combination of using complementing imaging techniques that is expected to deliver more quantitative and qualitative information about a given specimen as opposed to just using one of the applied techniques.

Most of the current activities in CLEM presented in this book are centered around three key topics. First, it is the development of CLEM probes that attracts the most attention in the field. Probes that allow the detection of a labeled structure by fluorescence and electron microscopy are desperately needed. Second, much research has been done on the preservation of fluorescence, such as GFP, in plastic-embedded samples for electron microscopy. Again, these activities are mainly driven by the need to correlate specific signals with cellular ultrastructure. Third, it is the analysis of the correlative data that comes more and more into focus. Automated registration algorithms will be increasingly important to accurately correlate LM and EM data. All the techniques presented in this book "fall" into one or more of these three key topics.

In parallel to this second volume, we will continue to teach techniques for CLEM during a biannual practical course sponsored by EMBO. It is certainly our continued hope that both the practical course and this publication will help to further motivate cross talk and collaboration between all imaging worlds.

Thomas Müller-Reichert and Paul Verkade
Dresden and Bristol
July 1, 2014

ACKNOWLEDGMENTS

The authors would like to thank Sarah Lay (Elsevier) for her help in bringing this second volume on Correlative Light and Electron Microscopy to completion.

CHAPTER

1

Retracing in Correlative Light Electron Microscopy: Where is My Object of Interest?

Lorna Hodgson*, David Nam†, Judith Mantell‡, Alin Achim†, Paul Verkade*,‡,§

**School of Biochemistry, University of Bristol, Bristol, United Kingdom*

†Visual Information Laboratory, University of Bristol, Bristol, United Kingdom

‡Wolfson Bioimaging Facility, University of Bristol, Bristol, United Kingdom

§School of Physiology & Pharmacology, University of Bristol, Bristol, United Kingdom

CHAPTER OUTLINE

Abstract

Correlative light electron microscopy (CLEM) combines the strengths of light and electron microscopy in a single experiment. There are many ways to perform a CLEM experiment and a variety of microscopy modalities can be combined either on separate instruments or as completely integrated solutions. In general, however, a CLEM experiment can be divided into three parts: probes, processing, and analysis. Most of the existing technologies are focussed around the development and use of probes or describe processing methodologies that

Methods in Cell Biology, Volume 124, ISSN 0091-679X, http://dx.doi.org/10.1016/B978-0-12-801075-4.00001-X

explain or circumvent some of the compromises that need to be made when performing both light and electron microscopy on the same sample. So far, relatively little attention has been paid to the analysis part of CLEM experiments. Although it is an essential part of each CLEM experiment, it is usually a cumbersome manual process. Here, we briefly discuss each of the three above-mentioned steps, with a focus on the analysis part. We will also introduce an automated registration algorithm that can be applied to the analysis stage to enable the accurate registration of LM and EM images. This facilitates tracing back the right cell/object seen in the light microscope in the EM.

INTRODUCTION

The combination of light and electron microscopy using a correlative light electron microscopy (CLEM) approach is a powerful tool for the study of dynamic intracellular membrane trafficking events at high resolution (Brown et al., 2012; Kukulski et al., 2012; van Weering et al., 2010). Light microscopy (LM) involves the visualization of both live and fixed cells at low resolution using fluorescent markers, while electron microscopy (EM) can examine fixed cells at near atomic-level spatial resolution, revealing the detailed ultrastructure of the cell and the three-dimensional intracellular architecture. When utilized together in a single CLEM experiment, these two imaging modalities provide greater insight into subcellular protein localization and trafficking than that, which is achievable by each technique alone.

In order to localize proteins, they usually have to be marked with a specific label. Fluorescent proteins such as GFP and probes such as the Alexa Fluor dyes can be employed in fluorescence light microscopy (FLM) to reveal the localization and trafficking of a particular protein of interest within the cell. The movement of this protein and its colocalization with other labeled proteins can be examined to reveal information about its function. However, only fluorescently labeled proteins and/or cellular compartments are visible using FLM and, therefore, this technique is unable to reveal details about the cellular environment in which the protein is located (van Weering et al., 2010). Furthermore, FLM and indeed many conventional LM imaging techniques are limited by the resolution capabilities of the microscope and thus cannot resolve structures closer than ~150 nm.

The diffraction limitations of conventional LMs can be overcome using super-resolution imaging techniques, such as photoactivated localization microscopy, stochastic optical reconstruction microscopy, and 4Pi microscopy (Betzig et al., 2006; Hell & Stelzer, 1992; Huang, Babcock, & Zhuang, 2010; Rust, Bates, & Zhuang, 2006). The increased resolution afforded by these super-resolution techniques enable the study of single molecules and protein–protein interactions in live cells. However, even these techniques are incapable of visualizing process that occur at the nanometer scale and lack much needed information about the cellular environment and nature of the surrounding unlabeled structures (Betzig et al., 2006; Gustafsson, 2005; Schönle, Glatz, & Hell, 2000). In addition, although constantly improving, the live-imaging capabilities of most super-resolution LM techniques are currently limited.

By contrast, EM is a high-resolution technique capable of revealing the precise location of a labeled protein and the ultrastructure and morphology of the compartment in which the protein is localized (Slot & Geuze, 2007). However, due to sample preparation and properties of the transmission electron microscope (TEM), it is not possible to use this technique to study dynamic and/or transient trafficking events in live cells, other than to analyze many time-resolved snapshots and reconstruct the event retrospectively (Kukulski et al., 2012). In addition, due to the small sample size and thickness required for penetration by the electron beam, it is somewhat challenging and time consuming to identify rare events within a collection of cells.

CLEM thus bridges the gap between light and electron microscopy and can overcome many of the limitations of these individual techniques. It allows first for the identification of the protein of interest at the LM stage and for instance movement or approximate colocalization and then enables information about "reference space" to be gathered at the EM level. In general, CLEM experiments first involve the imaging of a cell labeled with a fluorescently tagged protein or probe using the LM followed by sample processing, embedding, and sectioning, before finally retracing and analyzing the same cell using the EM. These three steps: probes, processing, and analysis in the CLEM process are necessary for almost any CLEM experiment and we will briefly discuss the current status for each. For the analysis part, we will focus on CLEM using transmission EM. Relocation methods for the SEM are described in some of the other chapters in this book.

1.1 STEPS IN THE CLEM WORKFLOW

1.1.1 PROBES

A variety of CLEM-specific probes and markers have been developed over recent years to study protein trafficking and localization in a range of different applications (reviewed in Giepmans, 2008) (Fig. 1.1). In general, a suitable CLEM probe consists of a fluorescent component that is visible by LM and an electron-dense particle that is detectable using EM.

When designing a CLEM experiment, it is important to establish which CLEM probe is the most effective for the specific scientific question being asked. Ideally, both the fluorescent and electron-dense components should be linked to a single molecule to provide the most accurate correlation between the light and electron microscopy images. However, this is not always possible due to issues concerning fluorophore quenching by the gold particle, probe aggregation, and mistargeting (Brown & Verkade, 2010; Kandela & Albrecht, 2007). Therefore, a combinatorial approach is sometimes taken using two separate probes that interact together, most commonly a fluorescently tagged antibody and a colloidal gold marker.

CLEM probes can be grouped into two broad categories: invasive and noninvasive probes. Noninvasive probes either contain both LM and EM visible components conjugated to an antibody or ligand or a sequence encoding a fluorescent moiety that can be converted into an electron-dense particle. By contrast, invasive probes often

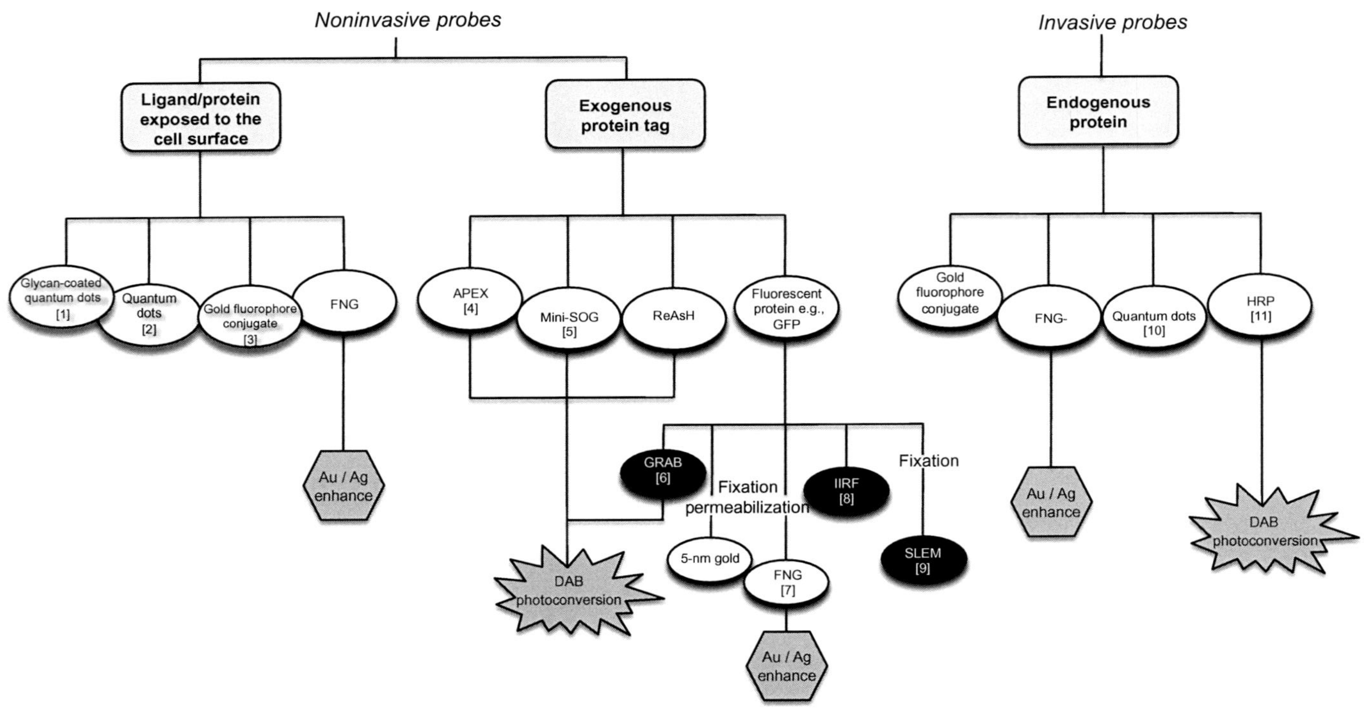

FIGURE 1.1

Flow diagram of available CLEM probes including noninvasive and invasive probes. Abbreviations: APEX, ascorbate peroxidase; DAB, 3,3′-diaminobenzidine; IRF, in-resin fluorescence; mini-SOG, mini singlet oxygen generator; GRAB, GFP recognition after bleaching; FNG, FluoroNanoGold; SLEM, sectioned light electron microscopy; HRP, horseradish peroxidase. [1] Benito-Alifonso (2014); [2] van Weering et al. (2010); [3] Brown and Verkade (2010); [4] Martell et al. (2012); [5] Shu et al. (2011); [6] Grabenbauer et al. (2005); [7] Polishchuk et al. (2012); [8] Peddie et al. (2014); [9] Hodgson et al. (2014); [10] Giepmans et al. (2005); and [11] Polishchuk et al. (2000).

contain separate fluorescent and electron-dense elements and require permeabilization of the cellular membrane, post fixation, to enable penetration into the cell interior. However, permeabilization of cellular membranes can alter the ultrastructure of the cell when examined using the EM, therefore, noninvasive probes are more desirable.

Noninvasive probes containing both LM and EM visible elements, such as quantum dots (QDs) can be attached to a ligand or antibody that binds to an extracellular tag. QDs are an exciting noninvasive probe that have generated much interest as suitable candidates for use in live labeling CLEM studies (Giepmans, Deerinck, Smarr, Jones, and Ellisman, 2005). They consist of a fluorescent, electron-dense crystal core containing cadmium telluride or cadmium selenide that is visible at both the LM and EM stage (Giepmans et al., 2005). QDs can be internalized into the cell when coupled to an antibody or targeting molecule which is exposed to cell surface, and although care must be taken (and controls included), they traffic in a similar manner to the uncoupled endogenous protein (Giepmans, 2006).

Alternatively, noninvasive probes can be linked to an endogenous protein through a genetic tag or linkage, such as the FlAsH/ReAsH system and mini singlet oxygen generator (mini-SOG). The fluorescent flavoprotein, mini-SOG, engineered from the blue-light photoreceptor, *Arabidopsis* phototropin 2 is a useful noninvasive biosynthetic CLEM probe for labeling endogenous proteins (Shu et al., 2011). It functions in a similar manner to the biarsenical derivative of resorufin, ReAsH (Gaietta et al., 2002) by generating singlet oxygen species that catalyze the polymerization of diaminobenzidine (DAB), following photooxidation. Mini-SOG is genetically tagged to the protein of interest, thus providing a true correlation between the light and electron microscopy images.

Furthermore, endogenous proteins can be directly tagged with a fluorescent moiety, such as GFP, which can be converted into an electron-dense precipitate by photooxidation of DAB (Grabenbauer et al., 2005; Meisslitzer-Ruppitsch et al., 2008). Alternatively, ectopically expressed proteins can be fused to GFP and a tag, such as the peroxidase protein, APEX or the metal-binding protein metallothionein, which are detectable by EM through photooxidation of DAB or through binding to gold clusters, respectively (Martell et al., 2012; Risco et al., 2012). These noninvasive approaches are useful for correlative studies because the EM component is essentially generated from the LM-visible tag or the LM and EM components are directly linked to the protein of interest, thus enabling true correlation of LM and EM images.

As described above, probes are often linked to a ligand or antibody that recognizes a protein or exofacial tag that is exposed to the extracellular milieu and subsequently internalized into the cell. Alternatively, when the antigen or protein of interest remains intracellular and the probe is membrane impermeable, a noninvasive probe must be added following fixation, permeabilization, and often ultrathin sectioning. An antibody linked to FluoroNanoGold is often used following chemical permeabilization of the cellular membrane by detergents or inorganic solvents (Robinson & Takizawa, 2009). Following permeabilization, this probe penetrates through the membrane into the cell due to its small size, but requires silver enhancement in order to visualize the gold in the EM.

1.1.2 PROCESSING

There are multiple different ways to perform a correlative experiment and a number of different microscopy modalities can be employed at both the LM and EM stages (Fig. 1.2). The experimental approach once again taken depends on the scientific question that is being asked and the CLEM probe that is being used.

If a transient or dynamic intracellular trafficking event such as the fusion of two compartments or the redistribution of a protein from one compartment to another is being studied, then a live-cell imaging approach may be taken at the LM stage. This involves visualizing and acquiring serial stacks of cells labeled with a fluorescently tagged protein, such as GFP, or a CLEM probe with a fluorescent moiety using a confocal microscope. Once the event of interest has been identified and recorded, the sample is then chemically fixed or immobilized by freezing, processed for EM and the same cell is relocated in the TEM.

When studying transient endocytic trafficking events, the rapid fixation of the specimen is imperative. Consequently, live-cell imaging combined with high-pressure freezing (HPF) or plunge freezing is desirable and retains good ultrastructure of the specimen (Brown, Mantell, Carter, Tilly, & Verkade, 2009; McDonald, Benavides, & Yacaman, 1998; Verkade, 2008). Chemical fixation is restricted by the speed of the diffusion of aldehydes and the formation of cross-links within the sample and may introduce artifacts into the sample (Kellenberger et al., 1992). It is, however, often used when further immunolabeling is required postembedding and sectioning.

An alternative CLEM approach may be taken when examining rare events within a population of cells. In this instance, the LM stage is required to identify the cell expressing the particular protein or phenotype and this can be done following EM processing and sectioning and does not require live-cell imaging. This is routinely used in combination with Tokuyasu immunolabeling of cryosections (Tokuyasu, 1973). The ultrathin sections are incubated with antibodies and gold markers and then imaged first by LM and then transferred to the TEM microscope (Hodgson, Tavare, & Verkade, 2014). The correlation of LM and EM images using this technique is much simpler as both imaging modalities are performed on the same sections and therefore any alterations in the appearance of the cells following embedding and sectioning will not interfere with the image registration. Furthermore, this approach allows for the examination of multiple cells and proteins within a sample. This technique has also been adapted to use in combination with live-cell imaging and to study cells *in situ*, using a flat embedding protocol (van Rijnsoever, Oorschot, & Klumperman, 2008).

1.1.3 ANALYSIS

The final stage of the CLEM protocol involves the analysis and relocation of the same cell identified using the LM, in the EM. This relocation is most frequently done manually, aided by finder grids, fiducial markers, and the surrounding cell features. The sample is first analyzed at low magnification in order to identify the particular cell of interest, before increasing the magnification to visualize the ultrastructure

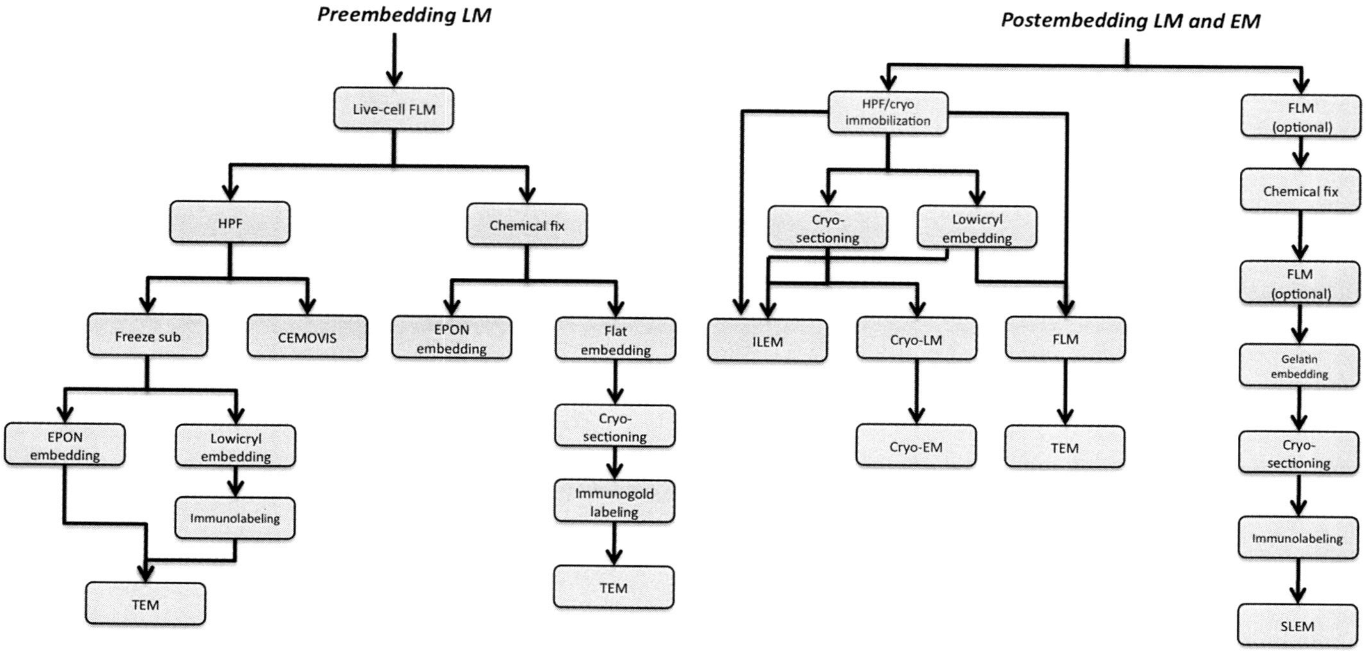

FIGURE 1.2

Workflow of different CLEM techniques that are available to use in combination with noninvasive and invasive probes. Abbreviations: FLM, fluorescence microscopy; HPF, high-pressure freezing; CEMOVIS, cryo-electron microscopy of vitreous section; TEM, transmission electron microscopy; ILEM, integrated light electron microscopy; SLEM, sectioned light electron microscopy.

within that cell and the precise location of the CLEM probe. The LM and EM images are finally registered by manually creating image overlays as described previously (Keene, Tufa, Lunstrum, Holden, & Horton, 2008).

The manual relocation and registration of LM and EM images can be challenging and time consuming and, therefore, it is important that accurate automated registration tools are developed. Distortions in the section created during the processing and sectioning of the specimen can result in differences in cellular shape, orientation, and features making the manual image registration extremely difficult. In addition, the delay between LM imaging of the event of interest and fixation or freezing of the sample may also affect the correlation of the images.

Many of these problems can be circumvented by embedding and sectioning samples prior to LM imaging, as done during a number of different studies including sectioned light electron microscopy (Hodgson et al., 2014), cryo-EM (Chapter 10) and lowicryl embedding of HPF samples (Kukulski et al., 2011; Peddie et al., 2014). Recent advancements have led to development of automated image registration processes for some of these techniques, which are discussed below.

Sectioned light electron microscopy using the Tokuyasu technique for labeling ultrathin sections has been successfully used in a variety of CLEM studies (Koster & Klumperman, 2003; Polishchuk et al., 2000). In this technique, sections are picked up on coated finder grids, which are visible in both the LM and EM and immunostained with various primary antibodies, fluorescent secondary antibodies, and gold markers. The ultrathin sections are then examined by FLM and the position of particular cells of interest on the finder grid are recorded and relocated back in the EM. Alternatively, GFP-expressing cells can be labeled with an anti-GFP primary antibody and gold-conjugated secondary (Hodgson et al., 2014). This method overcomes the issues of artifacts or alterations introduced during specimen processing and sectioning interfering with the registration of the two images, but still requires manual searching for the cell of interest and correlation of the EM and LM images.

LM imaging following sample preparation has also been applied to study vitrified samples in cryo-LM and EM using a specialized cryo-stage attachment (Sartori et al., 2007; van Driel, Valentijn, Valentijn, Koning, & Koster, 2009). In this method, cells are frozen and retained at cryogenic temperatures throughout imaging at LM and EM stages. This ensures that the samples remain unaltered between LM and EM imaging and no artifacts are introduced during the processing of the sample, such as during the fixation, dehydration or staining (Lucic et al., 2007). Furthermore, cryo-immobilization and processing for cryo-EM retains samples in a hydrated state preserves both fluorescence and ultrastructure (see also chapters 9 and 10). Recent studies have combined cryo-LM and EM with an automated 2D coordinate system in order to identify the cell of interest (Fukuda et al., 2013). This technique first involves the manual rough correlation of the LM image with a low-magnification EM image using grid bars to find the region of interest. Finally, reference points, such as holes in the support film and grid bar corners within the sample are identified and used to create a coordinate transform between the two images to identify the cell of interest in the EM.

Hydrophilic resins, such as lowicryl have been shown to allow preservation of GFP fluorescence following embedding and therefore are useful for FLM imaging after sample preparation (Keene et al., 2008). HPF samples are freeze-substituted and then embedded in lowicryl resin. This method has been successfully utilized to correlate fluorescence with ultrastructural information to gain insight into a variety of different biological systems and processes, including HIV particle attachment to mammalian cells, zebrafish embryo development, and endocytosis in budding yeast of HPF samples (Kukulski et al., 2011; Nixon et al., 2009). In these aforementioned studies, fluorescent microspheres that are visible in both LM and EM were used as fiducial markers in combination with an automated correlation procedure to identify the cell of interest from the LM images (Kukulski et al., 2011). The samples are incubated with the microspheres prior to imaging and the location of each fiducial marker is measured and used to calculate an optical transform between the EM and LM images. The optimal transform can then be applied to the areas of fluorescence in the LM image to identify the corresponding region in the EM. This automated process dramatically decreases the time taken to identify the cell of interest and provides near nanometer registration accuracy between the two images (Schorb & Briggs, 2013).

The techniques listed above overcome the issues of fixation and dehydration artifacts and sectioning distortions that can alter the appearance of the cell following LM imaging. However, they still require transfer of samples from LM to EM, which can lead to sample damage and increase time taken for relocation of the region of interest.

An integrated light and electron microscope (ILEM) allows for the imaging of LM and EM samples at cryogenic temperatures in a single microscope (Agronskaia et al., 2008; Faas et al., 2013). The integration of the two microscope modalities enables the imaging of the sample at LM followed immediately by the imaging at EM and removes the need for complicated correlation procedures. Regions of interest from the LM images can be used to calibrate the microscope allowing for precise and accurate location within the EM (<1 nm) (Faas et al., 2013). The drawback of using an integrated system is that the user cannot choose or tailor the imaging modalities to the experimental question as seen for a two-step imaging approach, thus making this procedure unsuitable for some applications where a particular microscopy technique (e.g., live imaging to determine the origin of an event) is required. In this particular instance, the reduced resolution of the LM in the ILEM (around 500 nm) may limit the identification of smaller structures. But it is an ideal tool to find objects of interest in a cryo-TEM where limited electron doses are essential (chapter 10).

As such, integrated microscopy but also LM imaging following embedding are less well suited to study dynamic and transient intracellular trafficking events as neither technique allows for live-cell FLM imaging. Furthermore, if an event is rare within a population of cells, it is often easier to identify this event first at the LM stage before sample processing. In general, the cell of interest is imaged live at the LM stage, the sample processed and the location of that same cell of interest recorded and retained throughout the sample processing before relocation in the EM. This is somewhat challenging when you consider that the LM samples are generally considerably larger than those imaged in the EM, corresponding to whole monolayers of cells and ultrathin sections of a small region of those cells, respectively. Therefore, the LM samples must be trimmed and sectioned prior to the EM relocation. It is not possible to use finder grids or fiducial markers to relocate the cell of interest in the EM and register the LM and EM images following live-cell imaging and sample processing. The identification of the cell of interest is instead achieved by examining the surrounding cell shapes, orientation, and nucleus location and comparing these to the LM images acquired before embedding and sectioning (see Fig. 1.3). This can be an extremely challenging and time-consuming process.

To overcome some of the challenges described above, we have developed algorithms that aid with retracing. In the remainder of this chapter, we focus on a CLEM workflow for cultured cells to retrace (very) rare events (see Fig. 1.4). We describe an automated image registration process that can be utilized to enable the accurate registration of LM and EM images. In the last part, we will describe the current status of this method and how it can be developed further.

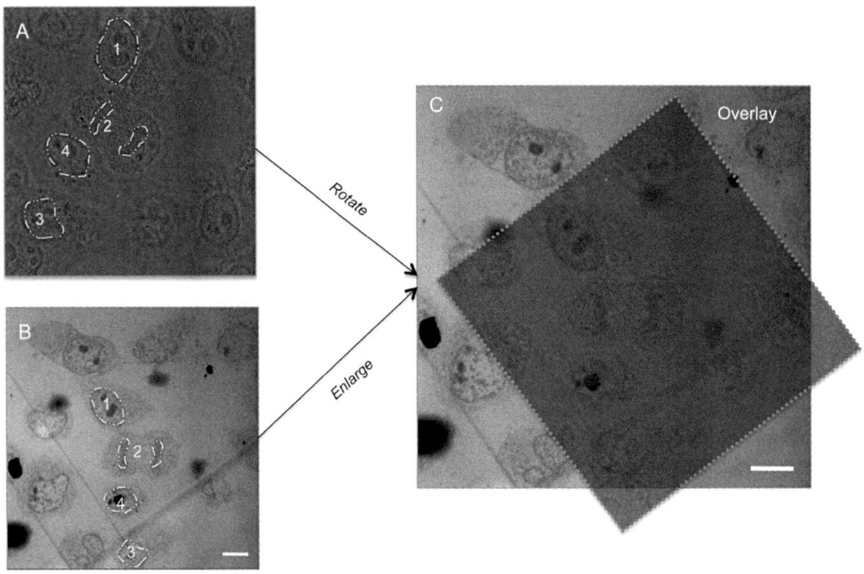

FIGURE 1.3

LM and EM manual overlay. (A) Bright-field LM images and (B) EM images can be (C) manually aligned post image acquisition. The shape and position of the nuclei (manually drawn in yellow (white in print version)) within each cell and the position of cells in relation to each other are used to orientate and identify the cell of interest. The two images are then overlayed by altering the transparency of the LM image, rotating the LM image and enlarging the EM image. Scale bar LM images and EM images = 10 μm.

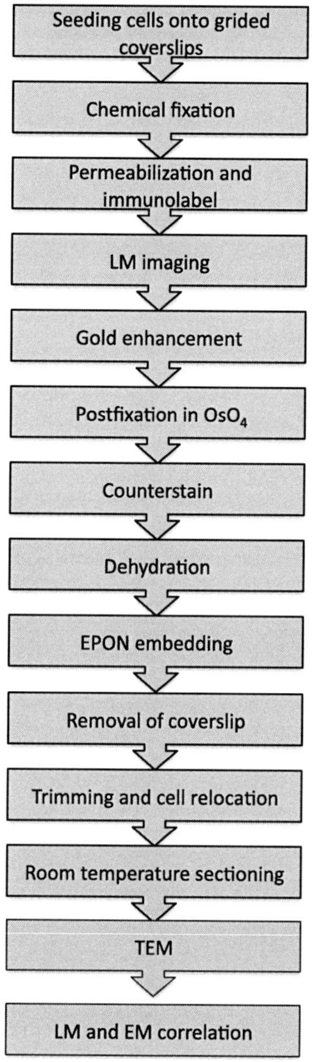

FIGURE 1.4

Experimental workflow.

1.2 METHODS

1.2.1 CELL CULTURE AND IMMUNOSTAINING

Glass-bottomed dishes (Mattek, P35G-2-14-C-GRID) or live-cell carriers (Brown et al., 2012; Verkade, 2008) containing gridded coverslips are routinely used to assist in the trimming and relocation of the cell of interest throughout sample processing.

For chemically fixed samples, this involves growing cells in an imaging dish containing a glass coverslip embossed with a finder grid pattern (van Weering et al., 2010), while cells for HPF-CLEM are grown on sapphire discs inside a live-cell carrier containing a carbon-coated finder pattern (Brown et al., 2012).

HeLa cells are seeded into a glass-bottomed imaging dish containing an embossed finder grid pattern at 400,000 cells per dish and left to adhere overnight. Rare events of interest such as cell division or specific stages within that process may be identified by fluorescence alone if detailed information of the ultrastructure is not required. If in addition, the localization of a protein is required, an immunogold-labeling step is included. Dishes are chemically fixed in 3% PFA containing 0.05% GA, permeabilized in 0.1% saponin before immunostaining with a primary antibody followed by FluoroNanoGold (Polishchuk, Polishchuk, Luini, MullerReichert, & Verkade, 2012; Polishchuk et al., 2000). Nuclear stains such as DAPI can be added to the cells before FLM imaging to visualize the nucleus which will later aid in the registration of LM and EM images.

Alternatively, if the antigen of interest is exposed to the extracellular media then cells may be labeled live using a noninvasive CLEM probe containing both the fluorescent tag and EM visible particle before fixation.

1.2.2 LIGHT MICROSCOPY IMAGING

Fluorescence LM imaging is performed using a confocal scanning light microscope. The cell of interest is first identified and then bright-field images of the position of the cell on the grid and fluorescence images of the nucleus and the localization of the CLEM probe are acquired (Fig. 1.5). Z-stack slices are then taken to reveal the localization of the fluorescent CLEM probe throughout the cell. Finally, an overview of the grid pattern is taken either by switching to a lower magnification lens or by acquiring multiple bright-field images of the area surrounding the cell of interest and stitching a mosaic together using the MosaicJ plug-in in FiJi imaging software. This final step is crucial when trimming the sample to size before sectioning and for retracing the cell of interest back in the EM.

1.2.3 SAMPLE PROCESSING FOR EM—COARSE ALIGNMENT

Following imaging the sample is processed and embedded in plastic resin. However, first the FluoroNanoGold particle must be enlarged so it is more visible by EM. This involves the enhancement of the 1.4 nm FluoroNanoGold using commercially available Gold or Silver enhancement kits (Polishchuk et al., 2012). The sample is post-fixed in 2% osmium tetroxide in phosphate buffer before washing with buffer and water. It is possible to store the samples overnight at 4 °C in H_2O at this point without significant alteration of the ultrastructure. Next, the samples are stained with 3% uranyl acetate in H_2O and dehydrated with a graded series of ethanol (70%, 80%, 90%, 96%, 100%, and 100%). The dish is then covered in EPON resin and left at room

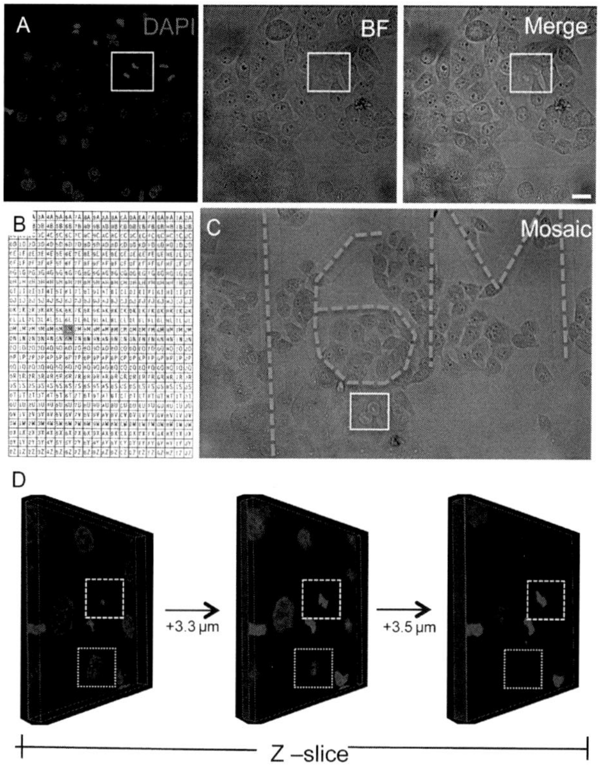

FIGURE 1.5

LM imaging of sample. (See the color plate.)

temperature for at least 2 h. The EPON is removed and replaced with fresh resin, which is left on the dish overnight at 60 °C. Finally, fresh resin is added to the center of the imaging dish and a resin-stub placed on top. The samples are then left at 60 °C for 24–48 h to allow the resin to polymerize.

Next, the coverslip and imaging dish must be removed before trimming and sectioning (Fig. 1.6). This is achieved by rapidly transferring the embedded dish between liquid nitrogen and boiling water to dissolve the glue holding the coverslip in place. The coverslip can then be removed using a razor blade, leaving an imprint of the finder grid in the hardened resin block. This imprint can act as a guide during trimming of the block when correlated with the LM images in order to relocate the cell of interest and reduce the size of the sample prior to serial sectioning (Polishchuk et al., 2000). The imaging dish is cut and excess resin is removed using pliers and then the resin block is trimmed to a pyramid, before cutting 70–300 nm serial sections and collecting the section ribbons on pioloform-coated EM slot grids.

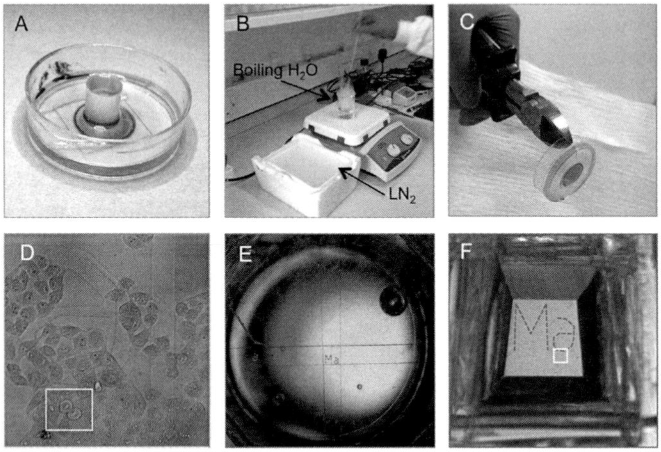

FIGURE 1.6

Example of trimming the block to size using finder grid. (See the color plate.)

1.2.4 AUTOMATED ALIGNMENT

Once the resin block has been trimmed and serial sections cut (either 70 nm for standard imaging or 300 nm for electron tomography), the sample is then examined in the TEM. The finder grid imprint is lost after the first section of the block is cut so the grid is no longer present to assist in the relocation of the cell of interest. The relocation is performed manually by comparing the orientation, position, and shape of the cells from the LM image, as demonstrated previously (see Fig. 1.3). The first few sections cut from the block should have retained the grid bar indent; therefore, it is often easier to find the cell of interest in these sections using the grid coordinates as a reference. Once the position of the cell of interest has been found in the first section on the grid, it is possible to use this as a guide to find the cell of interest in the other sections within the same ribbon. However, the imprint is not always obvious in the EM sections.

Thus, as discussed previously, finding back the cell and/or organelle of interest observed by LM in the EM can be a real challenge, especially when groups of cells have been imaged. Hence, we set out to develop automated solutions that would aid us in this process (Fig. 1.7). The method is based on the feature that when cells are seeded they distribute in unique patterns each time. Following image acquisition, the EM micrograph and LM images are automatically aligned using a feature-based image registration algorithm (Nam et al., 2014). This algorithm registers the EM image with the bright-field LM image, using the geometric centers (cell centroids) of cells in both images.

First, cell centroids are detected. Centroid detection is achieved by creating a voting image, using a cone-shaped voting kernel. The vertex of the kernel is oriented in the negative gradient direction (at specific key points from the image). The voting

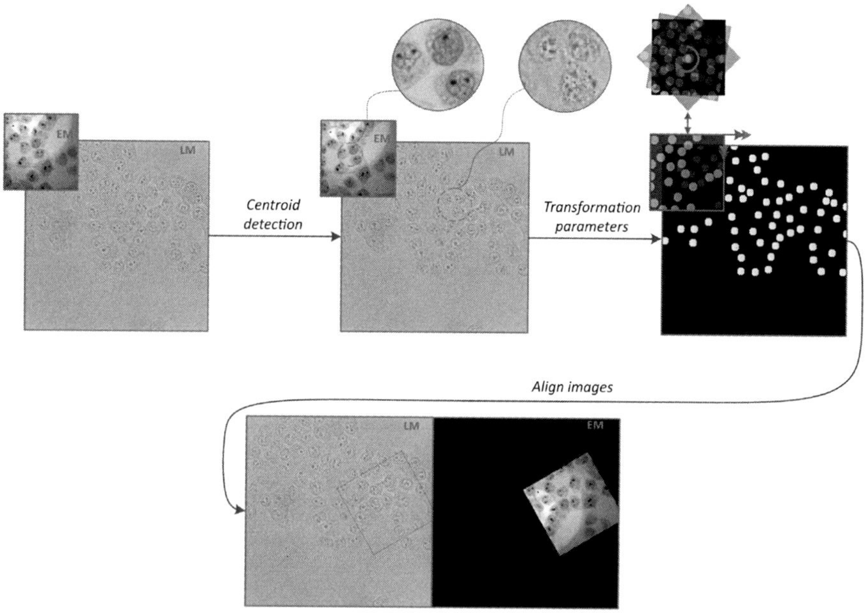

FIGURE 1.7

Workflow automated retracing. (See the color plate.)

image is then a summation of kernel contributions from each key point (Qi, Xing, Foran, & Yang, 2012). Three parameters define the voting area: r_{min}, r_{max}, and θ. r_{min} and r_{max} define the length of the voting kernel. Ideally, the radius of the cell should be halfway between r_{min} and r_{max}. r_{min} is set to be half the length of the cell radius and r_{max} to be 1.5 the length of the radius. The distance between these two values can be reduced to give a sharper voting image, but only if there is a lot of consistency in the observed cell size, and the cells appear very circular. Likewise, if there are larger variations in cell shape and size, we recommend increasing the distance between these two values. The angle θ is the angle at the vertex of the cone. We used 30°, while this value is suitable for most cases, decreasing its value can result in a sharper voting image (but with decreased robustness). Importantly, this method accounts for variation in the contrast between EM and LM images and section artifacts—such as dirt particles that would otherwise skew the calculation of the cell centroid.

The registration algorithm next calculates the transformation parameters using the cell centroid positions so that the EM and LM images align resulting in an LM and EM image overlay that is comparable to manually aligned images, but has been achieved more quickly and easily (Fig. 1.5). This is achieved by fitting disks onto cell center positions—a little smaller than the size of a cell, to add robustness—and then rotating and shifting the EM image across the LM image. The position with best overlap is used to obtain the correct parameters. The process is also repeated for the flipped EM image. Most importantly, the method is also able to account for the appearance or disappearance of nuclei in serial sections of the cell. This is a common occurrence in cultured cell monolayers, as some cells are tightly attached to the surface, while others loosely adhere and thus appear in a higher plane.

Automatically aligned images were compared to manually registered images to test the accuracy of the method. The manually registered images were created using Fiji by manually selecting and subsequently aligning corresponding points in the LM and EM image. The differences between these manually aligned images and those done with our method were found to be negligible.

1.3 MATERIALS

1. 35-mm glass-bottomed imaging dishes containing embossed finder grid (MatTek Corporation, P35G-2-14-C-GRID).
2. PHEM buffer: 60 mM PIPES, 25 mM HEPES, 10 mM EGTA, 2 mM $MgCl_2$ (pH 6.9).
3. Fixative: 3% Paraformaldehyde and 0.05% glutaraldehyde (GA) in PHEM buffer (pH 6.9).
4. Permeabilization solution: 0.1% saponin in PHEM buffer.
5. Blocking solution: 1% bovine serum albumin, 0.1% saponin in PHEM buffer.
6. FluoroNanoGold: Alexa Fluor® 594 FluoroNanoGold™ Fab' Fragment goat anti rabbit IgG (Molecular Probes, #A24923).
7. 5 μg/ml 4′,6-diamidino-2-phenylindole dihydrochloride (DAPI) (Molecular Probes, #D1306).
8. R-GENT Silver Enhancement-EM (Aurion, #500.044).
9. 200 mM Phosphate buffer (PB), pH 7.4.
10. 1% Osmium tetroxide (OsO_4) in phosphate buffer.
11. 3% Uranyl acetate ($UO_2(CH_3COO)_2 \cdot 2H_2O$) in H_2O.
12. 812 EPON resin (TAAB).
13. Confocal microscope: Leica SP5 confocal imaging system.
14. Ultratome: Leica UC6 ultramicrotome.
15. EM microscope: FEI 120 kV BioTwin Spirit TEM.

CONCLUSIONS

We have developed a method for the automated alignment of LM and EM images. Currently, the method helps us to confidently identify a cell of interest inside a group of cells. The proposed method is a feature-based image registration algorithm, which first detects cell centers and then calculates the transformation parameters to register the two images together during the final analysis stage of a CLEM experiment. The registration of LM and EM images is typically done manually by altering the transparency of one of the images, rescaling and rotating that image so that the two images

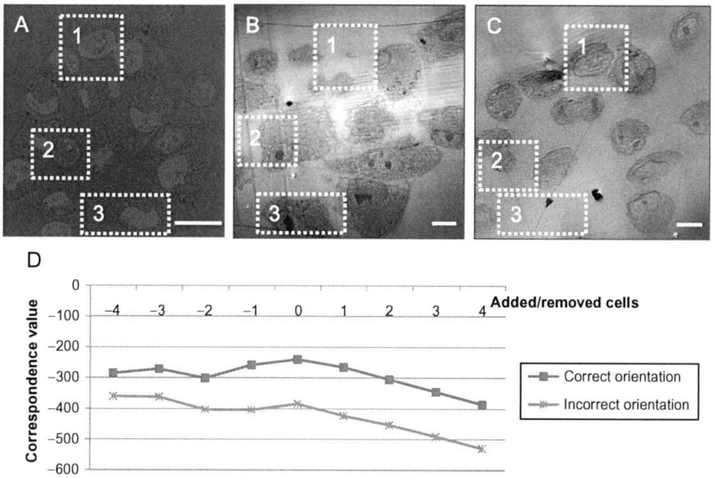

FIGURE 1.8

Removal of cells.

align (Keene et al., 2008). This automated procedure greatly reduces the time taken to overlay EM and LM images and provides a reliable and accurate alternative to manually aligned images. Furthermore, this algorithm is capable of correctly registering images even after removal or addition of cells from the EM image (Fig. 1.8). This is key as cells may appear and/or disappear as you move through the serial slices in EM. Since the method focuses on detection of center location rather than for instance shape identification it also avoids misidentification of EM images taken from deeper inside the block. Close to the block surface, where the cell attachment sites are, the cell also has the shape that corresponds most to the bright-field image (which does not contain z-information as seen in fluorescent confocal images). However, deeper inside the block, cell images generally become smaller and may also change shape in the EM. Finally, this method is capable of registering images of confluent cells. It is currently extremely challenging if not impossible to identify a cell within a tightly packed and overpopulated region in the EM as many of the visual cues used for registration of the images, such as gaps in the section where there are no cells, are not present. Therefore, this method enables the identification of the cell of interest within a cluster of cells, which is often not possible by eye. This feature is of much interest as we have previously been trying to trace back rare events (less than 1 in 100 cells) without the use of this method with very little success.

At present, the algorithm is capable of registering bright-field LM images with EM images. The theory could be developed further to allow the correlation of fluorescent CLEM probes with EM images. In addition, it could be applied to EM imaging software and the automated coordinate stage to pinpoint the cell of interest and zoom in to a particular region of the cell at the point of imaging, thus removing the need for manual searching and identification of the cell using the microscope.

Finally, this technique could be combined with super-resolution FLM procedures and electron tomography to reveal the detailed and precise 3D localization of a particular protein within a collection of cells but also within the cell.

This method greatly improves the registration of EM and LM images and thus facilitates in the study of protein dynamics, localization and interactions at high resolution. This CLEM-specific algorithm is an extremely useful tool for biologists, further bridging the gap between the two imaging modalities and allowing the researcher to take advantage of all strengths offered by each technique.

In future, it would be extremely useful to implement and develop this technology further for use in TEM. This would involve, first importing the LM image into the TEM image acquisition software and then creating a stitched image providing an overview of the grid containing the cell sections and the location of the individual cells and nuclei (Faas et al., 2012). Finally, the software could then automatically retrace the cell or group of cells imaged in the LM and as a last step could acquire either a standard TEM image or even acquire a tomogram of the area of interest. These developments could significantly shorten a CLEM experiment and provide more data to be analyzed than currently possible.

REFERENCES

Agronskaia, A., Valentijn, J., van Driel, L., Schneijdenberg, C., Humbel, B., Henegouwen, P., et al. (2008). Integrated fluorescence and transmission electron microscopy. *Journal of Structural Biology*, *164*, 183–189.

Benito-Alifonso, D., Tremel, S., Hou, B., Lockyear, H., Mantell, J., Fermin, D. J., Verkade, P., Berry, M., & Galan, M. C. (2014). Lactose as a "trojan horse" for quantum dot cell transport. *Angewandte Chemie (International Ed. in English)*, *53*, 810–814.

Betzig, E., Patterson, G., Sougrat, R., Lindwasser, O., Olenych, S., Bonifacino, J., et al. (2006). Imaging intracellular fluorescent proteins at nanometer resolution. *Science*, *313*, 1642–1645.

Brown, E., Mantell, J., Carter, D., Tilly, G., & Verkade, P. (2009). Studying intracellular transport using high-pressure freezing and Correlative Light Electron Microscopy. *Seminars in Cell & Developmental Biology*, *20*, 910–919.

Brown, E., van Weering, J., Sharp, T., Mantell, J., Verkade, P., & Mullerreichert, T. (2012). Capturing endocytic segregation events with HPF-CLEM. *Correlative Light and Electron Microscopy*, *111*, 175–201.

Brown, E., & Verkade, P. (2010). The use of markers for correlative light electron microscopy. *Protoplasma*, *244*, 91–97.

Faas, F. G., Avramut, M. C., Van Den Berg, B. M., Mommaas, A. M., Koster, A. J., & Ravelli, R. B. (2012). Virtual nanoscopy: Generation of ultra-large high resolution electron microscopy maps. *The Journal of Cell Biology*, *198*, 457–469.

Faas, F., Barcena, M., Agronskaia, A., Gerritsen, H., Moscicka, K., Diebolder, C., et al. (2013). Localization of fluorescently labeled structures in frozen-hydrated samples using integrated light electron microscopy. *Journal of Structural Biology*, *181*, 283–290.

Fukuda, Y., Schrod, N., Schaffer, M., Feng, L. R., Baumeister, W., & Lucic, V. (2013). Coordinate transformation based cryo-correlative methods for electron tomography and focused ion beam milling. *Ultramicroscopy*, *143*, 15–23.

Gaietta, G., Deerinck, T. J., Adams, S. R., Bouwer, J., Tour, O., Laird, D. W., et al. (2002). Multicolor and electron microscopic imaging of connexin trafficking. *Science*, *296*, 503–507.

Giepmans, B. (2006). Role of connexin43-interacting proteins at gap junctions. *Cardiovascular Gap Junctions*, *42*, 41–56.

Giepmans, B. (2008). Bridging fluorescence microscopy and electron microscopy. *Histochemistry and Cell Biology*, *130*, 211–217.

Giepmans, B. N., Deerinck, T. J., Smarr, B. L., Jones, Y. Z., & Ellisman, M. H. (2005). Correlated light and electron microscopic imaging of multiple endogenous proteins using Quantum dots. *Nature Methods*, *2*, 743–749.

Grabenbauer, M., Geerts, W., Fernadez-Rodriguez, J., Hoenger, A., Koster, A., & Nilsson, T. (2005). Correlative microscopy and electron tomography of GFP through photooxidation. *Nature Methods*, *2*, 857–862.

Gustafsson, M. G. (2005). Nonlinear structured-illumination microscopy: Wide-field fluorescence imaging with theoretically unlimited resolution. *Proceedings of the National Academy of Sciences of the United States of America*, *102*, 13081–13086.

Hell, S., & Stelzer, E. (1992). Properties of a 4PI confocal fluorescence microscope. *Journal of the Optical Society of America A-Optics Image Science and Vision*, *9*, 2159–2166.

Hodgson, L., Tavare, J. T., & Verkade, P. (2014). Development of a quantitative Correlative Light Electron Microscopy technique to study GLUT4 trafficking. *Protoplasma*, *251*, 403–416.

Huang, B., Babcock, H., & Zhuang, X. (2010). Breaking the diffraction barrier: Super-resolution imaging of cells. *Cell*, *143*, 1047–1058.

Kandela, I., & Albrecht, R. (2007). Fluorescence quenching by colloidal heavy metals nanoparticles: Implications for correlative fluorescence and electron microscopy studies. *Scanning*, *29*, 152–161.

Keene, D., Tufa, S., Lunstrum, G., Holden, P., & Horton, W. (2008). Confocal/TEM overlay microscopy: A simple method for correlating confocal and electron microscopy of cells expressing GFP/YFP fusion proteins. *Microscopy and Microanalysis*, *14*, 342–348.

Kellenberger, E., Johansen, R., Maeder, M., Bohrmann, B., Stauffer, E., & Villiger, W. (1992). Artifacts and morphological-changes during chemical fixation. *Journal of Microscopy-Oxford*, *168*, 181–201.

Koster, A. J., & Klumperman, J. (2003). Electron microscopy in cell biology: Integrating structure and function. *Nature Reviews Molecular Cell Biology*, (Suppl.), SS6–SS10.

Kukulski, W., Schorb, M., Welsch, S., Picco, A., Kaksonen, M., & Briggs, J. (2011). Correlated fluorescence and 3D electron microscopy with high sensitivity and spatial precision. *Journal of Cell Biology*, *192*, 111–119.

Kukulski, W., Schorb, M., Welsch, S., Picco, A., Kaksonen, M., & Briggs, J. A. (2012). Precise, correlated fluorescence microscopy and electron tomography of lowicryl sections using fluorescent fiducial markers. *Methods in Cell Biology*, *111*, 235–257.

Lucic, V., Kossel, A., Yang, T., Bonhoeffer, T., Baumeister, W., & Sartori, A. (2007). Multiscale imaging of neurons grown in culture: From light microscopy to cryo-electron tomography. *Journal of Structural Biology*, *160*, 146–156.

Martell, J., Deerinck, T., Sancak, Y., Poulos, T., Mootha, V., Sosinsky, G., et al. (2012). Engineered ascorbate peroxidase as a genetically encoded reporter for electron microscopy. *Nature Biotechnology*, *30*, 1143–1148.

McDonald, K., Benavides, H., & Yacaman, M. (1998). High pressure freezing freeze substitution of some key organisms used in molecular genetic studies. *Electron Microscopy*, *4*, 283–284.

Meisslitzer-Ruppitsch, C., Vetterlein, M., Stangl, H., Maier, S., Neumuller, J., Freissmuth, M., et al. (2008). Electron microscopic visualization of fluorescent signals in cellular compartments and organelles by means of DAB-photoconversion. *Histochemistry and Cell Biology*, *130*, 407–419.

Nam, D., Mantell, J., Hodgson, L., Bull, D., Verkade, P., & Achim (2014). Feature-based registration for correlative light and electron microscopy images. *Proceedings on IEEE Image Processing,* (in press).

Nixon, S., Webb, R., Floetenmeyer, M., Schieber, N., Lo, H., & Parton, R. (2009). A single method for cryofixation and correlative light, electron microscopy and tomography of zebrafish embryos. *Traffic*, *10*, 131–136.

Peddie, C., Blight, K., Wilson, E., Melia, C., Marrison, J., Carzaniga, R., et al. (2014). Correlative and integrated light and electron microscopy of in-resin GFP fluorescence, used to localise diacylglycerol in mammalian cells. *Ultramicroscopy*, *143*, 3–14.

Polishchuk, R., Polishchuk, E., Luini, A., Mullerreichert, T., & Verkade, P. (2012). Visualizing live dynamics and ultrastructure of intracellular organelles with preembedding correlative light-electron microscopy. *Correlative Light and Electron Microscopy*, *111*, 21–35.

Polishchuk, R., Polishchuk, E., Marra, P., Alberti, S., Buccione, R., Luini, A., et al. (2000). Correlative light-electron microscopy reveals the tabular-saccular ultrastructure of carriers operating between Golgi apparatus and plasma membrane. *Journal of Cell Biology*, *148*, 45–58.

Qi, X., Xing, F., Foran, D. J., & Yang, L. (2012). Robust segmentation of overlapping cells in histopathology specimens using parallel seed detection and repulsive level set. *IEEE Transactions on Biomedical Engineering*, *59*, 754–765.

Risco, C., Sanmartin-Conesa, E., Tzeng, W., Frey, T., Seybold, V., & De Groot, R. (2012). Specific, sensitive, high-resolution detection of protein molecules in eukaryotic cells using metal-tagging transmission electron microscopy. *Structure*, *20*, 759–766.

Robinson, J., & Takizawa, T. (2009). Correlative fluorescence and electron microscopy in tissues: Immunocytochemistry. *Journal of Microscopy-Oxford*, *235*, 259–272.

Rust, M., Bates, M., & Zhuang, X. (2006). Sub-diffraction-limit imaging by stochastic optical reconstruction microscopy (STORM). *Nature Methods*, *3*, 793–795.

Sartori, A., Gatz, R., Beck, F., Rigort, A., Baumeister, W., & Plitzko, J. (2007). Correlative microscopy: Bridging the gap between fluorescence light microscopy and cryo-electron tomography. *Journal of Structural Biology*, *160*, 135–145.

Schönle, A., Glatz, M., & Hell, S. W. (2000). Four-dimensional multiphoton microscopy with time-correlated single-photon counting. *Applied Optics*, *39*, 6306–6311.

Schorb, M., & Briggs, J. A. G. (2013). Correlated cryo-fluorescence and cryo-electron microscopy with high spatial precision and improved sensitivity. *Ultramicroscopy*, *143*, 24–32.

Shu, X., Lev-Ram, V., Deerinck, T., Qi, Y., Ramko, E., Davidson, M., et al. (2011). A genetically encoded Tag for correlated light and electron microscopy of intact cells, tissues, and organisms. *PLoS Biology*, *9*, e1001041.

Slot, J., & Geuze, H. (2007). Cryosectioning and immunolabeling. *Nature Protocols*, *2*, 2480–2491.

Tokuyasu, K. (1973). Technique for ultracryotomy of cell suspensions and tissues. *Journal of Cell Biology*, *57*, 551–565.

van Driel, L., Valentijn, J., Valentijn, K., Koning, R., & Koster, A. (2009). Tools for correlative cryo-fluorescence microscopy and cryo-electron tomography applied to whole mitochondria in human endothelial cells. *European Journal of Cell Biology*, *88*, 669–684.

van Rijnsoever, C., Oorschot, V., & Klumperman, J. (2008). Correlative light-electron microscopy (CLEM) combining live-cell imaging and immunolabeling of ultrathin cryosections. *Nature Methods*, *5*, 973–980.

van Weering, J., Brown, E., Sharp, T., Mantell, J., Cullen, P., Verkade, P., et al. (2010). Intracellular membrane traffic at high resolution. *Electron Microscopy of Model Systems*, *96*, 619–648.

Verkade, P. (2008). Moving EM: The rapid transfer system as a new tool for correlative light and electron microscopy and high throughput for high-pressure freezing. *Journal of Microscopy*, *230*, 317–328.

CHAPTER

Fluorescing the Electron: Strategies in Correlative Experimental Design

2

Kimberley H. Gibson, Daniela Vorkel, Jana Meissner, Jean-Marc Verbavatz
Max Planck Institute of Molecular Cell Biology and Genetics, Dresden, Germany

CHAPTER OUTLINE

Methods in Cell Biology, Volume 124, ISSN 0091-679X, http://dx.doi.org/10.1016/B978-0-12-801075-4.00002-1

Abstract

Correlative light and electron microscopy (CLEM) encompasses a growing number of imaging techniques aiming to combine the benefits of light microscopy, which allows routine labeling of molecules and live-cell imaging of fluorescently tagged proteins with the resolution and ultrastructural detail provided by electron microscopy (EM). Here we review three different strategies that are commonly used in CLEM and we illustrate each approach with one detailed example of their application. The focus is on different options for sample preparation with their respective benefits as well as on the imaging workflows that can be used. The three strategies cover: (1) the combination of live-cell imaging with the high resolution of EM (time-resolved CLEM), (2) the need to identify a fluorescent cell of interest for further exploration by EM (cell sorting), and (3) the subcellular correlation of a fluorescent feature in a cell with its associated ultrastructural features (spatial CLEM). Finally, we discuss future directions for CLEM exploring the possibilities for combining super-resolution microscopy with EM.

Abbreviations

CLEM correlative light and electron microscopy
EM electron microscopy
fP fusion protein
FP fluorescent protein
GFP green fluorescent protein
LM light microscopy
PALM photoactivated localization microscopy
ROI region of interest
SEM scanning electron microscopy
STED stimulated emission depletion microscopy
STORM stochastic optical reconstruction microscopy
TEM transmission electron microscopy
TIRF total internal reflection fluorescence microscopy
YFP yellow fluorescent protein

INTRODUCTION

Within the field of biology, a myriad of imaging technologies and techniques have arisen in pursuit of illuminating and identifying specific structures and molecules of interest. Research studies depend more and more on light microscopy (LM) to locate fluorescently tagged molecules like proteins, RNA, DNA, and lipids. While useful and applicable to many purposes, these techniques are limited in that they can only reveal a small selection of tagged or targeted molecules within a cell or a tissue without providing insight into the relationship or context that molecule or associated structure may have with its surroundings. Combining electron microscopy (EM) with the attributes of fluorescence microscopy allows us to provide an ultrastructural context in studying relationships between structure, localization, and function for a wide variety of molecules (Müller-Reichert & Verkade, 2012).

Both LM and EM have their advantages and disadvantages. Fluorescence microscopy can be employed to study fluorescently tagged molecules in live or fixed biological samples. In diffraction-limited LM, however, the resolution for fluorescent markers and endogenously expressed proteins is approximately 200 nm (*x*- and *y*-axes) which is incapable of resolving smaller biological structures like vesicles, microtubules, or organelle subcompartments (Betzig et al., 2006; Huang, Babcock, & Zhuang, 2010; Patterson, Davidson, Manley, & Lippincott-Schwartz, 2010; Robinson, Takizawa, Pombo, & Cook, 2001). In contrast, high resolutions approaching 1 nm can be achieved with EM using conventional staining methods allowing for better identification of structural details. However, transmission electron microscopy (TEM) depends on the passage of highly accelerated electrons through a very thin film of sample necessitating the use of fixatives and resin embedding or freezing to stabilize the sample so that it can be finely sectioned. Imaging of biological samples with an electron beam also requires *en bloc* staining during fixation steps, or poststaining after sectioning. Both depend on the use of high atomic number heavy metals like osmium, lead, and uranium. Specific labeling of molecules (proteins, RNA, DNA, etc.) by EM is challenging and live-cell imaging is nearly impossible.

Correlative light and electron microscopy (CLEM) incorporates a growing number of techniques evolving out of the need to combine the advantages of both LM (usually fluorescence microscopy) and EM. As is true for any research study, techniques and protocols must be adopted and often adapted to answer the question at hand. Various CLEM techniques have been proposed in the literature. Here, we are considering three standard but general CLEM strategies (Fig. 2.1) that, depending on the exact aim, can be employed and modified to fit a specific purpose. Integral to each strategy is the processing step until which fluorescence properties of the fluorescent probe used must be retained. It should be noted that various types of fixation and preparation steps for EM can denature or bleach fluorophores in a sample (Tables 2.1–2.3). Another issue to consider is the ability to spatially relocate the same cell, structure, or region of interest (ROI) in EM after fluorescence imaging at the desired resolution for correlation.

The first strategy that we consider is where dynamic processes are being observed, or where specific time points in development must be recorded. Cells and tissues have to be in a living state during fluorescence imaging (Polishchuk et al., 2000; Müller-Reichert, Srayko, Hyman, O'Toole, & McDonald, 2007; Woog, White, Büchner, Srayko, & Müller-Reichert, 2010; Fig. 2.1A). For time-resolved correlation, the sample must then be rapidly fixed or cryo-immobilized (i.e., through high-pressure freezing or freeze-plunging) at that critical time point of interest before processing for EM (Guizetti et al., 2011; Pelletier, O'Toole, Schwager, Hyman, & Müller-Reichert, 2006; van Rijnsoever et al., 2008). Chemically fixed cells may be embedded in resin (Polishchuk et al., 2000) or they can be further processed for Tokuyasu cryo-sectioning after fixation if desired (Cortese, Diaspro, & Tacchetti, 2009; Tokuyasu, 1973; Robinson et al., 2001; van Rijnsoever et al., 2008). High-pressure frozen samples can be freeze-substituted with lower concentrations of fluorescence-affecting chemicals like osmium, potassium ferrocyanide, or uranyl acetate (Table 2.1) before resin embedding, if fluorescence needs to be preserved further (Kukulski et al., 2011; Petrovska et al., 2014; Watanabe & Jorgensen, 2012; Watanabe et al., 2011).

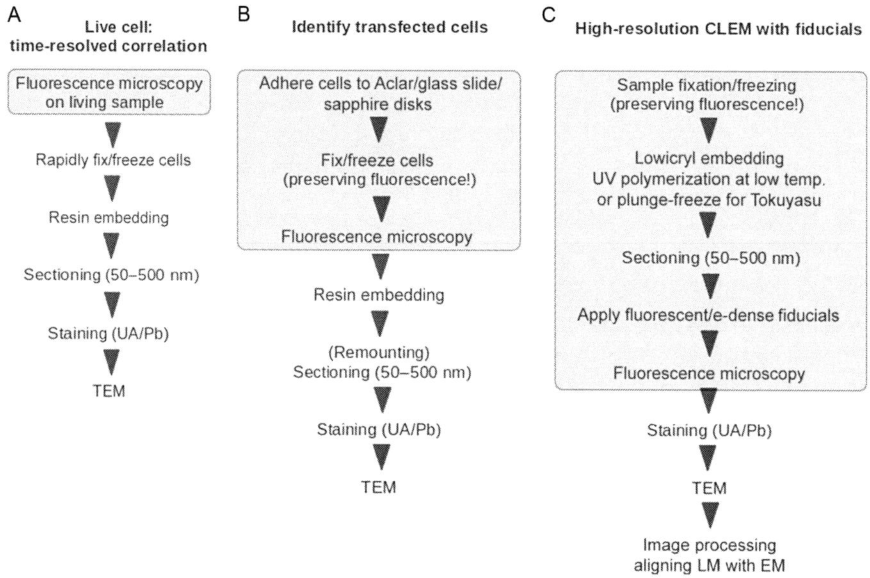

FIGURE 2.1

Three general strategies when considering CLEM in experimental design, where dynamic or development-specific processes must be observed via fluorescence microscopy prior to sample immobilization. (A) LM imaging is performed on living samples after which TEM processing can proceed in samples where transfection rates tend to be lower. (B) Specific fluorescent cells or regions can be detected postfixation via LM before final resin embedding, sectioning, and TEM imaging occur where it is necessary to localize specific subregions within a cell. (C) High-resolution CLEM can be adopted as means of accurately aligning LM with TEM images using fiducials in correlating fluorescently tagged substrates with structures observed in EM. Boxes denote the processing steps where sample fluorescence must be preserved for light microscopy imaging.

A second example of a CLEM strategy is applicable for cell cultures with low transfection rates of a fluorescently tagged protein, where only a fraction of cells may express the protein of interest. Cells exhibiting a phenotype or feature of interest can be identified pre- or postfixation by fluorescence microscopy. Such problematic samples can be adapted for CLEM by adhering cells or tissues to a substrate, such as ACLAR® plastic sheets or glass slides (Guizetti, Mäntler, Müller-Reichert, & Gerlich, 2010; Liv et al., 2013; Polishchuk et al., 2000, van Rijnsoever et al., 2008) for fluorophore-retaining chemical fixation, or on sapphire disks for high-pressure freezing followed directly by fluorescence imaging (Fig. 2.1B). After the fluorescence-identification of transformed cells, the samples can be resin-embedded, then polymerized followed by separation from the adhering substrate, remounted, and then sectioned for TEM imaging.

Table 2.1 List of Chemicals, Their Properties, and Fluorescence-Affecting Concentrations Employed During Chemical Fixation or Freeze-Substitution in CLEM

Chemical or fixative	Fluorescent protein (FP)	Publication	Notable features
Paraformaldehyde (PFA) Aldehyde, cross-linking Negligible autofluorescence	GFP, mCherry, RFP, mEos2, Dronpa, Kaede	Betzig et al. (2006), Hodgson, Tavaré, and Verkade (2014), van Rijnsoever, Oorschot, and Klumperman (2008), Brown, Fetter, Tkachuk, and Clayton (2010), this study	1–5% PFA retains fluorescence in Tokuyasu cryo-sectioning or for low-temperature acrylic resin embedding
Glutaraldehyde (GA) Aldehyde, cross-linking Retains FP fluorescence, but causes autofluorescence Excellent structural preservation	mEos2, Dronpa	Kopek, Shtengel, Xu, Clayton, and Hess (2012), Brown et al. (2010), Watanabe et al. (2011)	0.1–2% GA for excellent structural preservation, but causes autofluorescence that can be quenched using sodium borohydride without quenching FP fluorescence
Osmium tetroxide Very oxidative Membrane contrasting, structural preservation Quenches fluorescence	tdEos, Citrine, GFP, Dendra	Watanabe et al. (2011)	0.001% OsO_4 (with 0.1% $KMnO_4$) in freeze-substitution retains fluorescence, preserving membrane structure; >0.1% OsO_4 in freeze-substitution destroys fluorescence
Potassium permanganate Oxidizer Membrane contrasting, structural preservation	tdEos, Citrine, GFP, Dendra	Watanabe et al. (2011)	0.1% $KMnO_4$ in freeze-substitution retains fluorescence and preserves structure; use of 0.001% OsO_4 with 0.1% $KMnO_4$ preserved both fluorescence and structure
Uranyl acetate (UA) Heavy metal stain	mEos2, Dronpa, GFP, mCherry	Brown et al. (2010), Kukulski et al. (2011), Yuan, Li, Hong, and Hong (2014), Petrovska et al. (2014)	0.1–0.5% UA in freeze-substitution retains FP fluorescence

Continued

Table 2.1 List of Chemicals, Their Properties, and Fluorescence-Affecting Concentrations Employed During Chemical Fixation or Freeze-Substitution in CLEM—cont'd

Chemical or fixative	Fluorescent protein (FP)	Publication	Notable features
Acetone Solvent Mild fixative Scavenges free radicals	GFP, tdEos, Citrine, Dendra, mCherry	Watanabe et al. (2011), this study	100% Acetone quenches fluorescence; recommend adding water (i.e., 5% H_2O) during infiltration, or freeze-substitution. Acetone impedes resin polymerization, so optimally transition into ethanol before resin infiltration
Water Polar solvent Hydration necessary to prevent quenching of fluorescent proteins	GFP, Citrine, Dendra, tdEos	Watanabe et al. (2011)	Inclusion of ~3–5% H_2O during freeze-substitution, infiltration, and acrylic resin embedding helps to retain fluorescence

A third example is applied to the correlation of a fluorescent signal with notable ultrastructural features at the EM level. This issue has often been addressed by employing immunogold labeling in EM in combination with (immuno)fluorescence (Agronskaia et al., 2008; Buser & McDonald, 2010; Cortese, Diaspro, & Tacchetti, 2009; Karreman et al., 2012; Polishchuk et al., 2000; Pranke et al., 2011; Robinson et al., 2001). However, this is not an option for proteins or antigens of interest if no suitable antibody exists. Sometimes, even when antibodies are available, the antigen is not sufficiently abundant at the surface of EM sections to enable detection. The use of semi-thick sections (~150–400 nm) in electron tomography or thicker sections (500–700 nm) in scanning electron microscopy (SEM; Kopek et al., 2012) defeats the point of on-section immunolabeling, which depends on binding of antibodies to antigens exposed only on the surface of a section. However, combining the advantages of fluorescence microscopy with that of EM while superseding the caveats of either technique opens new avenues into illuminating cell ultrastructure and answering many biological queries. To that aim, fluorescently tagged proteins or other fluorescent molecules, like endocytic cargo, can be imaged by fluorescence microscopy, and the fluorescence localization can then be used to reidentify those same subcellular areas of interest with higher-resolution EM (Fig. 2.1C). Such high-resolution CLEM enables us to precisely locate structures or areas within a cell or tissue where a molecule of interest is targeted. Structural data from EM can then be compared with fluorescence localization information by overlaying the LM and EM images. This strategy requires that we preserve sample fluorescence until after sectioning, when fluorescence imaging can be performed. Sustaining fluorescence in standard EM protocols is tricky as chemical fixation (Table 2.1), freeze-substitution (Tables 2.1 and 2.2), *en bloc* staining (Tables 2.1 and 2.2), and resin embedding (Table 2.3) steps must be designed to minimize cross-linking, oxidation, and photobleaching of the fluorophore without compromising structural preservation. Importantly, certain fluorophores will be more robust after successive chemical treatments than others. An excellent option for fluorescence preservation is to prepare cells or tissues for Tokuyasu or cryo-sectioning technique (Tokuyasu, 1973), which does not require resin embedding.

Given the body of literature on CLEM, we are fortunate that a number of fluorescent proteins, such as green fluorescent protein (GFP), mCherry, mCitrine, Eos, and others, have proven amenably resilient to a wide array of standard and modified

Table 2.2 Cryo-immobilization techniques effectively used in CLEM

Fixation or cryo-immobilization technique	Fluorescent protein (FP)	Publication	Notable features and reagents
Cryo-EM: plunge-freezing (PF)	GFP	Schorb and Briggs (2013)	Cryo-fluorescence imaging on p22 bacteriophages post-FP, followed by cryo-EM
	GFP, mRFP	Duke et al. (2013)	Cryo-fluorescence imaging followed by cryo-soft X-ray tomography
High-pressure freezing (HPF)	EGFP	Watanabe et al. (2011)	GFP fluorescence after HPF followed by direct thawing showed no change in fluorescence level in *C. elegans* Pmyo-2::GFP
HPF, then freeze-substitution	tdEos, Citrine	Watanabe et al. (2011)	0.001% OsO_4 with 0.1% $KMnO_4$ in 95% acetone, ~5% H_2O
	EGFP, mCherry	Kukulski et al. (2011), Yuan et al. (2014), Petrovska et al. (2014)	0.1% Uranyl acetate, in acetone
Chemical fixation (excluding Tokuyasu cryo-prepared samples)	EGFP, mCherry, RFP	This study	3% PFA; 0.2% GA in 1 × PBS buffer; no osmium or uranyl acetate postfixation. Samples embedded in HM20

Table 2.3 List of embedding resins or mediums and their effects on structure and fluorescence

Embedding medium	Fluorescent protein (FP)	Publication	Notable features/ issues
Lowicryl HM20 Nonpolar acrylic resin Low-temperature infiltration Low-temperature UV polymerization Hydrophobic	EGFP; mCherry	Kukulski et al. (2011), Yuan et al. (2014)	Fluorescence is stable after UV polymerization
	Dendra2	Petrovska et al. (2014), this study	Fluorescence bleached due to photoactivation during UV polymerization
Lowicryl K4M Polar, acrylic resin Low-temperature infiltration Low-temperature UV polymerization Hydrophilic	tdEos; mCitrine	Watanabe et al. (2011)	Hydrophilic resin; the use of ~5% H_2O in resin to maintain fluorophore hydration interfered with polymerization, affecting sectioning quality
LR White Acrylic resin High-temperature or low-temperature polymerization UV polymerization optional Hydrophilic	Dendra; tdEos; mEosFP	Watanabe et al. (2011)	Fluorescence quenched due to acidity at pH 5.5; fluorescence retained by stabilizing pH of LR White with ethanolamine; *C. elegans* expressing H2B-tdEos or Dendra used in PALM and then SEM
	mEos2; Dronpa	Kopek et al. (2012), Brown et al. (2010)	Fluorescence retained, but poor mitochondrial structure
LR Gold Acrylic resin Low-temperature polymerization UV polymerization optional Hydrophilic	tdEos; mCitrine	Watanabe et al. (2011)	Poor tissue penetration resulting in poor sectioning
	GFP-immunolabeling only	Buser and McDonald (2010)	Good structural preservation; GFP fluorescence postembedding was not determined
Glycol methacrylate Acrylic (GMA) resin Low-temperature polymerization Hydrophilic	tdEos, mCitrine; Dendra	Watanabe et al. (2011), Watanabe and Jorgensen (2012)	3% H_2O in resin; 70% fluorescence retained; fusion proteins with mCitrine used STED microscopy; tdEos and Dendra were used in PALM, then SEM; inconsistent sectioning due to poor cross-linking

Table 2.3 List of embedding resins or mediums and their effects on structure and fluorescence—cont'd

Embedding medium	Fluorescent protein (FP)	Publication	Notable features/ issues
			of resin with sample exterior
Methylcellulose (Tokuyasu/cryo-sectioning) Chemical fixation Gelatin embedding Sucrose infiltration Plunge-freezing Cryo-sectioning Sections stabilized with methylcellulose	dEosFP, Kaede	Betzig et al. (2006)	0.1% GA/4% PFA fixation, cryo-sectioned, PALM imaging, then embedded in methylcellulose for TEM
	EGFP	Hodgson, Tavaré, and Verkade (2014)	2% PFA/0.2% GA fixation, cryo-sectioning, compare immunolabeling with CLEM
	mGFP; labeled with dextran–Alexa568; or transferrin–Alexa588	van Rijnsoever et al. (2008)	Live-cell fluorescence imaging, cells fixed with 4% PFA/0.05% GA *in situ*, cryo-sectioned, imaged via TEM
	mEos	Kopek et al. (2012)	2% GA; autofluorescence quenched postcryo-sectioning with sodium borohydride
	mEos2; Dronpa	Brown et al. (2010)	No EM, 0.1% glutaraldehyde/4% paraformaldehyde; cryo-embedding/sectioning, then sodium borohydride quenching before fluorescence imaging

EM sample processing techniques, as outlined in Tables 2.1–2.3. From this basis, it is possible to develop multiple strategies to investigate biological questions using a correlative approach.

Once fluorescence-friendly EM protocols have been established, the utility of the method is dependent on our ability to correlate fluorescence signals with precise EM structures for high-resolution CLEM. Tissues and cells are often densely packed with a cornucopia of organelles, undulating membranes, cisternae, fibrillar structures, proteinaceous complexes or arrangements, or budding and blebbing vesicular bodies. If the goal is to isolate fluorescence to any one of these substructures, it must be done with conviction. Not only do we need to relocate the same fluorescence region in EM that we identified via LM, we also need to accurately align and superimpose the fluorescence signals on the EM structural data to make a comparison. Optimizing the alignment of LM and TEM images requires that we utilize fiducial markers. For CLEM, minute spheres, between 100 and 200 nm in diameter (smaller than LM resolution limit), that are both fluorescent and electron dense for identification in both LM and TEM are excellent fiducials (Kukulski et al., 2011; Petrovska et al., 2014; Schellenberger et al., 2013; Schorb & Briggs, 2013). These can then be used in conjunction with imaging software as reference points to aid in aligning, warping, and overlaying LM and TEM images such that we can make accurate comparisons between the fluorescence localization in LM with the detailed structural data provided via TEM. With these tools in mind, we can consider the variable strategies through which we may further illuminate structures in molecular and cellular context.

2.1 STRATEGIES FOR CLEM

2.1.1 LIVE-CELL IMAGING AND TIME-RESOLVED CORRELATION BETWEEN LIGHT AND ELECTRON MICROSCOPY

Biological processes encompass dynamic events involving localization and translocation of organelles or protein complexes, molecular diffusion and suborganelle sequestration, fusion or fission of membranes, and of course, structural assembly or disassembly. Live-cell imaging allows us to follow specific processes that can occur within short time frames. Those cells or tissues identified at a specific desired stage of the process can be cryo-immobilized or chemically fixed to ensure that the stage in which it is transitioning is preserved. For example, to study Golgi subcompartmentalization and transport, Polishchuk et al. (2000) utilized COS-7 cells transfected with vesicular stomatitis virus G protein (VSVG) that was tagged with GFP. Using confocal microscopy, they observed the procession of the VSVG–GFP from the endoplasmic reticulum (ER) to *trans*-Golgi and post-Golgi after various chemical treatments or temperature shifts. When VSVG–GFP fluorescence dynamics were subcompartmentalized into Golgi-to-plasma membrane (PM) vesicles budding from the Golgi, the cells were then fixed with 4% paraformaldehyde and 0.05% glutaraldehyde while adhered to coverslips. Further confocal imaging followed cell fixation to image fluorescent structures within the *z*-axis. Cells on the coverslips were embedded in resin and later relocated and remounted for serial sectioning, followed by TEM imaging. Stacking the imaged serial sections in *z* followed by segmentation of resulting ER-to-Golgi, *trans*-Golgi, Golgi-to-PM, and also PM-fusion structures provided structural context at various crucial stages in cell trafficking.

Separately, Müller-Reichert et al. (2007) and Woog et al. (2010) embarked on a mission to characterize microtubules in early *Caenorhabditis elegans* embryos undergoing mitosis or meiosis. Live *C. elegans* embryos expressing β-tubulin-GFP, to label the microtubules, and H2B-mCherry, to visualize the chromatin, were observed via fluorescence microscopy (Woog et al., 2010). Embryos entering the desired stage were then rapidly high-pressure frozen (McDonald, Morphew, Verkade, & Müller-Reichert, 2007) as a means of cryo-preserving them while minimizing ice crystal formation. After freeze-substitution and embedding in resin, they were ensured that each embryo was worth serial sectioning and preparing for investigation via electron tomography. We have used this approach to study the organization of centrosomal microtubules in *C. elegans* embros during metaphase of mitosis. Figure 2.2A shows a fluorescence picture acquired immediately prior to high-pressure freezing. Dual-axis tilt series of serial sections through the centrosomal region (Fig. 2.2B, yellow box) were acquired, reconstructed, and combined (Fig. 2.2C) to allow for further segmentation (Fig. 2.2C and D) and analysis (Fig. 2.2E) of centriolar microtubules (Weber et al., 2012). Using this technique, we can determine with absolute certainty that every embryo sectioned is at the exact time point in the cell cycle, to follow through with further time-consuming EM tomography image acquisition, processing, and analyses.

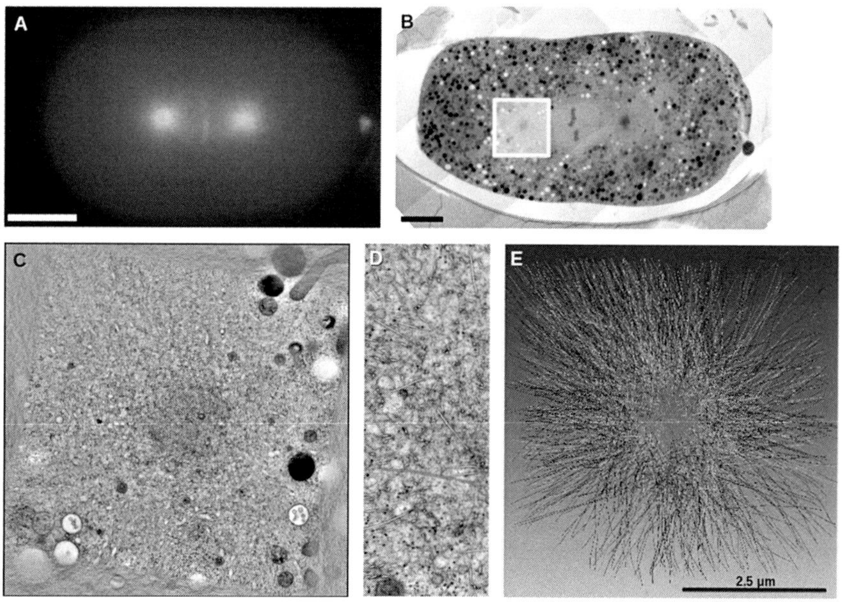

FIGURE 2.2

Imaging of a live *C. elegans* embryo during metaphase just prior to high-pressure freezing. (A) Mitotic spindles in blue, microtubules (MT) labeled by β-tubulin-GFP, and chromatin-labeled H2B-mCherry, red; (B) EM with centrosomes (blue) and chromatin (red) pseudo-colored; yellow box denotes centrosomal area selected for EM tomography; (C) combined dual-axis tomogram showing centrosome with segmented MTs (green lines), red line highlights one MT; (D) enlarged area showing selected MT from C (in red). (E) Three-dimensional model of segmentated microtubules generated by using the ZIBAmira software package. (See the color plate.)

Similar approaches were utilized by van Rijnsoever et al. (2008) to localize the lysosomal membrane protein LAMP-1-mGFP as expressed in human hepatoma HepG2 cells that were grown on Formvar-coated glass coverslips. The cells were chemically fixed on a coverslip during fluorescence imaging when cells reached the desired stage. Directly after chemical fixation, they proceeded to acquire ~200-nm step confocal image stacks through the *z*-axis of the cell to ease relocation of precise substructures for later TEM imaging. The cells on the coverslip were then processed using a modified Tokuyasu cryo-embedding protocol, followed by cryo-sectioning. The resulting 70-nm-thick serial sections were imaged via TEM and compared with confocal images to ensure subregions were relocated within the fluorescence *z*-stack of the same cell. Testing the utility of live-cell, EM correlated imaging, van Rijnsoever et al. (2008) were able to track and distinguish two differently labeled endosomes in live imaging; one endosome contained either dextran or transferrin-Alexa-568, while another carried both mGFP and either a dextran- or transferrin-bound Alexa568. After observing the motion dynamics of both endosomes, the cell was processed for cryo-immunogold labeling for GFP, which was then relocated via TEM. Here, van Rijnsoever et al. (2008) were able to show the possibility of isolating individual dynamic structures via fluorescence that could later be reidentified in EM for structural analysis.

2.1.2 THE NEEDLE IN A HAYSTACK—IDENTIFICATION OF TRANSFECTED CELLS

Fluorescence microscopy is often used to isolate cells of interest in cell cultures with low frequency, as is often the case for transfected primary culture cells where only a small proportion of cells may be transformed. Cells expressing those genes of interest may show anywhere from marginal to drastic molecular or structural differences, making identification of those details by electron microscopy akin to seeking out a needle in a haystack. Having a means of locating a specific cell transformed with a fluorescent transgene by fluorescence microscopy followed by structural examination with TEM can absolve the aforementioned problems.

We utilized this CLEM strategy in ascertaining the involvement of the vesicle-trafficking protein, Sec22 in ER–PM contact sites in HeLa cells (Petkovic et al., 2014) and neurons.

Neuron cells are not always efficiently transfected. To ensure that we could later relocate an identified fluorescent for TEM investigation, we grew transfected neuron cells on a thin, adherent grated ACLAR® sheet until ~70% confluence was reached (Fig. 2.3). Cells were fixed in place on the ACLAR® sheet with a low concentration of glutaraldehyde to retain fluorescence and preserve ultrastructure (Fig. 2.3A).

Before proceeding through further fixation and embedding steps, cells were prepared by adding a drop of buffer with a coverslip over top of the ACLAR® sheet. Any EGFP-Sec22-expressing cells were located in position on the ACLAR® sheet via epifluorescence microscopy (Fig. 2.3B). The location of the cells on the ACLAR® sheet was noted by imaging landmarks at lower (Fig. 2.3A) and higher magnification

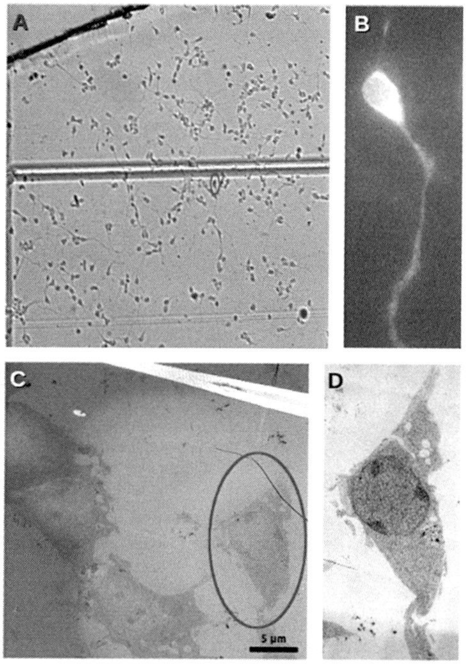

FIGURE 2.3

Locating a transfected neuron expressing GFP-Sec22 via fluorescence for TEM imaging. (A) A 10× bright field image showing aldehyde-fixed neural cells at ~70 confluence on an ACLAR sheet, green ellipse (light gray in the print version) denotes a single fluorescent cell near a grid line etched into the ACLAR sheet; (B) green fluorescence from the same GFP-expressing cell at 100× magnification; followed by (C) relocation of that cell at lower magnification TEM using linear etches as landmarks; before (D) closer examination via TEM of the same cell at higher magnification.

LM using DIC. After LM imaging, the entire ACLAR® sheet was retrieved for further fixation with 1% osmium and flat-embedded in epoxy resin. Throughout all steps described thus far, cells remained adhered in place on the ACLAR® sheet, allowing for ease of relocation after polymerization. The area of resin with identified fluorescent cells was excised from the ACLAR®, remounted on a block, sectioned, and mounted on grids for TEM imaging, ensuring that the cell imaged by TEM was indeed the same cell located in fluorescence microscopy. Here, we were able to relocate the same transfected neuron cell first via fluorescence microscopy for later TEM imaging (Fig. 2.3C), ensuring that the cell was indeed transfected, thereby promoting further examination of this cell (Fig. 2.3D) to ascertain whether any specific transgene-related structural changes could be noted. Following cell relocation, it was demonstrated by electron tomography that Sec22 mediates the apposition of ER and PMs at contact sites (Petkovic et al., 2014).

2.1.3 HIGH-RESOLUTION CLEM FOR PRECISE ALIGNMENT OF FLUORESCENCE AND EM DATA

In subcellular CLEM, the aim is to correlate the fluorescence signal to cell organelle ultrastructure. The optimal methods of preserving fluorescence and ultrastructure involve cryo-techniques. Several different cryo-immobilization methods have been successfully employed in CLEM, including high-pressure freezing, freeze-plunging, and Tokuyasu technique for cryo-sectioning (Table 2.2). High-pressure freezing is an excellent method to preserve fluorescence as it rapidly stabilizes fluorescent proteins without denaturing or photobleaching them (Watanabe et al., 2011). If the sample or tissue is too large to be frozen or not amenable to freezing, then chemical preservation is also an option. The use of chemicals, however, should be carefully considered to minimize damage to the fluorophore (see Table 2.1).

While high-pressure freezing itself does not result in loss of fluorescence (Schorb & Briggs, 2013; Watanabe et al., 2011), the subsequent freeze-substitution and resin-embedding steps can reduce fluorescence through photobleaching, pH changes, dehydration, oxidation, and chemical cross-linking (Katayama, Yamamoto, Mizushima, Yoshimori, & Miyawaki, 2008: Watanabe et al., 2011). The use of osmium tetroxide and glutaraldehyde should be minimized during freeze-substitution so that the ultrastructure is preserved without losing the fluorescent signal or creating too much background autofluorescence. Proteins that we have tested with success include GFP, RFP, and mCherry. Watanabe et al. (2011) preserved the fluorescence of mCitrine or tdEos fused to the histone, H2B and the mitochondrial outer membrane protein, TOM-20 (Watanabe & Jorgensen, 2012) after high-pressure freezing followed by freeze-substitution in 0.001% OsO_4, 0.1% $KMnO_4$ in 95% acetone containing ~5% H_2O. Osmium tetroxide and $KMnO_4$ do cross-link membranes; however, the lower concentrations of each reagent in this cocktail preserved both membrane structure and fluorescence sufficiently enough to enable fluorescence imaging.

It is not recommended to use Lowicryl/Unicryl or even low-temperature formulations of LR White (Brown et al., 2010) if a sample expresses endogenous UV-sensitive monomeric photoswitchable or photoactivatable fluorescence probes (i.e., Dendra2, Kaede, mEos, etc.) as these will likely be photoactivated and then photobleached during the extended 2–3 days of UV resin polymerization. We noted that a protein tagged with Dendra2 failed to retain any fluorescence after HM20 polymerization and sectioning, whereas the same protein tagged with mCherry was easily identified by epifluorescence microscopy. We presently have not tested this protocol on PA-GFP, EosFP, or other photoactivatable or -switchable fluorophores and cannot definitely state whether their fluorophores would survive UV polymerization. Table 2.3 documents effective methods of preserving photoactivatable fluorescent proteins for CLEM, providing alternatives for HM20 (Watanabe et al., 2011). Multimeric photoactivatable fluorophores may prove to be more resistant to photobleaching during UV polymerization as they are not as easily degraded as the more sensitive monomeric forms of some photoactivatable/switchable fluorophores like

Dendra2 or Kaede (Campbell & Davidson, 2010; Patterson et al., 2010; Watanabe et al., 2011). For samples expressing UV-sensitive fluorescent probes of interest, other acrylic resins should be considered. However, the alternative of 60 °C polymerization used for resins including LR Gold and LR White may also destroy some fluorescence.

Watanabe et al. (2011) tested several acrylic resins for their ability to oxidize, photobleach, or preserve fluorescent probes and found that glycol methacrylate (GMA), when mixed with an additional 3% H_2O, preserves fluorescence best in comparison to Lowicryl K4M, LR Gold, and LR White. They noted that adding water to the final GMA resin mixture renews fluorescence by up to 30% of that observed in nonhydrated resin mixtures, highlighting that dehydrated fluorophores are in an inactive nonfluorescent state. Hydration of fluorescent proteins plays an important role in maintaining fluorescence, as does pH. Many fluorescent proteins such as EGFP, Sapphire, and the dsRed derivative, mRFP (S146G) are acid-sensitive, quenching fluorescence between pH 4.3 and 6.2 (Katayama et al., 2008). GMA remains alkaline at pH 8 during polymerization. Another advantage of GMA is that it polymerizes at lower temperatures, between −4 °C and −20 °C without UV light. If samples possessing UV-sensitive fluorescent proteins (i.e., Dendra2, Kaede) are to be processed for fluorescence post-EM embedding, then it would be wise to consider infiltrating with GMA rather than Lowicryl HM20 or K4M.

If worries persist regarding sensitive fluorophores, other methods such as Tokuyasu cryo-ultramicrotomy (see above), with the possibility of including a lower 0.1–0.3% addition of glutaraldehyde, can be considered. If higher concentrations of glutaraldehyde are needed for optimal structural preservation, it is possible to quench autofluorescence postfixation and embedding by incubating sections with sodium borohydride prior to fluorescence acquisition (Brown et al., 2010; Kopek et al., 2012). Fluorescence maintenance in cryo-sections for a wide variety of fluorophores has been extensively documented and is adaptable for CLEM using epifluorescence, confocal, and super-resolution fluorescence microscopy (SRFM) techniques (Betzig et al., 2006; Hodgson, Tavaré, & Verkade, 2014; Kopek et al., 2012; van Rijnsoever et al., 2008). For example, Hodgson, Tavaré, & Verkade, 2014 effectively preserved and localized the EGFP-tagged GLUT4 to the PM in adipose cells prepared using the Tokuyasu method. Betzig et al. (2006), Brown et al. (2010), and Kopek et al. (2012) showed that cryo-sections are able to withstand the extensive imaging times used for SRFM techniques such as photoactivated localization microscopy (PALM).

To obtain fluorescence information useful for comparison with EM ultrastructure, we want to strategically align the fluorescently tagged protein or molecule within a more precise structural context observed in EM. The sample processing strategy must simultaneously prolong fluorescence and minimize sample shifting between fluorescence and EM imaging steps. We are no longer simply interested in whether a cell or tissue is expressing a protein of interest indicative of its transformed status. We now have to preserve fluorescence in our sample as we process it using standard EM methodologies, while also ensuring that we can relocate fluorescence-associated structures within ultrastructure-dense electron micrographs. To that aim, fiducial markers have been used. Fiducials are spherical points that can be applied to the sections on a grid that are then used to identify those regions on the grid at higher magnifications. CLEM alignment requires that the fiducials are both fluorescent and electron dense, so that we can see the same fiducials in LM and then later at higher TEM magnifications. For example, Blue FluoSphere® (Molecular Probes, Inc., Leiden, The Netherlands) fiducials were used by Kukulski et al. (2011). These can be used with green or red cellular fluorescent markers. Schellenberger et al. (2013) and Schorb and Briggs (2013) opted to use multichannel fluorescent 100-nm TetraSpeck™ microspheres (Life Technologies, Carlsbad, CA, USA) as fiducial markers allowing for fluorescence emission in red, green, and blue channels. These 100-nm beads are also sufficiently electron dense enough to be visualized by TEM. Both Schellenberger et al. (2013) and Schorb and Briggs (2013) tested fluorescent fiducial alignment procedures using fluorescently tagged viral particles that were plunge-frozen for cryo-fluorescence microscopy followed by cryo-electron tomography. Schellenberger et al. (2013) tested chromatic shifts or mechanical stage shifts in *x*- and *y*-axes when using multi-color (red–green–blue) emitting TetraSpeck™ beads. They found, on average, ~150-nm shift between the same beads imaged in green and blue channels. For the same beads imaged in red and blue channels, they noticed even higher average ~300–400-nm shift. For high-precision CLEM image alignment, especially when imaging closely associated structures in multichannel fluorescence imaging, it is recommended to first align images acquired in different channels to one another before proceeding to align fluorescence image data with that of EM images. Given the nature of Formvar and plastic thin sections, there is much room for stretching, shrinking, and warping of plastic surfaces between grid bars, between cells, and between LM and EM imaging. Using an alignment program that can account for elastic warping in distances between fiducial markers selected in LM images versus EM images is essential.

We are now challenged with translating suboptimal fluorescence resolution (~200 nm) onto an electron micrograph with resolutions varying between 1 and 20 nm (Kukulski et al., 2011; Schellenberger et al., 2013; Schorb & Briggs, 2013). This can be problematic if fluorescently tagged proteins localize to areas of a cell where many organelles are densely packed, interlinked, or compartmentalized, confusing attempts to correlate fluorescence localization with a specific organelle or region of the cell.

For suspension cultures, we have applied high-pressure freezing. We tested a modified freeze-substitution protocol adapted from Kukulski et al. (2011) to prepare yeast cells that had been transfected with an EGFP-tagged amphipathic lipid-sensing domain of synapsin previously described by Krabben et al. (2011) and explored further by Pranke et al. (2011). This domain interacts with highly curved membranes, such as those comprising vesicular surfaces, playing a role in binding and aggregating synaptic vesicles in neurons. Transformed yeast cells were freeze-substituted and embedded in HM20 resin. We found fluorescence difficult to detect in thin sections (i.e., <120 nm thick) and sections thicker than 150 nm will be difficult to resolve by TEM unless electron tomography is used. Therefore, we used 150-nm-thick sections, labeled with 200-nm diameter Blue FluoSpheres used as fiducial markers for correlation between LM and EM. Given that many of our proteins of interest were endogenously tagged with GFP or mCherry (Petrovska et al., 2014), having fluorescent fiducials that, when quenched, were mostly fluorescing in only the UV-blue channel was optimal. Second, the 200 nm polystyrene-based Blue FluoSpheres® were of sufficient electron density to be easily observed via electron microscopy rendering them useful in aligning areas of interest in TEM with LM images. We acquired images at either 10 × or using a 100 × (1.4 NA) oil immersion lens with an upright epifluorescence microscope at excitation/emission of 500/520 for blue, for the fiducials, and 488/514 for GFP (Fig. 2.4A). Images were also acquired in bright field DIC and overlaid on the low-magnification TEM image (Fig. 2.4C),

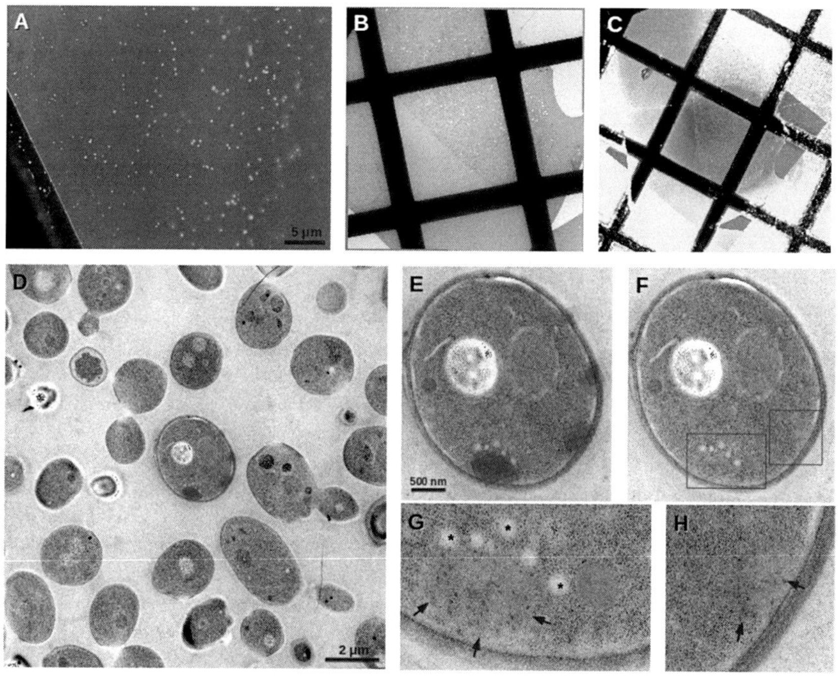

FIGURE 2.4

Investigation of structures associated with a synapsin-EGFP fusion protein in yeast. (A) 100 × LM images of synapsin-GFP, green fluorescence, overlayed with 200 nm Blue FluoSphere fiducials, blue fluorescence, on a DIC image; (B) 100 × TEM micrograph, 2 × 2 montage of the same ROI as identified by LM; (C) 100 × TEM image overlayed on 10 × DIC image; (D) GFP signal and blue fiducials correlated to a subregion in 11,000 × TEM micrograph; (E) TEM image of yeast cell with correlated GFP signal; (F) TEM image without fluorescent channels, GFP-correlated regions in boxes; (G) bottom-center subregion from (F) showing vesicle clusters in a ribosome-exclusion zone adjacent to the plasma membrane; (H) bottom-right subregion from (F) highlighting similar structure as seen in (G). Vesicle clusters: arrows; cortical ER: asterisks. (See the color plate.)

to visualize grid bars and landmarks on the Formvar surface. We relocated the same area where the fluorescent cell was located in TEM, first by finding and imaging the same ROI at lower 100 × TEM magnification (Fig. 2.4B). To distinguish between densely packed yeast cells, we performed a rough alignment using the lower magnification 10 × DIC image with the 100 × TEM image (Fig. 2.4C). After noting the subregion within the TEM image where our fluorescent cell was located, we acquired montages at higher magnifications (~ 1–2 nm pixel size) in TEM (Fig. 2.4D and E). Using the blue-electron-dense fiducials, we aligned the LM and TEM images such that we were able to identify the precise location of the synapsin-GFP chimera in the yeast cell (Fig. 2.4E). The GFP signal correlated exactly with clusters of vesicles within the yeast cell (Fig. 2.4F) that were always found laying adjacent the PM (Fig. 2.4G and H). For the alignment of fiducials, we used the Amira imaging software package (FEI Company, the Netherlands; Fig. 2.5).

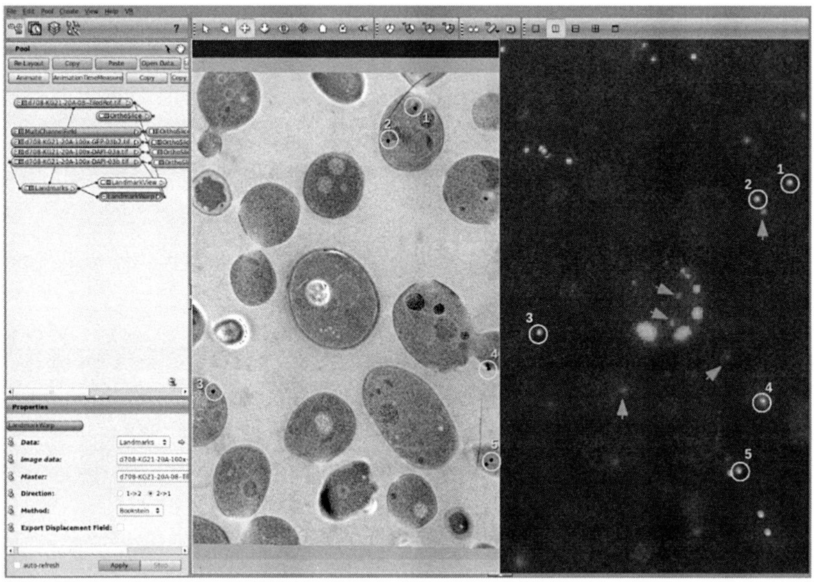

FIGURE 2.5

Screen capture of the Amira imaging software interface with imported TEM image (center panel) and fluorescence (right panel) images. Using the MultiChannelField, the epifluorescence alpha-channel images of GFP (green) and Blue FluoSpheres (DAPI channel, blue) are pseudo-colored. Using the LandMark (2 sets) module in Amira, electron-dense spheres corresponding to FluoSphere fiducials are selected as landmark points consecutively in first, the EM image, and then the fluorescence image (yellow circles surrounding pale blue selected "landmark" points). When a sufficient number of points have been selected, the fluorescence image is LandMarkWarped (registered) to the EM image. Red arrows denote FluoSpheres that detached between fluorescence and TEM imaging. (See the color plate.)

While this provides exceptionally powerful data regarding fluorescent protein localization, there were a few problems arising with the alignment procedure. Between fluorescence and TEM imaging, there were several steps through which the grid was treated with water or other reagents, in rinsing away the glass slide mounting medium or during staining with lead citrate. During these steps, some of the FluoSpheres dislodged from the Formvar-section surface and were either rinsed away (blue fluorescent "shadows" in Figs. 2.4D and E and 2.5, right panel) or relocated elsewhere on the grid. Most of the fiducials remained in place, still allowing for accurate alignment between LM and EM.

2.1.4 STAINING FOR TEM

Once samples have been imaged for fluorescence, then poststaining with either or both lead citrate and uranyl acetate can be performed. Karreman et al. (2012) tested the resilience of the fluorescent dye, TRITC after poststaining with 0.2% uranyl acetate and found a 68% reduction in fluorescence intensity while fluorescence lifetime was lowered by 25%. Consider that, due to interactions between the heavy metal stains and the fluorophore, the option to acquire any further fluorescence data after poststaining the grid is not worthwhile. Standard staining procedures employed for uranyl acetate and lead citrate are applicable at this stage of CLEM procedure.

2.2 METHODS

2.2.1 CRYO-IMMOBILIZATION AND SEGMENTATION OF MICROTUBULES IN MITOTIC SPINDLES IN *C. elegans* EMBRYOS BY ELECTRON TOMOGRAPHY

The protocols for fluorescence imaging, high-pressure freezing and freeze-substitution followed by electron tomography, and tilt-series acquisition were documented in detail by Müller-Reichert et al. (2007) and Woog et al. (2010). Computational tools for microtubule segmentation and analysis are described further by Torsney-Weir et al. (2011), Weber et al. (2012), and Redemann et al. (2014).

2.2.2 LOCALIZATION OF EGFP-SEC22-TRANSFECTED NEURON CELLS

1. Etch a grid-like design onto an ACLAR sheet using a razor blade or fine-gauged syringe needle. This grid will be used as a reference guide to aid in later relocation of transfected cells.
2. Coat the etched-surface of the ACLAR sheet with 0.1% poly-L-lysine to aid in culture cell adherence. Allow solution to dry.
3. Place the ACLAR sheet, etch-side facing up, onto the bottom of a culture dish. We use 6-well plates.
4. Cortical neuron cells were dissected and dissociated before being plated in MEM loading medium on ACLAR sheets within a 6-well cell culture plate.

After 2 h of attachment to the ACLAR sheets, the culture medium is changed to neurobasal medium (with B27 and glutamine). Cells are then transfected at DIVI with GFP or GFP-Sec22b (Petkovic et al., 2014) and allowed to incubate for 24 h allowing for expression.

5. While adhered to the ACLAR sheets, cells are fixed with 0.1% glutaraldehyde and 1% paraformaldehyde in 1 × PBS buffer for 30 min to 1 h. The sheets are rinsed three times with 1 × PBS buffer.
6. A droplet of 1 × PBS is applied to the ACLAR sheet with preserved cells, and a coverslip is laid on top of the ACLAR for fluorescence imaging.
7. Cells expressing EGFP-tagged Sec22p are identified and located via fluorescence microscopy followed by bright field microscopy so as to allow for later relocation of nonfluorescent cells, structural details, and razor-blade etches on the ACLAR sheet.
8. Cells are imaged by LM at low (4 × or 10 ×) and higher (63 × dry, or 100 ×, 1.4 NA, oil immersion) magnification.
9. After LM imaging, the coverslip is gently pried from the ACLAR surface. A few droplets of filtered water are added at the edge of the ACLAR sheet and coverslip. The tip of a forceps is wedged between the two surfaces to help break surface tension so as to not dislodge cells. The ACLAR sheet is washed with ddH_2O for approximately 20 min.
10. Cells on the ACLAR sheet are incubated with 1% OsO_4 reduced with 0.15% potassium ferrocyanide for 60–120 min at room temperature.
11. The ACLAR sheet is washed three times with ddH_2O.
12. Dehydration series are performed: 50% ethanol, 70%, 80%, 90%, 96% followed by three changes in 100% anhydrous ethanol with 10 min per step at room temperature.
13. The ACLAR sheet is transitioned from 100% ethanol into increasing concentrations of Epon Lx112 or your epoxy resin of choice. Three transition steps comprised of 1:3 (v/v), followed by 1:1 (v/v), and then 2:3 (v/v) of ethanol-to-Lx112 resin are performed for 1 h per incubation.
14. Cells are infiltrated for ~4 h or overnight with 100% Epon Lx112.
15. The ACLAR sheet is then embedded in 100% Epon Lx112 and laid in a thin layer of resin followed by polymerization at 60 °C for ~48 h.
16. Due to the strong adherence of the neuron cells to the ACLAR sheet, cells have remained in their original positions throughout the fluorescence imaging and subsequent fixation, dehydration, and resin-embedding steps. Surface details, grid patterns, and lines drawn on the ACLAR sheet can be identified using a dissection microscope to relocate general areas where fluorescent cells of interest were identified prior to osmium fixation.
17. Areas where cells of interest were located are sliced out of the ACLAR sheet using a razor blade. Areas with cells of interest are remounted on an epoxy-based dummy block using a two-phase Epoxy 907 Adhesive System to stably hold the sample in place.

18. Section the area with cell(s) of interest. Here, we use a Leica UCT Ultramicrotome and transfer 70-nm-thick (thin sections) or 300-nm-thick (tomography) sections onto Formvar-coated copper slot grids.
19. Grids are stained with uranyl acetate and lead citrate and then imaged with a Tecnai-12 BioTwin TEM (FEI Company, Eindhoven, the Netherlands) for thin sections or a Tecnai F30 TEM (FEI Company) for electron tomography:
    - Cells originally identified by fluorescence microscopy are located at lower/medium TEM magnifications (between 300× through 3000×) by comparing the arrangement of cells adjacent, ACLAR sheet grid marks or etches, and other notable features.
    - The cell(s) of interest are imaged at higher TEM magnifications.

2.2.3 HIGH-RESOLUTION CLEM: SAMPLE PREPARATION, SECTIONING, IMAGE ACQUISITION, AND ALIGNMENT

1. *Saccharomyces cerevisiae* cells transformed with a synapsin-GFP fusion protein (fP) are inoculated into an $S_{Glucose-Ura+Doxycycline}$ (as a repressor) culture and grown overnight at 30 °C. Cells are backdiluted into fresh $S_{Glucose-Ura}$ medium containing no Doxycycline and grown until log phase ($OD_{600}=0.3$–0.5).
2. Concentrate the cell culture by gentle vacuum filtration with 0.45-m HV Durapore® Membrane Filters (Merck Millipore, Millerica, MA, USA) in a Vacuum Filtering Flask (Merck Millipore) with a 125-mL capacity Glass Microanalysis Filter Holder (Merck Millipore).
3. After filtration, concentrated cells are kept moist on the filter paper by readdition of 30–50 μL unfiltered growth medium every 15 min to prevent cell dehydration before high-pressure freezing.
4. For high-pressure freezing, we use 0.1 × 1.5-mm membrane carriers (Leica Microsystems) in concert with the Leica EM PACT2 high-pressure freezing machine with a rapid transfer system (Leica Microsystems). Before freezing, 0.5 μL of 20% bovine serum albumin (BSA) dissolved in H_2O is transferred to the membrane carrier to act as a cryo-protectant.
5. Scrape a small blob of filtered yeast from the filter paper using a toothpick. Mix the yeast cells with the 20% BSA solution in the membrane carrier. Immediately transfer the membrane carrier into the EM PACT2 for high-pressure freezing.
6. After freezing, membrane carriers are maintained under liquid N_2 to prevent ice crystal formation until freeze-substitution can be performed.
7. Freeze-substitution is carried out using a Leica AFS2 (Leica Microsystems) machine. Membrane carriers are transferred into a freeze-substitution cocktail containing 0.1% uranyl acetate, 2–4% H_2O, in distilled acetone. The premade cocktail is stored in 1 mL aliquots in cryo-vials under liquid N_2 until needed.

8. Sample processing initiates with a −90 °C step lasting 50–58 h before transitioning to −45 °C over 9 h. Within the 2 h −45 °C step, the cocktail is rinsed twice with prechilled 100% acetone (distilled) before transitioning into prechilled 100% anhydrous ethanol. For yeast or other cells cultured in 0.5–2% glucose-based mediums, it is best to use ethanol to dissolve remaining cytosolic glucose, which impedes acrylic and/or epoxy resin infiltration (Buser & McDonald, 2010).
9. Within the AFS2 machine, sample membrane carriers are transferred from the cryo-vials and positioned within a Leica Flow-through ring in a Reagent bath (Leica Microsystems) container to ease reagent handling and subsequent polymerization of samples. Insert the membrane carriers with the sample facing upward. Prechilled anhydrous ethanol is added to the Flow-through ring to ensure that sample carriers remain moist during transfer.
10. For fluorescence preservation, freeze-substituted samples are embedded in low-temperature Lowicryl HM-20 acrylic resin. This resin can be infiltrated into samples between −45 °C and −25 °C, followed by UV polymerization at −25 °C for 48–56 h. Due to the toxicity and volatility of HM-20, it is recommended to install the AFS2 machine in a walk-in fumehood, or adapt a mobile fumehood with air filtration system over top of the AFS2 sample chamber. Beginning at −45 °C, sample carriers are infiltrated with 10% HM-20 in anhydrous ethanol followed by 25% HM-20, then 50%, 75% and then 100% HM-20 over a 10–12 h period until reaching −-25 °C. Three changes with 100% HM-20 are made over a ~24 h period.
11. To facilitate HM-20 polymerization under hypoxic conditions, an ACLAR foil (Leica Microsystems) is laid over top of the Flow-through ring (Leica Microsystems) after filling the Flow-through ring and Reagent bath with fresh HM-20. The samples are polymerized in the AFS2 machine with the UV LED lamphousing (Leica Microsystems) at −25 ºC for 48–56 h.
12. After polymerization, sample cylinder blocks with membrane carriers still attached are sliced from the plastic Flow-through ring using a razor blade.
13. Membrane carriers are separated from the sample cylinder blocks by shearing HM-20 resin from the base of the membrane carrier and then gently prying the carrier from the base of the cylinder, leaving the yeast sample embedded in the cylinder. The entire cylinder with the embedded yeast cells can now be mounted into an ultramicrotome chuck for sectioning.

2.2.3.1 Sectioning and fiducial application

1. Sections are cut with an Ultracut UCT Microtome (Leica) set to the ~150 nm thickness range and then mounted on Formvar-coated 100-mesh copper grids. To facilitate section relocation in LM and TEM, try to place sections in the center of the grid. Ensure that grids have some feature allowing for relocation of grid spaces in LM and EM. For example, often square Cu grids have an "arrow" in the center. Do not use slot grids, as Formvar films tend to break when retrieving the

grid from the glass slide mounting medium after fluorescence microscopy imaging. Slot grids also lack useful features necessary for relocation and orientation of ROIs for CLEM.

2. Apply fluorescent fiduciary markers to the section surface. We use Invitrogen Molecular Probes™ Blue FluoSphere® fiducials, 200 nm diameter. It is recommended to quench the Blue FluoSpheres® as the spheres are highly fluorescent in the Blue/UV/Cyan/Green epifluorescence channels and can often obfuscate identification of any fluorescently tagged protein. This can be done with 0.1% Tween 20 for 10 min, followed by two washes in PBS (Kukulski et al., 2011). To prevent clustering, sonicate the solution immediately prior to use.

2.2.3.2 Fluorescence microscopy

1. Mount the grid, section side up, on a glass slide in 30 μL of nonhardening mounting medium with coverslip (we use Vectashield® Vector Laboratories, Peterborough, UK).
2. Find the center of grid at low magnification (i.e., 10 ×) in bright field, noting orientation of grid on a sheet of paper. Drawing a diagram or printing a large picture of your grid from the manufacturer's Web site upon which you can sketch locations will help relocate ROIs later in TEM.
3. Proceed to higher magnification lens (i.e., 100 ×, 1.4 NA, oil immersion) using only transmitted light to avoid photobleaching the sample, recheck the grid diagram for orientation (i.e., find central "arrow"). Avoid selecting and imaging areas near the edge of the grid. These peripheral areas will be more difficult to reacquire later in TEM due to mechanical limitations in *x*/*y*-axis TEM stage motion.
4. Isolate fluorescently tagged structures on sections in fluorescent channel(s) specific to fluorescent marker (i.e., GFP, dsRed, mCherry, etc.).
5. Once an ROI has been located, acquire an image in the appropriate channel(s). Without moving the stage, image the same ROI under a Blue filter to acquire fiducial fluorescence. Given that regions between grid bars where Formvar/sections are unsupported and are therefore, not flat, often only part of the ROI will be in focus. Try to acquire several images in the same ROI in the same epifluorescence channel using different focusing steps to ensure that as many areas within the ROI are in focus. Ensure that you have acquired all relevant fluorescent channels before moving the stage to seek out a new ROI.
6. When removing the grid from the glass slide/coverslip, use a syringe filled with ddH_2O and 0.2-μm filter with needle attached to apply water near the edge of the coverslip and slide. It is important to avoid damaging the Formvar film and sections on the grid.
7. Wedge the tip of forceps between the slip and the slide to gently and slowly pry the coverslip away from the slide. The water should seep in between the slip and slide to break the cohesion-tension of the Vectashield® mounting medium.
8. Pick the grid up with forceps, clamping the grid in place by applying tension in the forceps with a rubber O-ring. Holding the grid vertically over a surface,

rinse the grid with filtered ddH_2O by allowing single droplets to fall onto the grid from the filter–syringe–needle. Draw excess water off with a wedge of filter paper and allow the grid to air dry.

9. Post-stain with lead citrate for 3 min.

2.2.3.3 TEM imaging

For CLEM, a setup that can acquire montaged images encompassing a larger field of view is optimal to simplify correlation of EM with LM images. First, we perform an initial LM-to-EM alignment using a large field of view TEM image in order to ascertain which cells are actually worth imaging at higher magnification. This can be done at low magnification.

1. Take EM micrographs at lower magnification to acquire an ROI (montaging if necessary) approximating the fluorescence microscopy field of view at 100 × magnification. The magnification should be high enough to identify the fiducials: use a pixel size of ~10 nm. The ROI should show notable features, such as cell orientation, shape, size, with respect to section edges or to grid bars.
2. Do a rough alignment of the montaged image with the fluorescence images (Blue FluoSphere fiducials and GFP/mCherry, etc.), using appropriate software Fiji/ImageJ or Amira. The rough CLEM overlayed image is used to identify any specific subregions within the TEM image that are worth acquiring at higher TEM magnifications.
3. When interesting subregions within the ROI are identified using the aligned LM and low-magnification TEM image, return to the TEM to acquire higher magnification images at the final resolution. Montaging can again be used to cover the desired field of view. One possibility is to directly transfer the LM image data (Fluorescence and DIC/Widefield, etc.) to the TEM computer: open the LM data in SerialEM (http://bio3d.colorado.edu/SerialEM/) and use the Registration Points option to align selected areas in the LM images to low/medium-magnification TEM images acquired with SerialEM (Schorb & Briggs, 2013).
4. Once interesting subregions have been located in roughly correlated low-magnification LM/TEM images, high-magnification TEM images can immediately be acquired.

2.2.3.4 High-resolution CLEM image alignment

For LM/TEM alignment purposes, several programs can be used to align or overlay fluorescence and TEM images using fiducials. Section warping due to stretching and shrinking of plastic embedding resin incurred during sectioning, grid handling, staining, and electron beam damage will affect the alignment procedure. It is important to choose a program that can account for minute elastic shifts in subregions of an ROI between LM and TEM imaging. To this end, we have employed the LandMarkWarp module or subprogram within the Amira (FEI Company, Eindhoven, the Netherlands) imaging software (Fig. 2.5).

1. Open both fluorescence LM and EM images in Amira and display them side by side in two display windows.
2. Using the LandMarkWarp function, select fiducials pairs (one in LM and one in EM) that correspond to those same fluorescent sphere.
3. Warp and overlay the LM image onto the TEM image with confidence that our fluorescence data actually correspond to specific structures observed in TEM. Other options include using the Landmark and Registration (elastic stack alignment) plugins in Fiji/ImageJ (http://fiji.sc/Fiji).

2.3 MATERIALS

2.3.1 REAGENTS

Chemicals from EMS, Science Services GmbH, München, Germany: 100% acetone, distilled; 16% paraformaldehyde (aqueous EM grade); 25% glutaraldehyde (EM grade, distillation, purified); Formvar (15/95 Resin powder); lead citrate and Epon LX112 Epoxy resin kit; Cryo-Lok, cryogenic vials (2.0 mL); Lowicryl HM-20 kit, ACLAR® fluoropolymer film, copper slot grids (2 mm × 1 mm slot), 100-mesh square copper grids, and uranyl acetate (Polysciences, Inc., Eppelheim, Germany). Potassium hexacyanoferrate (III) trihydrate (Sigma-Aldrich Chemie GmbH, Munich, Germany). 6-Well cell culture plate (Merck Millipore, Millerica, MA, USA). Two-Part Epoxy Adhesive Kit, Epoxy 907 Adhesive System (Miller-Stephenson Chemical Company, Inc., Sylmar, CA, USA). Bovine serum albumin, Fraction V (Sigma-Aldrich Chemie GmbH, Munich, Germany). The fluorescent fiducials described were 0.2-μm diameter Blue (365/415) FluoSphere® Fluorescent Microspheres (Molecular Probes, Inc., Leiden, The Netherlands). For LM, grids were mounted in Vectashield® H-1000 mounting medium (Vector Laboratories, Inc., Burlingame, CA, USA).

2.3.2 CELL CULTURES

S. cerevisiae BY4742 transformed with a construct encoding the amphipathic synaptic vesicle targeting region of synapsin (amino acids 69–96) (Krabben et al., 2011) fused to a portion of the coiled coil region of GMAP210 (amino acids 39–375), followed by yEGFP. Cells were cultured in S-medium minus Ura, with 2% Glucose. Mouse cortical neuron primary cells were initially grown in loading medium (MEM, fetal calf serum, glutamine, and glucose) and then in neurobasal medium supplemented with B27 and glutamine. Culturing conditions and GFP-Sec22-related constructs are described in more detail by Petkovic et al. (2014).

2.3.3 LIGHT MICROSCOPY

LM images were acquired with an upright Zeiss Axioplan2 epifluorescence microscope (Carl Zeiss, Oberkochen, Germany), with Metaview imaging software (Visitron Systems, Puchheim, Germany).

2.3.4 ELECTRON MICROSCOPY AND IMAGE PROCESSING

High-pressure freezing, and freeze-substitution supplies were from Leica Microsystems GmbH, Wetzlar, Germany: EMPACT2 high-pressure freezing machine, membrane carrier 1.5 mm × 0.1 mm, AFS2 freeze-substitution machine, UV lamphousing with LEDs, Reagent bath, Flow-through rings, and ACLAR® foils. Sectioning was performed using an Ultracut UCT Microtome (Leica). We employed a Tecnai-12 BioTwin TEM operated at 100 kV (FEI Company, Eindhoven, NL), with an axial Tietz F214A CCD camera (TVIPS GmbH, Gauting, Germany). With this camera, we have the option of using either EM-Menu 4.0 (TVIPS) software or SerialEM (http://bio3d.colorado.edu/SerialEM/index.html) to acquire TEM images. Either software can produce montaged or tiled images. Dual-axis electron tomography tilt series were acquired on a 300 kV Tecnai G2 F30 Twin TEM (FEI Company) with a bottom-mounted Gatan US 1000 CCD camera (Roper Scientific) with SerialEM (http://bio3d.colorado.edu/SerialEM/). Tilt series were reconstructed and the two axes were combined in IMOD (http://bio3d.colorado.edu/imod/). Centrosomal microtubules were segmented using the automated MicrotubuleTracing module and Segmentation Editor (Redemann et al., 2014; Torsney-Weir et al., 2011; Weber et al., 2012) within the ZIBAmira imaging software package (Zuse Institute Berlin, Germany). The automated MicrotubuleTracing software is also available online: http://amitube.mpi-cbg.de/. Image processing for the alignment of fiducials between LM and EM was also done in ZIBAmira.

2.4 DISCUSSION

Here we have reviewed and described three general directions. In the first approach, we have combined live-cell imaging and electron microscopy to study the 3D organization of microtubules during cell division. The purpose is to use EM resolution to trace individual microtubules in a densely crowded environment at a very precise time point in the cell cycle. This approach requires rapid fixing of the sample at the time point of interest. To that aim, we have used high-pressure freezing, but chemical fixation is also possible. The advantage of this time-resolved approach is that we know the exact state of the cell before processing it for EM. However, there is no direct spatial correlation between LM and EM images of the sample. In the second example, we used fluorescence microscopy to initially locate a cell of interest among hundreds for further examination at the EM level. This form of "cell sorting" allows us to address the issue of phenotypic variability in transfected cell cultures to definitively identify transfected cells from untransfected cells. Further EM processing and subsequent EM-level imaging can then be performed on the fluorescence-identified cells of interest. The third approach described herein involves spatial correlation of LM and EM data and is more challenging as it requires higher precision correlation.

Instead of separating the LM and TEM imaging steps, one could employ the new Tecnai™ iCorr TEM (FEI Company, Eindhoven, the Netherlands) which is adapted from a standard FEI Tecnai TEM platform, where an additional 15 × objective lens with fluorescence excitation ranging from 460 to 500 nm has been integrated into the TEM column, allowing for rapid and consecutive acquisition of fluorescence and TEM image data on the same EM grid, within the same apparatus. Interestingly, the iCorr™ system can be adapted for use in cryo-electron microscopy allowing for investigation of fluorescence-tagged proteins retained after freeze-plunging followed by acquisition within the same TEM. The cryo-adaptability of the iCorr™ system is especially useful given that antibodies cannot be applied to frozen sections, preventing the use of immunolabeling. For cryo-EM, only endogenously labeled samples can be used. For resin-embedded samples, Agronskaia et al. (2008) used a precursor to the iCorr system, the integrated Laser and Electron Microscope, to localize Alexa488-labeled catalase in peroxisomes of rat liver cells. To further the use of this technology, Karreman et al. (2012) considered the effects of low-vacuum conditions on various dyes, noting that certain dyes were optimally fluorescent when hydrated. They tested the fluorescence intensity of Alexa488, Alexa532, Cy2, Cy3, FITC, and TRITC in water or N_2 gas (to mimic low-vacuum, TEM conditions), finding that Cy2 and TRITC were the most stable, along with Cy3 and FITC, in N_2.

We have described our protocol with the intent of examining our samples via TEM. However, there is no reason that the protocols described herein could not be similarly examined using SEM. Serial sections of the same cell identified by LM could then be reexamined by SEM. Liv et al. (2013) integrated a 100 ×, 1.4 NA objective lens inside the vacuum chamber of a SEM, below the sample in an inverted epifluorescence LM configuration. A 470 nm LED provided illumination for fluorophore excitation. To study cell-to-cell interactions, they labeled actin in adenocarcinoma cells with phalloidin–Alexa488 or cortactin–Alexa488 and were able to identify higher concentrations of the fluorescent markers in extrusions or fine lamellar outgrowths. Provided poststaining of the sample does not interfere with the fluorescence signal, this approach could be used for either low- or high-resolution correlation between LM and EM, possible as an alternative to using fiducials.

Finally, it should be noted that all the methods described above are compatible with 3D imaging in EM, using for instance electron tomography (Guizetti et al., 2011; Pelletier et al., 2006; Petkovic et al., 2014; Swulius & Jensen, 2012) or SEM (Kopek et al., 2012).

Some limitations of the approaches described here should also be considered. For instance, fluorescent protein fusions have previously been noted to contribute to mislocalization, misfolding, and dimerization, resulting in structural artifacts when tagged to proteins of interest (Coumans et al., 2013; Katayama et al., 2008; Quattrocchino, Spelt, & Koes, 2013). Some fluorescent proteins, notably multimeric DsRed and GFP and other *Aequorea victoria*-derived fluorescent variants, can dimerize at higher concentrations in a cell (Von Stette, Noireclerc-Savoye, Goedhard, Gadella, & Royant, 2012) which can interfere with localization or substrate binding, thereby raising issues with regard to fluorescence imaging and quantification. Von Stette et al. (2012) developed an obligately monomeric version of cyan fluorescent protein, called mTurquise presenting an A206K mutation; other researchers have used this same mutation to render yellow fluorescent protein (YFP) into mCitrine, GFP into m(E)GFP and mVenus, and the red mCherry. Many transgenic systems involve overexpression of said fluorescent fPs also resulting in structural or biochemical changes not at all representative of the biological system under study (Comley et al., 2011; Coumans et al., 2013). Untargeted DsRed and its derivatives, mRFP and zFP506, were found to accumulate in lysosomes while retaining fluorescence (Katayama et al., 2008). Merely expressing untagged GFP in breast cancer cells induced a myriad of proteomic changes resulting in altered expression of proteins involved in protein folding, immune response, and cytoskeleton regulation (Coumans et al., 2013). YFP expression was also implicated in increasing gene expression for proteins involved in oxidative/metabolic stress, inflammation, and apoptosis signaling neurons of transgenic mice (Comley et al., 2011). Swulius and Jensen (2012) reported that the helical cytoskeleton described in previous studies utilizing YFP tagged to the actin-homolog MreB in *Escherichia coli* MC1000/pLE7 is an artifact of the YFP tag. Using cryo-electron tomography, they examined and compared untransformed wild-type *E. coli* pLE7 with cells expressing native MreB, to MreB-YFP, and MreB tagged with mCherry and found that only YFP-tagged cells exhibited filamentous structures adjacent the inner membrane of the cells. In lieu of this finding, they suggested that much of the literature surrounding helical cytoskeleton studies with YFP-MreB fPs in bacteria ranging from *Bacillus subtilis*, *Caulobacter crescentus*, and *Vibrio cholerae* should be reconsidered.

Similarly, we encountered fluorescent protein-induced structural alterations when studying the aggregative capabilities of the glutamine synthesis, mCherry fP (Petrovska et al., 2014). In considering potential structural artifacts, we compared fP-mCherry-transformed yeast cells with those expressing the photoactivatable Dendra2-fP or those with untagged fP. We found striking differences in the cytosolic filaments found in the mCherry fusion versus the Dendra2-fused and nonfused Gln1. In spite of these structural differences, our CLEM strategy could not be pursued using Dendra2-tagged protein because it did not survive the final extended 48 h of UV-light exposure required to polymerize the Lowicryl HM20 resin in which the yeast cells were embedded. In lieu of this, we could have embedded cells in a different resin, such as GMA, described earlier.

These artifacts are difficult to avoid, but they can be detected by testing multiple approaches and comparing the results. By preserving, at minimum, two samples: the first expressing the protein of interest with a fluorescent tag and the second expressing the protein of interest, alone, we can test the degree of structural interference induced by that fluorescent probe using electron microscopy. Given that not all fluorescent tags are alike, comparing fluorescence patterns of different fluorescent protein fusions (i.e., GFP, YFP, mCitrine, mCherry, DsRed, Dendra, mEosFP, etc.) followed by TEM or SEM examination is useful to affirm the degree to which a particular fluorescent probe interferes with the structure and localization of a given protein of interest.

2.4.1 TOWARD SUPER-RESOLUTION CLEM

While high-resolution CLEM overcomes many previous barriers in localizing specific fluorescent signals to specific structures in EM, the best fluorescence resolution that can be obtained with standard epifluorescence, confocal, or spinning disk LM is approximately ~200 nm in the *x*-/*y*-axes and ~400–600 nm in the *z*-axis. When comparing closely associated structures at the EM level using correlative fluorescence signal for that locale, the lower LM resolution limits the degree to which we can definitively state that a fluorescent signal corresponds to a specific structure or subregion within an aligned electron micrograph. Even by fluorescence microscopy standards, the 200 nm limit in the *x*/*y*-resolution impedes that ability to precisely localize or colocalize fluorescently tagged molecules, even when they target to different subcompartments or trafficking pathways. This limitation is well recognized and over the past decade, it has been giving rise to the development of several technologies and techniques. SRFM depends on the ability to separate individual fluorescent probes or molecules not only in space but also over time, so that individual molecules can be imaged. Enabling many SRFM techniques like PALM, STORM, or stimulated emission depletion microscopy (STED) is the employment of photoactivatable, photoswitchable, or photoblinking fluorescent probes that can be repeatedly, through targeted doses of fluorescent light, activated and deactivated before finally photobleaching (Betzig et al., 2006; Herbert, Soares, Zimmer, & Henriques, 2012; Patterson et al., 2010; Watanabe & Jorgensen, 2012). These comprise some of the most demanding imaging techniques involving extended fluorophore excitation and acquisition times. Already, SRFM techniques have been employed in CLEM techniques, rendering the fluorescence resolution comparable to EM resolution in ways that provide more meaningful correlative structural localization data than previously attainable by more established epifluorescence or confocal microscopy methods. Betzig et al. (2006) demonstrated localization of cytochrome *c* oxidase tagged with dEosFP using total internal reflection fluorescence microscopy (TIRF) and PALM on cryo-prepared thin-section COS-7 cells. They relocated and aligned the PALM image with a TEM image, showing sublocalization of cytochrome *c* oxidase in mitochondrial matrices. Watanabe and Jorgensen (2012) localized a tdEos-tagged TOM-20 in *C. elegans* mitochondria using PALM followed by SEM on thin sections. Watanabe et al. (2011) further compared resolution and fluorescence localization in confocal, STED, TIRF, and PALM using the histone, H2B and outer mitochondrial membrane protein, TOM20 fused to either Citrine or tdEos. Fluorescent patterns identified in the highest resolution fluorescence imaging techniques, STED and PALM were relocated and correlated in EM, with a precision exceeding that of standard epifluorescence, confocal, and TIRF microscopy. Furthering the exploration of mitochondrial subcompartmental localization, Kopek et al. (2012) tagged TFAM, a mitochondrial DNA-binding protein, with mEos2 in mouse 3T3sw fibroblast cells. They coupled PALM fluorescence data with correlational focused ion beam SEM 3D imaging on Tokuyasu cryo-sections, isolating TFAM in mitochondrial nucleoids. Expanding on these examples, we can imagine a myriad of applications for super-resolution in future CLEM studies.

ACKNOWLEDGMENTS

The authors wish to thank Dr. Cathy Jackson for permitting the use of yeast cells as well as Simon Alberti, Doris Richter, Ivana Petrovska, Matthias Munder, and Liliana Malinovska for aiding with yeast culturing methods, reagents, and discussions.

REFERENCES

Agronskaia, A. V., Valentijn, J. A., van Driel, L. F., Schneijdenberg, C. T. W. M., Humbel, B. M., van Bergen en Henegouwen, P. M. P., et al. (2008). Integrated fluorescence and transmission electron microscopy. *Journal of Structural Biology*, *164*, 183–189.

Betzig, E., Patterson, G. H., Sougrat, R., Lindwasser, O. W., Olenych, S., Bonifacino, J. S., et al. (2006). Imaging intracellular fluorescent proteins at nanometer resolution. *Science*, *313*, 1642–1645.

Brown, T. A., Fetter, R. D., Tkachuk, A. N., & Clayton, D. A. (2010). Approaches to super-resolution fluorescence imaging of mitochondrial proteins using PALM. *Methods*, *51*, 458–463.

Buser, C., & McDonald, K. (2010). Correlative GFP-immunolocalization microscopy in yeast. *Methods in Enzymology*, *470*, 603–618.

Campbell, R. E., & Davidson, M. W. (2010). Part I. Types of imaging reporter genes. Chapter 1. Fluorescent reporter proteins. In S. S. Gambhir, & S. S. Yaghoubi (Eds.), *Molecular imaging with reporter genes* (pp. 3–40). New York: Cambridge University Press.

Comley, L. H., Wishart, T. M., Baxter, B., Murray, L. M., Nimmo, A., Thomson, D., et al. (2011). Induction of stress in neurons from transgenic mice expressing yellow fluorescent protein: Implications for neurodegeneration research. *PLoS One*, *6*(3), e17639.

Cortese, K., Diaspro, A., & Tacchetti, C. (2009). Advanced correlative light/electron microscopy: Current methods and new developments using Tokuyasu cryosections. *Journal of Histochemistry & Cytochemistry*, *57*(12), 1103–1112.

Coumans, J. V. E., Gau, D., Poljak, A., Wasinger, V., Roy, P., & Moens, P. (2013). Green fluorescent protein expression triggers proteome changes in breast cancer cells. *Experimental Cell Research*, *320*(2014), 33–45.

Duke, E. M. H., Razi, M., Weston, A., Guttman, P., Werner, S., Henzler, K., et al. (2013). Imaging endosomes and autophagosomes in whole mammalian cells using correlative cryo-fluorescence and cryo-soft X-ray microscopy (cryo-CLXM). *Ultramicroscopy*, *143*, 77–87. http://dx.doi.org/10.1016/j.ultramic.2013.10.006.

Guizetti, J., Mäntler, J., Müller-Reichert, T., & Gerlich, D. W. (2010). Correlative time-lapse imaging and electron microscopy to study abscission in HeLa cells. *Methods in Cell Biology*, *96*, 591–601.

Guizetti, J., Schermelleh, L., Mäntler, J., Maar, S., Poser, I., Leonhardt, H., et al. (2011). Cortical constriction during abscission involves helices of ESCRT-III-dependent filaments. *Science*, *331*, 1616–1620.

Herbert, S., Soares, H., Zimmer, C., & Henriques, R. (2012). Single-molecule localization super-resolution microscopy: Deeper and faster. *Microscopy and Microanalysis*, *18*(6), 1419–1429.

Hodgson, L., Tavaré, J., & Verkade, P. (2014). Development of a quantitative correlative light electron microscopy technique to study GLUT4 trafficking. *Protoplasma*, *251*(2), 403–416.

Huang, B., Babcock, H., & Zhuang, X. (2010). Breaking the diffraction barrier: Super-resolution imaging of cells. *Cell*, *143*, 1047–1058.

Karreman, M. A., Agronskaia, A. V., van Donselaar, E. G., Vocking, K., Fereidouni, F., Humbel, B. M., et al. (2012). Optimizing immuno-labelling for correlative fluorescence and electron microscopy on a single specimen. *Journal of Structural Biology*, *180*, 382–386.

Katayama, H., Yamamoto, A., Mizushima, N., Yoshimori, T., & Miyawaki, A. (2008). GFP-like proteins stably accumulate in lysosomes. *Cell Structure and Function*, *33*, 1–12.

Kopek, B. G., Shtengel, G., Xu, C. S., Clayton, D. A., & Hess, H. F. (2012). Correlative 3D superresolution fluorescence and electron microscopy reveal the relationship of mitochondrial nucleoids to membranes. *Proceedings of the National Academy of Sciences of the United States of America*, *109*(16), 6136–6141.

Krabben, L., Fassio, A., Bhatia, V. K., Pechstein, A., Onofri, F., Fadda, M., et al. (2011). Synapsin I senses membrane curvature by an amphipathic lipid packing sensor motif. *Journal of Neuroscience*, *31*, 18149–18154.

Kukulski, W., Schorb, M., Welsch, S., Picco, A., Kaksonen, M., & Briggs, J. A. (2011). Correlated fluorescence and 3D electron microscopy with high sensitivity and spatial precision. *Journal of Cell Biology*, *192*(1), 111–119.

Liv, N., Zonnevylle, A. C., Narvaez, A. C., Effting, A. P. J., Voorneveld, P. W., Lucas, M. S., et al. (2013). Simultaneous correlative scanning electron and high NA fluorescence microscopy. *PLoS One*, *8*(2), e55707.

McDonald, K. L., Morphew, M., Verkade, P., & Müller-Reichert, T. (2007). Recent advances in high-pressure freezing. In J. Kuo (Ed.), *Electron microscopy, methods and protocols: 369. Methods in molecular biology* (pp. 143–173) (2nd ed.). Totowa, NJ: Humana Press Inc. (Chapter 8).

Müller-Reichert, T., Srayko, M., Hyman, A., O'Toole, E. T., & McDonald, K. (2007). Correlative light and electron microscopy of early *Caenorhabditis elegans* embryos in mitosis. *Methods in Cell Biology*, *79*, 101–119.

Müller-Reichert, T., & Verkade, P. (Eds.), (2012). Correlative light and electron microscopy. In *Methods in cell biology: Vol. 111*. (pp. 1–460). Elsevier Inc., Oxford, UK: Academic Press.

Patterson, G., Davidson, M., Manley, S., & Lippincott-Schwartz, J. (2010). Superresolution imaging using single-molecule localization. *Annual Review of Physical Chemistry*, *61*, 345–367.

Pelletier, L., O'Toole, E., Schwager, A., Hyman, A. A., & Müller-Reichert, T. (2006). Centriole assembly in *Caenorhabditis elegans*. *Nature*, *444*, 619–623.

Petkovic, M., Jemaiel, A., Daste, F., Specht, C. G., Izeddin, I., Vorkel, D., et al. (2014). The SNARE Sec22b has a non-fusogenic function in plasma membrane expansion. *Nature Cell Biology*, *16*(5), 434–444.

Petrovska, I., Nüske, E., Munder, M. C., Kulasegaran, G., Malinovska, L., Kroschwald, S., et al. (2014). Filament formation by metabolic enzymes is a specific adaptation to an advanced state of cellular starvation. *eLife*, *3*, e02409.

Polishchuk, R. S., Polishchuk, E. V., Marra, P., Alberti, S., Buccione, R., Luini, A., et al. (2000). Correlative light-electron microscopy reveals the tubular-saccular ultrastructure of carriers operating between Golgi apparatus and plasma membrane. *The Journal of Cell Biology*, *148*(1), 45–58.

Pranke, I. M., Morello, V., Bigay, J., Gibson, K., Verbavatz, J. M., Antonny, B., et al. (2011). α-Synuclein and ALPS motifs are membrane curvature sensors whose contrasting chemistry mediates selective vesicle binding. *Journal of Cell Biology*, *194*(1), 89–103.

Quattrocchino, F. M., Spelt, C., & Koes, R. (2013). Transgenes and protein localization: Myths and legends. *Trends in Plant Science*, *18*(9), 473–476.

Redemann, S., Weber, B., Möller, M., Verbavatz, J. M., Hyman, A. A., Baum, D., et al. (2014). The segmentation of microtubules in electron tomograms using amira. *Methods in Molecular Biology*, *1136*, 261–278.

Robinson, J. M., Takizawa, T., Pombo, A., & Cook, P. R. (2001). Correlative fluorescence and electron microscopy on ultrathin cryo-sections: Bridging the resolution gap. *Journal of Histochemistry and Cytochemistry*, *49*, 803–808.

Schellenberger, P., Kaufman, R., Siebert, C. A., Hagen, C., Wodrich, H., & Grünewald, K. (2013). High-precision correlative fluorescence and electron cryo microscopy using two independent alignment markers. *Ultramicroscopy*, *143*, 41–51. http://dx.doi.org/10.1016/j.ultramic.2013.10.011.

Schorb, M., & Briggs, J. A. G. (2013). Correlated cryo-fluorescence and cryo-electron microscope with high spatial precision and improved sensitivity. *Ultramicroscopy*, *143*, 24–32, pii: S0304-3991(13)00291-X.

Swulius, M. T., & Jensen, G. J. (2012). The helical MreB cytoskeleton in Escherichia coli MC1000/pLE7 is an artifact of the N-terminal yellow fluorescent protein tag. *Journal of Bacteriology*, *194*(23), 6382–6386.

Tokuyasu, K. (1973). Technique for ultracryotomy of cell suspensions and tissues. *Journal of Cell Biology*, *57*, 551–565.

Torsney-Weir, T., Saad, A., Möller, T., Weber, B., Hege, H. C., Verbavatz, J. M., et al. (2011). Tuner: Principled parameter finding for image segmentation algorithms using visual response surface exploration. *IEEE Transactions on Visualization and Computer Graphics*, *17*(12), 1892–1901.

van Rijnsoever, C., Oorschot, V., & Klumperman, J. (2008). Correlative light-electron microscopy (CLEM) combining live-cell imaging and immunolabelling of ultrathin cryosections. *Nature Methods*, *5*(11), 973–980.

Von Stette, D., Noireclerc-Savoye, M., Goedhard, J., Gadella, T. W. J., Jr., & Royant, A. (2012). Structure of a fluorescent protein from *Aequorea victoria* bearing the obligate-monomer mutation A206K. *Acta Crystallographica*, *F68*, 878–882.

Watanabe, S., & Jorgensen, E. M. (2012). Visualizing proteins in electron micrographs at nanometer resolution. *Methods in Cell Biology*, *111*, 298–306 (Chapter 15).

Watanabe, S., Punge, A., Hollopeter, G., Willig, K., Hobson, R. J., Davis, M. W., et al. (2011). Protein localization in electron micrographs using fluorescence microscopy. *Nature Methods*, *8*(1), 80–84.

Weber, B., Greenan, G., Prohaska, S., Baum, D., Hege, H. C., Müller-Reichert, T., et al. (2012). Automated tracing of microtubules in electron tomograms of plastic embedded samples in *Caenorhabditis elegans* embryos. *Journal of Structural Biology*, *178*(2), 129–138.

Woog, I., White, S., Büchner, M., Srayko, M., & Müller-Reichert, T. (2010). Correlative light and electron microscopy of intermediate stages of meiotic spindle assembly in the early *Caenorhabditis elegans* embryo. *Methods in Cell Biology*, *111*, 223–234 (Chapter 12).

Yuan, Y., Li, M., Hong, N., & Hong, Y. (2014). Correlative light and electron microscopic analyses of mitochondrial distribution in blastomeres of early fish embryos. *The FASEB Journal*, *28*, 577–585.

CHAPTER

Metallothioneins for Correlative Light and Electron Microscopy

3

Isabel Fernández de Castro, Laura Sanz-Sánchez, Cristina Risco

Cell Structure Laboratory, Centro Nacional de Biotecnología, Consejo Superior de Investigaciones Científicas (CNB-CSIC), Cantoblanco, Madrid, Spain

CHAPTER OUTLINE

Abstract

Structural biologists have been working for decades on new strategies to identify proteins in cells unambiguously. We recently explored the possibilities of using the small metal-binding protein, metallothionein (MT), as a tag to detect proteins in transmission electron microscopy. It had been reported that, when fused with a protein of interest and treated *in vitro* with gold salts, a single MT tag will build an electron-dense gold cluster ~1 nm in diameter; we provided proof of this principle by demonstrating that MT can be used to detect intracellular proteins in bacteria and eukaryotic cells. The method, which is compatible with a variety of sample processing techniques, allows specific detection of proteins in cells with exceptional sensitivity. We illustrated the applicability of the technique in a series of studies to visualize the

Methods in Cell Biology, Volume 124, ISSN 0091-679X, http://dx.doi.org/10.1016/B978-0-12-801075-4.00003-3

intracellular distribution of bacterial and viral proteins. Immunogold labeling was fundamental to confirm the specificity of the MT-gold method. When proteins were double-tagged with green fluorescent protein and MT, direct correlative light and electron microscopy allowed visualization of the same macromolecular complexes with different spatial resolutions. MT-gold tagging might also become a useful tool for mapping proteins into the 3D-density maps produced by (cryo)-electron tomography. New protocols will be needed for double or multiple labeling of proteins, using different versions of MT with fluorophores of different colors. Further research is also necessary to render the MT-gold labeling procedure compatible with immunogold labeling on Tokuyasu cryosections and with cryo-electron microscopy of vitreous sections.

INTRODUCTION

The functional organization of the cell is the result of the coordinated interactions of numerous biomolecules. As many different macromolecular complexes exist within the cell, the development of methods to identify specific molecules in the crowded environment of the cell interior will open up new possibilities for transmission electron microscopy (TEM) and further our understanding of how cells compartmentalize and regulate their activities. For this purpose, specialists in electron microscopy are searching for genetically clonable tags that could match the applications of the green fluorescent protein (GFP) in light microscopy.

A few years ago, Mercogliano and DeRosier proposed the use of the small metal-binding protein, metallothionein (MT), to develop a clonable tag for TEM (Mercogliano & DeRosier, 2006, 2007). MTs are metal-ion chelators that capture a variety of metals in different ionic states and at various stoichiometric ratios (Furey et al., 1986; Nielson, Atkin, & Winge, 1985). MT is present in some prokaryotes and eukaryotes, in which overexpression is induced in response to toxic metals and agents that induce oxidative stress. Nonetheless, the main function of MT is to control the storage and interchange of biologically essential metals such as zinc and copper (Chan, Huang, Merrifield, Salgado, & Stillman, 2002).

Isoform 1 of murine MT is a small, 6 kDa, 61-amino acid protein that comprises 20 cysteine residues. These residues bind gold atoms very efficiently in a reaction that builds an electron-dense gold nanocluster of $\sim$1 nm, easily visualized by a method we have termed metal-tagging TEM or METTEM (Fig. 3.1). With initial generous support from Drs. David DeRosier and Christopher Mercogliano, we explored the possibilities of using MT as a tag for TEM of proteins in cells, and demonstrated that MT can be used to detect intracellular proteins in bacteria (Diestra, Cayrol, Arluison, & Risco, 2009; Diestra, Fontana, Guichard, Marco, & Risco, 2009), mammalian cell lines (Risco et al., 2012), yeasts (Barajas, Fernández de Castro Martín, Pogany, Risco, & Nagy, 2014), and viruses (Cristina Risco, unpublished data).

Adjusting the procedure to eukaryotic cells required careful revision of the protocols initially developed for bacteria, due to questions concerning the toxicity of gold salts, the efficiency of gold uptake, and potential interference from endogenous MT.

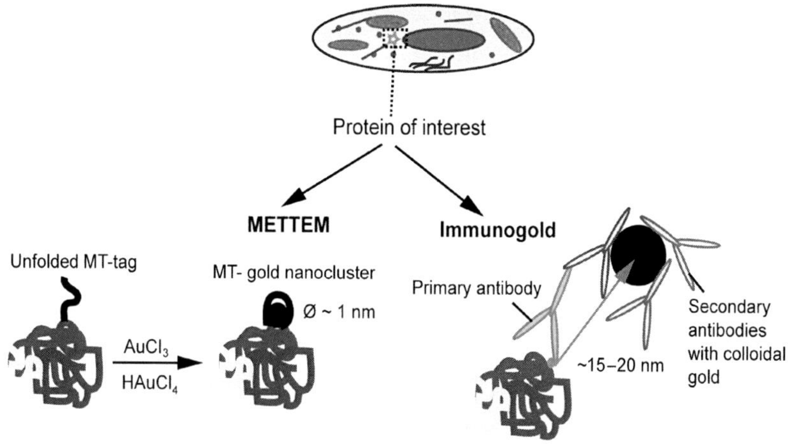

FIGURE 3.1

Detection of MT-gold-tagged proteins in cells. The scheme summarizes the principles of metal-tagging transmission electron microscopy (METTEM; left) and immunogold labeling (right). (See the color plate.)

Our studies showed that MT-gold nanoclusters formed within minutes in cytosol and within organelles of mammalian cell lines, long before the first signs of cytotoxicity could be detected. Moreover, background from endogenous cellular MT was negligible. Gold nanoclusters were readily detected by conventional TEM in unstained thin sections. In stained sections, nanoclusters were visualized by electron spectroscopic imaging (ESI), a method that increased detection sensitivity of MT-gold-tagged proteins. Immunogold labeling was fundamental to confirm the specificity of the MT-gold method. The specificity of immunogold labeling and the sensitivity of MT tagging can be combined for simultaneous detection of several proteins. A number of studies from different laboratories have now shown interesting results with MT tagging for TEM of isolated macromolecular complexes and proteins in cells (Ariyasu, Onoda, Sakamoto, & Yamamura, 2009; Barajas et al., 2014; Bouchet-Marquis, Pagratis, Kirmse, & Hoenger, 2012; Delebecque, Lindner, Silver, & Aldaye, 2011; Fukunaga et al., 2007, 2012; Zhou, Sun, Tai, & Sui, 2012). In our experience, successful results with MT tagging need specific adjustments of the protocols for each cell type and protein studied. Our investigations also established the first CLEM protocols for proteins fused with both GFP and MT (Risco et al., 2012).

3.1 RATIONALE

This chapter presents a detailed description of the protocols for detecting MT-tagged proteins with TEM and CLEM methods, from cell treatment with gold salts to visualization in 2D and 3D (Barajas et al., 2014; Diestra, Cayrol, et al., 2009; Diestra, Fontana, et al., 2009; Risco et al., 2012). We describe the controls necessary to confirm the specificity of labeling results, as well as some considerations regarding procedure resolution and sensitivity. We will discuss our experience with the method in a variety of biological specimens, as well as some important elements to be taken into account when facing the analysis of a new protein in a new cell type. A discussion of future challenges concerning the procedure is included and new protocols for CLEM of MT-tagged proteins are proposed.

3.2 METHODS

3.2.1 GENERAL TECHNICAL CONSIDERATIONS

In our protocols developed for bacteria, we used gold chloride I (AuCl). Live bacteria are incubated 1–3 h with 10 mM AuCl in optimal conditions for nanocluster formation in MT-tagged proteins. Further work with yeast and mammalian cells showed that gold salt concentration and incubation time can be markedly reduced, which solved the question of potential toxicity of gold salts in eukaryotic cells. We also tested two other gold salts, gold chloride III ($AuCl_3$) and chloroauric acid ($HAuCl_4$), which provided the best results due to their higher solubility and stability in water solution. Samples for TEM must be processed in conditions that do not mask small ($\sim$1 nm) gold nanoclusters. This means that no contrast agents or staining can be used, as they mask the MT-gold particles. For thin-sectioning, cells must be embedded in acrylic resins, which are more transparent than epoxy resins. As is routine for any other tags, when MT fusions with target proteins are being planned, a preliminary study must be carried out to determine where the MT tag can be inserted in the protein sequence without affecting protein folding and function. MT is five times smaller than GFP; our experience shows that most proteins that can be fused successfully with GFP can also be fused with MT, and with both MT and GFP. Nonetheless, those Cys residues that are critical for protein folding and/or activity can be affected by gold binding. Some of these cases can be solved satisfactorily; our experience in this matter will be discussed below.

3.2.2 MT TAGGING FOR BACTERIA

Escherichia coli was an ideal specimen for developing the MT-gold tagging method because bacteria can be adapted to grow in the presence of heavy metals. This was done by a controlled progressive increase in metal concentration in the culture medium (Diestra, Cayrol, et al., 2009; Diestra, Fontana, et al., 2009). Moreover, *E. coli* has no endogenous MT. A number of controls were developed to determine how gold salts interact with live bacteria, alone or with exogenously expressed MT (Fig. 3.2). Ultrathin sections of bacteria that had not been incubated with gold salts showed the diffuse electron density of the cell contents (Fig. 3.2A). When incubated with gold salts, bacteria were filled with gold deposits of variable size (Fig. 3.2B). When bacteria were transformed to express MT (not fused to a protein) and treated with gold salts, most gold atoms were captured by cytosolic MT, which built small nanoclusters (Fig. 3.2C). It is important to note that, to be visible by conventional TEM,

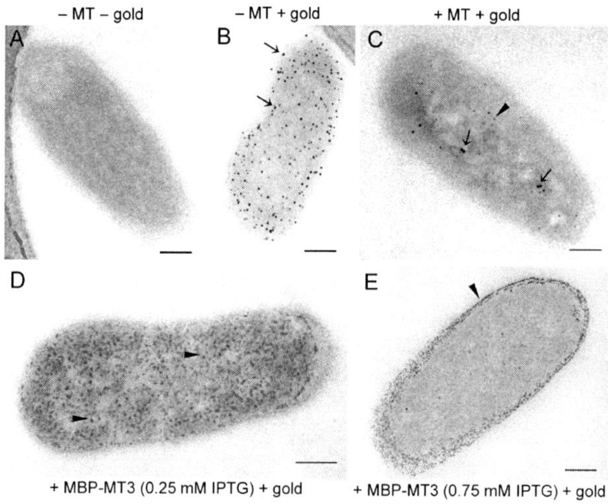

FIGURE 3.2

MT in bacteria (controls). (A) Ultrathin section of a control untransformed *Escherichia coli* cell grown without gold salts and embedded in acrylic resin in the absence of contrast agents. (B) Control cell treated with gold salts showing electron-dense aggregates of variable size (arrows). (C) Transformed bacterium expressing MT and treated with gold salts. Electron-dense particles ~1 nm in diameter (arrowhead) and larger gold aggregates (arrows) are visible. (D) Transformed cell after treatment with 0.25 mM IPTG and incubation with gold salts, expressing a cytosolic version of maltose-binding protein tagged with three concatenated copies of MT (MBP-MT3). Gold nanoclusters associated to MBP-MT3 show a ribosome-like pattern (arrowheads). (E) Bacterium treated with 0.75 mM IPTG. Gold nanoclusters associated to MBP-MT3 are larger and are seen as aggregates at the cell periphery (arrowhead). Scale bars are 100 nm.

Panels (A, B, D and E): reproduced from Diestra, Fontana, et al. (2009) with permission.

MT-gold nanoclusters must have a minimum number of gold atoms, estimated at 20–40 (Mercogliano & DeRosier, 2006, 2007). When cytosolic MT was expressed in sufficient amounts, the protein captured the most gold atoms. If gold amounts exceeded MT chelation capacity, some larger gold deposits formed (Fig. 3.2C). This equilibrium must be achieved for correct visualization of MT-tagged proteins in cells, and testing must be specific for each case (see examples in Fig. 3.2D and E). When we used a cytosolic version of the maltose-binding protein (MBP) fused to three concatenated copies of MT, chimera expression showed very different intracellular localization. At low levels (obtained with 0.25 mM IPTG), MBP-MT3 molecules were seen in the vicinity of ribosomes (Fig. 3.2D). When the protein was expressed in larger amounts (with 0.75 mM IPTG), however, it formed aggregates or inclusion bodies that were displaced toward the cell periphery (Fig. 3.2E).

Two additional bacterial proteins were fused to MT and studied by METTEM (Figs. 3.3 and 3.4). When AmiC, a component of the *E. coli* division ring, was fused with a single copy of MT and expressed in cells, the protein was visualized

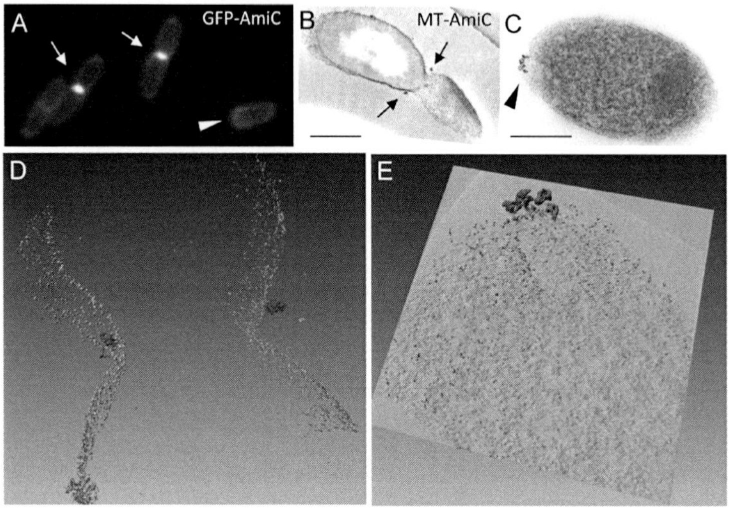

FIGURE 3.3

Visualization of AmiC in transformed *E. coli.* (A) Fluorescence microscopy of *E. coli* cells expressing GFP-AmiC. The protein accumulates in the periplasm of nondividing cells (arrowhead) and in the periplasm and division ring of dividing cells (arrows). (B and C) TEM of ultrathin sections of dividing (B) and nondividing (C) cells expressing MT-AmiC and treated with gold salts. Gold nanoclusters are seen in the periplasm, poles, and the constriction of the ring. (D and E) 3D electron tomography visualization of MT-AmiC molecules in cells. Scale bars are 200 nm. (See the color plate.)

Image in (A) was kindly provided by Dr. Piet A. J. de Boer. Panels (B–E): modified from Diestra, Fontana, et al. (2009) with permission.

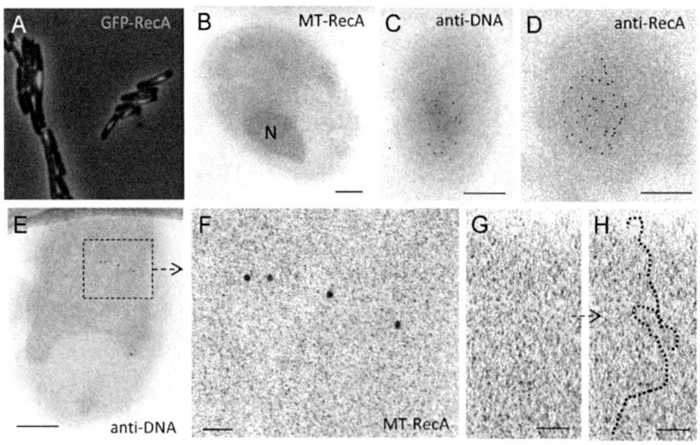

FIGURE 3.4

Distribution of RecA in transformed *E. coli.* (A) Fluorescence microscopy of *E. coli* cells expressing GFP-RecA. The protein accumulates in cytosolic nucleoid-like zones. (B–E) Cells expressing MT-RecA and incubated with gold salts present a dense nucleoid-like body (N) that reacts with anti-DNA (C and E) and anti-RecA (D) antibodies in immunogold assays. (F, G and H) Enlargements of dense nucleoids show small ~1 nm nanoclusters with a fibrous-like pattern. (G and H) show the same image; in (H) a dotted line has been added to some fibers to assist visualization. Scale bars are 100 nm in (B–E), 25 nm in (F), and 10 nm in (G) and (H).

Image in (A) was kindly provided by Dr. Steven J. Sandler. Panels (B–F): reproduced from Diestra, Fontana, et al. (2009) with permission.

both in the periplasm and at the poles of nondividing cells, while in dividing cells, the protein was seen in the septal ring and in the periplasm (Diestra, Fontana, et al., 2009). These subcellular sites were identical to those shown by light microscopy of cells expressing GFP-fused AmiC (Fig. 3.3A–C). Electron tomography showed three-dimensional fine details of AmiC–MT macromolecular complexes; protein molecules exhibited diverse aggregation states in the periplasm, the poles, and the constriction ring (Fig. 3.3D and E). A third protein, the DNA-binding protein RecA, was included in the study. When this protein was fused with a single copy of MT and expressed in *E. coli*, small MT-gold nanoclusters accumulated in the bacterial nucleoid, with a pattern identical to that seen by light microscopy of RecA-GFP-expressing bacteria (Fig. 3.4 A–D). High magnification views of these electron-dense nucleoids showed small nanoclusters arranged in a fiber-like pattern, probably a result of RecA-MT binding to chromosomal DNA (Fig. 3.4E–H).

3.2.3 MAMMALIAN CELL LINES

When we transferred the METTEM protocol to mammalian cell lines, gold salt concentration and incubation times were notably reduced. Our studies showed that MT tagging did not affect intracellular trafficking or function of rubella virus (RUBV) proteins (Risco et al., 2012). We studied the intracellular distribution of the P150 subunit of the RUBV replicase, which had been fused with MT and GFP (Risco et al., 2012). Two main protocols were developed, one involving permeabilization with the bacterial exotoxin streptolysin O (SLO) to facilitate uptake of gold atoms while avoiding toxicity of the gold salt (Fig. 3.5) and a second approach for living cells (Fig. 3.6); cytotoxicity tests were performed in the second case. When live cells were treated with 1 mM AuCl or 0.5 mM $AuCl_3$ for up to 1 h, there were no changes in viability as determined by trypan blue exclusion, in actin network integrity as studied by fluorescence microscopy, or in ultrastructural morphology as visualized by TEM. Cytopathic effects such as partial actin depolymerization were seen only after 2 h. Based on these observations, we restricted incubation with gold salts to 15–30 min. We confirmed that these time periods caused no appreciable cytotoxicity, but were sufficient for intracellular MT-tagged proteins to build 1-nm gold nanoclusters at different intracellular locations such as plasma membrane, lysosomes, and perinuclear organelles.

One of our concerns when adjusting the protocols to mammalian cell lines was the potential interference of endogenous MT (Chan et al., 2002), which are readily detectable by Western blot analysis in cell lysates and by immunogold labeling in cryosections (Risco et al., 2012). To our surprise, endogenous MT did not appear to induce formation of gold clusters. This might be due to the fact that MT expression is strictly controlled in eukaryotes. Cellular MT is fully metallated, with very little or no free MT, and no free Cu and Zn in the cell (Beyersmann & Haase, 2001; Rae, Schmidt, Pufahl, Culotta, & O'Halloran, 1999). As these metals are only partially displaced by gold atoms when bound to MT (Schmitz, Minkel, Gingrich, &

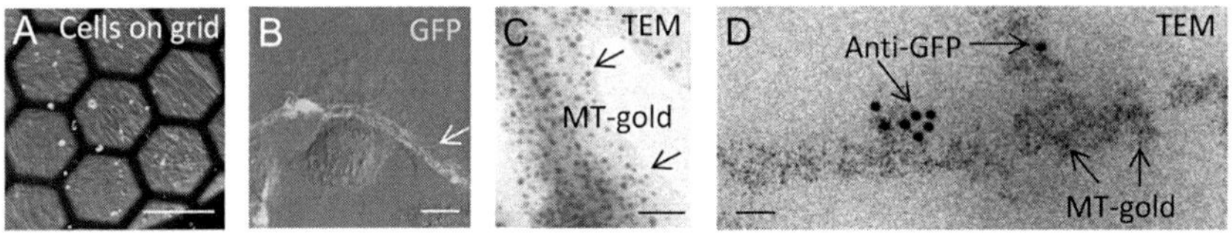

FIGURE 3.5

Correlative GFP-fluorescence and metal-tagging TEM of a GFP-MT dual-tagged protein in transfected mammalian cells. Cells grown on grids were permeabilized with SLO before incubation with gold salts. (A) Fluorescence microscopy shows transfected cells that express the RUBV P150 protein fused with GFP and MT. (B) Differential interference contrast and GFP-fluorescence microscopy show the protein arranged in filament bundles. The area marked with a white arrow in (B) corresponds to the TEM projection image in (C). (D) Correlative immunogold and METTEM. Cells treated as in (C) were embedded in acrylic resin. Ultrathin sections were incubated with anti-GFP antibody and a 10-nm colloidal gold conjugate before TEM. Immunogold signal colocalizes with MT-gold nanoclusters. Scale bars are 250 μm in (A), 10 μm in (B), and 25 nm in (C) and (D). (See the color plate.)

Modified from Risco et al. (2012) with permission.

Shaw, 1980), the endogenous MT would be unable to form gold clusters with the minimum number of gold atoms necessary for TEM detection.

We established three CLEM protocols for viral proteins dual tagged with MT/GFP and expressed in mammalian cell lines. When expressed in the absence of other replicase components, the RUBV P150 replicase builds filament bundles (Fig. 3.5). Cells expressing MT-GFP-tagged P150 were grown on EM grids and visualized by fluorescence microscopy and DIC (differential interference contrast) (Fig. 3.5A). After sample permeabilization with SLO and incubation with gold salts, we performed direct correlative GFP and TEM (Fig. 3.5B and C). Cells were dried on the grid and studied by TEM. Electron microscopy of the thin peripheral areas of "green" cells showed that the intracellular green filaments had built numerous electron-dense nanoclusters (Fig. 3.5C). Another protocol was developed for correlative immunogold and METTEM of the cell monolayers. Cells treated with SLO and gold salt were chemically fixed with aldehydes and embedded in acrylic resin. Ultrathin sections were incubated with an anti-GFP antibody and 10-nm colloidal gold conjugates, and visualized by TEM (Fig. 3.5D). Colocalization of immunogold particles and MT-induced gold nanoclusters confirmed that P150-GFP-MT had built ~1 nm particles in the green filament bundles. The third CLEM protocol was established for nonpermeabilized, live cells (Fig. 3.6). When coexpressed with the P90 subunit of RUBV replicase, P150-GFP-MT is incorporated into perinuclear active replication organelles, which are modified lysosomes and synthesize viral RNA (Fontana et al., 2010) (Fig. 3.6A). For both light and electron microscopy imaging, we used antibodies specific for double-stranded RNA (dsRNA), an intermediate of the viral replication process, to identify those "mature" replication organelles in which the viral replicase became active (Fernández de Castro, et al., 2014; Fontana, López-Montero,

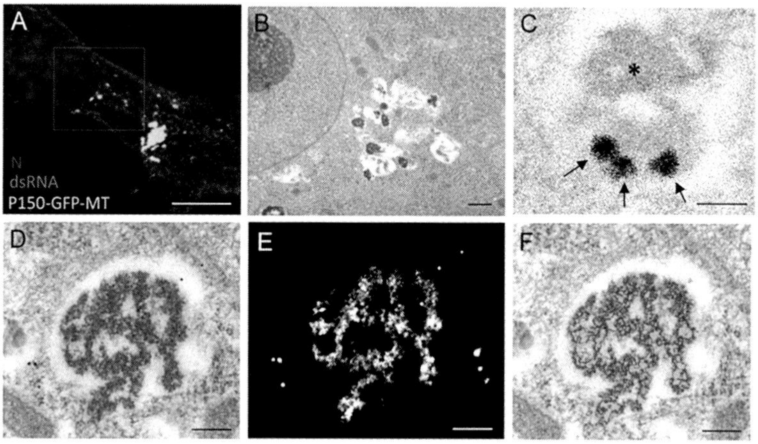

FIGURE 3.6

METTEM and electron spectroscopic imaging (ESI) of a GFP-MT-gold-tagged protein in mammalian cells. (A) Fluorescence microscopy imaging of cells with RUBV P150 protein fused with GFP and MT, and expressed as part of an active viral replicon. The protein incorporates into active perinuclear replication organelles that are labeled with antibodies specific for double-stranded RNA (dsRNA, red). (B) Stained ultrathin section of the perinuclear region of a transfected cell with replication organelles. (C) Unstained ultrathin section of a replication organelle containing several aggregates (arrows) of P150-GFP-MT-gold from an unpermeabilized cell treated with gold salts. Part of the organelle looks dense (asterisk), but gold nanoclusters are not distinguished. (D–F) Elemental distribution of gold in a stained ultrathin section of a replication organelle containing P150-GFP-MT-gold. (D) Ultrastructure of the organelle visualized by elastic bright field microscopy. (E) Net gold elemental distribution obtained by ESI and background subtraction. (F) Gold signal superimposed on the elastic bright field microscopy image, showing the precise distribution of MT-gold-tagged P150 molecules in the subdomains of the replication organelle. N, nucleus. Scale bars are 10 μm in (A), 0.5 μm in (B), and 200 nm in (C) to (F). (See the color plate.)

Panels (C–F): reproduced from Risco et al. (2012) with permission.

Elliott, Fernández, & Risco, 2008; Fontana et al., 2007; Risco et al., 2012). Transfected cells were treated *in vivo* with gold salts, chemically fixed with aldehydes, and embedded in acrylic resin. Semi-thick sections were studied by fluorescence microscopy and selected green cells analyzed in ultrathin sections by TEM (Fig. 3.6B and C). Replication organelles contained densely packed complexes of MT-gold P150 molecules, as seen in unstained sections (Fig. 3.6C). Although some areas within the organelles looked dense, they did not contain distinguishable gold nanoclusters (Fig. 3.6C, asterisk). Elemental gold distribution, as imaged by ESI and after background subtraction, showed P150-MT-gold molecules in most organelle membranes (Fig. 3.6D–F). With the pronounced increase in detection sensitivity, an important advantage of ESI is that it can be used in stained samples, which allows simultaneous observation of ultrastructural details and the molecular location of MT-tagged proteins. In this study, MT-tagged RUBV capsid protein was also visualized by METTEM. The analysis showed capsid-MT molecules in the cytosol and in active replication organelles, with patterns very similar to those shown by the untagged form of the protein and by a capsid-RFP fusion protein (Risco et al., 2012).

3.2.4 YEAST

Protocols for METTEM visualization of MT-tagged proteins in yeast had to be designed to overcome the difficulties imposed by the thick cell wall, an obstacle when processing yeasts for TEM. Partial elimination of the wall was the best solution for efficient infiltration of gold salts, fixatives, and resins, to allow optimal formation of MT-gold nanoclusters and visualization of intracellular compartments (Barajas et al., 2014). To remove the cell wall, yeast was incubated with zymolyase in medium A (1 × yeast nitrogen base, 2% (w/v) glucose, 1 × amino acids, 1 M sorbitol, 20 mM Tris–Cl, pH 7.5). Spheroplasts were then incubated (75 min) with 2 mM $HAuCl_4$ in the same medium, which was also maintained during fixation. For TEM, we studied ultrathin sections of spheroplasts from yeast strains that express MT-tagged viral proteins. We imaged the intracellular distribution of MT-tagged P33 protein, the replicase of the Tombusvirus *Tomato Bushy Stunt Virus*. METTEM showed the arrangement of protein molecules when incorporated into active complexes in replication organelles (Fig. 3.7). As with mammalian cells, untransformed yeast showed no nonspecific background when treated with gold salts (Barajas et al., 2014).

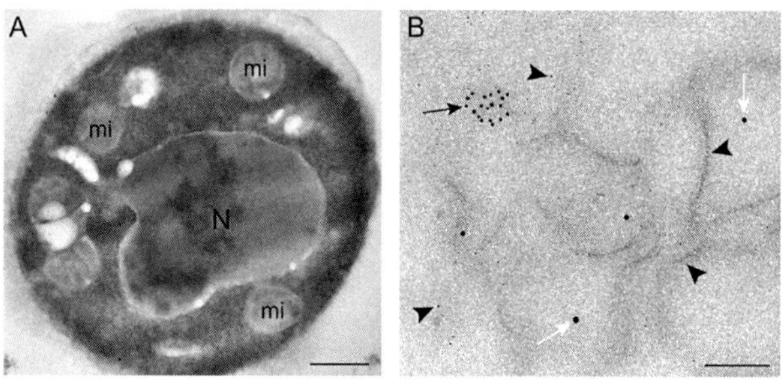

FIGURE 3.7

Visualization of MT-tagged proteins in yeast. The Tombusvirus replicase p33 was tagged with MT and expressed in yeast as part of an active replicon. (A) Stained ultrathin section of a spheroplast expressing p33-MT, treated with gold salts and embedded in epoxy resin. N, nucleus; mi, mitochondria. (B) METTEM of p33-MT-gold. Nanoclusters concentrate in a peripheral membranous compartment (black arrowheads). Immunogold labeling with dsRNA-specific antibodies show the site in the compartment at which the p33-MT replicase molecules become active (black arrow); the membranes of this compartment contain the RER marker PDI as shown by immunogold labeling (white arrows). Scale bars are 0.5 μm (A) and 100 nm (B).

3.2.5 A GENERAL PROTOCOL: HOW TO GET STARTED

Our experience with TEM visualization of MT-gold-tagged proteins in bacteria, yeast, and mammalian cells can be summarized in a general procedure that might serve as a starting point. Specific adjustments must be made depending on the protein and cell type. The main steps are

1. Identify the best sites for inserting tags into the protein sequence and test that the MT-(GFP)-tagged protein activity is not affected.
2. Become familiar with the ultrastructure of the cell type being studied.
3. Prepare gold salt stocks in distilled water and store them at 4 °C in the dark because they precipitate when exposed to light. Before adding gold to cells, mix the stock with the cell culture medium to the desired final concentration and confirm that gold does not precipitate in the medium. During incubation with gold, samples must be kept in the dark.
4. Study gold salt-treated control cells by TEM to confirm lack of nonspecific background.
5. Transfect/transform the target cell type for optimal expression of the MT-tagged protein under study.
6. Treat transfected cells with 0.2 mM $HAuCl_4$ (30 min) in the dark.
7. Process cells for embedding in acrylic resin in the absence of osmium or staining agents.
8. Use immunogold labeling to confirm signal specificity and activity of MT-tagged proteins *in situ*.
9. For correlative microscopy of GFP-MT-dual tagged proteins, cells can be studied by fluorescence microscopy of semi-thick sections. Fluorescent cells are photographed and selected for ultrathin sectioning and TEM.
10. TEM of serial sections or electron tomography of semi-thick sections is used to study the 3D arrangement of MT-tagged proteins in different intracellular locations.

3.3 DISCUSSION

Our studies have shown that MT tagging is a very sensitive method for labeling proteins in cells. The procedure is compatible with a variety of sample processing techniques for TEM and CLEM, although there are some important questions to consider before starting. For example, cysteine residues in active sites of proteins can be inactivated by gold binding. This is the case of the three cysteines in the *E. coli* RecA DNA-binding protein; when gold atoms bind to these Cys residues, protein capacity to form filaments on DNA is impaired (Mercogliano & DeRosier, 2006). Incubation with the metal chelator penicillamine resolved this difficulty and similar treatments can be applied to other Cys-containing proteins (Diestra, Fontana, et al., 2009; Mercogliano & DeRosier, 2006). Proteins with transmembrane domains can also be problematic. The NSm scaffolding protein participates in Bunyamwera virus assembly. NSm does not admit any tags in its N- or C-terminal domains, and the only possible insertion site is in one of its three transmembrane domains. Active NSm-GFP chimeras and recombinant viruses were obtained using this site (Shi et al., 2006), but MT-tagged NSm molecules accumulated in the rough endoplasmic reticulum, probably due to incorrect folding, and were unable to reach the viral assembly sites in Golgi membranes (Noelia López-Montero, Richard M. Elliott, & Cristina Risco, unpublished data). Nonetheless, MT tagging did not interfere with the folding, localization, or function of most membrane proteins we have tested so far (Barajas et al., 2014; Risco et al., 2012). Additional considerations regarding the method are discussed in the following sections.

3.3.1 RESOLUTION AND SENSITIVITY

The MT-tagging method allows highly sensitive labeling of cell proteins. After the first results were obtained in bacteria, the question was raised as to whether the MT-gold nanoclusters are indeed individual protein molecules. Our experience leads us to consider that, in most cases, they probably are (Diestra, Fontana, et al., 2009; Mercogliano & DeRosier, 2007; Risco et al., 2012). We have also observed that when several MT-tagged protein molecules are in close proximity, a single gold cluster can be built of several MT. This is our interpretation, for example, of the larger gold clusters in bacteria when MBP-MT3 was expressed at high levels, aggregated, and formed inclusion bodies (Fig. 3.2E). Higher resolution studies in simple systems will be necessary to determine how MT-gold nanoclusters form *in situ*. For example, recombinant viruses with a defined number of MT-tagged proteins can be assembled, treated with gold salts, and studied by cryo-TEM. MT-gold nanoclusters can be counted to analyze whether each particle corresponds to a single protein molecule. Promising preliminary data point to a notable increase in sensitivity of the MT-tagging method when combined with ESI/elemental gold mapping (Fig. 3.6). More testing is needed to quantify the sensitivity achieved by METTEM alone or with ESI.

3.3.2 MULTIPLE LABELING

It is not yet clear whether MT tagging will allow labeling of two or more proteins. Based on our results in bacteria, different-sized MT peptides (e.g., 6 kDa vs. 18 kDa) will produce nanoclusters of different diameters. If possible, MT tags designed with distinct affinities for different metals would be another interesting strategy. We nonetheless consider the most promising approach to be MT tagging combined with immunogold labeling on Tokuyasu cryosections (Hurbain & Sachse, 2011; Slot & Geuze, 2007). A protocol that combines these methods would be ideal for simultaneous labeling of two or more protein species. We observed that small MT-gold nanoclusters are masked by the methylcellulose layers used to stabilize Tokuyasu cryosections (Diestra, Cayrol, et al., 2009) (Fig. 3.8). Protocols to compatibilize the techniques will thus have to be designed. Increasing MT-gold nanocluster diameter with silver enhancement might allow these two specific, highly sensitive labeling techniques to be combined.

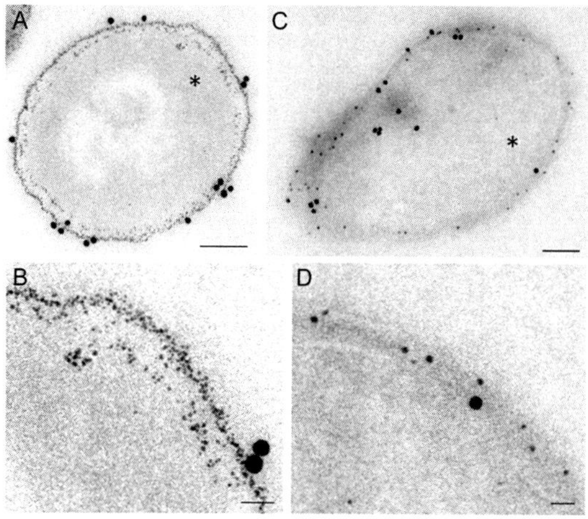

FIGURE 3.8

MT-gold tagging, Tokuyasu cryosections and immunogold. *Hfq E. coli* cells expressing the bacterial protein Hfq fused with MT and treated with gold salts. (A) Ultrathin section of a bacterium embedded in acrylic resin. The smallest dense particles correspond to Hfq-MT-gold complexes of ~1 nm that localize in both the inner and outer bacterial cell membranes. The 15-nm colloidal gold particles are bound to anti-LamB antibodies that label the outer bacterial membrane. (B) Enlargement of the region marked by an asterisk in (A). (C) Immunogold labeling on a thawed cryosection of *Hfq E. coli* cells expressing Hfq-MT and treated with gold salts. Double immunogold labeling was performed with a rabbit anti-Hfq polyclonal antibody, followed by goat anti-rabbit secondary antibody conjugated with 15 nm colloidal gold particles, as well as a mouse anti-MT monoclonal antibody followed by goat anti-mouse secondary antibody conjugated with 10 nm colloidal gold particles. Signals coexist in the peripheral membranes where small MT-gold nanoclusters of ~1 nm are no longer distinguishable. (D) Enlargement of the area marked by an asterisk in (C). The image has been rotated. Scale bars are 100 nm in (A) and (C), and 25 nm in (B) and (D).

Modified from Diestra, Cayrol, et al. (2009) with permission.

3.3.3 NEW PROTOCOLS FOR CLEM

The use of fluorophores smaller than GFP, such as mini-SOG (Shu et al., 2011), in combination with smaller MT peptides could be an interesting solution for proteins that do not tolerate the insertion of large tags. Proteins fused to the beta domain of mouse MT isoform 1 have been expressed in bacteria and nanoclusters <1 nm were detected (Elia Diestra & Cristina Risco, unpublished data). Different versions of MT with fluorophores of different colors might also provide a new means for multiple labeling in CLEM. For strict correlation and visualization of the same molecules in light and electron microscopy, our future studies will combine our current experience in CLEM (Risco et al., 2012; Sanz-Sánchez & Risco, 2013) with the strategies using fiducial markers as described by Briggs (Kukulski et al., 2011).

3.3.4 MT FOR THREE-DIMENSIONAL ELECTRON MICROSCOPY

MT tagging is a technique originally designed to localize proteins in the 3D-density maps produced by (cryo)-electron tomography of cells, and in volumes obtained by 3D reconstructions of serial sections. Indeed, as shown above, MT-gold nanoclusters can be visualized in vitrified isolated macromolecular complexes (Bouchet-Marquis et al., 2012; Mercogliano & DeRosier, 2007; Zhou et al., 2012), in frozen bacteria (Diestra, Fontana, et al., 2009), and in thick-sections of bacteria and eukaryotic cells visualized in 3D by electron tomography (Diestra, Fontana, et al., 2009; Isabel

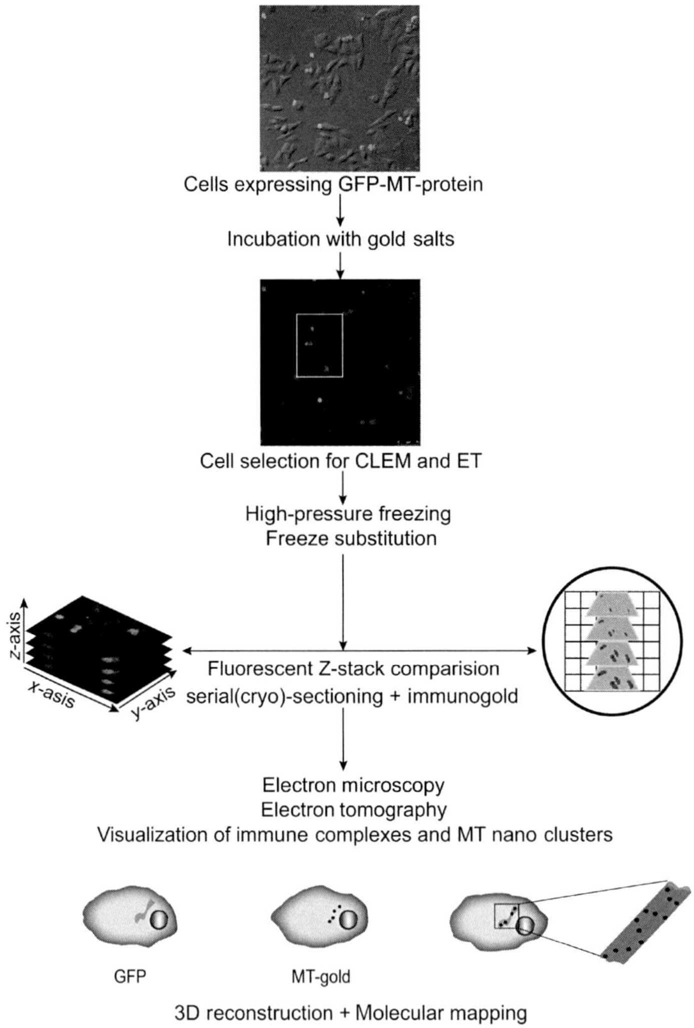

FIGURE 3.9

Scheme summarizing the workflow to study GFP-MT-tagged proteins in mammalian cells, with the technical steps for processes from the live cell to CLEM and 3D EM.

Fernández de Castro, Daniel Barajas, Peter D. Nagy, & Cristina Risco, unpublished results). Further studies will be necessary to combine the MT-gold labeling method with cryo-electron microscopy of vitreous sections (CEMOVIS). A scheme summarizing our proposed workflow from live cells to 3D EM of GFP-MT-tagged proteins is shown in Fig. 3.9.

ACKNOWLEDGMENTS

We would like to express our gratitude to Prof. David DeRosier and Dr. Christopher P. Mercogliano for their initial support with the MT-tagging project. Our gratitude to Dr. Martin Sachse for critically reading the chapter, to Drs. Piet A. J. de Boer, Steven J. Sandler, Elia Diestra, and Sergio Marco, and to Ms. Eva Sanmartín for providing images, to Drs. Veronique Arluison and Christopher P. Mercogliano for bacterial stains, to Dr. Peter D. Nagy for yeast strains, and to Catherine Mark for excellent editorial assistance.

This work was funded by an FPI fellowship (to I. F. C.) and research grants BIO2009-07255 and BIO2012-33314 from the Spanish Ministry of Economy and Competitiveness, and grant PIF200620F0024 from the "Proyectos Intramurales de Frontera" program of the Spanish National Research Council (CSIC) (to C. R.).

REFERENCES

Ariyasu, S., Onoda, A., Sakamoto, R., & Yamamura, T. (2009). Alignment of gold clusters on DNA via a DNA-recognizing zinc finger-metallothionein fusion protein. *Bioconjugate Chemistry*, *20*, 2278–2285.

Barajas, D., Fernández de Castro, I., Pogany, J., Risco, C., & Nagy, P. D. (2014). Noncanonical role for the host Vps4 AAA+ ATPase ESCRT protein in the formation of the Tomato bushy stunt virus replicase. *PLoS Pathogens*, *10*, e1004087.

Beyersmann, D., & Haase, H. (2001). Functions of zinc in signaling, proliferation and differentiation of mammalian cells. *Biometals*, *14*, 331–341.

Bouchet-Marquis, C., Pagratis, M., Kirmse, R., & Hoenger, A. (2012). Metallothionein as a clonable high-density marker for cryo-electron microscopy. *Journal of Structural Biology*, *177*, 119–127.

Chan, J., Huang, Z., Merrifield, M. E., Salgado, M. T., & Stillman, M. J. (2002). Studies of metal binding reactions in metallothioneins by spectroscopic, molecular biology, and molecular modelling techniques. *Coordination Chemistry Reviews*, *233–234*, 319–339.

Delebecque, C. J., Lindner, A. B., Silver, P. A., & Aldaye, F. A. (2011). Organization of intracellular reactions with rationally designed RNA assemblies. *Science*, *333*, 470–474.

Diestra, E., Cayrol, B., Arluison, V., & Risco, C. (2009). Cellular electron microscopy imaging reveals membrane localization of the bacterial Hfq protein. *PLoS One*, *4*, e8301.

Diestra, E., Fontana, J., Guichard, P., Marco, S., & Risco, C. (2009). Visualization of proteins in intact cells with a clonable tag for electron microscopy. *Journal of Structural Biology*, *165*, 157–168.

Fernández de Castro, I., Zamora, P. F., Ooms, L., Fernández, J. J., Lai, C. M.-H., Mainou, B. A., et al. (2014). Reovirus forms neo-organelles for progeny particle assembly within reorganized cell membranes. *mBio*, *5*, e00931-13.

Fontana, J., López-Iglesias, C., Tzeng, W.-P., Frey, T. K., Fernández, J. J., & Risco, C. (2010). Three-dimensional structure of Rubella virus factories. *Virology*, *405*, 579–591.

Fontana, J., López-Montero, N., Elliott, R. M., Fernández, J. J., & Risco, C. (2008). The unique architecture of Bunyamwera virus factories around the Golgi complex. *Cellular Microbiology*, *10*, 2012–2028.

Fontana, J., Tzeng, W.-P., Calderita, G., Fraile-Ramos, A., Frey, T. K., & Risco, C. (2007). Novel replication complex architecture in rubella replicon-transfected cells. *Cellular Microbiology*, *9*, 875–890.

Fukunaga, Y., Higashihara, A., Nishino, Y., Yasunaga, T., Jin, M., & Miyazawa, A. (2012). Enhanced detection efficiency of genetically encoded tag allows the visualization of monomeric proteins by electron microscopy. *Journal of Electron Microscopy*, *61*, 229–236.

Fukunaga, Y., Hirase, A., Kim, H., Wada, N., Nishino, Y., & Miyazawa, A. (2007). Electron microscopic analysis of a fusion protein of postsynaptic density-95 and metallothionein in cultured hippocampal neurons. *Journal of Electron Microscopy*, *56*, 119–129.

Furey, W. F., Robbins, A. H., Clancy, L. L., Winge, D. R., Wang, B. C., & Stout, C. D. (1986). Crystal structure of Cd, Zn metallothionein. *Science*, *23*, 704–710.

Hurbain, I., & Sachse, M. (2011). The future is cold: Cryo-preparation methods for transmission electron microscopy of cells. *Biology of the Cell*, *103*, 405–420.

Kukulski, W., Schorb, M., Welsch, S., Picco, A., Kaksonen, M., & Briggs, J. A. (2011). Correlated fluorescence and 3D electron microscopy with high sensitivity and spatial precision. *Journal of Cell Biology*, *192*, 111–119.

Mercogliano, C. P., & DeRosier, D. J. (2006). Gold nanocluster formation using metallothionein: Mass spectrometry and electron microscopy. *Journal of Molecular Biology*, *355*, 211–223.

Mercogliano, C. P., & DeRosier, D. J. (2007). Concatenated metallothionein as a clonable gold label for electron microscopy. *Journal of Structural Biology*, *160*, 70–82.

Nielson, K. B., Atkin, C. L., & Winge, D. R. (1985). Distinct metal-binding configurations in metallothionein. *Journal of Biological Chemistry*, *260*, 5342–5350.

Rae, T. D., Schmidt, P. J., Pufahl, R. A., Culotta, V. C., & O'Halloran, T. V. (1999). Undetectable intracellular free copper: The requirement of a copper chaperone for superoxide dismutase. *Science*, *284*, 805–808.

Risco, C., Sanmartín-Conesa, E., Tzeng, W.-P., Frey, T. K., Seybold, V., & de Groot, R. J. (2012). Specific, sensitive, high-resolution detection of protein molecules in eukaryotic cells using metal-tagging transmission electron microscopy. *Structure*, *20*, 759–766.

Sanz-Sánchez, L., & Risco, C. (2013). Multilamellar structures and filament bundles are found on the cell surface during bunyavirus egress. *PLoS One*, *8*, e65526.

Schmitz, G., Minkel, D. T., Gingrich, D., & Shaw, C. F., 3rd. (1980). The binding of Gold(I) to metallothionein. *Journal of Inorganic Biochemistry*, *112*, 293–306.

Shi, X., Kohl, A., Leonard, V. H., Li, P., McLees, A., & Elliott, R. M. (2006). Requirement of the N-terminal region of orthobunyavirus nonstructural protein NSm for virus assembly and morphogenesis. *Journal of Virology*, *80*, 8089–8099.

Shu, X., Lev-Ram, V., Deerinck, T. J., Qi, Y., Ramko, E. B., Davidson, M. W., et al. (2011). A genetically encoded tag for correlated light and electron microscopy of intact cells. *PLoS Biology*, *9*, e10011041.

Slot, J. W., & Geuze, H. J. (2007). Cryosectioning and immunolabeling. *Nature Protocols*, *2*, 2480–2491.

Zhou, Q., Sun, S., Tai, P., & Sui, S.-F. (2012). Structural characterization of the complex of SecB and metallothionein-labeled proOmpA by cryo-electron microscopy. *PLoS One*, *7*, e47015.

CHAPTER

Correlative Light Microscopy and Electron Tomography to Study Von Willebrand Factor Exocytosis from Vascular Endothelial Cells

4

Marjon J. Mourik*, Frank G.A. Faas*, Karine M. Valentijn*, Jack A. Valentijn*, Jeroen C. Eikenboom[†], Abraham J. Koster*

**Department of Molecular Cell Biology, Section Electron Microscopy, Leiden University Medical Center, Leiden, The Netherlands*

[†]Department of Thrombosis and Hemostasis, Einthoven Laboratory for Experimental Vascular Medicine, Leiden University Medical Center, Leiden, The Netherlands

CHAPTER OUTLINE

Methods in Cell Biology, Volume 124, ISSN 0091-679X, http://dx.doi.org/10.1016/B978-0-12-801075-4.00004-5

Abstract

Revealing the ultrastructure and function of fluorescently labeled cellular components by correlative light and electron microscopy (CLEM) facilitates the study of structure–function relationships in complex biological processes. Given the diversity of available fluorescent tags, light microscopy is ideal for monitoring dynamic cellular processes, while electron microscopy reveals the morphological context of structures at high resolution.

Endothelial cells lining the blood vessel wall contain storage organelles called Weibel–Palade bodies (WPBs), which contain tubules of densely packed helical spirals of the blood coagulation protein Von Willebrand factor (VWF). Exocytosis of WPBs is triggered upon vascular damage and results in the transformation of stored tubular VWF into secreted extracellular VWF. Upon exocytosis, VWF rearranges into long filamentous strings to recruit platelets from the blood. During this secretion process, large intracellular VWF exocytosis structures are formed called secretory pods. Here, we describe a CLEM method used to study the relationship between the secretory pod and secreted VWF where confocal microscopy on whole cells was combined with serial electron tomography on chemically fixed, plastic-embedded sections. We show that the combination of these two well-established microscopy modalities provides a robust and generic CLEM method suitable for the characterization of VWF secretion sites.

INTRODUCTION

Light microscopy (LM) was the first techniques used to study small organisms and single cells that were otherwise invisible to the naked eye. Since the development of the light microscope, biologists have strived to visualize biological specimens in ever-greater detail. To this end, technical advances in LM have generated a variety of sophisticated imaging systems designed to improve the resolution (Garini, Vermolen, & Young, 2005). Also numerous labeling strategies have been developed such as fluorescent dyes, antibodies, and fusion proteins which enable real-time imaging of specific proteins in dynamic biological processes (Suzuki, Matsuzaki, Hagiwara, Aoki, & Takata, 2007). The benefit of using these fluorescent labeling techniques is that they can provide specific, real-time information on the location of the labeled structure. However, the morphological context remains unobserved.

The advent of electron microscopy (EM) made it possible to visualize cellular morphology beyond the resolution attainable by LM. EM has contributed significantly as a tool to image tissue architecture, cellular morphology, and the structure of subcellular organelles (Koster & Klumperman, 2003). Labeling specific structures in EM samples is usually achieved through the use of antibody gold conjugates. However, this technique is susceptible to loss of antigenicity during sample preparation, and also depends upon antibody specificity.

The need for morphological information of fluorescently labeled structures in LM and the need for additional labeling techniques in EM stimulated the development of correlative light and electron microscopy (CLEM). This technique facilitates the study of structure–function relationships in biological processes by combining LM and EM imaging of the same sample, where the fluorescence signal is later used to localize specific structures in the electron microscope. However, the differences in specimen preparation for EM and LM make the combination of these imaging modalities challenging. Preparation of samples for LM often requires mild fixation and permeabilization of the cells, which diminish ultrastructural preservation, whereas conventional EM sample preparation often involves strong glutaraldehyde and osmium fixation which obscures epitopes for antibody labeling and quenches fluorescence. Therefore, the fluorescent labeling and EM sample preparation steps for CLEM are strategically chosen or adapted to avoid these complications. Some CLEM methods require fluorescence imaging prior to EM sample preparation (Brown, Mantell, Carter, Tilly, & Verkade, 2009; Kobayashi et al., 2012; Loussert, Forestier, & Humbel, 2012). Others have adapted the EM sample preparation protocol by reducing the amount of metal staining to preserve fluorescence in the EM sample or to enable fluorescent labeling with antibodies on EM sections (Karreman et al., 2012; Kukulski et al., 2011). In addition to the sample preparation for CLEM, relocating regions of interest is also challenging when using successive imaging modalities. To facilitate the relocation process, a letter or number coded meshwork is generally used. This meshwork can be engraved or carbon coated onto cell culture coverslips (including sapphire disks) (Brown et al., 2009; Kobayashi et al., 2012; Loussert et al., 2012; Polishchuk, Polishchuk, & Luini, 2012) or is provided by EM finder grids (Ren, Kruhlak, & Bazett-Jones, 2003; van Driel, Valentijn, Valentijn, Koning, & Koster, 2009). This approach for navigation only provides a rough indication of the location of corresponding structures. Overlaying the LM and EM images is often sufficient to locate the correlating structures. To achieve more accurate overlays, dedicated CLEM probes can be used that provide both fluorescence and EM labeling. Examples of such bimodal labels are quantum dots, fluorescent antibodies combined with small gold particles (Brown & Verkade, 2010), or the genetic fluorescent tag miniSOG which can precipitate osmiophilic diaminobenzidine (Shu et al., 2011).

The chosen strategy for CLEM often depends on the studied biological material, the resolution required to answer the particular research question, and on the combination of LM and EM techniques used for correlation. As a consequence, many different approaches for CLEM have been developed. Here, we will exemplify a more generic method for CLEM which was used to characterize the exocytosis site of the blood coagulation protein Von Willebrand factor (VWF).

Endothelial cells that line the inner surface of blood vessels are able to respond very rapidly to vascular damage by secreting specialized storage organelles called Weibel–Palade bodies (WPBs) (Aird, 2007; Weibel & Palade, 1964). These storage organelles are formed at the *trans*-Golgi network where VWF multimerizes and is packed into long helical tubules (Berriman et al., 2009; Huang et al., 2008). The tubules of VWF drive granule formation and give the WPB, its distinct elongated cigar shape (Haberichter, Jozwiak, Rosenberg, Christopherson, & Montgomery, 2002; Michaux et al., 2003; Voorberg et al., 1993). Upon vascular damage, the endothelial cells are stimulated to provoke WPBs exocytosis, thereby releasing VWF into the blood (Rondaij, Bierings, Kragt, van Mourik, & Voorberg, 2006). This secreted VWF subsequently forms long filamentous strings of VWF which bind to exposed collagen and captures platelets to initiate the formation of a primary platelet plug (Ruggeri, 2003). Deficiencies caused by mutations in the VWF gene can disturb the amount and quality of secreted VWF, leading to the most common hereditary bleeding disorder, Von Willebrand disease (VWD) (Sadler et al., 2000). The mutations can also result in less functional VWF or disturb normal biogenesis, storage in WPBs, and secretion. To obtain more information on VWF secretion on a structural level, we studied WPB exocytosis in cultured endothelial cells. By LM, we identified a novel structure involved in VWF exocytosis, the secretory pod. This structure is only present in stimulated endothelial cells and appears as a large VWF positive spherical-shaped structure that associates with WPBs and VWF strings (Mourik et al., 2013; Valentijn et al., 2010). Transmission electron microscopy (TEM) revealed that stimulated endothelial cells contained large membrane-enclosed organelles of similar dimensions. The membrane-enclosed organelles contained lacy material and were often in close proximity with WPBs. To show that the structures observed by TEM contained VWF and represented, the VWF-positive secretory pods seen by LM, CLEM was performed (Mourik et al., 2013; Valentijn et al., 2010). We combined confocal laser scanning microscopy with conventional TEM and electron tomography to reveal the structures in three dimensions using serial sections.

4.1 RATIONALE

The benefit of CLEM is that it provides a platform to find the metaphorical "needle in a hay stack." Other methods that were explored to determine the relationship between the membrane delimited structure observed by TEM and VWF secretion were unsuccessful and required extensive searching in TEM samples. With CLEM, the secretory pods were specifically targeted, using the fluorescent labeling as a guide to find and identify the corresponding ultrastructure by TEM.

The main challenge in performing CLEM on secretory pods was to define a workflow to fluorescently label VWF and to retrace the observed fluorescent spots to specific structures observed in the TEM sections. Transient expression of a WPB marker fused with green fluorescent protein appeared to result in a low number of transfected endothelial cells. Furthermore, those cells contained fewer labeled WPBs compared to nontransfected cells. To increase the number of detectable targets for CLEM, we therefore used native nontransfected cells and an extracellular antibody to label secreted VWF. Double labeling experiments, in which extracellular VWF was labeled prior to cell permeabilization for subsequent intracellular labeling, showed that large extracellular dots of VWF often colocalized with fused secretory pods. By imaging

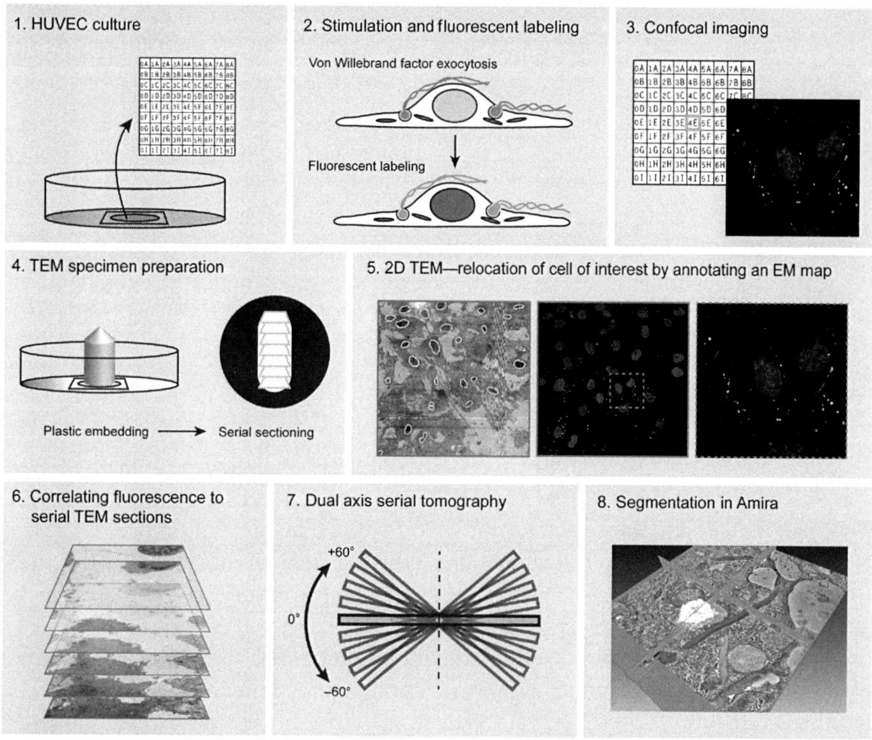

FIGURE 4.1

Schematic overview of the sequential steps of the CLEM procedure. (See the color plate.)

the typical extracellular patterns of secreted fluorescently labeled VWF for CLEM, we aimed to correlate the secreted VWF with fused secretory pods. We developed a correlative method for adherent cells by combining confocal microscopy with conventional TEM sample preparation. To study the VWF exocytosis, we preferred to work with a confluent layer of cells to have a maximal secretion response upon cellular stimulation, and to increase the number of targets for CLEM. To overcome the technological and practical challenges encountered when working with a confluent cell layer, several strategies were designed for both LM and EM to track and trace the region of interest. An overview of the described CLEM method is presented in Fig. 4.1.

4.2 METHODS

4.2.1 CELL CULTURE

To study VWF exocytosis from endothelial cells, a confluent monolayer of cells was required to induce a good cellular response upon stimulation. When confluence was reached, the cells were kept in culture for one additional day as this further boosted the stimulation response due to increased numbers of WPBs per cell. Pooled cryopreserved human umbilical vein endothelial cells (HUVECs) were thawed and seeded at a cell density of 5000 cells/cm^2 in complete endothelial growth medium-2 (EGM-2) in endothelial cell attachment factor (ECAF) coated T25 tissue culture flasks. Cells were incubated at 37 °C and 5% CO_2 and medium was refreshed every other day. When confluence was reached cells were subcultured and seeded at a density of 3000 cells/cm^2 in the 35 mm grid-engraved MatTek glass bottom petri dishes (P35G-2-14-C-GRID) coated with ECAF to promote cell adhesion. HUVECs were used for experiments until passage 5, as these primary cells gradually dedifferentiate when cultured for longer periods *in vitro*.

4.2.2 REGULATED WPB EXOCYTOSIS

To induce WPB exocytosis and VWF secretion, cells were stimulated with phorbol 13-myristate 12-acetate (PMA), which is a synthetic diacylglycerol analog. PMA was diluted in dimethyl sulfoxide (DMSO) to create a stock solution of 20 mM. To create the stimulation medium, the 20 mM PMA stock solution was diluted to a final concentration of 80 nM in endothelial cell incubation medium (ECIM). Control medium was supplemented with a similar amount of DMSO as was present in the stimulation medium. The cells were preincubated with ECIM for 10 min at 37 °C 5% CO_2 and subsequently stimulated under similar conditions with ECIM containing 80 nM PMA (or DMSO for controls) for 20 min. After stimulation, cells were fixed with ice-cold solution of 3% paraformaldehyde and 0.5% glutaraldehyde in 0.1 M phosphate buffer, pH 7.4 for 30 min at 4 °C.

4.2.3 FLUORESCENT LABELING

To label exocytosed VWF, cells were rinsed for 5 min with 50 mM Tris–HCl, blocked with phosphate-buffered saline (PBS) and 5% normal goat serum (NGS) for 5 min, and then incubated with fluorescein isothiocyanate (FITC) conjugated sheep anti-VWF antibody diluted 1:200 in PBS 5% NGS for 1 h. After several washing steps with PBS, the cells were incubated in 3% paraformaldehyde and 0.5% glutaraldehyde in 0.1 M phosphate buffer, pH 7.4, containing 300 nM of the nuclear stain 4′,6-diamidino-2-phenylindole (DAPI). Although the addition of glutaraldehyde in the fixative slightly quenches fluorescence, preservation of cell ultrastructure is significantly improved compared to paraformaldehyde fixation only.

4.2.4 CONFOCAL MICROSCOPY

Imaging secreted VWF was performed on an inverted confocal laser scanning microscope. The engraved grid in the glass bottom dish was used to locate target regions for CLEM. However, a confluent layer of cells obscures the visibility of the engraved grid in the MatTek dish, which makes localization of regions of interest technically more demanding than for nonconfluent layers.

- For confocal imaging, we started with a 10 × objective and focused in bright field (BF) to visualize the engraved grid in the petri dish. *Tip*: When focus is not easily found in BF mode, the DAPI signal can be used instead. After focusing on the DAPI, we returned to BF to adjust the light intensity and the aperture diaphragm for more contrast.
- The dish was rotated such that the letter–number coding in the grid was straight up and easily legible.
- The objective lens was then changed to 63 × oil immersion for imaging regions of interest for CLEM. We specifically searched for areas displaying fluorescently labeled VWF strings connected to large VWF dots.
- At regions of interest, a *z*-stack was recorded. The top focus plane and bottom focus plane were selected of the 3D volume of VWF-FITC-labeled structures. The *z*-stack for VWF-FITC and DAPI were sequentially acquired at 4 × (Fig. 4.2A) digital zoom with frame average of 4.
- We also collected single images of VWF-FITC and DAPI at 2 × digital zoom (Fig. 4.2B) and 1 × digital zoom. Furthermore, a phase contrast or a differential interference contrast (DIC) image was acquired to visualize the letter–number coordinates in the bottom of the petri dish.
- If the letter–number code in the bottom of the petri dish was not recognized using the 63 × objective, lower magnification objective lenses (such as

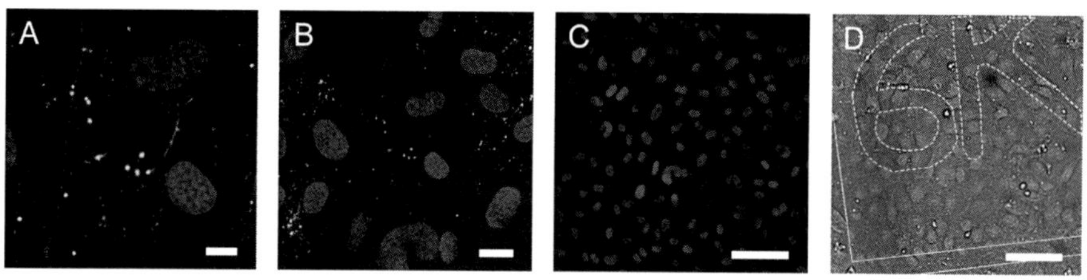

FIGURE 4.2

Fluorescence imaging for CLEM. Before image acquisition, the orientation of the engraved glass bottom dish was adjusted to ensure the letter–number code in the engraved glass bottom dish was right-side-up. (A) High magnification image of exocytosed VWF (VWF-FITC, green channel) and surrounding nuclei (DAPI, blue channel). Regions containing VWF strings (arrows) connected to VWF dots (arrowheads) were of particular interest. Image are maximum intensity projection of a *z*-stack acquired using a 63 × oil objective NA 1.4 and 4 × digital zoom. Scale bar, 10 μm. (B) Overview at intermediate magnification of the imaged VWF and nuclei shown in panel (A). Images were acquired using a 63 × oil objective NA 1.4 and 2 × digital zoom. Scale bar, 20 μm. (C and D) Identification of the number-letter code of the imaged region using a 40 × oil objective lens, NA 1.25, 1 × digital zoom. (C) DAPI image of the nuclei in the region on interest. (D) Overlay of a differential interference contrast (DIC) image and the DAPI image shown in panel (C). The number-letter code is delineated by a white-dashed line. The solid line highlights the borders of the imaged square. Scale bar panel (C and D), 100 μm. (See the color plate.)

This figure was partially reproduced from Mourik et al. (2013).

40 × or 20 ×) were used. We then collected single DAPI and DIC images (Fig. 4.2C and D).

Tip: If oil has to be removed from the dish to image with dry objectives, use the nuclear pattern from the DAPI stain to relocate the region of interest.

- Once the letter–number code nearest the cell of interest was recognized, a note was made of its location.
- Subsequently, the previously acquired *z*-stacks were converted to maximum intensity projections. In the Leica application suite advanced fluorescence (LAS AF) software, this is done in *Process—3D projection*. Set angles at 0. Select *maximum intensity projection* and press *Apply*. Finally, the images were exported as TIFF files.

4.2.5 SAMPLE PREPARATION FOR EM

Once the fluorescence imaging step is completed, the cells were immediately placed in 2% glutaraldehyde in 0.1 M sodium cacodylate buffer, pH 7.4 for at least 1 h for subsequent specimen preparation for EM. Before continuing with sample preparation, regions of interest were marked on the bottom of the petri dish. Marking the dish prior to the postfixation step with osmium is important given that a confluent, osmium-fixed layer of cells will completely obscure the engraved grid in the bottom of the dish. Marking this area on the dish is crucial for a later step in the preparation protocol in which a resin-filled capsule is positioned over the region of interest.

After marking the region of interest, the cells were washed several times with 0.1 M sodium cacodylate buffer, pH 7.4, and postfixed for 1 h with 1% osmium tetroxide in 0.1 M sodium cacodylate buffer, pH 7.4 at room temperature. The cells were then incubated with 1% tannic acid in 0.1 M sodium cacodylate buffer, pH 7.4 for 30 min. Cells were washed with 0.1 M sodium cacodylate buffer, pH 7.4, and dehydrated with a gradient series of ethanol: 5 min 70% and 70% ethanol overnight, two times 5 min 80% and 90% ethanol, followed by 5 min and 1 h 100% ethanol. Infiltration of the cells with resin was achieved by incubating the cells with a 1:1 mixture of 100% ethanol and Epon (LX-112) for 1.5 h at room temperature, which was followed by an incubation in 100% Epon for 2 h at room temperature and a 10 min incubation with fresh Epon at 60 °C to remove trace ethanol. Excess Epon was removed from the dishes and a large BEEM® capsule filled with Epon was placed on top of the cells on the marked region (Fig. 4.3A). Large size 00 BEEM® capsules were used to ensure the capsule covered the region of interest, as capsules can become displaced in the first hours of polymerization. Resin blocks were polymerized for 2 days at 60 °C.

To remove the BEEM® capsule with the resin embedded cells from the dishes, the dishes were placed first in liquid nitrogen, and then on a hot plate. This freezing and warming of the block were repeated until the resin block detached from the dish. The BEEM® capsule was then removed and the resin block was placed in a holder for sectioning (Fig. 4.3B).

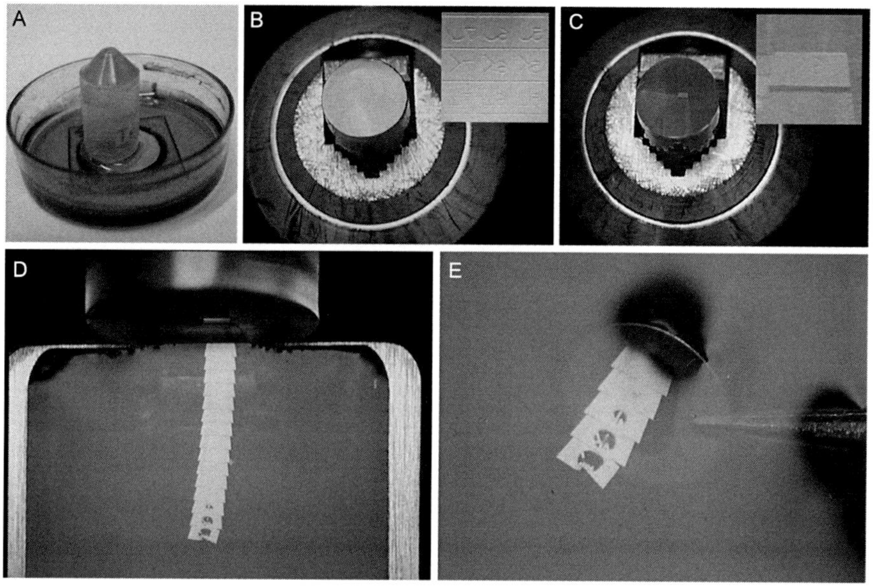

FIGURE 4.3

Serial sectioning. (A) A resin-filled BEEM® capsule on top of the flat embedded cells of interest. (B) After removing the resin block from the petri dish, the block is placed in a holder for sectioning. On the surface of the resin block a (mirrored) representation of the glass engraved number-letter grid is visible. The insert shows a magnified part of the block surface. (C) The resin around the region of interest is trimmed to form a pyramid for sectioning. The insert shows a higher magnification image of the pyramid, where the coordinate corresponding to the cell of interest (6K) is still visible. (D) A ribbon of serial sections floating on water in the boat of a diamond knife. On one side the sections are still attached to the edge of the diamond knife. (E) Collecting a short ribbon of sections on a copper single-slot grid. The single-slot grids were glow discharged before use and contain a carbon coated Formvar® support film. (See the color plate.)

An imprint of the engraved grid was retained in the surface of the resin block, which facilitates precise localization of the region previously imaged by LM (Fig. 4.3B, insert). The material around the region of interest was trimmed using a razor blade and a glass knife to create a pyramid suitable for sectioning at the region of interest of about 1 mm in width and 500 μm in height (Fig. 4.3C). In this way, the section covers the whole slot of an EM grid. To section the block for EM, a diamond sectioning knife was mounted on a microtome. The knife was then moved toward the surface of the resin block until a band of light appeared between the knife and the block surface. In our experience, it helps to switch off the top light from the microtome to sharpen the band of light observed between the knife edge and block surface. We adjusted the knife and block angle until the block and knife were roughly aligned to each other, i.e., the distance between knife and block was consistent over the whole pyramid surface. The knife was then moved closer to the block and the alignment was refined. The boat of the sectioning knife was filled with distilled water and serial sections of 100 nm were cut (Fig. 4.3D). Sections were picked up with glow discharged L2 × 1 single-slot copper grids, which were coated with Formvar® film and carbon (Fig. 4.3E). Since the support film on single-slot grids tends to break easily, a low number of sections (five to six) were collected on each grid to spread the risk of losing the sectioned material. To increase the contrast of the sections, they were incubated with 7% uranyl acetate in water for 20 min, washed extensively on droplets of water, and incubated with Reynolds lead citrate for 10 min. Finally, the sections were washed on droplets of water, blotted, and stored in a grid box.

4.2.6 CORRELATING LM TO TEM SECTIONS

During the specimen preparation for TEM, the entire 3D volume of the confluent layer of cells was sliced into a series of sections with a thickness of 100 nm. However, relocating a cell previously imaged by fluorescence LM in a single TEM section can be challenging in a confluent layer of cells. The cells appear as a layer of dense material in the TEM at low magnification (Fig. 4.4A), which impedes region-of-interest relocation on the basis of gross cell morphology. To overcome this limitation, and to relocate the area of interest, we recorded a mosaic of multiple TEM images of a single EM section (Fig. 4.4C, left) using the software MyMesh/MyStitch (version 5) as presented by Faas et al. (2012). Mosaic data collection was performed as follows:

- The sections were preirradiated for approximately 10 min at low magnification (about 100 ×) with a medium beam intensity (spotsize 2). Preirradiation is required to ensure homogeneous shrinkage of the sample in the *z*-axis prior to high-resolution data collection and avoids large variations in image intensity during data collection.
- The sample was adjusted to eucentric height and focused. The magnification was set to 860 × to visualize the nuclei of the cells in the section.
- The MyMesh software was opened (Fig. 4.4B). We selected a 25% overlap and entered a base name (e.g., Gridbox number and grid position) for each image to be acquired.
- The region for the overview scan was selected by providing MyMesh with the stage coordinates of the upper left- and lower right-corner of the quadrant we wanted to image (Fig. 4.4B). Clicking *Calculate* and *Add to queue* adds the region to the list of meshes for acquisition. The scan is started by pressing *Collect all data*. *Note*: Although the software provides an autofocus function, this is generally not required for low magnification imaging.

After acquisition of the mosaic, the program MyStitch was used to create a stitched compound file. In MyStitch, the contrast and brightness of the complete stitch was adjusted based on the histogram of individual images. Additionally, the option to correct for differences in beam intensity and gain was selected before starting the stitching process. Aperio ImageScope software was used to view and annotate the stitched file interactively at low and high magnification. To simplify relocation of

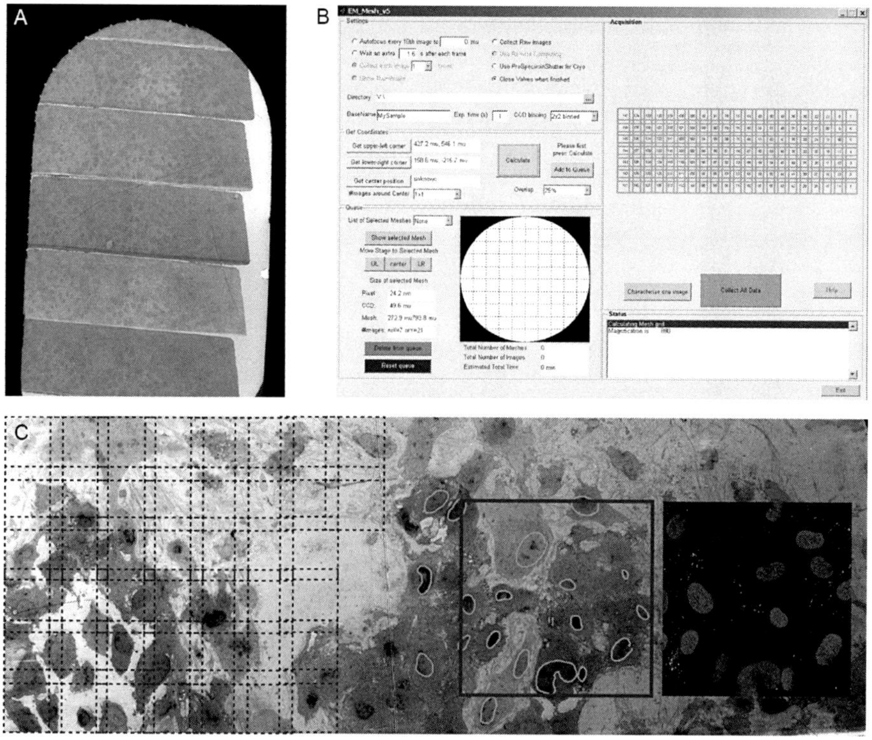

FIGURE 4.4

Relocating the region of interest in TEM sections. (A) Serial sections at low magnification at the TEM. (B) The graphical user interface of the MyMesh software to collect a scan of a whole TEM section. (C) Scan of a TEM section that is composed of multiple images. The scan can be viewed at low and high magnification in Aperio ImageScope and annotations can be added to simplify relocation of the area of interest. (See the color plate.)

the cells previously imaged with LM, we delineated the nuclei manually in the TEM stitch (Fig. 4.4C, center). We compared the nuclear pattern and shape of the cells in the TEM section and DAPI image (Fig. 4.4C, right) and compensated for changes in the orientation of the TEM section by rotating and flipping the TEM overview stitch in ImageScope. Once the region of interest is relocated in the stitch, navigation back to this position at the TEM is straightforward, and additional data can be recorded at higher magnifications.

To correlate the previously imaged fluorescence of secreted VWF to structures in the TEM sections, 2D images of the correlating cell were acquired in consecutive sections to create an overlay of the LM and EM data:

- In Adobe Photoshop™, EM images of consecutive sections were placed in separate layers. The EM images were aligned to each other using the nuclei as reference. The layers were made temporarily transparent to allow manual alignment of the layers.

- The maximum intensity projections of the individual LM channels (DAPI and VWF-FITC, in gray scale) were added to two new layers in the Photoshop file.
- Two additional layers were created. One layer was colored green and the other blue using the *Paint Bucket Tool* (*Shift+g*).
- The layer with the VWF-FITC LM image was then selected and copied (*Crtl+a*, *Ctrl+c*).
- In the green-colored layer, the *Quick Mask Mode* (*Q*) was entered and the VWF-FITC image was added in the green layer (*Ctrl+v*).
- The image was inverted (*Ctrl+i*) and the *Quick Mask Mode* was switched off by clicking again on the icon or by pressing *Q*. The remaining selected material was removed by pressing the *delete* button.
- This process was repeated for the DAPI channel in the blue layer.
- The correlation between the masked fluorescence and original LM image was checked.
- The layers containing the original fluorescent images were switched off so that only the masked versions are visible.
- Then, the layers containing fluorescent images (including invisible layers) were selected and simultaneously adapted using *free transform* (or *Ctrl+t*). The layers were scaled and rotated manually until the DAPI stain fitted best within the nuclei in the EM images. Structures found outside the cell, designating secreted VWF, were also used to fine tune the overlay.

  Note: To accommodate for shrinking in the EM, the LM images were sometimes also slightly adjusted with a different scaling in x and y.

Once the overlay was constructed, the fluorescent spots (Fig. 4.5A and B) were analyzed by turning the visibility of the layers of the consecutive EM sections off and on. An important consideration is that the fluorescence projection image contains information of the whole 3D cell, while the EM sections represent only 100 nm. Therefore, correlating structures were often found in multiple sections (Fig. 4.5C). The fluorescence of secreted VWF (Fig. 4.5A and B) was found to correlate with a secretory pod, a membrane delineated structure, with a WPB in close proximity. Analysis of consecutive sections showed that the secretory pod contained dense material which was connected with an extracellular globular mass of secreted VWF (Fig. 4.5C). We experienced that when an overlay between LM and EM was made, differences in correlation between the fluorescence and individual EM sections occurred which is possibly due to nonuniform shrinkage of the sections by the electron beam. In our case, this was not problematic as the investigated structures were relatively large.

4.2.7 SERIAL TOMOGRAPHY OF CORRELATING STRUCTURES

To obtain morphological 3D information of the structures targeted by LM, electron tomography on serial sections was performed (McIntosh, Nicastro, & Mastronarde, 2005). To align the images of the tilt series for 3D reconstructions gold fiducials were

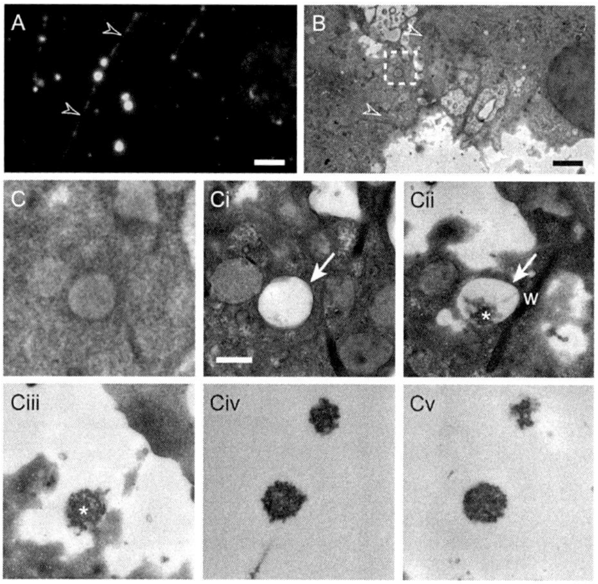

FIGURE 4.5

Correlating fluorescence to serial TEM sections. (A) Fluorescence of secreted dots of VWF connected to VWF strings (VWF-FITC, green) and the nucleus of a surrounding cell (DAPI, blue). Scale bar, 5 μm. (B) Overlay of the fluorescence image shown in panel (A) and the electron micrograph of the corresponding region in a TEM section. Scale bar, 5 μm. (C) High magnification of the boxed area in panel (B). (Ci–Cv) Consecutive TEM sections shown in structures that were found to correlate with fluorescently labeled VWF. (Ci) A membrane delineated secretory pod (arrow) containing little electron-dense material. (Cii) Secretory pod (arrow)-containing dense material of VWF (asterisk). A Weibel–Palade body (w) was also observed next to the secretory pod. (Ciii–Cv) Extracellular globular-shaped VWF. Scale bar, 500 nm. (See the color plate.)

This figure was partially reproduced from Mourik et al. (2013).

used. After recording high-resolution 2D images, 10 nm gold fiducials were added to the back side of the grid (nonsection side). The grid was then placed into a dual-axis tomography holder (Fischione, model 2040) for dual-axis image collection. In the TEM, the regions with the correlating structures were easily recognized due to previous 2D imaging. On these preirradiated positions the resin had become thinner and therefore appeared brighter in the TEM. For each of the five sections on the grid, the stage coordinates of the correlating structures were stored. Using SerialEM (version 2.8.6) (Mastronarde, 2005), dual-axis tilt series were collected for each of these recorded positions. Tomographic tilt series were recorded with tilt angles from +60° to −60° in 1° intervals, an exposure time of 1 s, at 11,000× magnification, using defocus −1 μm, spotsize 2 and binning 2, resulting in a 1.98-nm pixel size. Autofocus was performed automatically every 10° and at every tilt angle below

−56 or above 56°. The beam intensity was adjusted by the software to maintain the mean intensity at the camera at 17,000 counts for each recorded image. For all stored positions, the *A*-axis was collected first. When the *A*-axis was finished, two images are generated by the program which provide two options for a 90° rotation for collection of the *B*-axis. To keep the reference images for the alignment of the *B*-axis, it was best to make a screenshot to save the reference images in a separate file. SerialEM did not save the reference images. Once the *A*-axis for each position was acquired, the sample was rotated and the same regions were relocated using the saved reference image. Finally, the *B*-axis was collected for all the stored positions.

4.2.8 RECONSTRUCTION AND JOINING OF SERIAL TOMOGRAMS

Using the dual-axis tilt series collected from five serial sections, the 3D structures of extra- and intracellular VWF organization at the site of VWF exocytosis was reconstructed. The tomography reconstruction interface of eTOMO from the IMOD software suite (Kremer, Mastronarde, & McIntosh, 1996) was used. On the IMOD website (http://bio3d.colorado.edu/imod/), a detailed description of the reconstruction software is provided as well as a tutorial for dual-axis tomography reconstructions. Using this software, the *A*- and *B*-axis tilt series were reconstructed separately by resolution-weighted back projection and combined into a single tomogram (Fig. 4.6A; Mastronarde, 1997). During tomographic reconstruction, the gold particles were computationally erased to reduce the ray artifacts caused by these markers.

Next, the tomograms were joined into one volume using the joining software from IMOD. In the joining software, the top and bottom of consecutive tomograms were defined and manually aligned to each other. There were also options in the software for automatic alignment but this was not applicable in our case since the overlapping features in the consecutive tomograms were not so abundant. A detailed tutorial and additional information for joining tomograms can be found on the IMOD website (http://bio3d.colorado.edu/imod/).

4.2.9 MODELING AND VISUALIZATION

Generating a three-dimensional model of the data can provide a clearer understanding of the relevant structures within the tomogram. To this end, segmentation of the final tomogram was performed to highlight the exocytosis event of VWF secretion from the secretory pod (Fig. 4.5B). Before segmentation, the volume of the combined tomograms was binned by a factor of 2 in all directions to reduce the file size. We also filtered the volume by nonlinear anisotropic diffusion (*K* value 10 and 10 iterations) to enhance the contrast in the tomogram. Both steps were performed using IMOD (Kremer et al., 1996).

Segmentation was performed using the Amira software package. One of the first steps for segmentation is to assign specific subvolumes of the tomogram to a specific feature (e.g., cell membrane, WPBs, VWF). Depending on the material, this was achieved by manually masking regions once every 5–10 slices. Interpolation

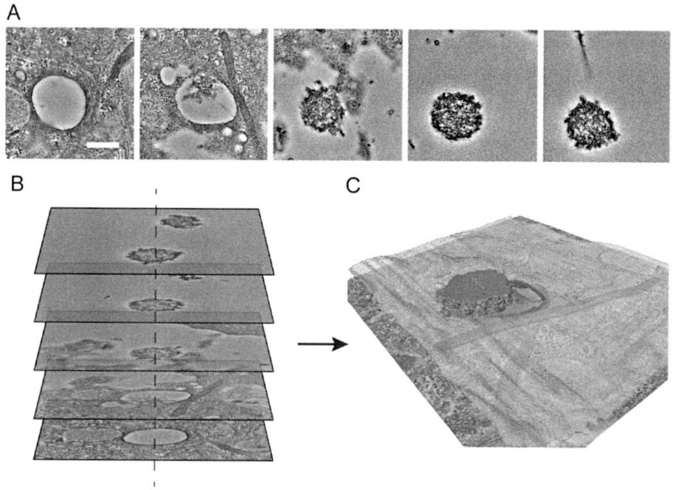

FIGURE 4.6

Serial tomography and segmentation. (A) Tomographic slices from five serial tomograms. Scale bar, 500 nm. (B) Schematic representation of the segmentation in joined serial tomograms. (C) The segmented three-dimensional model of VWF exocytosis from a secretory pod which was obtained from five combined serial tomograms (orange, secreted VWF; pink, secretory pod membrane; blue, docked WPB; transparent yellow, plasma membrane; green, clathrin-coated pits). See also supplementary movie S1 (http://dx.doi.org/10.1016/B978-0-12-801075-4.00004-5). (See the color plate.)

This figure was partially reproduced from Mourik et al. (2013).

between these slices yielded a volumetric segmentation for each feature. To refine the selection, the threshold function was used to ensure only voxels with intensity values corresponding to those of the feature of interest were included. Although voxel patches surrounding the feature with similar intensity values may be incorrectly included in the selection, these can be removed with the "*remove islets*" function. In addition to this palette of functions, Amira also provides options to shrink, grow, or smooth the selected volume.

The segmentation of the combined volume shows the 3D organization of the correlated secreted VWF expelled by a secretory pod. In addition, the structures associated with the secreted VWF were modeled such as the secretory pod membrane, a nearby WPB, and the cell membrane with several clathrin-coated pits (Fig. 4.5C). Finally, a movie from the tomogram and the segmentation was made using the DemoMaker and MovieMaker options in Amira. The tomogram alone, with the 3D model and finally the rotating model are shown in Supplementary movie S1 (http://dx.doi.org/10.1016/B978-0-12-801075-4.00004-5). A detailed description of Amira can be found in the Amira user guide.

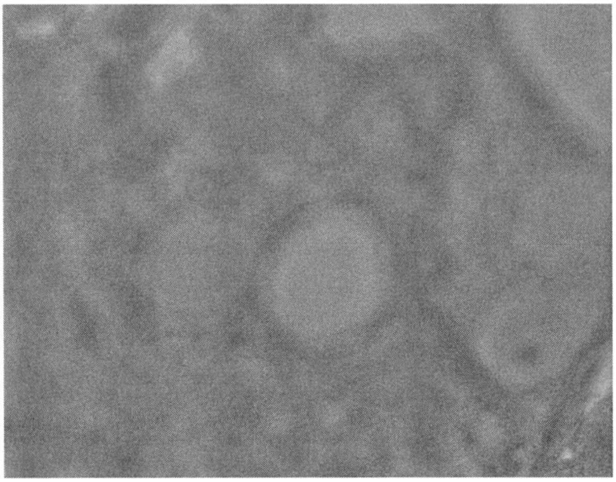

SUPPLEMENTARY MOVIE S1

Segmentation of five joined tomograms collected in consecutive sections showing Von Willebrand factor exocytosis from an endothelial cell. Colors represent: orange, secreted VWF; pink, secretory pod membrane; blue, docked WPB; yellow, plasma membrane; green, clathrin-coated pits.

This movie is reproduced from Mourik et al. (2013).

4.3 MATERIALS AND INSTRUMENTATION

4.3.1 CELL CULTURE

Instrumentation: Hettich Rotanta P centrifuge, CleanAir flow cabinet Class II type EF/B, Marienfeld Fuchs-Rosenthal cell counting chamber.

Materials: Pooled HUVEC cryo-ampule (Lonza, Walkersville, MD, C2517A), T25 culture flasks (Greiner Bio-One, Frickenhausen Germany, 690175), 5 mm glass bottom dishes with engraved grid (MatTek, Ashland, MA, P35G-2-14-C-GRID).

Reagents: EGM-2 BulletKit (Lonza, CC-3162), subculturing reagents Reagent-Pack™ (Lonza, CC-5034), ECAF (Sigma–Aldrich, St. Louis, MO, E9765).

4.3.2 WPB EXOCYTOSIS AND FIXATION

Instrumentation: Grant water bath, New Brunswick Scientific Innova CO-48 incubator.

Reagents: ECIM consisting of 20 mM HEPES sodium salt (Sigma–Aldrich, H3784), 112 mM sodium chloride (Sigma–Aldrich, S5886), 6 mM potassium chloride (Sigma–Aldrich, P5405), 1 mM magnesium sulfate (Sigma–Aldrich, M7774), 1 mM calcium chloride (Merck, White House Station, NJ, 1725701000), 1.2 mM dipotassium hydrogen phosphate (Mallinckrodt Baker, CAS# 7758-11-4), 5.6 mM D-glucose (Sigma–Aldrich, G7021), and 2% human serum albumin (Sigma–Aldrich,

A1653). The pH was adjusted to 7.4 using 1 N HCl. Eight percentage of paraformaldehyde (EMS, Hatfield, PA, 157-8), 8% glutaraldehyde (EMS, 16019), sodium phosphate monobasic monohydrate (Sigma–Aldrich, S6938), sodium phosphate dibasic dehydrate (Sigma–Aldrich, 71643).

4.3.3 FLUORESCENCE LABELING

Reagents: Trizma® base (Sigma–Aldrich, T6066), PBS, NGS (Dako, Glostrup, Denmark, X090710), anti-VWF-FITC (Abcam, Cambridge, UK, ab8822), 4′,6-diamidino-2-phenylindole dihydrochloride (DAPI) (Invitrogen, Grand Island, NY, D1306).

4.3.4 LIGHT MICROSCOPY

Instrumentation: Leica TCS SP5 confocal laser scanning microscope (Vienna, Austria) equipped with a motorized stage, a 405-nm laser diode, a tunable argon laser and a multiphoton laser. The objectives on the microscope were 10 × (NA 0.3), 20 × (NA 0.7) dry lenses, 40 × (NA 1.25), and 63 × (NA 1.4) oil immersion lenses.

Reagents: Leica immersion oil type-F (refraction index 1.52, cat. no. 115 1359).

Software: LAS AF.

4.3.5 ELECTRON MICROSCOPY

Instrumentation: Emitech K950 carbon coater with K350 glow discharge unit, Memmert heating oven, Leica Ultracut-6 microtome, FEI Tecnai 12 Biotwin 120 kV (FEI, Einthoven, The Netherlands) equipped with an FEI Eagle 4 × 4k CCD camera and compustage, FEI single tilt holder, dual-axis tilt holder (Fischione Instruments, Export, PA, model 2040).

Materials: BEEM® capsules size 00 pyramidal (EMS, 70000-B), Diatome ultra 45® sectioning knife (Hatfield, PA), copper slot grids L2 × 1 (Storck Veco B.V., Eerbeek, The Netherlands, OD841130), Dumont tweezers #5 (EMS, 72700-D).

Reagents: Sodium cacodylate (Sigma–Aldrich, C0250), osmium tetroxide (EMS, 19150), tannic acid (EMS, 21710), ethanol (Mallickrodt Baker, 8006.2500), Epon Lx-112 kit (Ladd, Williston, VT, 5-21210), Formvar® 1595 E (Merck, 12164), uranyl acetate (Fluka, Sigma–Aldrich, 73943), Reynolds lead citrate, 10 nm colloidal gold fiducials (Cell Microscopy Center, Utrecht, The Netherlands).

Software: FEI TEM server TIA, MyTEM/MyMesh/MyStitch version 5 (Faas et al., 2012), SerialEM version 2.8.6 (Mastronarde, 2005).

4.3.6 DATA PROCESSING

Instrumentation: High end 64-bit computer with two Intel® Xeon® processors CPU 2.67 GHz, and 16-GB RAM running Windows 7 Professional.

Software: Adobe Photoshop™ CS3, ImageJ version 1.45s (NIH, USA), Aperio ImageScope version 11.1.2, IMOD version 4.1.10 (Kremer et al., 1996), Amira visualization package version 2.4.5 (Visage Imaging).

4.4 DISCUSSION

The CLEM strategy described in this chapter was designed to study the secretion of VWF from endothelial cells. This CLEM technique of imaging fluorescently labeled structures by confocal microscopy prior to resin embedding for EM is relatively simple, robust, and generally applicable. To relocate the targeted areas imaged by fluorescence LM in serial TEM sections, glass bottom cell culture dishes containing an engraved letter–number coded grid were used in combination with the MyTEM/MyStitch scanning/stitching software to create an EM map for navigation (Faas et al., 2012). The EM map is especially beneficial when performing CLEM on a confluent layer of cells as the region of interest is difficult to recognize based on the shape of the cells alone.

The advantage of the approach described earlier is that it combines conventional methods that are commonly used, and does not require highly advanced sample preparation steps or imaging equipment. Moreover, this method can be adapted straightforwardly to every type of fluorescent label (e.g., chemical stains, antibody labeling, or fused fluorescent proteins) and to a variety of cell types (Kobayashi et al., 2012; Loussert et al., 2012; Polishchuk et al., 2012; Valentijn et al., 2010). The engraved grid in the glass bottom dish provides a large search area which is beneficial for studies that investigate rare cellular events or for studies that work with biological systems that cope with a low transfection efficiency of fluorescently labeled proteins. As shown by Polishchuk et al., this type of CLEM can also be combined with live-cell imaging which can be performed prior to fixation (Polishchuk et al., 2012).

The most challenging step in successfully performing CLEM is tracking the region of interest throughout the workflow. The current commercially available glass bottom dish with the engraved localization system is only one of many methods available. Other methods include the creation of a localization pattern by placing a finder grid on top of the cells (Ren et al., 2003) or by carbon coating the meshwork of finder grids on a glass bottom dishes before seeding cells (Kobayashi et al., 2012). These localization methods are used primarily as a navigation tool for ultramicrotomy of the TEM sample, to ensure that the area selected for sectioning contains the region of interest. Once the sections are placed on a grid and visualized in the TEM, relocating the cell of interest is usually achieved by visual recognition of the patterns of cells. When working with low cell densities, this approach has a high success rate. However, this becomes more challenging when the number of cells in the section increases or when the 2D profile of the cell changes significantly upon sectioning. As shown by Polishchuk et al., it is also possible to use one of the first sections as a navigation map (Polishchuk et al., 2012). The first sections retain the imprint of the glass engraved grid of the petri dish which was also visible on the surface of the resin block (Fig. 4.3D and E). In our experience, this method still has limitations, particularly when the correlating structure resides on a subsequent grid. Others reported the use of toluidine blue-stained sections for navigation in the EM (Kobayashi et al., 2012). We also explored this approach but found that preparation and imaging of toluidine blue-stained sections was as time consuming as collecting a stitch at the

TEM. The main advantage of the stitch is that it directly represents one of the sections on the TEM grid, making relocation and orientation much more convenient for subsequent TEM imaging. Once the same cell is relocated in the TEM section the correlation between the previously recorded fluorescence and the ultrastructure has to be examined.

To correlate the LM images to the EM images, we projected the fluorescence signal on top of the EM images. We used the nuclei in the EM images and DAPI stain to align the fluorescence to the EM. In our case, the extracellular VWF could also be used to refine the alignment as it appeared to function as a bimodal marker forming distinct patterns seen by both LM and EM. When the fluorescently labeled structure is relatively large (0.5–1 μm) relocating the correlating ultrastructure is relatively simple. However, when structures become smaller bimodal markers such as fluorescently labeled latex beads may be required to correct for nonuniform deformations. Using these bimodal markers, Kukulski et al. have demonstrated that the LM can be aligned to EM images with an accuracy better than 100 nm (Kukulski et al., 2011).

4.5 OUTLOOK

The choice for the most effective CLEM strategy depends on the biological question. New developments in instrumentation, sample preparation, and bimodal labels contribute to an expanding number of CLEM strategies, driving a progression toward less involved workflows that are faster, more accurate, and applicable to a wider variety of biological systems.

The knowledge of today, combined with the latest types of specimen preparation and imaging instrumentation, provides novel approaches to study VWF secretion sites by CLEM that were not available a few years ago.

One of these approaches, developed by Kukulski et al., enables fluorescence retention in plastic embedded biological material (Kukulski et al., 2011). Fluorescently labeled extracellular VWF is retained in the EM sections while providing excellent morphological preservation. An additional benefit of this approach is the possibility of subsequent on-section immunogold labeling, which could be used to target intracellular VWF.

Complementary to this approach, integrated LM–TEM systems such as the iLEM (Agronskaia et al., 2008; Faas et al., 2013) or an integrated LM in a scanning electron microscope (SEM) (Liv et al., 2013) could provide huge advances in imaging efficiency. Using these integrated systems correlations can be made instantly, without the time consuming relocation to the regions of interest required using conventional, separate LM and EM imaging steps.

Another potential CLEM method to study VWF secretion sites could be the combination of confocal microscopy with 3D volume focused ion beam SEM (FIB-SEM). The benefit of FIB-SEM is that the automated data collection can be performed relatively fast compared to TEM analysis of serial sections. Correlating multiple regions imaged by confocal microscopy with 3D volume SEM can be performed easily, vastly increasing the imaging throughput relative to other CLEM methods. Although it is more challenging to visualize small and delicate structures such as actin, FIB-SEM imaging also provides a good alternative for electron tomography as it can generate large 3D stacks with slices up to 3 nm in thickness (Narayan et al., 2013). For this, grid-engraved MatTek dishes could provide a good navigation tool for both LM and SEM. In addition, this CLEM approach can be easily combined with other fluorescent LM modalities such as live-cell imaging and super resolution LM techniques such as structured illumination microscopy and photoactivated localization microscopy (Requejo-Isidro, 2013).

These exciting advances in CLEM methodologies will enable the investigation of an ever-broader range of biological questions. Information obtained through structure–function studies will contribute both to our general understanding of diseases such as VWD and the cellular mechanisms underlying their development.

ACKNOWLEDGMENTS

The authors would like to thank Cristina Avramut for help in cell culture, Joop Wiegant for help with confocal microscopy, Lennard Voortman for help with Adobe Photoshop, Montserrat Barcena for help with SerialEM, and Roman Koning for critical reading of the manuscript, and for help with Amira. This work was financially supported by The Netherlands Organization for Scientific Research (NWO TOP, grant no. 91209006).

REFERENCES

Agronskaia, A. V., Valentijn, J. A., van Driel, L. F., Schneijdenberg, C. T., Humbel, B. M., van Bergen en Henegouwen, P. M., et al. (2008). Integrated fluorescence and transmission electron microscopy. *Journal of Structural Biology*, *164*(2), 183–189. http://dx.doi.org/10.1016/j.jsb.2008.07.003.

Aird, W. C. (2007). Phenotypic heterogeneity of the endothelium: I. Structure, function, and mechanisms. *Circulation Research*, *100*(2), 158–173. http://dx.doi.org/10.1161/01.RES.0000255691.76142.4a.

Berriman, J. A., Li, S., Hewlett, L. J., Wasilewski, S., Kiskin, F. N., Carter, T., et al. (2009). Structural organization of Weibel–Palade bodies revealed by cryo-EM of vitrified endothelial cells. *Proceedings of the National Academy of Sciences of the United States of America*, *106*(41), 17407–17412. http://dx.doi.org/10.1073/pnas.0902977106.

Brown, E., Mantell, J., Carter, D., Tilly, G., & Verkade, P. (2009). Studying intracellular transport using high-pressure freezing and correlative light electron microscopy. *Seminars in Cell and Developmental Biology*, *20*(8), 910–919. http://dx.doi.org/10.1016/j.semcdb.2009.07.006.

Brown, E., & Verkade, P. (2010). The use of markers for correlative light electron microscopy. *Protoplasma*, *244*(1–4), 91–97. http://dx.doi.org/10.1007/s00709-010-0165-1.

Faas, F. G. A., Avramut, M. C., van den Berg, B. M., Mommaas, A. M., Koster, A. J., & Ravelli, R. B. (2012). Virtual nanoscopy: Generation of ultra-large high resolution electron microscopy maps. *The Journal of Cell Biology*, *198*(3), 457–469. http://dx.doi.org/10.1083/jcb.201201140.

Faas, F. G. A., Barcena, M., Agronskaia, A. V., Gerritsen, H. C., Moscicka, K. B., Diebolder, C. A., et al. (2013). Localization of fluorescently labeled structures in frozen-hydrated samples using integrated light electron microscopy. *Journal of Structural Biology*, *181*(3), 283–290. http://dx.doi.org/10.1016/j.jsb.2012.12.004.

Garini, Y., Vermolen, B. J., & Young, I. T. (2005). From micro to nano: Recent advances in high-resolution microscopy. *Current Opinion in Biotechnology*, *16*(1), 3–12. http://dx.doi.org/10.1016/j.copbio.2005.01.003.

Haberichter, S. L., Jozwiak, M. A., Rosenberg, J. B., Christopherson, P. A., & Montgomery, R. R. (2002). The von Willebrand factor propeptide (VWFpp) traffics an unrelated protein to storage. *Arteriosclerosis, Thrombosis, and Vascular Biology*, *22*(6), 921–926. http://dx.doi.org/10.1161/01.atv.0000017063.36768.87.

Huang, R. H., Wang, Y., Roth, R., Yu, X., Purvis, A. R., Heuser, J. E., et al. (2008). Assembly of Weibel–Palade body-like tubules from N-terminal domains of von Willebrand factor. *Proceedings of the National Academy of Sciences of the United States of America*, *105*(2), 482–487. http://dx.doi.org/10.1073/pnas.0710079105.

Karreman, M. A., Agronskaia, A. V., van Donselaar, E. G., Vocking, K., Fereidouni, F., Humbel, B. M., et al. (2012). Optimizing immuno-labeling for correlative fluorescence and electron microscopy on a single specimen. *Journal of Structural Biology*, *180*(2), 382–386. http://dx.doi.org/10.1016/j.jsb.2012.09.002.

Kobayashi, K., Cheng, D., Huynh, M., Ratinac, K. R., Thordarson, P., & Braet, F. (2012). Imaging fluorescently labeled complexes by means of multidimensional correlative light and transmission electron microscopy: Practical considerations. *Methods in Cell Biology*, *111*, 1–20. http://dx.doi.org/10.1016/B978-0-12-416026-2.00001-7.

Koster, A. J., & Klumperman, J. (2003). Electron microscopy in cell biology: Integrating structure and function. *Nature Reviews Molecular Cell Biology*, *4*, SS6–SS10.

Kremer, J. R., Mastronarde, D. N., & McIntosh, J. R. (1996). Computer visualization of three-dimensional image data using IMOD. *Journal of Structural Biology*, *116*(1), 71–76. http://dx.doi.org/10.1006/jsbi.1996.0013.

Kukulski, W., Schorb, M., Welsch, S., Picco, A., Kaksonen, M., & Briggs, J. A. (2011). Correlated fluorescence and 3D electron microscopy with high sensitivity and spatial precision. *The Journal of Cell Biology*, *192*(1), 111–119. http://dx.doi.org/10.1083/jcb.201009037.

Liv, N., Zonnevylle, A. C., Narvaez, A. C., Effting, A. P., Voorneveld, P. W., Lucas, M. S., et al. (2013). Simultaneous correlative scanning electron and high-NA fluorescence microscopy. *PLoS ONE*, *8*(2), e55707. http://dx.doi.org/10.1371/journal.pone.0055707.

Loussert, C., Forestier, C. L., & Humbel, B. M. (2012). Correlative light and electron microscopy in parasite research. *Methods in Cell Biology*, *111*, 59–73. http://dx.doi.org/10.1016/B978-0-12-416026-2.00004-2.

Mastronarde, D. N. (1997). Dual-axis tomography: An approach with alignment methods that preserve resolution. *Journal of Structural Biology*, *120*(3), 343–352. http://dx.doi.org/10.1006/jsbi.1997.3919.

Mastronarde, D. N. (2005). Automated electron microscope tomography using robust prediction of specimen movements. *Journal of Structural Biology*, *152*(1), 36–51. http://dx.doi.org/10.1016/j.jsb.2005.07.007.

McIntosh, R., Nicastro, D., & Mastronarde, D. (2005). New views of cells in 3D: An introduction to electron tomography. *Trends in Cell Biology*, *15*(1), 43–51. http://dx.doi.org/10.1016/j.tcb.2004.11.009.

Michaux, G., Hewlett, L. J., Messenger, S. L., Goodeve, A. C., Peake, I. R., Daly, M. E., et al. (2003). Analysis of intracellular storage and regulated secretion of 3 von Willebrand disease-causing variants of von Willebrand factor. *Blood*, *102*(7), 2452–2458. http://dx.doi.org/10.1182/blood-2003-02-0599.

Mourik, M. J., Valentijn, J. A., Voorberg, J., Koster, A. J., Valentijn, K. M., & Eikenboom, J. (2013). von Willebrand factor remodeling during exocytosis from vascular endothelial cells. *Journal of Thrombosis and Haemostasis*, *11*(11), 2009–2019. http://dx.doi.org/10.1111/jth.12401.

Narayan, K., Danielson, C. M., Lagarec, K., Lowekamp, B. C., Coffman, P., Laquerre, A., et al. (2013). Multi-resolution correlative focused ion beam scanning electron microscopy: Applications to cell biology. *Journal of Structural Biology*, *185*, 278–284. http://dx.doi.org/10.1016/j.jsb.2013.11.008.

Polishchuk, R. S., Polishchuk, E. V., & Luini, A. (2012). Visualizing live dynamics and ultrastructure of intracellular organelles with preembedding correlative light-electron microscopy. *Methods in Cell Biology*, *111*, 21–35. http://dx.doi.org/10.1016/B978-0-12-416026-2.00002-9.

Ren, Y., Kruhlak, M. J., & Bazett-Jones, D. P. (2003). Same serial section correlative light and energy-filtered transmission electron microscopy. *The Journal of Histochemistry and Cytochemistry*, *51*(5), 605–612.

Requejo-Isidro, J. (2013). Fluorescence nanoscopy. Methods and applications. *Journal of Chemical Biology*, *6*(3), 97–120. http://dx.doi.org/10.1007/s12154-013-0096-3.

Rondaij, M. G., Bierings, R., Kragt, A., van Mourik, J. A., & Voorberg, J. (2006). Dynamics and plasticity of Weibel–Palade bodies in endothelial cells. *Arteriosclerosis, Thrombosis, and Vascular Biology*, *26*(5), 1002–1007. http://dx.doi.org/10.1161/01.ATV.0000209501.56852.6c.

Ruggeri, Z. M. (2003). Von Willebrand factor, platelets and endothelial cell interactions. *Journal of Thrombosis and Haemostasis*, *1*(7), 1335–1342.

Sadler, J. E., Mannucci, P. M., Berntorp, E., Bochkov, N., Boulyjenkov, V., Ginsburg, D., et al. (2000). Impact, diagnosis and treatment of von Willebrand disease. *Thrombosis and Haemostasis*, *84*(2), 160–174.

Shu, X., Lev-Ram, V., Deerinck, T. J., Qi, Y., Ramko, E. B., Davidson, M. W., et al. (2011). A genetically encoded tag for correlated light and electron microscopy of intact cells, tissues, and organisms. *PLoS Biology*, *9*(4), e1001041. http://dx.doi.org/10.1371/journal.pbio.1001041.

Suzuki, T., Matsuzaki, T., Hagiwara, H., Aoki, T., & Takata, K. (2007). Recent advances in fluorescent labeling techniques for fluorescence microscopy. *Acta Histochemica et Cytochemica*, *40*(5), 131–137. http://dx.doi.org/10.1267/ahc.07023.

Valentijn, K. M., van Driel, L. F., Mourik, M. J., Hendriks, G. J., Arends, T. J., Koster, A. J., et al. (2010). Multigranular exocytosis of Weibel–Palade bodies in vascular endothelial cells. *Blood*, *116*(10), 1807–1816. http://dx.doi.org/10.1182/blood-2010-03-274209.

van Driel, L. F., Valentijn, J. A., Valentijn, K. M., Koning, R. I., & Koster, A. J. (2009). Tools for correlative cryo-fluorescence microscopy and cryo-electron tomography applied to whole mitochondria in human endothelial cells. *European Journal of Cell Biology*, *88*(11), 669–684. http://dx.doi.org/10.1016/j.ejcb.2009.07.002.

Voorberg, J., Fontijn, R., Calafat, J., Janssen, H., van Mourik, J. A., & Pannekoek, H. (1993). Biogenesis of von Willebrand factor-containing organelles in heterologous transfected CV-1 cells. *The EMBO Journal*, *12*(2), 749–758.

Weibel, E. R., & Palade, G. E. (1964). New cytoplasmic components in arterial endothelia. *The Journal of Cell Biology*, *23*(1), 101–112. http://dx.doi.org/10.1083/Jcb.23.1.101.

CHAPTER

A Simple Procedure to Analyze Positions of Interest in Infectious Cell Cultures by Correlative Light and Electron Microscopy

5

Kazimierz Madela*, Sebastian Banhart†, Anja Zimmermann‡, Janett Piesker*, Norbert Bannert‡, Michael Laue*

**Advanced Light and Electron Microscopy (ZBS 4), Robert Koch Institute, Berlin, Germany*

†Junior Research Group Sexually Transmitted Bacterial Pathogens (NG 5), Robert Koch Institute, Berlin, Germany

‡Centre for HIV and other Retroviruses (FG 18), Robert Koch Institute, Berlin, Germany

CHAPTER OUTLINE

Abstract

Plastic cell culture dishes that contain a thin bottom of highest optical quality including an imprinted finder grid (μ-Dish Grid-500) are optimally suited for routine correlative light and electron microscopy using chemical fixation. Such dishes allow high-resolution fluorescence and bright-field imaging using fixed and living cells and are compatible with standard protocols for scanning and transmission electron microscopy. Ease of use during cell culture and imaging,

Methods in Cell Biology, Volume 124, ISSN 0091-679X, http://dx.doi.org/10.1016/B978-0-12-801075-4.00005-7

as well as a tight cover render the dishes particularly suitable for working with infectious organisms up to the highest biosafety level. Detailed protocols are provided and demonstrated by showing two examples: monitoring the production of virus-like particles of the Human Endogenous Retrovirus HERV-K(HML-2) by HeLa cells and investigation of Rab11-positive membrane-compartments of HeLa cells after infection with *Chlamydia trachomatis*.

INTRODUCTION AND RATIONALE

Cell cultures are one of the most frequently used tools to study infectious organisms, mainly because cell culture cells are the smallest viable units that can be infected. Moreover cell cultures are needed to propagate at least some infectious organisms, like for instance viruses or certain obligate intracellular bacteria, such as *Chlamydia*. An important advantage of working with cell cultures is their compatibility with most of the biosafety conditions which are necessary to avoid unintended infection of animals and humans. With the automation of light microscopes, live-cell imaging became a valuable approach for the investigation of infectious organisms. Applications range from simple observations of the cytopathic effect which a virus is causing over time to the analysis of virus entry, assembly, and egress using fluorescent virus proteins (e.g., Brandenburg & Zhuang, 2007; Schudt, Kolesnikova, Dolnik, Sodeik, & Becker, 2013). Since most infectious organisms are rather small and in case of viruses usually smaller than the resolution of conventional diffraction-limited light microscopy, imaging at highest spatial resolution is frequently necessary. Electron microscopy is able to visualize all infectious organisms known to date without using specific labeling and is therefore still used as a generic diagnostic tool (Curry, Appleton, & Dowsett, 2006; Goldsmith et al., 2013). Moreover, electron microscopy can reveal other, yet unexpected structures associated with infectious organisms even within the cellular context (which is also coined as the "reference space"; Brown, Mantell, Carter, Tilly, & Verkade, 2009) and thereby provides a more complete picture of the ultrastructure than fluorescence microscopy, including all super-resolution techniques that are currently available. However, to get maximum information, advantages of the different microscopy modes (i.e., resolution, specificity, dynamics) must be combined. Correlative light and electron microscopy (CLEM) allows to analyze corresponding positions of interest by light and electron microscopy in a successive process and to correlate directly the information obtained by the different microscopy modes.

Many procedures have been described to study cells and other biological systems by CLEM (compare this volume; Müller-Reichert & Verkade, 2012). In contrast to other more specialized approaches, we urged to develop a rather universal method, which can be routinely used to locate positions of interest in infectious cell cultures by light microscopy for a relocalization in the electron microscope. Importantly, such a method should be compatible with common biosafety standards up to the highest biosafety levels (BSL-3 and -4). While rapid freezing methods are rather complex to handle and most importantly inactivation of microorganisms by the complete process (freezing and the follow-up procedures) has still to be assessed in more detail (Vanhecke, Zuber, Brugger, & Studer, 2012), chemical fixation has proven that it is stabilizing (Laue, 2010) and inactivating (Rubo, Gardner, & Webb, 1967) infectious organisms efficiently. Moreover, chemical fixation is easily applied and usually needs only liquid handling, like it is done for medium exchange. However, while it is clear that many questions relevant in ultrastructural research can be answered by using chemical preparation methods (Griffiths, 1993, chapter 2.1), highest resolution and minimal artifacts can only be achieved by employing cryo-preparation techniques (e.g., Eltsov & Zuber, 2006).

Since a few years cell culture dishes (μ-Dishes; ibidi GmbH) are available for imaging which are entirely produced out of plastic, including a particular plastic at the bottom which guarantees highest optical performance in light microscopy. They are particularly useful for experiments carried out in biosafety laboratories where glass should be avoided or is not allowed because it can damage personal protection (e.g., gloves and suit). A variant of such dishes is equipped with a labeled grid pattern that is imprinted in the bottom allowing localization of positions of interest by light microscopy. We have selected these dishes (with or without a grid pattern) as our standard dish for live-cell imaging because they are ready to use (sterile and pretreated by charging) and similar in size to other dishes used for conducting cell culture experiments (e.g., six-well plates) which allows to switch from screening experiments to correlative microscopy without changing cell culture parameters. Another advantage is that the μ-Dish plastic is resistant against most chemicals used for standard preparation of electron microscopy.

In this chapter, we describe a method for correlative imaging of positions of interests in cells cultivated on μ-Dishes by using live-cell microscopy followed by either ultrathin-section transmission electron microscopy or scanning electron microscopy (SEM). A few studies already have used the μ-Dishes for correlative imaging of cells by light and electron microscopy (e.g., Katsen-Globa et al., 2014; Koppensteiner, Banning, Schneider, Hohenberg, & Schindler, 2012), but without extensive description of the methodology. We show results from two representative models as examples to demonstrate feasibility of our protocols. In one model, transfection of cell cultures was used for the generation of virus-like particles (VLPs) of a reconstituted Human Endogenous Retrovirus type K (HERV-K[HML-2]) (Chudak et al., 2013; George et al., 2011). A major goal of such experiments was to study the assembly and release of these viral particles from human cells transfected with expression plasmids encoding large parts of the reconstituted ancient retroviral genome. HERV-K(HML-2) proteins and viral particles are expressed in teratocarcinomas and several other human cancers and have been associated with carcinogenesis (Bannert & Kurth, 2006). The VLPs were fluorescent labeled by incorporation of a Gag-Cherry fusion protein into the regular Gag lattice during virion assembly. In pilot experiments, a 1:5 ratio of Gag-Cherry to Gag expression plasmids was found to give best fluorescent labeling, thus allowing efficient release of morphologically intact VLPs. To achieve a higher expression of the retroviral proteins, large parts of the coding sequences were codon optimized (George et al., 2011). In the other model, cells were infected with *Chlamydia trachomatis*, a pathogenic sexually transmitted bacterium, that invades human cells and replicates in a membrane-bound intracellular compartment termed "inclusion" (Moulder, 1991). It is important to study the formation and maintenance of these inclusions to understand the propagation and pathogenicity of *Chlamydia*. In particular, reorganization of the membrane compartments, such as the Golgi apparatus (Heuer et al., 2009), and acquisition of host cell lipids (Hackstadt, Rockey, Heinzen, & Scidmore, 1996; Rejman Lipinski et al., 2009) are issues for a detailed analysis of cellular trafficking processes which also involves live-cell imaging using various labeled cell compartments (e.g., Rab11-positive vesicles).

5.1 METHODS

5.1.1 CELL CULTURE, INFECTION AND TRANSFECTION

Culture dishes from ibidi company (μ-Dish; 35 mm in diameter, 12 mm high, with lid) were used for cell culture. Dishes possess a thin (170 μm) plastic bottom in which a labeled grid was imprinted (Fig. 5.1). Cultivation of cells was performed identically to standard procedures (six-well plates are perfect for testing of the cultivation conditions, because the individual wells have almost the same size than a μ-Dish). A pretreatment of the plastic surface is usually not necessary, if physically modified dishes (ibiTreat) are used to promote cell adhesion.

Transfection of Hela cells with HERV-K(HML-2) cDNA constructs

1. HeLa cells were seeded on μ-Dishes (150,000 cells/dish) and cultivated in Dulbecco's modified Eagle Medium (D-MEM) supplemented with 10% fetal calf serum (plus 50 units/ml penicillin, 50 μg/ml streptomycin and 2 mM

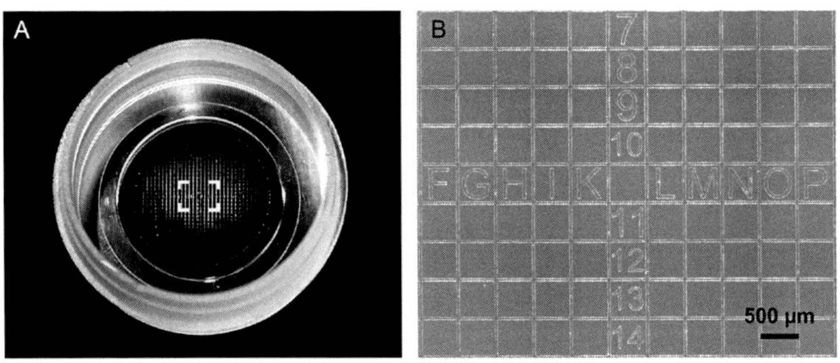

FIGURE 5.1

Cell culture dish (μ-Dish) with imprinted and labeled grid pattern. (A) Overview of the entire dish (without lid) which is 35 mm in diameter. The effective diameter for imaging is 21 mm, while the grid pattern covers a central area of 10 x 10 mm. Brackets indicate the central part of the grid pattern which is shown in (B) in detail by using light microscopy. One field has a size of 500 x 500 μm, i.e., the entire grid pattern is formed by 20 x 20 fields.

L-glutamine). Cultivation was done overnight at 37 °C, 5% carbon dioxide and 98% humidity.
2. Transfection was performed using PolyFect reagent according to instructions of the manufacturer and a mixture (1:5) of the plasmids pcDNAoricoGagCherry (Gag-Cherry) and pcDNAoricoHERV-K_GagProPol (George et al., 2011). After incubation overnight, reagents were removed and replaced by optimized D-MEM (Opti-MEM, without phenol red). For live-cell imaging Trolox, a vitamin E analogue, was added (100 μM) to reduce photo damage by free radicals.

Infection of Hela cells expressing Rab11A/eGFP with Chlamydia
1. HeLa cells were cultured in RPMI 1640 medium supplemented with 10% fetal calf serum, 1 mM sodium pyruvate, and 2 mM L-glutamine at 37 °C and 5% CO_2 in a humidified incubator.
2. *C. trachomatis* was propagated and purified from HeLa cell monolayers. Infections were performed in D-MEM (with 4.5 g/l glucose) supplemented with 5% fetal calf serum, 1 mM sodium pyruvate, and 2 mM L-glutamine (infection medium).
3. For live-cell imaging, approximately 100,000 HeLa cells were seeded in a single μ-Dish and cultivated overnight. The next day, cells were washed twice with infection medium and incubated with *C. trachomatis* L2 using a multiplicity of infection of approximately 0.5 at 35 °C and 5% CO_2 in a humidified incubator. After 2 h, cells were again washed with infection medium and then incubated at 35 °C. At 8 h *post infectionem* (p.i.), cells were transfected with pEGFP-Rab11A using Lipofectamine 2000 transfection reagent (Invitrogen GmbH, Germany) according to the manufacturer's protocol. Live-cell imaging was performed 24 h p.i. in infection medium (without phenol red).

5.1.2 CORRELATIVE LIVE-CELL IMAGING AND SEM

Visualization of HERV-K(HML-2) assembly and confirmation of fluorescent labeled HERV-K(HML-2) VLPs at the surface of HeLa cells

To optimize experimental conditions for studying the assembly and release of HERV-K (HML-2) VLPs, we have conducted correlative live-cell imaging and SEM. While live-cell imaging gave us information about the expression and distribution of the fluorescent labeled retrovirus protein (Gag-Cherry) over time, SEM provided the end-point control for the final formation of VLPs at the cell surface. The entire cell population was monitored over time by live-cell imaging and individual cells were selected for a detailed analysis when punctate fluorescence significantly concentrated near their cell surface. Position of cells relative to the grid pattern was noted, overview images were recorded and detailed inspection at higher magnification followed. After live-cell imaging, cells were immediately fixed by glutaraldehyde. The dish bottom was removed and processed for SEM. Recorded positions were easily relocalized relative to the grid pattern at low magnification in the SEM and surface of cells were inspected at higher magnification for the presence of virus particles (Fig. 5.2).

1. *Live-cell imaging*: Transfected Hela cells were inspected with an automated live-cell imaging microscope (Nikon Eclipse TE200-E; Nikon Corp., Japan). Imaging was done using a Plan-Fluor (20x, NA 0.5, DIC) or a Plan-Apochromat (60x, NA 1.4, DIC, oil) lens and an EMCCD camera (at a resolution of 512×512 pixels, 16 bit) for recording. All fields of interest were inspected using differential interference contrast (DIC) and fluorescent mode (excitation filter: 485 nm; emission filter: 526 nm). During imaging cells were maintained at a temperature of 37 °C and in a moistened atmosphere containing 5% carbon dioxide.
2. *Chemical fixation*: At the end of the observation period (usually up to a few hours), cells were fixed by removing the medium and replacing it with the fixative (2.5% glutaraldehyde in 0.05 M Hepes, pH 7.2). Incubation was at least for 30 min at 37 °C. Usually, samples were stored at 4 °C in fixative for a few days until follow-up preparation for SEM was performed.
3. *Sample preparation for SEM*: The thin bottom of the dish was extracted by using a sharp scalpel, avoiding touching the cells and dehydrated with ethanol (30%, 50%, 70%, 90%, 96%, each for 15 min; absolute ethanol, for 60 min, one change). Drying was performed according to the critical point drying (CPD) method using carbon dioxide and a critical-point dryer (CPD 030, Bal-Tec, Liechtenstein). Dried samples were mounted on a sample support (stub) with adhesive carbon tabs with the cells facing up. Sample surface was coated with a layer (approximately 5 nm thick) of gold/palladium in a sputter coater.
4. *SEM*: Imaging was done using a field-emission scanning electron microscope (Leo 1530 Gemini; Carl Zeiss Microscopy, Germany) at high vacuum conditions and the in-lens secondary electron detector at 5 kV acceleration voltage.

5.1.3 CORRELATIVE LIVE-CELL IMAGING AND THIN SECTION TRANSMISSION ELECTRON MICROSCOPY

Analysis of Rab11/eGFP-positive membrane compartments in *Chlamydia* infected HeLa cells

Infection of cells with the pathogenic bacterium *C. trachomatis* results in the formation of so-called intracellular inclusions, in which bacteria survive and grow (Moulder, 1991). Formation and maintenance of the inclusion is accompanied with a reorganization of the cellular metabolism including membrane trafficking. In particular, the Golgi apparatus is fragmented into mini-stacks (Heuer et al., 2009), a process which is depending on functional Rab6 and Rab11 (Rejman Lipinski et al., 2009). To study this reorganization of membranes in more detail HeLa cells were infected with *Chlamydia* and subsequently transfected with Rab11/eGFP. Expression of Rab11/eGFP and formation of inclusions were observed in a confocal live-cell imaging microscope (LSM 780). Cells which displayed

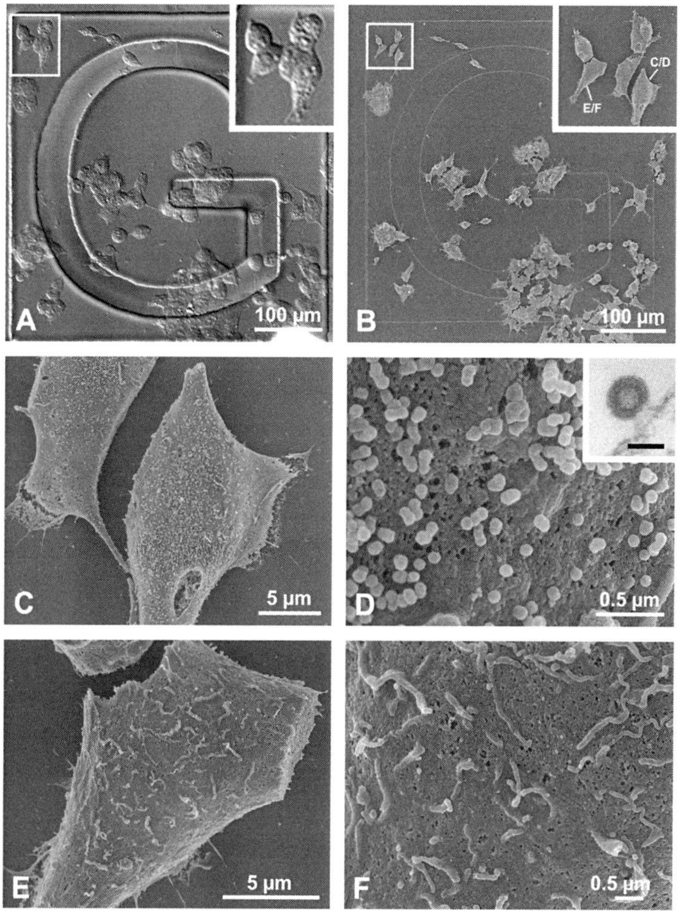

FIGURE 5.2

Correlative live-cell fluorescence and scanning electron microscopy of HeLa cells which are producing fluorescently labeled virus-like particles of the Human Endogenous Retrovirus HERV-K(HML-2). (A) Overlay of images from fluorescence and bright-field light microscopy of a field labeled "G" in which several cells show red fluorescence from expressed virus fusion proteins (Gag-Cherry). The inset shows a magnified view of the field marked with a white rectangle. (B) Scanning electron microscopy (SEM) of the same observation field as shown in (A). Inset: An increased magnification of the region marked with the rectangle in (A) and (B). Labeled cells are shown at higher magnification in (C) and (D) or (E) and (F), respectively. (C–F) SEM of selected cells as indicated in the inset in (B). (C and D) A cell which shows red fluorescence in (A) and several dot-like particles at the cell surface (C). At higher magnification (D) particles exhibit uniform size of about 100 nm which corresponds to the particle size of retroviruses. The inset shows a single budding retrovirus in an ultrathin section from a different preparation visualized by transmission electron microscopy. (E and F) Another cell which shows, in contrast to the cell in (C) and (D), no fluorescence in (A) and no dot-like particles at the cells surface. At higher magnification (F) the cell surface reveals only typical surface structures that are mainly microvilli of different lengths. (See the color plate.)

inclusions and significant Rab11/eGFP expression were analyzed in more detail by using higher magnification and recording of z-stacks to reveal distribution of fluorescence in 3D. Phase-contrast imaging was used for visualization of inclusions and bacteria which appeared as granules. Position of cells relative to the grid pattern was noted and documented by taken overview images. After chemical fixation, dehydration and embedding, positions of interest were relocalized and trimmed for serial thin sectioning.

Cells of interest were identified in sections at the transmission-electron microscope using the overview images taken by light microscopy from each position of interest. Different sections of a series were inspected and cells were photographed for registration with the respective images from z-stacks recorded with the confocal microscope. The result allowed a rough correlation of the fluorescent signal pattern from a z-slice with the structures visible on images taken from serial sections of the plastic bloc (Fig. 5.4C–E). Regions which showed significant fluorescence were analyzed in more detail at higher magnification using the transmission electron microscope. The analysis revealed the presence of small membrane-bound compartments, i.e., vesicles and tubules of different size (Fig. 5.4F), indicating that Rab11/eGFP positive membrane-bound compartments are recruited to the inclusions at this stage of infection.

1. *Live-cell imaging*: *Chlamydia*-infected and Rab11/eGFP-transfected Hela cells were investigated with an automated confocal laser-scanning microscope (LSM 780; Carl Zeiss Microscopy, Germany). Imaging of fluorescence was done with an excitation wavelength of 488 nm and emission was detected between 500 and 530 nm. Bright-field images were recorded in scanning mode using one of the photo multipliers (transmitted-light path). All investigations were done at 35 °C and in an atmosphere that was supplemented with 5% carbon dioxide. The lid of the culture dish was not firmly closed to allow diffusion of gas.
2. *Chemical fixation*: At the end of the live-cell imaging, culture medium was removed and cells were fixed either with 4% paraformaldehyde or with 2.5% glutaraldehyde in 0.05 M Hepes for at least 3 h at room temperature. Paraformaldehyde-fixed samples were again analyzed by fluorescence microscopy, before they were postfixed with 2.5% glutaraldehyde in Hepes buffer. This procedure allowed a rapid arrest of cellular events by paraformaldehyde fixation and stabilized them for a further analysis by fluorescence microscopy (e.g., to find regions of interest and to record overview images) without significant interference with the imaging. The additional postfixation with glutaraldehyde was necessary to provide a further stabilization of the cells which is beneficial for preservation of the ultrastructure. Samples were stored at 4–8 °C until further processing.
3. *Postfixation and en-bloc staining*: All incubations were done at room temperature, if not stated otherwise. All fluids were removed from the dish and immediately replaced by fresh fluids. Aldehyde fixatives were removed by washing twice in Hepes buffer (0.05 M, pH 7.2) for at least 10 min each, before

cells were incubated with 1% osmium tetroxide in distilled water for 1 h. After several washes with distilled water, samples were incubated with tannic acid (0.1% in 0.05 M Hepes, pH 7.2; for 30 min), followed by two washes (10 min each) with sodium sulfate (Na_2SO_4, 1%) and distilled water (3 × for 10 min each). Block contrasting with uranyl acetate (2% in distilled water) was performed for 1 h.

4. *Dehydration and embedding*: Dehydration of samples was performed in ethanol (70%, 80%, 90%, 100%, 100% ethanol; 15 min each). Infiltration with the resin (Epon) started with incubation in acetone as an intermedium (3 × for 15 min each) followed by mixtures of acetone and resin (3:1, 1:1, 1:3; for 1 h each) and pure resin (overnight). Finally, resin was replaced for fresh resin and samples were cured in an oven at 60 °C for at least 24 h. Note that the μ-Dish is not compatible with propylene oxide as intermedium.
5. *Relocation of positions*: Polymerized samples were inspected with an inverted light microscope (Nikon Eclipse TE2000-E) using a Plan-Fluor 20x (NA 0.5, DIC M/N2) lens and standard bright field imaging or an upright microscope (Zeiss Axiophot) using a Plan-Neofluar 10x (NA 0.3, Ph1) and phase-contrast imaging. Recorded positions were identified and inspected if cells were still present or not (Figs. 5.3 and 5.4A and B). Positions of interest were roughly labeled with a fine marker pen. The mark was used to label cutting lines for extraction of the respective regions by using a saw or a motorized cutting tool. Extracted plastic blocs were directly mounted into a holder of the ultramicrotome or onto a "dummy" plastic bloc.

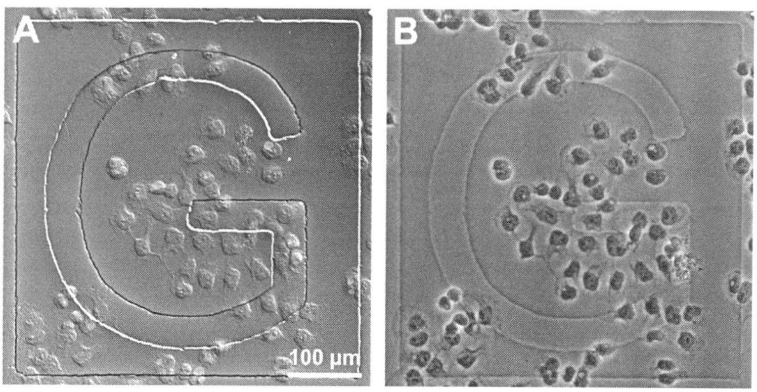

FIGURE 5.3

Correlative live-cell fluorescence and transmission electron microscopy. Comparison of the cell locations before and after embedding in resin (Epon) to check the presence of relevant cells and to select particular positions. (A) Overlay of fluorescence and differential interference contrast mode images to show positions of HeLa cells, infected with *Chlamydia trachomatis*, and expression of Rab11/eGFP (recorded with the Nikon Eclipse TE2000-E). (B) Phase contrast image of corresponding region after embedding in resin. (See the color plate.)

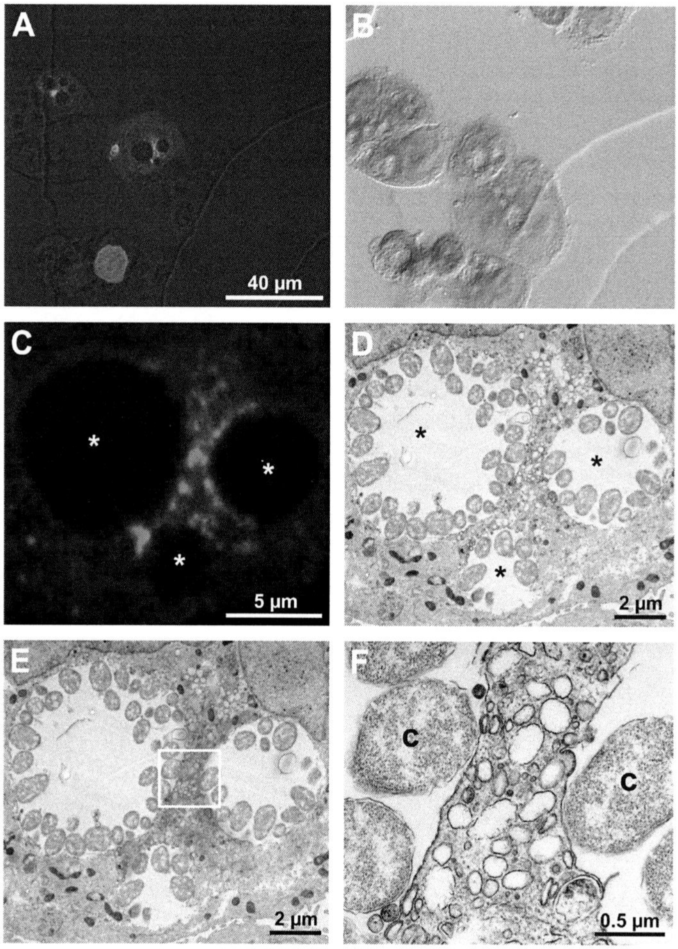

FIGURE 5.4

Correlative live-cell fluorescence and transmission electron microscopy (TEM) of HeLa cells which expressed Rab11/eGFP and which were infected with *Chlamydia trachomatis*.
(A) Confocal fluorescence microscopy of the Rab11/eGFP signal (overlay with bright-field transmission channel). (B) Relocalization of the same cells in the resin-embedded sample by using light microscopy and bright-field imaging. (C) Confocal fluorescence microscopy of a relevant region of one cell localized in the center of (A). The fluorescent signal from Rab11/eGFP shows a distinct distribution between the three inclusions (*) visible in the cytoplasm. (D) Ultrastructure of the region corresponding to (C) in ultrathin sections visualized by TEM. (E) Overlay of the manual registration of fluorescence and ultrastructure that is shown in (C) and (D). (F) Ultrastructural detail of the boxed region in (E) which shows a variety of small membrane compartments that are visible between two of the inclusions. *Chlamydia* (c) are localized close to the bounding membrane of the inclusions. (See the color plate.)

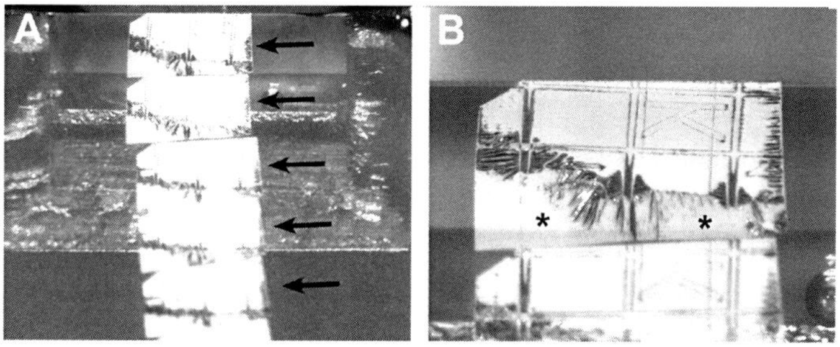

FIGURE 5.5

Ultrathin sections taken parallel to the bottom surface of an embedded μ-Dish. (A) A ribbon of sections (arrows) is floating on the water surface of the knife boat (blue (appearing dark grey in the lower part of the image in print version)). (B) Magnified view of a section which clearly shows the grid pattern and a label ("K") in the upper part that corresponds to the thin plastic of the dish bottom. The lower part of the section (*) corresponds to the resin above the dish bottom which contains the cells. Note that the inscriptions are partially visible by transmission electron microscopy in the very first sections above the dish bottom.

6. *Ultramicrotomy*: Under the dissection microscope, sample blocs were trimmed laterally to a width of approximately 1 mm keeping regions of interest in the middle of the bloc face. The plastic of the dish bottom was removed either by using a precise motorized sample trimming tool or the ultramicrotome and a diamond for semithin sectioning. With the latter approach maximum control of the depth of plastic removal was achieved. The transition between the dish and the plastic embedded sample was indicated by appearance of the grid pattern in the sections that were swimming on the trough of the diamond knife (Fig. 5.5). Serial ultrathin sections (60–80 nm thickness) were collected on copper slot grids which are covered by a film of pioloform.
7. *Transmission electron microscopy*: Sections were inspected in a transmission electron microscope (Tecnai12 BioTwin) at 120 kV without section staining or occasionally after staining sections with 2% uranyl acetate and 0.1% lead citrate.
8. *Image registration*: Registration of z-images from confocal fluorescence microscopy with images from transmission electron microscopy of physical sections was done manually using the Photoshop software (Adobe Inc., USA). Shape and distribution pattern of inclusions present in the cytoplasm facilitated registration.

5.2 MATERIALS

Cell culture

1. *Cells*: HeLa cervical epithelial cells (American Type Culture Collection [ATCC], No. CCL-2).
2. *Bacteria*: *C. trachomatis* Lymphogranuloma venereum biovar strain L2/434/Bu (ATCC No. VR-902B).

3. *Cell culture dishes (μ-Dish)*: 35 mm in diameter, with grid pattern (500 μm pitch; μ-Dish 35 mm, high Grid-500, sterile, ibiTreat; ibidi GmbH, Germany). *Note*: μ-Dishes with glass bottom and grid pattern (500 and 50 μm pitch) are also available. To allow proper imaging by DIC, particular lids which possess a glass insert (DIC-lid) were used. For further details compare information provided by ibidi GmbH on their website (www.ibidi.com).
4. *Medium, according to the cells used*: Dulbecco's modified Eagle Medium (D-MEM; Gibco/Invitrogen GmbH, Germany); RPMI 1640 medium (Gibco/Invitrogen GmbH, Germany); Opti-MEM I (reduced serum, without phenol red; Gibco/Invitrogen GmbH, Germany).

Infection and transfection

1. *HERV-K expression plasmids*: Plasmid pcDNAoricoHERV-K_GagProPol has been described previously (George et al., 2011). It allows CMV-promoter driven expression of reconstituted original Gag, Gag-Pro and Gag-Pro-Pol proteins of HERV-K(HML-2). These proteins do assemble into VLPs that bud from the cell surface (George et al., 2011). Plasmid pcDNAoricoGag-Cherry encodes the Gag with a C-terminally fused Cherry protein (Chudak et al., 2013).
2. *RAB11 expression plasmid*: RAB11A transcript variant 1 ligated into pEGFP-C1 (pEGFP-RAB11A; Clontech Laboratories, Inc., USA).
3. *Transfection reagents*: PolyFect (Qiagen GmbH, Germany); Lipofectamine 2000 (Invitrogen GmbH, Germany).

Live-cell imaging

1. *Nikon Eclipse TE2000-E*: Motorized microscope (Nikon Corp., Japan) with inverted optics and a Xenon illumination system (175 W) that is able to rapidly switch emission wave length (Lambda DG4, Sutter Instrument Company, USA) for epi-fluorescence imaging. Installed optical contrast modes: phase and DIC. Image recording was done using a color CCD-camera (DS-2Mv; Nikon Corp., Japan) or an EMCCD camera (Cascade 512B; Roper Scientific, USA). The system was equipped with an automatic focusing system (Perfect-Focus System, T-PFS; Nikon Corp., Japan) and a cultivation chamber that was heated and supplied with moistened artificial air (supplemented with 5% carbon dioxide).
2. *Zeiss LSM 780*: Motorized and digitally controlled confocal laser scanning microscope with inverted optics (Axio Observer Z1; Carl Zeiss Microscopy GmbH, Germany) that was equipped with three laser systems: a diode laser (405 nm emission line), a multiple-line Argon-laser (458, 488, 514 nm), a HeNe-laser (543, 594, 633 nm). Detection of fluorescence was done using two parallel photomultipliers or a 32 channel gallium arsenide spectral detector (QUASAR GaAsP 32). Bright-field images were recorded in scanning mode using one of the photo multipliers (transmitted-light path). Sample and optics were surrounded by a cultivation chamber (XL S1 LSM 710; Carl Zeiss Microscopy GmbH, Germany) that was heated and supplied with carbon dioxide (final concentration: 5%).

Scanning electron microscopy

1. *Fixative*: glutaraldehyde (25%; TAAB Laboratories Equipment Ltd., United Kingdom).
2. *Critical point dryer* (CPD 030, Bal-Tec, Liechtenstein) supplied with liquid carbon dioxide.
3. *Sputter coater* (E5100, Polaron, United Kingdom) equipped with an Au/Pd target.
4. *Field-emission scanning electron microscope*, Leo 1530 Gemini (Carl Zeiss Microscopy, Germany).

Thin section transmission electron microscopy

1. *Fixatives*: Paraformaldehyde (Merck, Germany); glutaraldehyde (25%; TAAB Laboratories Equipment Ltd, United Kingdom); osmium tetroxide (Carl Roth, Germany); tannic acid (e.g., Electron Microscopy Services, No. 21710).
2. *Epon resin*: 23.52 g glycidyl ether 100, 12.35 g dodenyl succinic anhydride, 14.13 g nadic methyl anhydride, 0.65 g dimethyl phthalate (Serva, Electrophoresis, Heidelberg, Germany).
3. *Ultramicrotome* (e.g., UC7; Leica Microsystems, Germany).
4. *Sample trimming tool* (EM RAPID; Leica Microsystems, Germany).
5. *Diamond knife* (e.g., Histo or Ultra 45°; Diatome, Switzerland).
6. *Section and en bloc staining*: uranyl acetate (2% in distilled water; e.g., Merck, Germany); lead citrate (e.g., 0.1% in distilled water; Serva, Germany).
7. *Transmission electron microscope* (e.g., Tecnai12 BioTwin; FEI Corp., The Netherlands), equipped with a digital CCD camera (MegaViewIII; Olympus Soft Imaging Solutions, Germany) operated at a resolution of 1376 x 1014 pixels, 14 bit.

5.3 DISCUSSION

Cell culture dishes that possess a plastic bottom of highest optically quality and an imprinted grid pattern (μ-Dishes) for localization of interesting positions are highly suitable for routinely performed CLEM. They are delivered ready-to-use and usually can be applied without any change of the cultivation conditions which are used for cultivation of cells in other plastic containers (e.g., a six-well plate). The μ-Dishes can be easily fitted to most light microscopes and stable live-cell imaging can be performed. After analysis by light microscopy the μ-Dishes allow relocalization of positions of interest for convenient imaging by either routine scanning or transmission electron microscopy.

CLEM using SEM has the advantage of combining two techniques with a comparable large field of view. This provides representative information of a sample and facilitates detection of rare events or structures. Several applications may profit from this application, e.g., the proof of pathogen uptake or release in a subset of cells. CLEM using SEM provides a very economical approach to check many positions in a single sample, which increases throughput and improves data quality by increasing the number of cells/samples inspected. SEM, in most cases, is restricted to information collected from outside of the cells and intracellular events are usually not detectable. However, using uncoated or slightly carbon-coated samples, internal structures may be visible by tuning the voltage from low voltages (surface information) to higher voltages (depth information) and/or energy filtering of the backscattered electrons if structures of interest contain heavy metals (Rohde, Branitzki-Heinemann, & Talay, 2013).

Different cultivation systems or procedures for CLEM using SEM were published before. Most of them use particular culture dishes with a glass bottom (e.g., from MaTek Corp.) or coverslips that possess particular grids or labels at their surface (e.g., Larson, Johnson, Webb, & Vogt, 2005). While glass is usually optically perfect and resistant against many chemicals, removal from the dish may be cumbersome. Coverslips have to be inserted in a dish or container for live-cell imaging which might impair with the imaging due to diffraction at the different layers. Another interesting procedure for CLEM in conjunction with SEM is to integrate a fluorescence microscope in the SEM (Liv et al., 2013) or to attach it to the SEM chamber (ClairScope; Morrison et al., 2012). However, both procedures need particular machines and either do not allow fluorescence imaging of living cells or use a rather unconventional SEM imaging of the material contrast which provides also information from inside the cells (ClairScope).

Intracellular structures are optimally visualized by using thin section electron microscopy. Still, the most commonly used instrument for analyzing ultrathin section is the transmission electron microscope. The major problem in CLEM using thin sections is the proper registration of positions acquired by the different imaging modes. While in many cases, a formal image registration is not needed, because only the position of the cell is relevant, in some cases a perfect match of positions is necessary. A couple of factors affect registration of images and especially in 3D, accurate registration is difficult to achieve (compare discussion and procedures in Kukulski, et al., 2012 and Padman, Bach, & Ramm, 2014). The image registration accuracy using the μ-Dish and procedures described in this chapter is not precisely known. It is definitely not better than the resolution of the light microscope which is around 250 nm in xy and much more in z. With confocal imaging of fluorescence a depth resolution of around 500 nm and above, depending on the pinhole size of the microscope, is expected, which is rather low in comparison to the depth resolution of a conventional ultrathin section (60–80 nm). Thus a perfect match between images obtained with both modes, even if serial optical and physical sectioning is applied, never would be achievable. As a consequence, in the example demonstrated, Rab11/eGFP-labeled vesicles could not be distinguished reliably from nonlabeled vesicles which are localized in close vicinity (Fig. 5.4E and F). However, immunogold labeling directed against the fluorescent protein provides a possibility to correlate localization of fluorescence signals with ultrastructural details at highest resolution (e.g., Mironov & Beznoussenko, 2013; Spiegelhalter et al., 2010). The μ-Dish seems to be compatible with pre-embedding and postembedding immunogold labeling. Preliminary experiments showed, that low-temperature embedding in Lowicryl K4M

(HM20 dissolves the bonding of the thin bottom to the plastic dish) allows postembedding immunolabeling for the green fluorescence protein (GFP) (M. Laue, unpublished data).

The CLEM procedure described in the present paper relies on chemical fixation of the cells, because high-pressure freezing of cells cultivated on the dish plastic did not result in a sufficient freezing quality (Laue, unpublished observations), most probably because thermal conductivity of the plastic is not good enough. In addition, working with infectious pathogens needs to follow particular biosafety rules, which are in many cases not compatible with high-pressure freezing, because pathogens may be accidentally aerosolized during sample handling or the freezing process. While this must be avoided with the highest possible likelihood for pathogens of the risk group 3 and 4 (pathogens that are causing diseases with a significant mortality that are extremely difficult to treat or untreatable), pathogens belonging to the risk group 2 (pathogens that cause diseases with no significant mortality and available medical treatment) might be fixed by high-pressure freezing. However, diseases that are caused by pathogens belonging to risk group 2 are in some cases far from being harmless and high-pressure freezing is not a sterilization procedure (compare Vanhecke et al., 2012). Researchers should consider the individual risks of their procedures and contact their local safety officers for further support to avoid spread of infections.

Stabilization of cells by chemical fixation is well established and is easily performed by a simple exchange of fluids within the culture dish, which can be done in a biosafety cabinet to increase biosafety. Inactivation properties of formaldehyde and glutaraldehyde are well documented for many pathogens (Rubo et al., 1967; Sabel, Hellman & McDade, 1969). In case of uncertainty, sterility controls can be performed which rules out that samples are still infectious, which is important to prove if samples must be transferred to laboratories of a lower biosafety class (e.g., an electron microscopy laboratory). Besides advantages for biosafety, stabilization of cell cultures by chemical fixation provides flexibility. At the end of an experiment, cell cultures can be routinely fixed and stored for a considerable time until decision of the follow up-processing, whether it might be electron microscopy or another light microscopy technique. To conserve fluorescence, initial formaldehyde fixation can be performed, as we did in some of the experiments with *Chlamydia*. Those samples can be reinspected under the light microscope after storage (e.g., to record more positions of interest) before starting with the follow-up preparation including further fixation steps to preserve structures for electron microscopy.

The major drawback of chemical fixation is that it is slow and incomplete which causes changes of the ultrastructure. In case of live-cell imaging experiments a considerable delay in the arrest of cellular events has to be expected in addition. In those cases, morphology and fluorescent signals should be checked after fixation. Prefixation with formaldehyde, as indicated earlier, and reinspection of cells of interest might be a way to solve this issue. If chemical fixation is not sufficient, high-pressure freezing might be applied, considering all of the biosafety aspects as discussed earlier. In many cases, a suitable alternative would be to work with attenuated or inactive variant of the infectious pathogen (e.g., Romero-Brey et al., 2012).

Alternative culture systems suitable for CLEM of chemically fixed cell cultures have been described in many papers before (e.g., Jiménez et al., 2009; Mironov & Beznoussenko, 2013). Most of them would allow to get the same information as obtained with the methods described in this chapter. However, the procedures described in this chapter most likely are much easier performed and more flexible than most of the other procedures described before which helps to analyze more samples in shorter time. Glass or plastic supports, like cover slips or Aclar, that are somehow labeled for recording positions of interest have to be inserted in culture containers for imaging which usually impairs high resolution imaging because of inappropriate thickness or optical properties (compare Padman et al., 2014). Moreover, they may move during live-cell imaging, if not fixed in the containers. Glass supports usually are more resistant toward chemicals, but may cause problems during or after embedding (e.g., separation of glass from resin bloc may impair cells).

SUMMARY

The μ-Dishes with an imprinted finder grid provide many advantages for CLEM: ready-to-use culture dish of a standard size; highest optical performance for fluorescence and bright-field imaging of living and fixed cells; compatibility with SEM and thin section transmission electron microscopy using chemical fixation and standard follow-up procedures; control of presence of cells at various steps of the preparation; compatibility with biosafety regulations up to the highest biosafety-level. Hence, CLEM using μ-Dishes is suited to answer routinely a variety of important basic questions relevant for the study of infectious microorganisms, for example, to search cell cultures for infected cells by live-cell imaging followed by electron microscopy to prove the presence of the pathogen within an individual cell, or, to investigate the ultrastructure of an infected cell at the end of a live-cell imaging experiment that has followed either fluorescent labeled pathogens or molecules related to infection.

ACKNOWLEDGMENTS

We would like to thank Christoph Schaudinn and Thomas Müller-Reichert for their valuable comments.

Declaration: The authors declare the absence of competing interests.

REFERENCES

Bannert, N., & Kurth, R. (2006). The evolutionary dynamics of human endogenous retroviral families. *Annual Review of Genomics and Human Genetics*, *7*, 49–173.

Brandenburg, B., & Zhuang, X. (2007). Virus trafficing—Learning from single-virus tracking. *Nature Reviews. Microbiology*, *5*, 197–208.

Brown, E., Mantell, J., Carter, D., Tilly, G., & Verkade, P. (2009). Studying intracellular transport using high-pressure freezing and correlative light and electron microscopy. *Seminar in Cell & Developmental Biology*, *20*, 910–919.

Chudak, C., Beimforde, N., George, M., Zimmermann, A., Lausch, V., Hanke, K., et al. (2013). Identification of late assembly domains of the human endogenous retrovirus-K(HML-2). *Retrovirology*, *10*, 140.

Curry, A., Appleton, H., & Dowsett, B. (2006). Application of transmission electron microscopy to the clinical study of viral and bacterial infections: Present and future. *Micron*, *37*, 91–106.

Eltsov, M., & Zuber, B. (2006). Transmission electron microscopy of the bacterial nucleoid. *Journal of Structural Biology*, *156*, 246–254.

George, M., Schwecke, T., Beimforde, N., Hohn, O., Chudak, C., Zimmermann, A., et al. (2011). Identification of the protease cleavage sites in a reconstituted Gag polyprotein of an HERV-K(HML-2) element. *Retrovirology*, *8*, 30.

Goldsmith, C. A., Ksiazek, T. G., Rollin, P. E., Comer, J. A., Nicholson, W. L., Peret, T. C. T., et al. (2013). Cell culture and electron microscopy for identifying viruses in diseases of unknown cause. *Emerging Infectious Diseases*, *19*, 886–891.

Griffiths, G. (1993). *Fine structure immunocytochemistry*. Berlin, Heidelberg: Springer.

Hackstadt, T., Rockey, D. D., Heinzen, R. A., & Scidmore, M. A. (1996). *Chlamydia trachomatis* interrupts an exocytic pathway to acquire endogenously synthesized sphingomyelin in transit from the Golgi apparatus to the plasma membrane. *EMBO Journal*, *15*, 964–977.

Heuer, D., Rejman Lipinski, A., Machuy, N., Karlas, A., Wehrens, A., Siedler, F., et al. (2009). Chlamydia causes fragmentation of the Golgi compartment to ensure reproduction. *Nature*, *457*, 731–735.

Jiménez, E. G., Van Donselaar, E. G., De Winter, D. A. M., Vocking, K., Verkleij, A. J., & Post, J. A. (2009). Gridded Aclar: preparation methods and use for correlative light and electron microscopy of cell monolayers, by TEM and FIB-SEM. *Journal of Microscopy*, *237*, 208–220.

Katsen-Globa, A., Meiser, I., Petrenko, Y. A., Ivanov, R. V., Lozinsky, V. I., Zimmermann, H., et al. (2014). Towards ready-to-use 3-D scaffolds for regenerative medicine: Adhesion-based cryopreservation of human mesenchymal stem cells attached and spread within alginate-gelatin cryogel scaffolds. *Journal of Material Science: Materials in Medicine*, *25*, 857–871.

Koppensteiner, H., Banning, C., Schneider, C., Hohenberg, H., & Schindler, M. (2012). Macrophage internal HIV-1 is protected from neutralizing antibodies. *Journal of Virology*, *86*, 2826–2836.

Kukulski, W., Schorb, M., Welsch, S., Picco, A., Kaksonen, M., & Briggs, J. A. G. (2012). Precise, correlated fluorescence microscopy and electron tomography of Lowicryl sections using fluorescent fiducial markers. *Methods in Cell Biology*, *111*, 235–257.

Larson, D. R., Johnson, M. C., Webb, W. W., & Vogt, M. V. (2005). Visualization of retrovirus budding with correlated light and electron microscopy. *Proceedings of the National Academy of Sciences of the United States of America*, *102*, 15453–15458.

Laue, M. (2010). Electron microscopy of viruses. *Methods in Cell Biology*, *96*, 1–20.

Liv, N., Zonnevylle, A. C., Narvaez, A. C., Effting, A. P. J., Voorneveld, P. W., Lucas, M. S., et al. (2013). Simultaneous correlative scanning electron and high- NA fluorescence microscopy. *PLoS One*, *8*, e55707.

Mironov, A. A., & Beznoussenko, G. V. (2013). Correlative microscopy. *Methods in Cell Biology*, *113*, 209–254.

Morrison, I. E. G., Dennison, C. L., Nishiyama, H., Suga, M., Sato, C., Yarwood, A., et al. (2012). Atmospheric scanning electron microscope for correlative microscopy. *Methods in Cell Biology*, *111*, 307–324.

Moulder, J. W. (1991). Interaction of Chlamydiae and host cells in vitro. *Microbiological Reviews*, *55*, 143–190.

Müller-Reichert, T., & Verkade, P. (2012). *Correlative light and electron microscopy*. Amsterdam: Elsevier (Methods in Cell Biology 111).

Padman, B. S., Bach, M., & Ramm, G. (2014). An improved procedure for subcellular spatial alignment during live-cell CLEM. *PLoS One*, *9*, e95967.

Rejman Lipinski, A., Heymann, J., Meissner, C., Karlas, A., Brinkmann, V., Meyer, T. F., et al. (2009). Rab6 and Rab11 regulate *Chlamydia trachomatis* development and golgin-84-dependent Golgi fragmentation. *PLoS Pathogens*, *5*, e1000615.

Rohde, M., Branitzki-Heinemann, K., & Talay, S. (2013). Kinetic studies in Streptococcal pathogenesis applying field emission scanning electron microscopy (FESEM). In R. Rachel (Ed.), *Proceedings of the microscopy conference 2013, Regensburg. Part II* (pp. 282–283).

Romero-Brey, I., Merz, A., Chiramel, A., Lee, J.-Y., Chlanda, P., Haselman, U., et al. (2012). Three-dimensional architecture and biogenesis of membrane structures associated with Hepatitis C virus replication. *PLoS Pathogens*, *8*, e1003056.

Rubo, S. D., Gardner, J. F., & Webb, R. L. (1967). Biocidal activities of glutaraldehyde and related compounds. *Journal of Applied Bacteriology*, *30*, 78–87.

Sabel, F., Hellman, A., & McDade, J. J. (1969). Glutaraldehyde inactivation of virus in tissue. *Applied Microbiology*, *17*, 645–646.

Schudt, G., Kolesnikova, L., Dolnik, O., Sodeik, B., & Becker, S. (2013). Live-cell imaging of Marburg virus-infected cells uncovers actin-dependent transport of nucleocapsids over long distances. *Proceedings of the National Academy of Sciences of the United States of America*, *110*, 14402–14407.

Spiegelhalter, C., Tosch, V., Hentsch, D., Koch, M., Kessler, P., Schwab, Y., et al. (2010). From dynamic live cell imaging to 3D ultrastructure: Novel integrated methods for high pressure freezing and correlative light-electron microscopy. *PLoS One*, *5*, e9014.

Vanhecke, D., Zuber, B., Brugger, S. D., & Studer, D. (2012). Safe high-pressure freezing of infectious micro-organisms. *Journal of Microscopy*, *246*, 124–128.

CHAPTER

6

Correlative Light and Electron Microscopy for a Free-Floating Spindle in *Xenopus laevis* Egg Extracts

Erin M. Tranfield[*,†], **Xavier Heiligenstein**[†,‡,§], **Ina Peristere**[†], **Claude Antony**[†,¶]

**Instituto Gulbenkian de Ciência, Rua da Quinta Grande 6, Oeiras, Portugal*

[†]European Molecular Biology Laboratory, Meyerhofstrasse 1, Heidelberg, Germany

[‡]Institut Curie, Centre de Recherche, Paris, France

[§]Structure and Membrane Compartments, CNRS UMR144, Paris, France

[¶]Institut de Génétique et Biologie Moléculaire et Cellulaire, 1 rue Laurent Fries, BP 10142, Illkirch, France

CHAPTER OUTLINE

Methods in Cell Biology, Volume 124, ISSN 0091-679X, http://dx.doi.org/10.1016/B978-0-12-801075-4.00006-9

Abstract

Cryoimmobilization is an optimal method of preserving sample ultrastructure in electron microscopy studies. However, cryoimmobilization is limited to thin samples and this limitation may necessitate the isolation of the structure of interest. For cellular structures that are found in low number, or only during certain phases of the cell cycle, an added benefit of isolation is the possibility to concentrate the structures. We developed a method to perform correlative light and electron microscopy on infrequent isolated subcellular structures. In this chapter, we will describe our protocol that uses a combination of existing techniques and new solutions for the isolation, identification, cryoimmobilization, targeted ultramicrotomy, and imaging of the free-floating meiotic spindles assembled in *Xenopus laevis* egg extract.

INTRODUCTION

Advanced imaging techniques such as electron tomography enable detailed three-dimensional studies of the ultrastructure of the cell at a resolution that was previously unachievable (Studer, Michel, & Muller, 1989). The improvements in three-dimensional resolution where supported by advancements in ultrastructural preservation by cryoimmobilization (Moor, Bellin, Sandri, & Akert, 1980). Cryoimmobilization of cellular components in vitreous ice followed by freeze-substitution processing is now accepted as a superior method of ultrastructure preservation when compared with conventional chemical fixation for room temperature electron microscopy (Studer et al., 1989). However, the need for sample vitrification limits cryoimmobilization to thin samples (Moor et al., 1980). In the case of high-pressure freezing, ideal samples have a maximum thickness of 200 μm (Studer, Humbel, & Chiquet, 2008), hence, restricting biological samples to small organisms, such as yeast or *Drosophila* embryos, thin cell monolayers, isolated tissue, or subcellular organelles. For structures that are found in low number, or only during selected phases of the cell cycle, the possibility to concentrate the structures at a higher density than found *in vivo* is of great value. Selection of isolated structures can be achieved by using correlative light and electron microscopy (CLEM). This approach begins by using fluorescence microscopy in combination with landmarks for the identification and registration of structures of interest. Following electron microscopy sample processing, targeted ultramicrotomy can be used for precise alignment and orientation of the sample prior to sample sectioning and electron microscopy imaging (Kolotuev, Bumbarger, Labouesse, & Schwab, 2012).

Our biological interest was focused on the three-dimensional ultrastructural organization of the meiotic spindle from the *Xenopus laevis* frog at microtubule protofilament resolution. The *X. laevis* fertilized egg is approximately 1 mm in size, which is too large to vitrify. Even if vitrification was possible, finding a single spindle within such a large volume would be almost impossible. An alternative was the well-established *X. laevis* egg extract method to assemble meiotic spindles at high number in a cytosol suspension (Lohka & Maller, 1985). The challenge of this model for CLEM was not only the lack of cell membrane protection around the spindle making it vulnerable to physical stresses (Salmon, Goode, Maugel, & Bonar, 1976) but also the spindle's unknown location within the suspension. To investigate our biological questions, we needed to know not only the location of the spindle within the volume of extract, but also its orientation and its size to accurately proceed with targeted ultramicrotomy.

In the pages that follow, we will describe our solutions to these challenges, which use a combination of existing techniques and new solutions for the fluorescence identification, cryoimmobilization and freeze-substitution sample preservation, targeted ultramicrotomy, and imaging of the *X. laevis* meiotic spindle. Although this protocol was developed to study the *X. laevis* meiotic spindle, we believe it can readily be adapted for any isolated or free-floating structure.

6.1 RATIONALE

Within this chapter, the modifications and new tools we developed for investigating free-floating cell structures by CLEM will be described in detail. In the development of this protocol, simple tools had to be constructed and procedures defined to maintain and protect the isolated spindles, to identify well-formed spindles as well as to follow a given spindle from the fluorescence microscope through to the electron microscope. Technical developments ranged from protecting the sample while imaging in the high-pressure freezer holder to the latest freeze-substitution steps after high-pressure freezing. The large spindle size and the need for 2 nm resolution required many compromises to achieve the resolution over such a large area.

6.2 METHODS

6.2.1 PREPARATION OF SPINDLES IN THE *X. laevis* EGG EXTRACT

To study the organization of the microtubules in the meiotic spindle, spindles were assembled in *X. laevis* egg extract (Lohka & Maller, 1985; Murray, 1991). We optimized the protocol previously explained in detail (Hannak & Heald, 2006) to obtain bipolar, isolated yet sufficiently concentrated spindles. Once an egg extract was confirmed by florescence microscopy to contain isolated, bipolar spindles, sapphire assemblies containing extract were constructed for high-pressure freezing as explained below. At no time did we chemically fix the extracts prior to high-pressure freezing.

6.2.2 IMAGING AND HIGH-PRESSURE FREEZING OF THE SPINDLE STRUCTURES

The self-assembled spindles in the extract are abundant, can easily be labeled fluorescently (Hannak & Heald, 2006) and are well within the optimal size range for cryoimmobilization. The main challenges of these isolated spindles for CLEM were (a) that they were free-floating making their location unpredictable and (b) that they

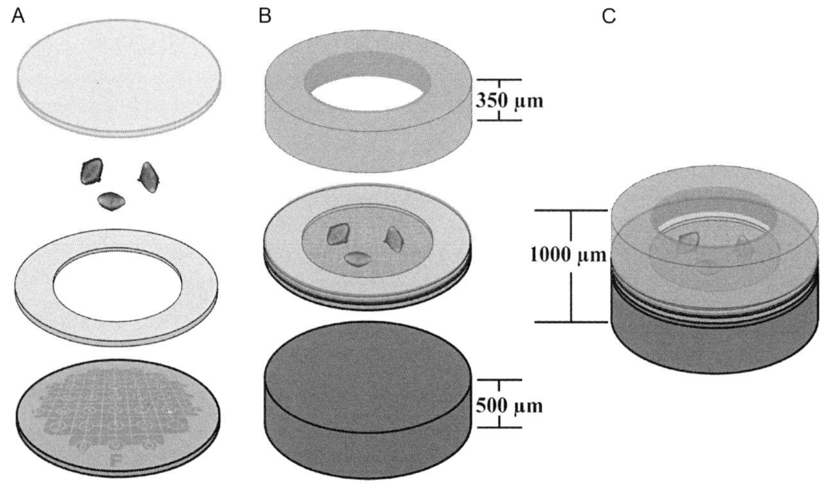

FIGURE 6.1

Schematic illustration of the sapphire assembly. (A) The assembly is composed of a carbon-coated sapphire, a gold spacer ring, the sample, and a second noncarbon-coated sapphire dipped in 1-Hexadecene. (B) The sapphire assembly was loaded into the tip of the HPM010 sitting on top of a "B" carrier oriented flat side up, and sitting under a ~350-μm PDMS ring. (C) Combined the stack was ~1000 μm fitting snugly into the tip of the HPM010 sample holder. Imaging of the structures within the sapphire assembly was done through the hole in the center of the PDMS ring. (See the color plate.)

were not protected by cell walls making them vulnerable (Salmon et al., 1976) and harder to contain for high-pressure freezing.

To address the first challenge, we contained and protected the spindles in a sapphire assembly (Fig. 6.1). The sapphire assembly was made of two sapphire discs spaced apart by a gold spacer ring (Fig. 6.1A). The extract was contained within the center of the gold spacer ring. Gold is less reactive and toxic than other material, such as copper (Moor et al., 1980). The sapphire assembly was built under a stereomicroscope. Bipolar, isolated spindle structures where identified and registered against the carbon landmarks within the sapphire assembly using an inverted epifluorescence microscope with long-working distance objectives. A carbon grid mask on the substrate sapphire disc allowed us to register the exact location of the spindles in the sapphire assembly. Following sample processing, the carbon mask readily transferred from the sapphire to the polymerized plastic resin thus keeping the landmarks with the samples throughout processing (McDonald et al., 2010).

The sapphire assembly was remarkably robust and easy to move around. However, after many failed attempts to relocate the spindles following electron microscopy processing, we grew concerned that if too much handling took place between imaging and freezing, the spindles may displace from their registered positions. Three potential sources of the problem were identified: first, the meiotic spindles were disassembling due to pressure applied during the closing of the sample holder of the Baltec HPM010 high pressure freezer (Hawes, Netherton, Mueller, Wileman, & Monaghan, 2007). Second, the unfixed specimens were drifting rapidly. Third, the samples were handled between registering the spindle location and cryoimmobilizing the spindle in the Baltec HPM010 high pressure freezer. To resolve spindle disassembling due to pressure, we created a soft ring out of polydimethylsiloxane (PDMS) to hold the sapphire assembly without pressing the specimen in the sample holder of the Baltec HPM010 high pressure freezer. To resolve the second and the third source of the problem, we imaged the sapphire assembly in the sample holder of the Baltec HPM010 high pressure freezer (Fig. 6.2A). After loading the tip with a "B" carrier, a sapphire assembly containing the sample and a PDMS ring (Fig. 6.1C), we closed and secured the sample holder with a clip (Fig. 6.2D and E). The clipped sample holder was then mounted on a newly designed imaging

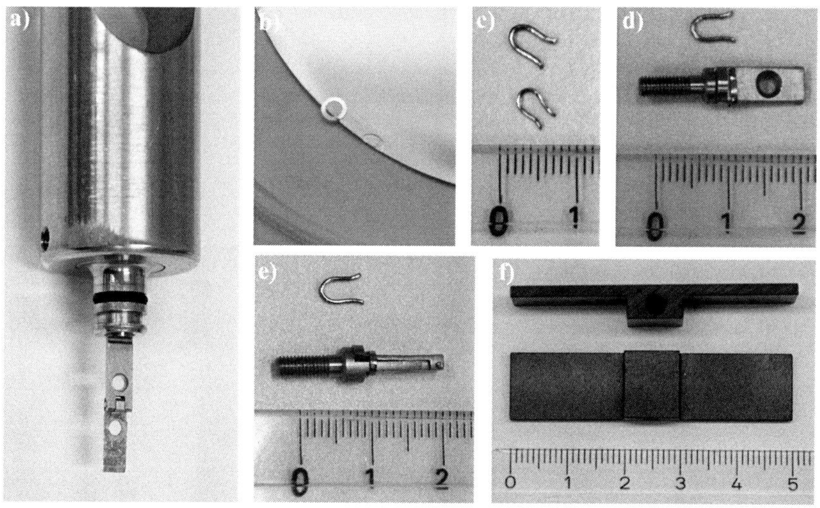

FIGURE 6.2

Imaging of the specimen in the tip of the holder for the BAL-TEC HPM010 high-pressure freezer. (A) The tip of the BAL-TEC HPM010 has been opened and checked to ensure there is no debris that could interfere with the loading and freezing process. (B) To facilitate building the sapphire assembly, the carbon-coated sapphire disc and a gold spacer ring (white in print version) are held down by a PDMS-coated black silicon wafer. On top of the sapphire disc rests the gold spacer ring (white in print version). The preparation awaits addition of the sample before a final sapphire is placed on top to enclose and protect the sample. The sapphire assembly is mounted into the holder tip and the tip is clamed closed. (C) To secure the tip a "U"-shaped clip was used. The tip of the BAL-TEC HPM010 clamped closed from the top (D) and from the side (E), respectively. In both images an additional clamp is at the top of the image. (F) Once the tip is clamped closed, it can be mounted onto an aluminum holder that fits on the stage of a light microscope for fluorescence imaging. A nut glued onto a microscopy glass side (not shown) can serve the same purpose as the aluminum holder.

platform (Fig. 6.2F), placed on the stage of an inverted fluorescence microscope and images were acquired through the whole in the center of the PDMS ring. From this platform to high-pressure freezing, the transfer time was reduced to 45–90 s and sample handling was minimized thereby reducing movement of or damage to the specimen between registration and cryoimmobilization. This combined approach dramatically increased the success at relocating the spindle following high-pressure freezing in the Baltec HPM010 and processing of the sample in a Leica AFS1. Below is the detailed protocol we followed:

Instruments

- Stereomicroscope
- Inverted epi-fluorescence light microscope (Axio Observer.Z1, Zeiss)
- Long-working distance objectives (LD Plan-Neofluar 20×/0.4 NA, 40×/0.6 NA, 63×/0.75 NA, Zeiss)
- High-pressure freezer (Baltec HPM010)
- Carbon coater (Baltec Med020)
- Glow-discharge machine (Homemade by EMBL)

Materials

- 90-mm filter paper, grade 50 (Whatman)
- Dumont Tweezers 5 Dumoxel
- P2 or P10 pipette (Gilson)
- D10 (10 μl) pipette tips (Gilson), ends cut with ~2 mm aperture
- Sapphire discs (outer ø=3 mm, ↕=50 μm; Wohlwend #405)
- Gold spacer ring (outer ø=3.05 mm, inner ø=2 mm, ↕=50 μm; Agar gold hole grid #AGG2620A)
- PDMS (Sylgard 184 Silicone Elastomer Kit, Dow Corning)
- 5-in. silicon wafer PDMS coated using a spin coater
- Petri dish, ø=10 cm
- B carriers (Wohlwend #242)
- Filter paper, grade 1, ø=90 mm (Whatman)
- 1-Hexadecene (Merck #8.22064.0500)
- PDMS rings (homemade, see below for protocol)
- Mextaform finder grids (#HF15)

6.2.2.1 How to prepare special materials

"B" carriers sitting on a filter paper soaked in a few drops of 1-Hexadecene—in a 10-cm petri dish place a filter paper, wet the paper with a small amount of 1-Hexadecene so that the 1-Hexadecene is forming small puddles on the filter paper. If too much 1-Hexadecene is used, it tends to spread and make a mess. Use as little as possible. You can use the same dish, with the same 1-Hexadecene until it starts to look dry, and/or dirty.

PDMS rings—using the PDMS kit from Dow Corning mix 10:1 (w/w) polymer to cross-linker in a 50-ml falcon tube. Place falcon tube on a rotator and mix for 1 h.

Placing mixture under vacuum will remove bubbles. Create a thin layer of PDMS—target is 330–350 μm thick. The most reproducible method we found was to have a workshop make wells of variable depth (330, 340, and 350 μm) in several aluminum plates (Fig. 6.3 pictures our 350-μm deep well). Use a second piece of flat aluminum to close the well. Polymerize the PDMS in a 60 °C oven for 2 h. Allow polymerized PDMS to cool before removing it from the aluminum mold. It is advisable to confirm the thickness of the polymerized PDMS using calipers before making the rings. To make the rings, use a 3-mm biopsy punch to create round discs. Use a 1.5-mm biopsy punch to create a hole in the center of the PDMS disc, resulting in a PDMS ring with an outer diameter of 3 mm and an inner diameter of 1.5 mm (Fig. 6.3C). The PDMS rings should be as round as possible and the internal hole as close to the center as possible. Over time the PDMS consistency will change thus rings older than 1 month should be discarded. Unpolymerized PDMS can be stored for a few weeks at −20 °C.

Carbon-coated, glow-discharged sapphire discs and glow-discharged gold spacer rings—on top of clean sapphire discs, place a finder grid with a pattern of your choosing. To keep the sapphire discs and finder grids from moving, we made a small tool (Fig. 6.4A). Load the grids in the carbon coater and generate a very thick carbon coating. We used eight twisted carbon threads in the BAL-TEC Med020. Following carbon coating, the carbon-coated sapphires, an equal number (plus 1 or 2 to ensure you always have enough) of clean sapphires and an equal number (plus a few more in case some are bent) of gold spacer rings where placed on a parafilm-wrapped glass slide (Fig. 6.4B). As soon as an extract sample was ready for high-pressure freezing, the slide was glow discharged under air for 5 s. Following glow discharge, the sapphires were used within 1 h. After 1 h a new preparation, as shown in Fig. 6.4B, were glow discharged and used.

Note: Repeated or more than 10 s of glow discharging caused the hydrophobic carbon layer to peal upon contact with water.

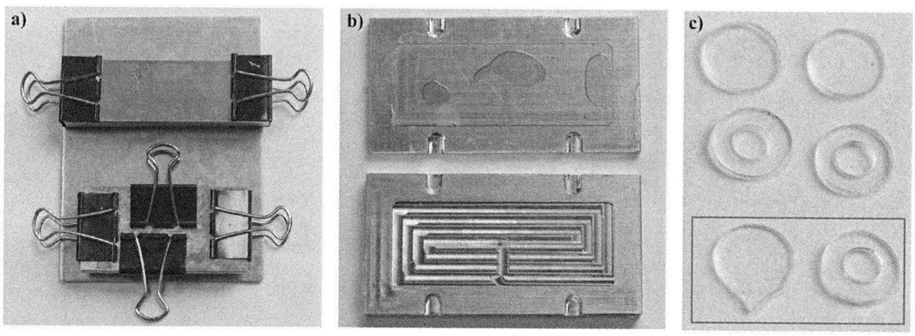

FIGURE 6.3

Preparation of PDMS rings. (A) PDMS molds clamped shut for polymerization. (B) Our 350 μm deep mold opened; the PDMS layer is resting on the lid (top), the 350 μm deep well is located in the bottom piece. The four insets on each piece facilitate opening the mold following polymerization. (C) Two PDMS discs (top), two PDMS rings (middle), and two rejected pieces that are incorrectly formed (bottom, in box) and should not be used.

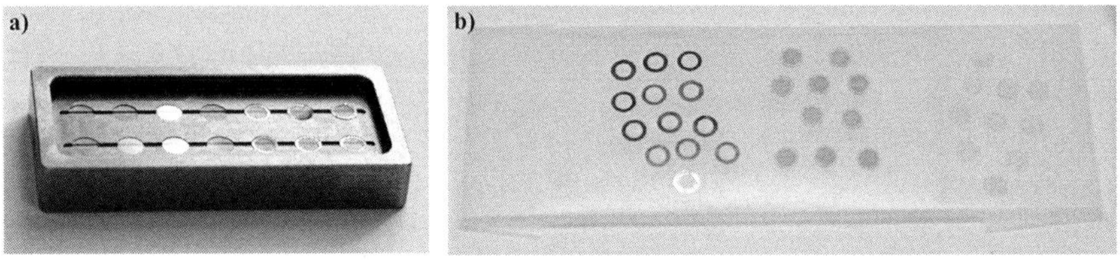

FIGURE 6.4

Carbon coating of sapphire discs. (A) To carbon coat the sapphire discs an aluminum dish with 14 3-mm insets was used. The sapphires could be mounted in the inset as shown for the left eight wells. On top of the sapphires, a finder grid was placed as shown in the right six wells. (B) In preparation for glow discharging, 10 carbon-coated sapphires, 11 clean sapphires (10+1 to ensure you always have enough), and 10+ (in case some are bent) gold sapphire rings (white in print version) where placed on a parafilm covered glass slide.

6.2.2.2 Sapphire assembly protocol

1. Prepare the holder of the Baltec HPM010 high-pressure freezer by ensuring the tip is open, clean, and ready for a sample (Fig. 6.2A).
2. Blot a "B" carrier soaked in 1-Hexadecene on filter paper and place it flat side up in the tip.
3. Put the holder to the side, but within easy reach.
4. Place a PDMS spin-coated wafer into a large petri dish with filter paper below.
5. Place a glow-discharged, carbon-coated sapphire disc half-way on the edge of the wafer with the carbon layer facing up. The PDMS wafer is sticky and holds the carbon-coated sapphire in place while you work.
6. Place the gold spacer ring on top of the carbon-coated sapphire (Fig. 6.2B, PDMS wafer is black with the sapphire and the gold spacer ring resting on top). Take care not to bend the spacer ring since it may result in air bubbles within the sapphire assembly.
7. Using 10-μm pipette tips with cut ends (which will reduce shearing and damage to the spindles), pipette ~1.2 μl of extract onto the sapphire in the center of the gold spacer ring.
8. Dip a clean, none carbon-coated sapphire in the 1-Hexadecene, blot on filter paper, and place on top of the gold spacer ring.
   - Coating the top sapphire with 1-Hexadecene helps the sample release (McDonald et al., 2010).
   - By placing the top sapphire disc on at an angle, less air bubbles will be between the sapphire discs.
9. Push the top sapphire down and close the sapphire assembly. The majority of the sample volume will be squeezed out.
10. The final volume between the sapphires will be ~0.16 μl; if any air bubbles are present, place the sapphire assembly in a tube destined for sapphire recycling and start over at step 1.
11. Pick up the entire sapphire assembly, blot off excess sample on filter paper, flip sapphire assembly over, and place into the tip for the high-pressure freezer.

The carbon sapphire should now be on the top away from the aluminum carrier to allow imaging of the sample on a fluorescence microscope.

12. Place a PDMS ring on top of the sapphire assembly (Fig. 6.1B); the entire assembly should be arranged as shown in Fig. 6.1C.
    - *Note*: the PDMS swells if in contact with 1-Hexadecene and will no longer fit into the HPM tip. Avoid contact with 1-Hexadecene.
13. Close the high-pressure freezing tip. Clip the tip closed using a tight fitting "U"-shaped, stainless steel clamp (Fig. 6.2C–E).
14. Check under the stereomicroscope to ensure the sapphire assembly is held in place by the PDMS ring, and that no bubbles have entered between the sapphires. If the sapphire assembly is moving in the tip, then the PDMS ring is not thick enough. Conversely, the PDMS ring should not be so thick that it is squished out the top.
15. Unscrew the clamped high-pressure freezing tip from the freezing handle. The clip will keep the tip closed and the sapphire assembly secure.
16. Screw the tip into the imaging holder (Fig. 6.2F) and place the imaging holder directly on the stage of an inverted fluorescence microscope. Ensure that the carbon sapphire is pointed down and close to the objective lens. Focus on the carbon grid mask, and acquire images in the appropriate fluorescence channel of your structure of interest.
17. An optimal sample contains at least one well-formed isolated structure. We defined isolated as no other structures within the same or neighboring square of the grid mask. Structures should not be positioned too close to the edge of the sapphire discs. If any small air bubbles are observed, then this sample needs to be placed in the sapphire recycling and sample preparation must return to step 1.
18. Once done imaging, remove the tip from the imaging holder and reattach it to the freezing holder. While observing through the stereomicroscope, carefully remove the "U"-shaped clip and make sure no bubbles appear and nothing seems to change.
19. Proceed as rapidly as possible with high-pressure freezing. Store high-pressure frozen samples in a labeled 0.5-ml Eppendorf Tube® until ready for freeze-substitution.

6.2.3 FREEZE-SUBSTITUTION OF THE SAMPLE AND EMBEDDING IN EPON

Cryoimmobilized samples were stored in liquid nitrogen for no longer than a week before being loaded into a Leica AFS 1 and taken through freeze-substitution.

We left the sapphire assembly unopened until the embedding steps to prevent damaging the samples.

The freeze-substitution protocol we used was optimized for our biological interest—the microtubule. We took guidance from the work of Giddings (2003) to optimize our protocol as follows:

6.2.3.1 Freeze-substitution protocol

Instruments

- Automatic freeze-substitution system (AFS-1, Leica)
- Stereomicroscope with darkfield illumination and air extraction.

Materials

- Anhydrous EM Grade Glutaraldehyde in Acetone (EMS #16530)
- Tannic acid (EMS #21700)

6.2.3.2 Detailed protocol

20. Place the entire frozen assembly into fixative cocktail #1: 2.5% dH_2O, 0.2% glutaraldehyde, 0.1% freshly prepared tannic acid in dried acetone. Hold at −90 °C for 8 h.
21. Increase temperature 5 °C per hour to −30 °C
22. Hold at −30 °C for 3 h
23. Remaining at −30 °C, rinse out fixation cocktail #1 with −30 °C dried acetone
24. Repeat acetone rinse 3 ×, 10 min per rinse
25. Remaining at −30 °C, replace acetone with fixative cocktail #2: 1% OsO_4+0.1% UA in acetone. Hold at −30 °C for 1 h
26. Increase temperature to 0 °C within 1 h
27. Hold at 0 °C for 30 min
28. Remaining at 0 °C, rinse out fixation cocktail #2 with 0 °C dried acetone.
29. Repeat acetone rinse 3 ×, 10 min per rinse
30. Remove samples from AFS1 for sapphire assembly opening and immediate embedding into Epon.

6.2.3.3 ACLAR® Flat embedding protocol

The following flat embedding procedure (Fig. 6.5) was used for two reasons: (a) the possibility to retrieve more than one spindle and (b) the specimen was located in a 50-μm layer, greatly simplifying the identification and sample orientation during data collection.

1. Cut five pieces of 0.2-mm thick ACLAR® 33C embedding film (EMS # 50425) the size of a glass slide.
2. Use double-sided tape to stick three of the sheets together. Cut the corner of the stack of ACLAR® so you know the left end from the right end (Fig. 6.5).
3. Use a sharp razor blade, or hole punch to cut up to five holes into the sheets (three holes in Fig. 6.5). These will become the sample wells. The smaller they are the better, but make sure your sample will fit easily.
4. Place one piece of intact ACLAR® on a glass slide.
5. Place the ACLAR® stack with holes on top of the intact ACLAR®.
6. Clamp them in place with paper clamps.
7. Add several drops of your resin of choice to the ACLAR® flat embedding wells

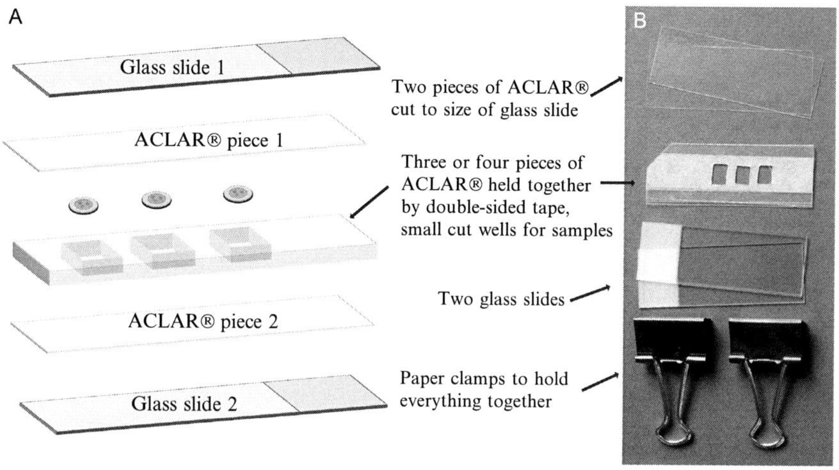

FIGURE 6.5

Schematic illustration of flat embedding. (A) An illustration of how the flat embedding molds were constructed. (B) An image of the material used for the flat embedding procedure.

8. Under a stereomicroscope with air extraction, in 100% acetone, in a 60-mm petri dish, remove the noncarbon-coated sapphire by sliding it parallel to the carbon-coated sapphire. Many times the noncarbon-coated sapphire has already detached during the freeze-substitution protocol. It is best to keep the gold spacer ring on, and the sample must remain attached to the carbon-coated sapphire so the carbon landmarks can later be used for orientation during sectioning.
9. Dip the sample in two small volumes of resin of choice; we filled the lid from a 0.5-ml Eppendorf Tube® for this procedure so we used about 400 μl of resin.
10. Place the carbon-coated sapphire down, the sample up in the ACLAR® wells (Fig. 6.5A). Push the sapphire down to the bottom of the well. Following resin polymerization, the carbon-coated sapphire will be closer to the edge of the resin making it easier to remove.
11. Carefully remove the clamps. Place the last piece of ACLAR® on top of the ACLAR® wells, try to avoid bubbles near the samples. Place the second glass slide on top. Clamp everything closed (Fig. 6.5B). Check that the samples are still in the wells.
12. Let this assembly rest at room temperature for 1–2 h allowing Epon infiltration.
13. Polymerize resin overnight at 60 °C

6.2.3.4 Removal of carbon-coated sapphire and re-embedding protocol

1. Remove the clamps, glass slides, and the bottom piece of ACLAR® (the side the carbon-coated sapphire discs are closest to). Use a razor blade to pry the slides apart if the resin has leaked.

2. While observing through a stereomicroscope, use a fine razor blade to carefully and thoroughly remove the Epon from the top and the edges of the carbon-coated sapphire

Important

- To make sure you are not on the sample side, ensure the carbon letters are inverted.
- If not enough Epon is removed the sapphire will most probably break during removal, which could leave sapphire dust on the sample, and subsequently damage the diamond knife.

3. Gently place the edge of a razor blade under the sapphire disc and lift the sapphire disc slowly. A rainbow wave under the sapphire should be visible as the sapphire releases from the Epon. This must be done slowly, with steady lifting. Lifting from multiple places will help. Eventually the sapphire will release (Fig. 6.6A).
4. If a sapphire breaks, just keep removing pieces. Try to get all small pieces off.
5. Once the sapphire is removed, also remove the gold spacer ring. A 2-mm diameter circle of sample should remain, with the carbon pattern clearly visible (Fig. 6.6A). The sample should be clearly visible when OsO_4 was used during the freeze-substitution and it should be clear when the sample has been lost (Fig. 6.6B).
6. Following the removal of the sapphire, the sample will be directly at the face of the block. Ultimately, the sample will be sectioned perpendicular to the exposed

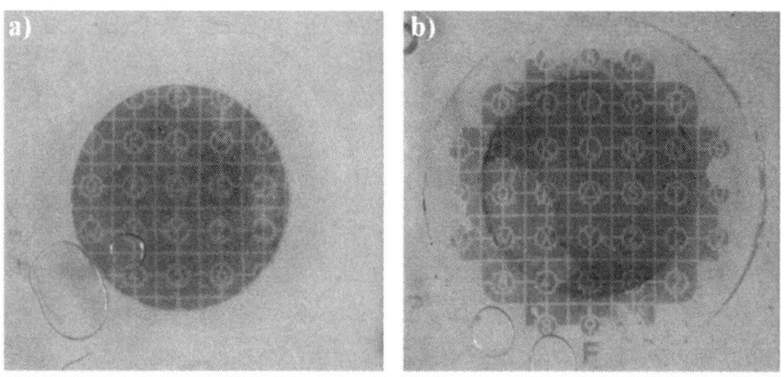

FIGURE 6.6

Two samples following processing and re-embedding. (A) An example of a sample that was embedded with the gold spacer ring (dark gray in print version). When the gold spacer ring (dark gray in print version) was removed just prior to re-embedding, a disc containing the sample was all that remained. By keeping the gold space ring (dark gray in print version) during the first embedding the preparation was very clean and the sample remained intact more often than not. (B) In contrast, this sample lost its gold spacer ring (dark gray in print version) before the first embedding, evident from the larger diameter of carbon embedded in the block. The loss of the sample over letters O, W, X, and Z is apparent because of the change of sample color. Sample loss was more common when the gold spacer ring (dark gray in print version) was separated from the sample in the early steps of sample processing.

block face therefore to protect the sample during sectioning the sample was re-embed in a mold made from ACLAR® sheets as done for the first embedding. The modifications to this are (a) the mold should be made from four layers of ACLAR® so the wells are deeper and (b) the wells should be cut larger to accommodate the sample.

7. Polymerize resin overnight at 60 °C

6.2.4 SAMPLE ALIGNMENT AND SERIAL SECTIONING

The OsO_4 and the glutaraldehyde in the freeze-substitution procedure and the embedding in Epon abolished the fluorescence properties of the dyes (Nixon et al., 2009). Therefore, we were no longer able to see the location of the spindles. All landmarking and orientation was done using the carbon grid mask against which the spindle had been imaged. The flat embedded sample was trimmed and serial sectioned perpendicular to the axis of the sapphire disc (Fig. 6.7). To use the carbon landmarks to find the location of the structure of interest in the block, while the block is mounted in the microtome, it was helpful to tilt the block so the carbon layer was approximately 45° relative to the vertical, and to use a piece of white paper underneath the block to homogenize the light source. In this position, a razor blade was used to mark the edges of the area to be sectioned. The block was tilted back into a vertical position and a glass knife was used to roughly trim the area containing the spindle. Using glass knives was a method to protect the diamond knives from any remnants of the sapphire disk remaining in the block. Once the majority of the trimming was done, then a 45° diamond trimming knife was used to trim the last edges and build a trapezoid (Bozzola & Russell, 1999). One to two hundred 300-nm serial sections were cut from the trapezoid using an ultra 35° diamond knife. Serial sections were collected on home-coated slot grids (protocol described below).

Instrument

- Ultramicrotome (UCT, Leica)

Materials

- Ultra 35° diamond knife for plastic microtomy (Diatome)
- Diamond 45° trimming knife (Diatome)
- Glass trimming knife (Homemade; Bozzola & Russell, 1999).
- 2×1 mm Cu/Pd 3.05 mm slot grids (Agar #G2500PD)
- 0.8–1.0% Formvar (EMS #15800) in chloroform
- Dumont Tweezers 5 Dumoxel
- Tools for sectioning—eyelashes, loops

Slot grid preparation—we prepared our own formvar-coated grids a few days before we planned to serial section the samples. The day before grid preparation make a 0.8–1.0% formvar in chloroform solution. Leave the solution to stir for several hours before being stored overnight. Rinse slot grids several times in chloroform and allow to dry before use. On the suggestion from Cindi Schwartz (Boulder, USA), each grid

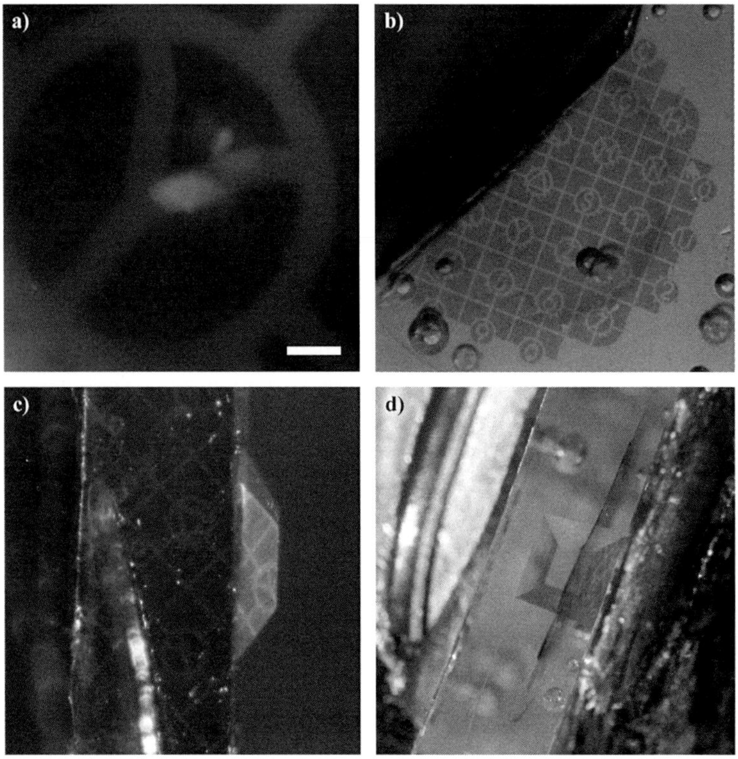

FIGURE 6.7

Sample registration critical for targeted ultramicrotomy. Using fluorescence images acquired just prior to cryoimmobilization (A, showing the letter "Y"), the block can be mounted in the holder of the ultramicrotome (B) and trimmed to target the structures of interest (C). When viewed from above a thin black sample is visible running through the polymerized resin (D). Scale bar = 30 μm. (See the color plate.)

should initially be predunked in a small volume of the formvar solution, placed face-up on a piece of filter paper and allowed to dry. We placed the Pd side up. Following dunking approximately 100 grids, the formvar solution should be discarded in the chloroform waste. Make silver (with a hint of gold) formvar films from the formvar solution, float the films on water and place the predunked grids face down (for us this was Pd side down) on the formvar film. When done placing grids, the film with the grids can be picked up on a glass slide covered with a clean, white office label. Place the grids in a 60-mm petri dish and warm them slightly (30–40 °C) to encourage drying. Warming during drying helps to have a tighter formvar film in the slot grids compared to grids dried at room temperature, this is particularly useful when the ambient humidity is high.

6.2.5 POSTCONTRASTING AND DATA ACQUISITION BY ELECTRON TOMOGRAPHY

Several combinations of poststaining were tested. We found that poststaining with Reynolds' lead citrate for 12 min was the best protocol for our samples. The most notable artifact was a gradient within the tomograms when a combination of uranyl acetate in 70% methanol and lead citrate were used to poststain the thick sections (Fig. 6.8). After extensive testing, we concluded that the best results were obtained by removing uranyl acetate from the poststaining.

Instruments

- An electron microscope equipped with a goniometer
- Rotation holder
- A large hard-drive to store data

Materials

- Renoylds' lead citrate (Reynolds, 1963)
- Dumont Tweezers 5 Dumoxel

The *X. laevis* spindle is approximately 15 μm in diameter and approximately 40 μm in length (Kronja, Kruljac-Letunic, Caudron-Herger, Bieling, & Karsenti, 2009). Protofilament resolution, which is about 2 nm, was needed to understand the three-dimensional microtubule organization of the meiotic spindle. To acquire an area that encapsulated the entire spindle and the region around the poles at 2 nm resolution would take approximately 36 h of continuous data acquisition. Attempts to do this failed due to the extensive nonlinear warping of the sections. Although we were never able to acquire the entire spindle from pole to pole in a single section, we did determine that the best method to reduce sample warping was to carbon coat

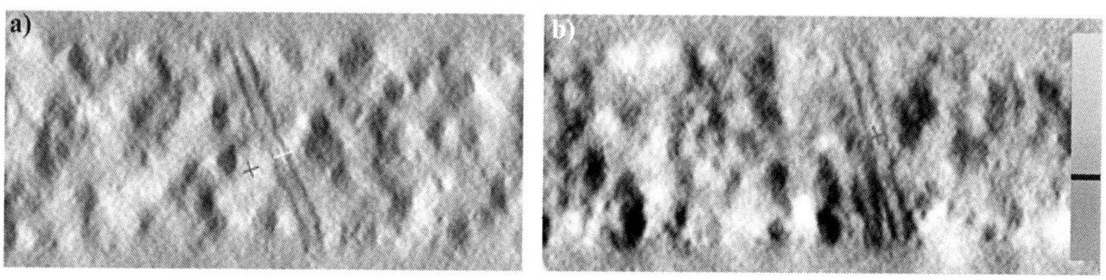

FIGURE 6.8

Variation in poststaining procedures visible in variable quality of electron tomograms. (A) Side view of a tomogram acquired from a sample poststained with only lead citrate. The staining is consistent throughout the depth of the section and the microtubule is evenly stained. (B) Side view of a tomogram acquired from a section poststained with uranyl acetate and lead citrate. The staining appears asymmetric as schematically shown on the right. The black bar marks the boundary between darker staining (bottom) and lighter staining (top). When the uranyl acetate was removed from the poststaining procedure this gradient was no longer encountered.

the slot grids. After testing a multitude of combinations of carbon coating and sectioning, the greatest reduction in warping was obtained by carbon coating both the front and the back of the grid with sections in a very thin, faintly visible layer of carbon. Carbon coating before section pick-up resulted in one of three challenges (a) increased breaking of Formvar films in the carbon coater; (b) the hydrophobic carbon layer on the grids pushed the section away from the grid during section pick-up, or (c) if the carbon-coated grid was glow discharged for 1–2 s, the grids became so hydrophilic that the grid was covered in water during section pick-up. All these challenges made serial sectioning very hard. We had the fewest challenges when the grids were carbon-coated following sectioning and staining. Not a single Formvar film over a slot grid was broken using this method. Although we arrived at this protocol independently, a similar approach has since been published (Walther, Schmid, & Hohn, 2013).

Tomography data was acquired on a FEI F30 using SerialEM and reconstructed using etomo (Kremer, Mastronarde, & McIntosh, 1996).

6.3 PERSPECTIVES AND FUTURE DIRECTIONS

Herein, we have presented the protocol we applied to the fluorescence identification, cryoimmobilization, processing, targeted ultramicrotomy, and imaging of the meiotic spindle assembled in the *X. laevis* egg extract (Fig. 6.9). Many of the steps were optimized for an isolated cell-free sample that needed to be protected, quality controlled, and whose size and location needed to be known throughout the entire sample processing procedure. Key steps in the method described herein are (a) the sapphire

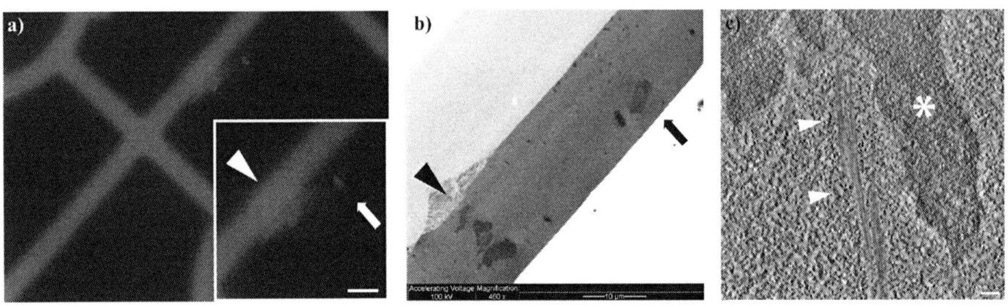

FIGURE 6.9

From light microscope to electron microscope. (A) An imperfect spindle can be seen by fluorescence microscopy. (B) Electron micrograph of the same imperfect spindle shown in panel A. The two areas of DNA, one large, one small (see matching arrows), are observed in both panel A and B confirming that this is the same sample and illustrating the success of our sample registration and targeted ultramicrotomy. (C) Electron tomography data of a microtubule bundle (arrowheads) visible near the large chromosome mass (asterisk) of the spindle shown in panel A and B. Scale bar = 15 µm (panel A), 10 µm (panel B), and 200 nm (panel C). (See the color plate.)

assembly held in place by a "B" carrier and a PDMS ring; (b) the "U"-shaped clip that held the high-pressure freezing tip closed and enabled fluorescence imaging of the sample mounted in the tip; and (c) the method of identifying the structures relative to the carbon grid mask and following them through sample processing and onto the electron microscope. Tools were developed and protocols adapted to enable a free-floating delicate sample to be taken from light microscopy through to the electron microscope.

ACKNOWLEDGMENTS

We thank Céline Pugieux for teaching E. M. T. the frog egg extract protocol. Kresimir Crnokic for maintaining the frog facility. Ivo A. Telley, Sarah Köhler, Mikhail Eltsov, Rachel Mellwig, Sabine Prugnaller, and Wim Hagen for their troubleshooting suggestions. We thank Sascha Blättel from the EMBL Mechanical Workshop for making and adapting the unique tools. François Nédélec and Jan Ellenberg for their constant support in this project and their fruitful protocol and experimental suggestions. Eileen O'Toole, Cindi Schwartz, Jean-Marc Verbavatz, and Thomas Müller-Reichert for protocol and troubleshooting suggestions. David Mastronarde for his never-ending help with the data reconstruction.

REFERENCES

Bozzola, J. J., & Russell, L. D. (1999). *Electron microscopy: Principles and techniques for biologists.* Bostan, MA: Jones and Bartlett Publishers.

Giddings, T. H. (2003). Freeze-substitution protocols for improved visualization of membranes in high-pressure frozen samples. *Journal of Microscopy, 212*(Pt 1), 53–61.

Hannak, E., & Heald, R. (2006). Investigating mitotic spindle assembly and function in vitro using Xenopus laevis egg extracts. *Nature Protocols, 1*(5), 2305–2314.

Hawes, P., Netherton, C. L., Mueller, M., Wileman, T., & Monaghan, P. (2007). Rapid freeze-substitution preserves membranes in high-pressure frozen tissue culture cells. *Journal of Microscopy, 226*(Pt 2), 182–189.

Kolotuev, I., Bumbarger, D. J., Labouesse, M., & Schwab, Y. (2012). Targeted ultramicrotomy: A valuable tool for correlated light and electron microscopy of small model organisms. *Methods in Cell Biology, 111*, 203–222.

Kremer, J. R., Mastronarde, D. N., & McIntosh, J. R. (1996). Computer visualization of three-dimensional image data using IMOD. *Journal of Structural Biology, 116*(1), 71–76.

Kronja, I., Kruljac-Letunic, A., Caudron-Herger, M., Bieling, P., & Karsenti, E. (2009). XMAP215-EB1 interaction is required for proper spindle assembly and chromosome segregation in Xenopus egg extract. *Molecular Biology of the Cell, 20*(11), 2684–2696.

Lohka, M. J., & Maller, J. L. (1985). Induction of nuclear envelope breakdown, chromosome condensation, and spindle formation in cell-free extracts. *The Journal of Cell Biology, 101*(2), 518–523.

McDonald, K., Schwarz, H., Muller-Reichert, T., Webb, R., Buser, C., & Morphew, M. (2010). "Tips and tricks" for high-pressure freezing of model systems. *Methods in Cell Biology, 96*, 671–693.

Moor, H., Bellin, G., Sandri, C., & Akert, K. (1980). The influence of high pressure freezing on mammalian nerve tissue. *Cell and Tissue Research, 209*(2), 201–216.

Murray, A. W. (1991). Cell cycle extracts. *Methods in Cell Biology*, *36*, 581–605.

Nixon, S. J., Webb, R. I., Floetenmeyer, M., Schieber, N., Lo, H. P., & Parton, R. G. (2009). A single method for cryofixation and correlative light, electron microscopy and tomography of zebrafish embryos. *Traffic*, *10*(2), 131–136.

Reynolds, E. S. (1963). The use of lead citrate at high pH as an electron-opaque stain in electron microscopy. *The Journal of Cell Biology*, *17*, 208–212.

Salmon, E. D., Goode, D., Maugel, T. K., & Bonar, D. B. (1976). Pressure-induced depolymerization of spindle microtubules. III. Differential stability in HeLa cells. *The Journal of Cell Biology*, *69*(2), 443–454.

Studer, D., Humbel, B. M., & Chiquet, M. (2008). Electron microscopy of high pressure frozen samples: Bridging the gap between cellular ultrastructure and atomic resolution. *Histochemistry and Cell Biology*, *130*(5), 877–889.

Studer, D., Michel, M., & Muller, M. (1989). High pressure freezing comes of age. *Scanning Microscopy*, *3*(Supplement), 253–268, discussion 268–259.

Walther, P., Schmid, E., & Hohn, K. (2013). High-pressure freezing for scanning transmission electron tomography analysis of cellular organelles. *Methods in Molecular Biology*, *931*, 525–535.

CHAPTER

7

Combining Wide-Field Super-Resolution Microscopy and Electron Tomography: Rendering Nanoscopic Correlative Arrays on Subcellular Architecture

Filip Braet[*,†,‡], Delfine Cheng[*], Minh Huynh[‡], Jeffrey Henriquez[*], Gerry Shami[*], Marko Lampe[§,¶]

[*]School of Medical Sciences (Discipline of Anatomy and Histology)—The Bosch Institute, The University of Sydney, Sydney, NSW, Australia

[†]Charles Perkins Centre, The University of Sydney, Sydney, NSW, Australia

[‡]Australian Centre for Microscopy & Microanalysis, The University of Sydney, Sydney, NSW, Australia

[§]European Molecular Biology Laboratory, Heidelberg, Germany

[¶]Department of Translational Pulmonology, Translational Lung Research Center (TLRC), University of Heidelberg, Heidelberg, Germany

CHAPTER OUTLINE

Methods in Cell Biology, Volume 124, ISSN 0091-679X, http://dx.doi.org/10.1016/B978-0-12-801075-4.00007-0

Abstract

In this chapter, the authors outline in full detail, an uncomplicated approach that enables the combination of wide-field fluorescence super-resolution microscopy with electron tomography, thereby providing an approach that affords the best possible confidence in the structures investigated. The methodical steps to obtain these high-throughput correlative nanoscopic arrays will be visually explored and outlined in detail. The authors will demonstrate the feasibility of the method on cultured Caco-2 colorectal cancer cells that are labeled for filamentous actin. The presented images, morphometric data, and generated models illustrate the strengths of our correlative approach for future advanced structural-biology-oriented questions. Correlative nanoscopy applications can be readily found in which there is a need to reveal biomolecular information at unprecedented resolution on subcellular behavior in various biological and pathobiological processes.

INTRODUCTION

Currently, we are at a very exciting point in time in the modern history of cellular microscopy. Recent developments in super-resolution microscopy are pushing spatial resolution far beyond the optical diffraction limit reaching resolving power close to electron microscopy (Schermelleh, Heintzmann, & Leonhardt, 2010). Although the theoretical concept of super-resolution microscopy was described as early as at the beginning of the previous century, we had to wait until the late 1990s before the first practical solutions became available, delivering near to single-molecule imaging in fluorescently labeled samples. Depending on the optical super-resolution technique used, measurements with lateral resolution of at least 60–80 nm, and even down to 20–25 nm can be achieved, opening up a unique view on cellular architecture at the supramolecular scale (Galbraith & Galbraith, 2011).

Several studies demonstrated the feasibility of imaging super-resolved subcellular detail via the use of fluorescent probes (reviewed in Fernández-Suárez & Ting, 2008). Typical lateral resolution values in the order of 25–75 nm have been reported. Localization microscopy with organic dyes became the technique with the highest achievable lateral resolution enhancement using standard biological probes. Therefore, in our laboratory, we used ground state depletion (GSD) followed by individual molecule (IM) return microscopy (Fölling et al., 2008)—also known as dSTORM (Heilemann et al., 2008)—for optical nanoscopy. This type of wide-field super-resolution microscopy generates super-resolved optical data at resolutions down to 25 nm via conventional fluorescent probes without the need of any further time-consuming reconstruction steps. In addition, key to the success of obtaining highly resolved super-resolution data is the choice of sample mounting conditions in order to collect a sufficient number of single molecule events during the course of the data collection. The type of buffer and nature of antifading reagents used herein are the most important contributors to longevity and brightness of the fluorophores, thereby creating a solid foundation to reconstitute a high-quality super-resolved image (Requejo-Isidro, 2013).

In this chapter, the decoration of filamentous actin with Alexa Fluor® 647 labeled phalloidin in cultured Caco-2 colorectal cancer cells was studied as an example to assess the different structural configurations and dimensions of filamentous actin via ground state depletion-super resolution (GSD-SR). In particular, this cytoskeleton protein forms an ideal subject of study to assess how the different imaging methods complement each other with respect to resolving power. It is well known that actin filaments can possess distinctive widths, ranging between 6 and 25 nm, pending on their structural arrangement (Chhabra & dos Remedios, 2008). Taking advantage of the opportunities offered by correlative light and electron microscopy (CLEM) (Kobayashi et al., 2012; Müller-Reichert & Verkade, 2012), we aimed to directly compare the optical data with electron tomography (ET) and this, on the same regions of interest (ROI) previously imaged via GSD-SR (Fig. 7.1). Typically, CLEM approaches allow to obtain complementary information about the same sample by means of two or more distinct types of microscopy and then merge those distinct streams of information together to obtain a far richer depth of insight into biological processes, that is, possible from a single mode of microscopy alone (Spiegelhalter, Laporte, & Schwab, 2014). This way combined microscopy approach undoubtedly adds a higher degree of confidence about the structures of interest as observations from the same cell can be directly compared across the different imaging methods (Jahn et al., 2012; Su et al., 2010).

7.1 RATIONALE

We present a straightforward approach in the sample preparation and data analysis for correlative super-resolution studies. The feasibility and ease of this approach will

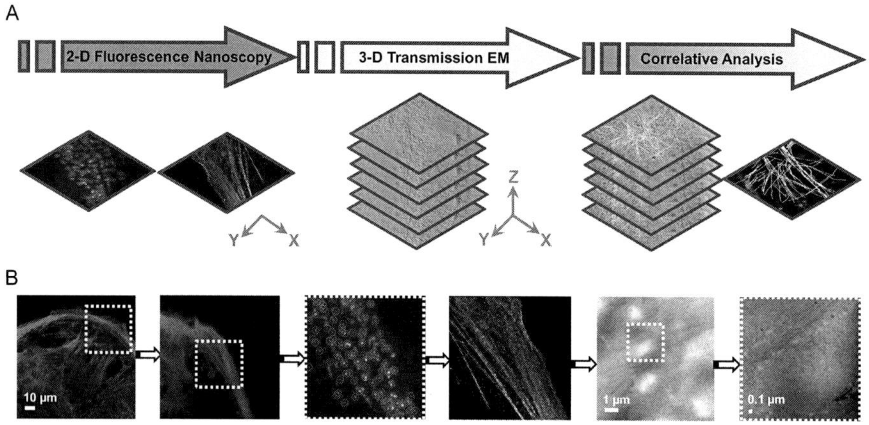

FIGURE 7.1

Schematic diagram outlining the correlated nanoscopy approach. (A) Principle of cross-correlative two-dimensional wide-field fluorescence super-resolution imaging and combined three-dimensional transmission electron microscopy on the same region of interest. Correlative analysis allows the exploration of electron tomography models together with fluorescent information retrieved at the single-molecule level. (B) Scheme depicting the different imaging steps involved in the acquisition of correlative ground state depletion data with transmission electron microscopy information. Note from left to right the remarkable length scales, i.e., "from ten's of microns down to a couple of ten's of nanometers," that get bridged. The white boxes indicate the region of interest and the ensuing areas analysed with the different imaging platforms. (See the color plate.)

be illustrated by presenting CLEM arrays of cellular actin obtained via GSD-SR and ET. This all will be supported by quantitative measurements in which GSD-SR data will be directly compared and extensively discussed with morphometric values extracted from the ET models. The different notes and tricks to facilitate relocation of the area of interest across the different imaging platforms will be documented throughout this chapter, including how to overlay and model the different microscope data sets that result in nanoscopic arrays. Of note, the CLEM approach described in this chapter has arisen from the combination of our past experience in this field (Braet et al., 2007; Jahn et al., 2009), the most recent advances in optical technologies and imaging instruments (Agrawal, Reilly, & Schroeder, 2013; Herbert, Soares, Zimmer, & Henriques, 2012), and the availability of the latest generation of fluorescent probes and the improvements in software for correlated data analysis (Sjollema, Schnell, Kuipers, Kalicharan, & Giepmans, 2012).

7.2 METHODS

The different methodological steps outlined under this section are graphically depicted in Fig. 7.2. The sequential key-steps to be undertaken in sample processing are given to successfully correlate single molecule fluorescence data with

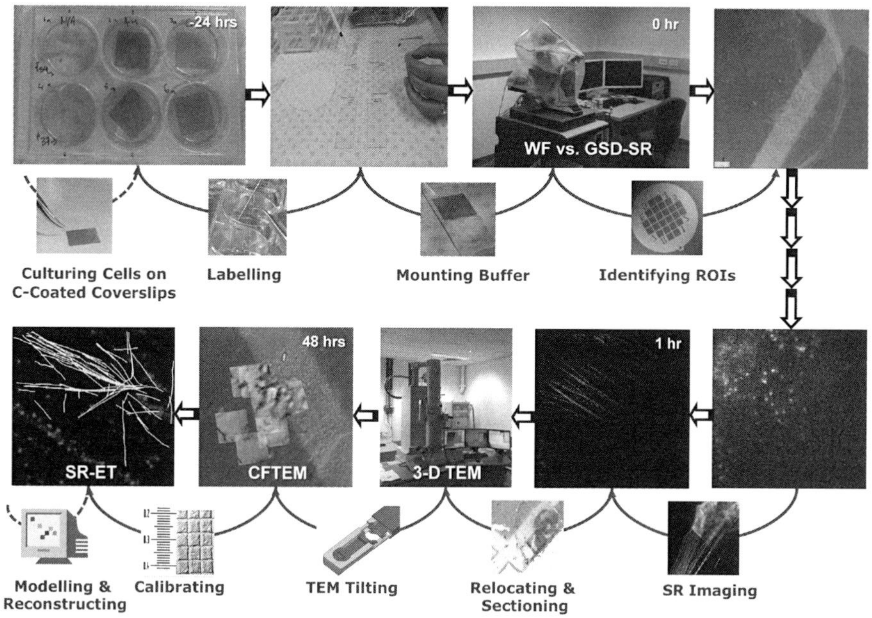

FIGURE 7.2

Outline of the different experimental steps involved to achieve correlated optical super-resolution transmission electron tomography data. *Note*: Read from left to right in the upper panel and right to left in the lower panel (black straight arrows); the curved arrows denote the critical steps involved for a successful relocation and subsequent image acquisition and data processing to achieve correlative nanoscopic arrays. Typically, the entire workflow from cell labeling up to correlative SR-ET data takes around 3 days of work. See also the supplementary information (http://dx.doi.org/10.1016/B978-0-12-801075-4.00007-0; i.e., SR_ET.wmv-movie) illustrating the main data collection steps in a time-lapse manner. Abbreviations: C, carbon; CFTEM, correlative fluorescence and transmission electron microscopy; GSD-SR, ground state depletion super-resolution microscopy; hrs, hours; WF, wide-field; SR-ET, optical super-resolution electron tomography; 3D TEM, three-dimensional transmission electron microscopy. (See the color plate.)

transmission electron microscopy (TEM) images. The approach is the result of the combination of previous approaches described by our team (Braet et al., 2002; Jahn & Braet, 2008; Kobayashi et al., 2012) and the latest advances reported regarding dedicated sample preparation steps for artifact-free super-resolution imaging (Curthoys, Mlodzianoski, Kim, & Hess, 2013; LeicaMicrosystems, 2014).

7.2.1 PRINTING RELOCATION AIDS ON CELL CULTURE GLASS COVERSLIPS

The following procedure describes the deposition of carbon-based relocation marks optimized for correlative super-resolution microscopy studies. This method is based on a protocol we described recently in much detail to correlate live-cell and/or confocal laser microscopy with classical sample preparation for transmission ET (Kobayashi et al., 2012):

1. Prewash square glass coverslips by soaking them in 0.1 M HCl for 1 h, followed by a few washes in distilled water. The coverslips are then subsequently soaked twice in 70% ethanol and in 100% ethanol, for 1 h each. This step has shown to enhance the adhesion of the carbon on the coverslip.
2. Carefully place two TEM finder grids, centered on the coverslip, one next to another, on each prewashed coverslip. Consistency in which side of the finder grid is placed on the coverslip facilitates future relocation.
3. Carbon-coat the coverslips to a thickness of 20–30 nm.
4. Remove finder grids: the coverslips are left with the finder grids patterns, which should be thick enough to be visible by eye.
5. Lightly scratch the coated side of the coverslip for reference.
6. Gently rinse the coverslips with 70% ethanol to remove unwanted deposits.
7. Bake the coverslips for at least 48 h at 60 °C: this step is crucial to retain the grid pattern throughout the remaining sample preparation procedures.

7.2.2 CELL CULTURE AND CONSIDERATIONS FOR SUBSEQUENT SUPER-RESOLUTION FLUORESCENT LABELING STUDIES

Once the glass coverslips are provided with the relocation prints one can start-up the routine cell-culture practice. The following standard procedure is applicable to all cell line-based experimentation with the exception of the growth media and cell culture supplements that might vary from cell to cell line. Below, we will outline a characteristic experimental design that encompasses all necessary sample labeling steps that leads to the successful acquisition of GSD-SR data. In this study, we used the Caco-2 colon cancer cell line as an example, experimental model for rendering nanoscopic arrays on actin filaments. The cell line was obtained from the American Type Culture Collection (ATCC).

1. Culture cells in 75 cm^2 tissue culture flasks and maintain in complete media consisting of advanced Dulbecco's Modified Eagle Medium (DMEM), FBS (10%), L-glutamine (2.5 mM), and antibiotic/antimycotic solution (100 U/ml penicillin, 100 μg/ml streptomycin, and 25 μg/ml amphotericin B).
2. Incubate the cells at 37 °C under 5% CO_2 and subculture to 70% confluency.
3. Place carbon-coated coverslips into 6-well tissue culture plates and UV sterilize for 20 min.
4. Wash the cells with phosphate buffered saline (PBS) and harvest the cells by applying trypsin-EDTA for 5 min at 37 °C before centrifugation at 2000 rpm for 3 min.
5. Resuspend the cells in complete media and place the cells into 6-well tissue culture plates containing carbon-coated coverslips at 70% confluency.

6. Allow the cells to attach to the coverslips for at least 24–48 h at 37 °C under 5% CO_2 before subsequent labeling with fluorophores (i.e., from step 8 onward).
7. Wash the cells twice with PBS pH 7.4 at 37 °C.
8. Fix the cells with 4% paraformaldehyde (PFA) solution in PBS for 10 min at room temperature.
9. Wash the cells 5 × with PBS at room temperature.
10. Make up a 6.6 μM (200 units/ml) solution of Alexa Fluor® 647 Phalloidin in PBS.
11. Place 200 μl of the staining solution per coverslip for 20 min at room temperature protected from light.
12. Wash the cells 5 × with PBS.
13. Mount coverslips onto single cavity slides with approximately 75 μl of PBS.
14. Seal coverslip edges with Picodent Twinsil® and store slides at 4 °C protected from light until required for imaging.

7.2.3 BRIGHT-FIELD, WIDE-FIELD FLUORESCENCE, AND SUPER-RESOLUTION MICROSCOPY

The following sequence of events allows (i) the registration of the cell(s) of interest for correlative purposes using bright-field imaging, (ii) the instant collection of fluorescent information under wide-field imaging conditions as reference to assess the labeling quality, (iii) followed by the subsequent acquisition of super-resolution image data to subsequently reconstruct the GSD-SR image of the selected ROI.

1. Capture transmitted light images (i.e., bright-field) of the whole grid area as a point of reference for relocation at the electron microscopy level. This step needs to be done prior to super-resolution imaging.
2. Immediately before GSD-SR imaging, remount sample in fresh 30 mM 2-mercaptoethylamine (MEA) (also known as cysteamine), reseal with Picodent Twinsil® and allow to set at room temperature for 10 min. Note, it is critical that there are no air bubbles in the MEA mountant as this will affect subsequent imaging.
3. Place the slide onto the microscope. The immersion oil should not contain any air bubbles as those can cause deflections/straylight of the excitation laser beam decreasing the image quality.
4. Identify the ROI using bright-field and (diffraction-limited) fluorescence microscopy and centered in the middle of the camera image. Note the sample rested for a couple of minutes before starting the first super-resolution image to ensure minimal drift.
5. Acquire wide-field image of the specimen in the super-resolution module (GSD-Wizard) followed by an image of the ROI in the center of the camera chip covering 18 μm × 18 μm in the sample plane. The super-resolution image was recorded in the central 18 μm × 18 μm ROI and the diffraction widefield image served as an initial comparison to the super-resolution image.

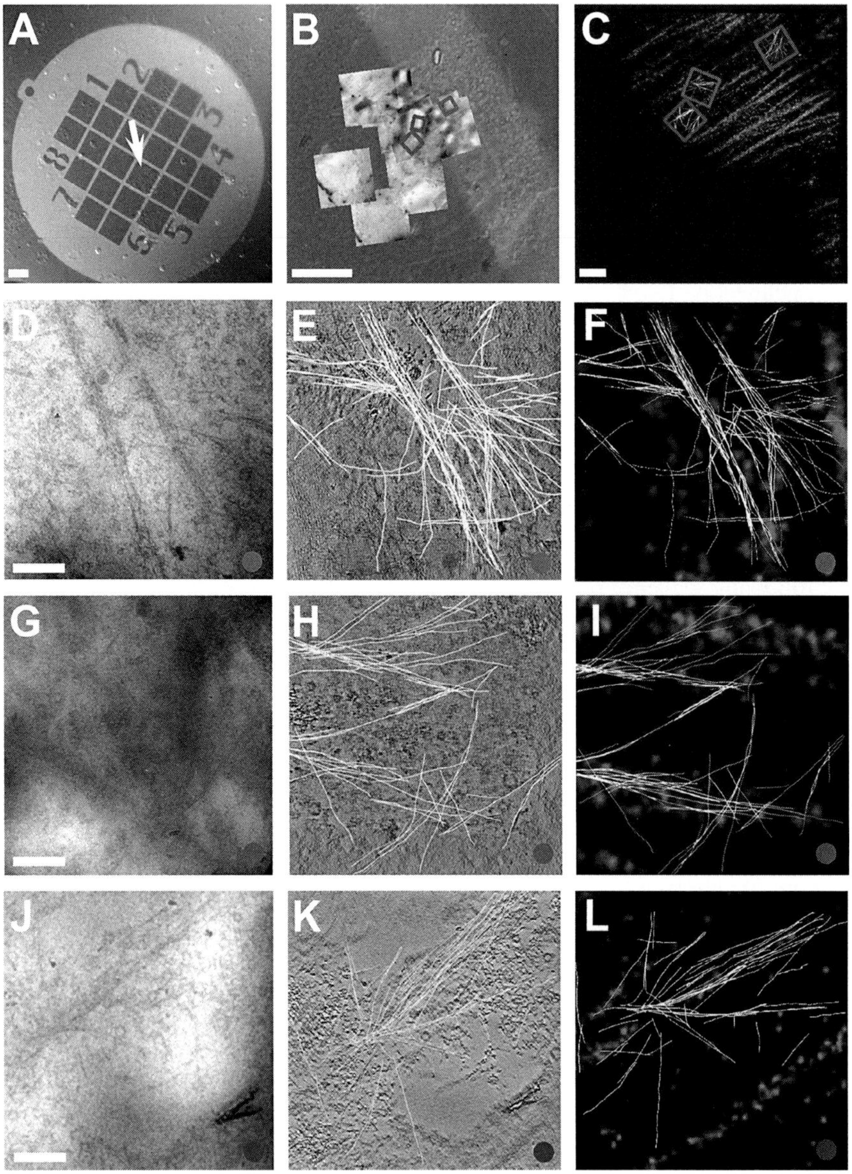

FIGURE 7.3

Combining bright-field, wide-field fluorescence, and super-resolution imaging with 3D transmission electron microscopy. Caco-2 colorectal cancer cells were labeled for filamentous actin. (A) The white arrow (A) denotes the cellular area of interest to be further analysed with wide-field fluorescence and ground stated depletion super-resolution microscopy. Note the clearly visible carbon relocation marks that facilitate further correlative

6. Next acquire the super-resolution image. This imaging step requires a two-step process of first transferring the fluorophores into the off-state (pumping) and then acquiring the single molecule blinking, which is immediately converted into a super-resolution image for live display. For "pumping" a higher laser power is applied than for imaging to ensure an efficient transfer of the molecules to the off-state. Therefore, it is not uncommon to use the full laser output power (up to 500 mW, measured before AOTF and fiber incoupling). For the images shown in Figs. 7.3 and 7.4 typically a laser power of up to 150 mW/~30% total laser power was used. The pumping step ends when single-well separated molecules are visible, with no or only a few fluorophores overlapping. The acquisition step generally requires an approximately two- to threefold lower laser power. Imaging was carried out at a frame rate of 100 frames per second (corresponding to 10 ms exposure time per frame) and concurrently the laser power was increased carefully during the course of the data collection to maximize the number of single molecule events per frame.
7. Stop the image acquisition manually when subsequent number of single molecule events was collected and/or the number of events per frame was dropping due to unavoidable bleaching to only a couple of events per frame (even when the 405 nm laser was at full power). We typically collected 100,000 images on the phalloidin labeled samples. Overlapping signals were rejected by the software (Leica SR GSD with LAS AF Version 2.6) and therefore not contributing to the final super-resolution image. We also wish to refer to Section 7.4.3 outlining additional notes and tricks to optimize the signal-to-noise ratio versus camera detection settings.

---

analysis. (B) Shows besides the wide-field information an example of initial correlative array mapping. Typically, we combine bright-field (e.g., gray transparent information of the entire image boundary) and wide-field fluorescence signals (red color) together with the 2D transmission electron microscopy data (e.g., seven small gray squares). This initial low-magnification correlative mapping exercise is extremely helpful in the later stages of correlative analysis and serves as a map (i.e., correlative cartography) that aids the generation of the correlative nanoscopy arrays (see the green, blue, and purple line boxes in B and C). (C) After the acquisition of the overall low magnification wide-field information, the region of interest is selected and super-resolution imaging can commence. Note the well-delineated actin structures obtained after GSD-SR imaging. The green, blue, and purple line boxes indicate the regions of interest that were next subjected to tomographic analysis (D–L). (D, G, and J) Represent the 2D transmission electron microscopy data acquired on 250 nm thick sections. (E, H, and K) The corresponding areas depicted under D, G, and L were next tilted, aligned, and modeled, resulting in a segmented dataset with actin as the structural feature of interest outlined (yellow color). (F, I, and L) Finally correlative optical super-resolution and 3D transmission electron microscopy models were generated in which the GSD-SR information shown in (C) was merged within the corresponding tomographic slices, showing the cross-correlation between light and electron microscopy information. Scale bars in (A) 100 μm, in (B) 10 μm, in (C) 2.5 μm, and in (D)–(L) 500 nm. (See the color plate.)

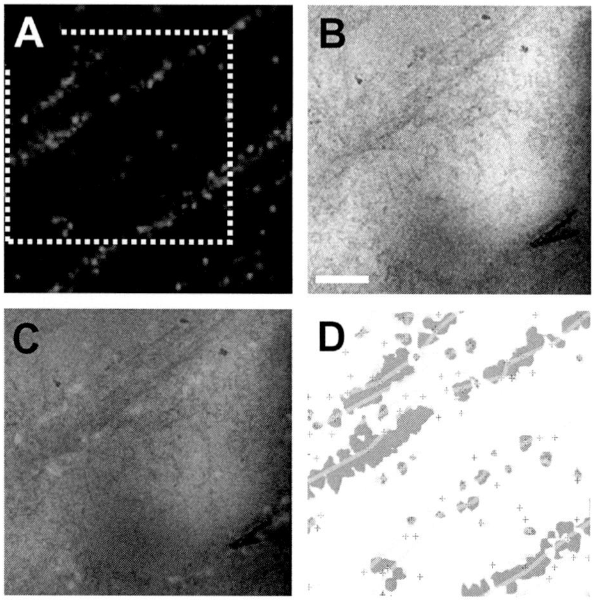

FIGURE 7.4

Image montage showing "predictive" simulation data using digital thresholding and employing near-neighboring algorithms on super-resolution images to "predict/calculate" electron microscopy morphology. (A) GSD-SR, (B) 2D TEM, (C) corresponding correlative GSD-SR and 2D TEM information, and (D) simulation outcome after applying a threshold filter (gray color) and connecting neighboring pixel information (crosses) via straight line connectors (yellow lines). Area analysed is indicated under (A) (white dotted line box under A). Notice the resemblance in structural information when comparing images (B)–(C) with (D). Scale bar, 500 nm. (See the color plate.)

7.2.4 SAMPLE PREPARATION FOR CORRELATIVE THREE-DIMENSIONAL TEM

After GSD-SR data had been collected, some careful sample preparation aspects have to be considered to guarantee subsequent relocation at the electron microscopy level. Critical to the success is the way to embed, trim the resin block, including how to section the sample in order to reduce the loss of the previous ROI and relocation marks. The following sequence of events is optimized for generating nanoscopy arrays on subcellular structure at the electron microscopy level.

1. Place the coverslip in a 6-well plate for the remaining steps after carefully peeling off the Twinsil® and separating the coverslip from the glass slide. Note that the use of Picodent Twinsil® allows the easy recovery of the glass coverslip containing the cells previously imaged under GSD-SR thereby avoiding any unwanted damage to the sample for further electron microscopy processing.

2. Gently wash the coverslip with warm PBS to remove the MEA.
3. Further fix the sample with 2.5% glutaraldehyde in PBS for 1 h and rinse with PBS twice for 10 min.
4. Stain with 1% osmium tetroxide in PBS for 20 min and wash 3 × 10 min with distilled water.
5. Dehydrate in ethanol at the following concentrations and times: 50%, 70%, 90% for 10 min each and 100% for 10 min twice.
6. Gradually infiltrate the sample with EPON resin.
7. The samples are then embedded using a site-specific and double-polymerisation method. At the end of the resin infiltration, apply a thin layer of resin to the bottom of the dish and carefully place two open-ended BEEM capsules centered with the grid patterns.
8. Place the well plate in an oven at 60 °C and allow the resin to polymerise for 3–5 h. The resin should not fully polymerise at this stage yet to prevent a separation in the final resin block.
9. Remove the well plate from the oven and fill up the embedding capsules with more resin, while the plate is still warm.
10. Return the well plate to the oven for final polymerisation overnight: the second polymerisation step produces a resin block of adequate size for handling and sectioning.
11. Retrieve the resin blocks by holding the embedding capsules tightly with a pair of tweezers and snapping blocks off the plate immediately after removing them from the oven or by warming up the plate on a hot plate: the monolayer of cells and the grid pattern remain on the surface of the resin block.
12. Relocate the ROI on the surface of the resin block with the aid of the grid pattern and the rest of the block is trimmed away using a razor blade.
13. Using an ultramicrotome, section ultra-thin (70 nm) or ultra-thick (180 nm) sections for conventional and tomography TEM imaging, respectively, and collect them onto a 3-window copper grid supported by a formvar film. Note, one can also section semithin sections of 300 nm, collect those on a glass slide and stain them with 0.5% toluidine blue for conventional light microscopy imaging. The section containing the information captured with the GSD-SR can then be retrieved for TEM by washing off the toluidine blue stain with drops of warm water and re-embedding the section by placing the glass slide on top of a BEEM capsule filled with fresh resin, centered with the section. The block containing the section of interest can be easily snapped off the glass slide after polymerisation and resectioned for TEM.
14. Poststain the grids with uranyl acetate (saturated solution in 50% ethanol) and Reynold's lead citrate for 10 min each.
15. Deposit for tomography purposes, after post-staining, deposit 10 nm colloidal gold, used as fiducial markers, by dipping the mounted and dried TEM grids in a 1:4 colloidal gold solution for 1 min, followed by three short washings in water. This step facilitates tracing and alignment of the individual images during the tilt series acquisition and tomogram alignment.

7.2.5 ET AND 3D-MODELING

The IMOD software package was used for tomogram reconstruction (eTomo) and segmentation (3dmod) (Boulder Laboratory for 3-D Electron Microscopy of Cells, University of Colorado, USA; Kremer, Mastronarde, & McIntosh, 1996). Individual actin filaments were manually segmented across the tomogram using the embedded drawing tools. Z-scale of the models were corrected to compensate for shrinkage during image acquisition.

1. Relocate ROIs with the aid of the corresponding DIC images by observing the morphology of the cellular landscape and the shape of the sections. Acquiring a mosaic of the ROI was also found very helpful as a larger area can be viewed and compared.
2. Record dual-axis tomography tilt series using the SerialEM software (Boulder Laboratory for 3-D Electron Microscopy of Cells, University of Colorado, USA) at 2° increments, from −60° to +60°, on the JEOL 2100 TEM at 200 kV.

7.2.6 CORRELATIVE ANALYSIS AND MORPHOMETRY

During the cross-correlation image analysis steps, it is crucial to ensure that the different pixel sizes are retained throughout all the image manipulation processes. In our hands, ImageJ (Abramoff, Magelhaes, & Ram, 2004) was used for image processing, correlative analysis and figure assembly (Fig. 7.3) at matching pixel resolution. The following procedure was followed to ensure accurate cross-correlative analysis:

1. Open the optical super-resolution and electron microscopy data in ImageJ.
2. Transfer and open all recorded end magnifications captured across the different imaging platforms into Microsoft Excel at the same time. This allows the calculation of the number pixels per length (e.g., WF vs. GSD-SR) and/or volume (e.g., EM) and hence allows to accurately correlate subcellular information.
3. Merge the optical and electron microscopy data—at matching pixel dimensions—via the "Image/Color/Merge-Channels" option under ImageJ. Next, final figure assembly can either be performed under ImageJ or Adobe Photoshop (Sedgewick, 2008) using the "replace color" and/or "layer/channel" tool.

7.3 INSTRUMENTATION AND MATERIALS

7.3.1 PRINTING RELOCATION AIDS ON CELL CULTURE GLASS COVERSLIPS

Instrumentation: Emitech K950C carbon coater; Heating oven (Binder WTC, Tubbingen, Germany).

Materials: Glass coverslips (22 mm × 22 mm) (ProSciTech, G422 No. 1); copper finder grids (Proscitech, GCU200F2).

Reagents: Ethanol (Merck, Cat No. 4.10230, CAS No. 64-17-5); hydrochloric acid (Analytical grade, Sigma Chemicals).

7.3.2 CELL CULTURE AND CONSIDERATIONS FOR SUBSEQUENT SUPER-RESOLUTION FLUORESCENT LABELING STUDIES

Instrumentation: Heraeus labofuge 200; Neubauer chamber.

Materials: Caco-2 cell line (American Type Culture Collection, ATCC, Item No. HTB-37); 75 cm^2 tissue culture flasks (In Vitro Technologies, COR430641); 6-well tissue culture plates (In Vitro Technologies, COR3516); tissue culture consumables (Corning Life Sciences).

Reagents: Ethanol (Merck, Cat No. 4.10230, CAS No. 64-17-5); Advanced DMEM media (Life technologies, 12491023); antibiotic/antimycotic (Life technologies, 15240104); heat-inactivated fetal bovine serum (FBS; Life technologies, 10100–147); L-glutamine (Sigma-Aldrich, G7513); trypsin-EDTA (Sigma-Aldrich, 59430C); Alexa Fluor® 647 Phalloidin (Life Technologies, A22287); paraformaldehyde (ProSciTech, C007); phosphate buffered saline (PBS; In Vitro Technologies, IVT3001302).

7.3.3 BRIGHT-FIELD, WIDE-FIELD FLUORESCENCE, AND SUPER-RESOLUTION MICROSCOPY IMAGING

Instrumentation: Leica SR GSD system consisting of an inverted DMI6000 B microscope equipped with the GSD/TIRF optical module and SuMo-stage; dedicated SR objective (Leica HCX PL APO 100/1.47); wide-field objectives HC PL APO 10x/0.40 and HCX PL FLUOTAR; iXon^{EM}+EMCCD camera (Andor Technology); TS-140 anti-vibration table (TS Table Stable Ltd®).

Materials: AF6000 Core Software Package (Leica Microsystems, LAS AF 11640800); GSDIM Software Module (Leica Microsystems, LAS AF 11888962); Single cavity microscope slides (ProSciTech, G341-12).

Reagents: Cysteamine (i.e., MEA) (Sigma-Aldrich, 30070); Picodent Twinsil® (Leica Microsystems, 11532691).

7.3.4 SAMPLE PREPARATION FOR CORRELATIVE THREE-DIMENSIONAL TEM

Instrumentation: heating oven (Binder WTC, Tubbingen, Germany); Leica Ultracut-7 ultramicrotome (Leica, Vienna, Europe).

Materials: standard embedding capsules (ProSciTech, RB001 BEEM®).

Reagents: copper grids, 200 mesh square (ProSciTech, GCu200); copper slot grid (Proscitech, GCU1M2M); copper grids, three slots (Proscitech, GCU2M3X1M); ethanol (see earlier); glutaraldehyde (ProSciTech, C001); lead nitrate (Sigma-Aldrich, 228621); osmium tetroxide (ProSciTech, C010); procure 812 embedding kit Epon resin (ProSciTech, C038); sodium citrate (ProSciTech, C077); sodium cacodylate (ProSciTech, C020); sodium hydroxide (Sigma-Aldrich, S8045); sucrose (Sigma-Aldrich, S9378); toluidine blue (ProSciTech, C078); uranyl acetate (ProSciTech, C079-F).

7.3.5 ET AND 3D-MODELING

Instrumentation: JEOL 2100 transmission electron microscope (JEOL, Tokyo, Japan).

Materials: DigitalMicrograph (v.1.82.366), SerialEM-software (v.3.1); IMOD full package (v.4.1, Boulder Laboratory for 3-Dimensional Electron Microscopy of Cells, University of Colorado, USA) (Kremer et al., 1996).

Reagents: Fiducial markers (Sigma Chemicals, 10 nm immunogold-labeled IgG).

7.3.6 CORRELATIVE ANALYSIS AND MORPHOMETRY

Instrumentation: Home-built 64-bit computer (16 GB RAM 1600 MHz, Intel i7-2700K Quad Core 3.5 GHz CPU) running Windows 7 Professional Software Package.

Materials: Microsoft Excel 2010 Professional Package; ImageJ (v1.47, NIH, USA) (Abramoff et al., 2004).

7.4 RESULTS AND DISCUSSION

7.4.1 RELOCATION: SAMPLE PREPARATION CONSIDERATIONS FOR CORRELATIVE SUPER-RESOLUTION AND TEM STUDIES

Essential to our approach is the provision of relocation of landmarks on the cell culture substrate allowing the swift relocation of a relative large number of cells within a given CLEM experiment (Brown, Mantell, Carter, Tilly, & Verkade, 2009). Evaporating carbon on a cell culture carrier is in our view superior, comes at low cost, and typically a sputter coater device is readily available in each facility dedicated to ultrastructural studies. Additionally, the glass coverslip used as cell culture carrier is a major parameter with high impact to super-resolution imaging. To obtain an optimal super-resolution image quality a uniform thin and flat coverslip is required. It is advised not to save on budget for this component of super-resolution imaging and hence to acquire the best commercially available glass coverslips.

Important to note, we found that the deposition of a gold or platinum-based sputter-coated footprint adversely affects GSD-SR imaging on the areas where the metal deposits are present (data not shown). This is most likely caused by the high-power lasers used under super-resolution microscopy resulting in heat conduction along the metal-based marks deposited onto the glass. Furthermore, we recommend the deposition of carbon marks in the center of the glass coverslip. This will beneficially affect super-resolution imaging quality. A better flatness and high stability is generally better achieved in the center of the glass coverslip. Furthermore, if desired, the carbon footprint is physically stable enough to allow subsequent coating with either 10 μg/mL collagen or 1 μg/mL fibronectin. Regardless of the extra coating the carbon layer is visibly thick enough ensuring relocation of the ROI under wide-field imaging conditions and within the ultramicrotome.

7.4.2 SAMPLE LABELING FOR WIDE-FIELD SUPER-RESOLUTION IMAGING AND SUBSEQUENT CORRELATIVE TEM

Various chemically synthesized and genetically encoded fluorophores are readily available for GSD-SR imaging. The success to attain a super-resolution image under GSD-SR conditions (i.e., by single molecule detection) depends on the inherent properties of each dye in combination with the imaging buffer of the fluorophore (Fölling et al., 2008; Heilemann et al., 2008). An updated listing of GSD-SR compatible fluorophores and their imaging environment can be found at the "Leica Microsystems-Life Sciences"-website (LeicaMicrosystems, 2014).

After the labeling procedure is finished, we prefer to mount our sample in PBS in the dark and store at 4 °C until SR-imaging. We noticed that glycerol-based mounting media usually adversely affect efficient single-molecule blinking and thereby super-resolution image quality. Depression slides are ideal to mount the sample in the presence of PBS for storage or PBS with MEA for imaging. Typically, 75 μL of buffer solution is sufficient to ensure that the sample is entirely submerged. Next, the careful and firm mounting of the sample is important to avoid any unwanted drift (movement) during GSD-SR data collection. For this Picodent Twinsil® is a preferred mean to keep the glass coverslip firmly in place. The two-component "glue" takes 10 min to set at room temperature.

Overnight storage of the sample in PBS after immunolabeling (and before super-resolution imaging) is uncritical in our hands. In-house test series showed that we can keep up samples for GSD-SR imaging upto 7–10 days without diminishing too-much fluorescence that blinking of the probes becomes impossible. However, for sparsely present structures, we do recommend imaging directly after the labeling procedure. We found that the detection of proteins attached to cytoskeletal fibers becomes more challenging. One way to circumvent this is to increase the concentration of the probe ensuring sufficiently stained structures.

The hardware solutions and stage compositions on offer for super-resolution by the different microscope suppliers are in general sufficient to control possible environmental-induced drift. In case, drift of the fluorescent data during the image acquisition is noticed (i.e., movement of fluorescent labeled structures) one should consider to include an additional fixation step with 4% PFA for 5 min after labeling. It is well known that for optical-based super-resolution microscopy, the cells should be fixed for long enough to avoid any unwanted movement during GSD-SR data collection. Typically, a drift in the order of 10–30 nm can be noticed in poorly fixed samples and will evidently adversely affect single molecule detection, and hence SR read-out.

To round up this section on labeling, we wish to underpin that besides the single labeling approach documented in this chapter the method can be easily extended on samples labeled for different structural components as no additional sample preparation steps are required that would affect the final outcome of the correlative arrays. In addition, the correlative labeling method outlined in this chapter concerns a noncombinatorial labeling approach, which allows to cross-correlate optical data from fluorescently labeled samples with electron microscopy data in the absence of electron-dense markers. This approach is generally quick and the samples are less inclined to sample-preparation artifacts since additional postlabeling or other sample preparation steps are avoided, or at least limited. Conversely, combinatorial labeling strategies permit the visualization of the tagged structure over the different microscopy platforms by applying additional label-rendering preparation steps for the respective beam-lines. For papers on correlative combinatorial versus noncombinatorial labeling approaches, we refer to the papers by Giepmans, Adams, Ellisman, and Tsien (2006), Giepmans (2008), Su et al. (2010), and Ellisman, Deerinck, Shu, and Sosinsky (2012).

7.4.3 THE WAY TO SUCCESSFUL GSD SUPER-RESOLUTION IMAGING

Any localization microscopy technique, such as GSD-SR, is based on single-molecule detection. The best possible image quality can be obtained by efficiently separating the active fluorophores and avoiding overlapping signals. Therefore, the major aim should be to match the fluorophore and embedding medium combination as well as the imaging parameters of the super-resolution system. Although the data acquisition of the super-resolution image is not the first step in the process, everything needs to be optimized for achieving the maximum performance in this step. The best "blinking" is derived discrete, nonoverlapping fluorophore spots, which are clearly distinguishable from any background. Nonblinking background can be to a certain extend excluded, but will still affect the localization precision. Therefore, the "pumping" step is of major importance and should be prolonged until the blinking is optimal. Switching too fast from the pumping to the imaging step and/or to a highly fluorescent background can even lead to the creation of some artifacts in the super-resolution image. If the signal intensity is too low, the fluorophores are hard to separate from the camera noise (low signal-to-noise ratio), Then, laser power, EMCCD gain and exposure time are the obvious parameters for optimization (see, *vide infra*). Please note that some samples do not need extensive pumping with high laser power and therefore a brief pumping step of few seconds with 5–15% laser power might be sufficient.

The camera pixels should not be saturated nor should the signals be too dim resulting in a low signal-to-noise ratio. Increasing the EMCCD gain (up to ~300 for an Andor 897) as well as matching the final laser intensity was the standard procedure to optimize the signal intensity. Typically a laser power of up to 30%/150 mW and an EMCCD gain of 300 is recommended. For precise localization, the software needs to distinguish background noise from real single molecule signals. Therefore, a major parameter is a threshold for background rejection, which can, e.g., be applied in photons per pixel. A typical threshold setting is 10 photons per pixel.

The correct determination of the threshold is a key step and can also and should be optimized in a postprocessing step while carefully monitoring the detection rate of single molecule signals in the raw data. Once determined the threshold can be used as a basis for subsequent images of the same sample type and only minor changes might be applied to accommodate the actual sample. We used a threshold around 10 photons per pixel for the super-resolution images shown in Figs. 7.3 and 7.4.

If a proper "blinking" cannot be achieved, it is in general not recommended to continue imaging and the focus should shift to the sample preparation part. Although, the sample was prepared from the start in a suitable fluorophore embedding, still a number of parameters can be optimized. The typical parameters are (i) concentration of components of the mounting medium; (ii) concentration of the antibody and/or probe; and (iii) the pH of the sample. As the staining and imaging parameters can vary significantly from cell to cell type, no universal set of parameters can be suggested. In our hands, the following optimization worked well and we recommend first the use of 30 mM MEA as a starting point. In case the detection of GSD-SR signal/blinking is "over-saturated" we tend to use a higher laser power and a higher concentration of MEA (i.e., 100 mM) to counter the effects. In the case of much lower events/blinking, we typically lower the concentration of MEA (i.e., 5–10 mM). This is also a particular situation that is valid during dual/double GSD-SR acquisition of a faintly and strongly fluorescent labeled structure. It is imperative that the cavity slide is entirely filled with MEA buffer and that there are no air bubbles present in the sample during mounting, as this will adversely affect the detection of single molecules. When imaging a dual labeled sample, it is advisable to use the longer wavelength laser line before imaging the shorter wavelength, i.e., 647 nm before 488 nm laser. If imaged in the reverse order, the signal of the longer wavelength fluorophore will be already bleached by the high power laser excitation of the shorter wavelength dye.

7.4.4 LINING UP CELL ARRAYS FOR SUBSEQUENT TEM MODELING STUDIES

ET allows the examination of 250 nm thick sections, instead of more routine 60 nm sections as used in conventional TEM. This three-dimensional imaging capability can image structures typically at 1–0.1 nm spatial resolution and with an axial resolving power in the order of 7–12 nm (Baumeister, 2002; Koster & Klumperman, 2003). Typically, the workflow includes the following image acquisition and data processing steps: (i) dual-axis tilting, (ii) alignment of the dual-axis tilting series, (iii) resolution-weighted back-projection of the tomograms, and (iv), modeling of the ROI(s) via computer-assisted tracking and/or manual segmenting within the tomograms. For this, we use the reliable SerialEM (recording), eTOMO (aligning), and 3dMOD (modeling) software packages developed primarily by David Mastronarde, Rick Gaudette, Sue Held, Jim Kremer, and Quanren Xiong at the Boulder Laboratory for 3-D Electron Microscopy of Cells (Kremer et al., 1996). The shareware software is readily available to all and comes with the different plugins for the different

Table 7.1 Direct comparison of actin filament thickness

Microscopy technique	Thickness ± SD (nm)
WF	387.3 ± 112.5 (n = 78)
SR	68.2 ± 21.3 (n = 186)
TEM	6.7 ± 1.1 (n = 100)

Morphometric data on actin filament (AF), thickness as determined by wide-field (WF), super-resolution (SR), and transmission electron microscopy (TEM). Data are derived from cross-correlative measurements on the same ROI and positions that were firstly imaged under WF conditions followed by SR microscopy and finally TEM. Note the tabulated data were obtained from the image set and nearby ROIs shown under Fig. 7.1B.
SD, standard deviation; n, *number of measurements; nm, nanometer.*

microscope manufacturers and/or computer operation system requirements. The software packages can be downloaded at http://bio3d.colorado.edu/imod/. The different packages undergo regular updates based on needs and user feedback.

7.4.5 CORRELATIVE ANALYSIS AND COMPLEMENTARY NANOSCOPIC INFORMATION

The final stage involves the successful translation (i.e., combining or merging) of the calibrated digital images from the different image capture platforms into one single "nanoscopic model" (Fig. 7.3). Key is the extensive documentation of the data capturing conditions for the different imaging platforms. For correlative image analysis, it is important that the digital data are entered at a matching pixel resolution before any further image processing steps can take place. In our hands, we recommend the professional image processing packages provided by ImageJ (Abramoff et al., 2004) or Adobe Photoshop (Sedgewick, 2008). The authors are aware that each laboratory or individual has preferred software suites to process data. The above listed packages, however, contain all required tools to adequately overlay data stacks without the risk of modifying information in the background during processing.

The data presented in Table 7.1 illustrate what correlative nanoscopy can bring to the researcher. Besides the obvious improvement in optical resolution attained, in which single fluorescent-labeled molecules can be seen decorating the bundles of actin filaments, the single actin filaments could be rendered visible only using electron microscopy. Those observations are underpinned by the measurements presented in Table 7.1. Moreover, when the two-dimensional optical arrays are translated into three-dimensional electron microscopy models one can start gathering balanced structural insights.

CONCLUSION

We present a straightforward method to combine LM-based super-resolution technology with TEM approaches to swiftly analyse cellular architecture in a cross-correlative manner within the nanometer regime (see also, SR_ET.wmv-movie on http://dx.doi.org/10.1016/B978-0-12-801075-4.00007-0). As such, the workflow outlined in this chapter allows one to generate correlative nanoscopic arrays in a high-throughput manner on biomolecular ultrastructure across length scales thereby offering an additional imaging tool set that permits the disclosure of novel structure–function information.

As discussed earlier, the nanoscopic arrays and data presented in Fig. 7.3 directly illustrate this point. The study of different actin configurations by means of correlative nanoscopy showed that the fluorescent signals were largely complementary to the information provided by the electron microscope. Evidently, once reaching 20–30 nm spatial resolution limit of the super-resolution microscope, better microscopy technology with a higher resolving power is required to beam-up the full make-up of the cell's interior. Especially, as per our example, where we could not resolve single actin filaments that have a typical thickness in the order of 6 nm (Table 7.1). In addition, in time we foresee that optical-based nanoscopy of single molecule events will be used to "forecast" electron microscopic information. In Fig. 7.4, for example, we endeavored to combine our correlative nanoscopy approach (Fig. 7.1) together with image data analysis software on single molecule data to connect near-neighboring events within our super-resolution images (Maji & Bruchez, 2012). Obviously, optical super-resolution imaging will never replace electron microscopy, but information as depicted in Fig. 7.4D may once result in the development of LM-EM software packages that aids cross-correlative analysis in an automated manner via pattern recognition of the data obtained at the single molecule optical level. Otherwise, researchers who do not have immediate access to electron microscopy may have the possibility to simulate a predictive EM model before considering the labor intensive and technical demanding exercise inherent to electron microscopy.

There is no better way to end this contribution than to quote Professor Stephan Hell during one of his conference presentations (Inventor photon-based super-resolution microscopy): "The ability to view life in nanometer dimensions via super-resolution technology opens the door to an understanding of intracellular life processes which was never thought possible before and which may lead to revolutionary discoveries on the subject of how diseases originate." There is no doubt that super-resolution imaging redefines the limits of optical microscopy. Moreover, no one will dispute that correlative fluorescence and electron microscopy technology permits one to bring the missing pieces of the puzzle on the sample's fine structure together in a way that is impossible with any one technique alone. And this is exactly what the current protocol has on offer: i.e., "bringing the best imaging and analytical capabilities of each microscope platform in unison."

ACKNOWLEDGMENTS

The authors acknowledge the facilities, as well as technical and administrative assistance from staff, of the AMMRF at the Australian Centre for Microscopy & Microanalysis of the

University of Sydney. The authors are also indebted to Leica Microsystems (Australia) and Jeol (Australasia) for their continuing support. Dr. M. L. acknowledges the support received by Leica Wetzlar. We wish to thank The University of Sydney (Research & Innovation) and the Charles Perkins Centre (Dr. M. Ainsworth—Chief Operating Officer) for funding some of the work reported herein. The authors are also indebted to Australian Research Council for granting support through the "Linkage Infrastructure, Equipment and Facilities" funding scheme (LE100100010).

REFERENCES

Abramoff, M. D., Magelhaes, P. J., & Ram, S. J. (2004). Image processing with ImageJ. *Biophotonics International*, *11*, 36–42.

Agrawal, U., Reilly, D. T., & Schroeder, C. M. (2013). Zooming in on biological processes with fluorescence nanoscopy. *Current Opinion in Biotechnology*, *24*, 646–653.

Baumeister, W. (2002). Electron tomography: Towards visualizing the molecular organization of the cytoplasm. *Current Opinion in Structural Biology*, *12*, 679–684.

Braet, F., Spector, I., Shochet, N. R., Crews, P., Higa, T., Menu, E., et al. (2002). The new anti-actin agent dihydrohalichondramide reveals fenestrae-forming centers in hepatic endothelial cells. *BMC Cell Biology*, *3*, 7.

Braet, F., Wisse, E., Bomans, P., Frederik, P., Geerts, W., Koster, A., et al. (2007). Contribution of high-resolution correlative imaging techniques in the study of the liver sieve in three-dimensions. *Microscopy Research and Technique*, *70*, 230–242.

Brown, E., Mantell, J., Carter, D., Tilly, G., & Verkade, P. (2009). Studying intracellular transport using high-pressure freezing and correlative light electron microscopy. *Seminars in Cell & Developmental Biology*, *20*, 910–919.

Chhabra, D., & dos Remedios, C. G. (2008). Actin: An overview of its structure and function. In C. G. dos Remedios, & D. Chhabra (Eds.), *Actin-binding proteins and disease: Vol. 8. Protein reviews* (pp. 1–15). New York: Springer.

Curthoys, N. M., Mlodzianoski, M. J., Kim, D., & Hess, S. T. (2013). Simultaneous multicolor imaging of biological structures with fluorescence photoactivation localization microscopy. *Journal of Visualized Experiments*, *82*, e50680.

Ellisman, M. H., Deerinck, T. J., Shu, X., & Sosinsky, G. E. (2012). Picking faces out of a crowd: Genetic labels for identification of proteins in correlated light and electron microscopy imaging. *Methods in Cell Biology*, *111*, 139–155.

Fernández-Suárez, M., & Ting, A. Y. (2008). Fluorescent probes for super-resolution imaging in living cells. *Nature Reviews. Molecular Cell Biology*, *12*, 929–943.

Fölling, J., Bossi, M., Bock, H., Medda, R., Wurm, C. A., Hein, B., et al. (2008). Fluorescence nanoscopy by ground-state depletion and single-molecule return. *Nature Methods*, *5*, 943–945.

Galbraith, C. G., & Galbraith, J. A. (2011). Super-resolution microscopy at a glance. *Journal of Cell Science*, *124*, 1607–1611.

Giepmans, B. N. (2008). Bridging fluorescence microscopy and electron microscopy. *Histochemistry and Cell Biology*, *130*, 211–217.

Giepmans, B. N., Adams, S. R., Ellisman, M. H., & Tsien, R. Y. (2006). The fluorescent toolbox for assessing protein location and function. *Science*, *312*, 217–224.

Heilemann, M., van de Linde, S., Schüttpelz, M., Kasper, R., Seefeldt, B., Mukherjee, A., et al. (2008). Subdiffraction-resolution fluorescence imaging with conventional fluorescent probes. *Angewandte Chemie (International Ed. in English)*, *47*, 6172–6176.

Herbert, S., Soares, H., Zimmer, C., & Henriques, R. (2012). Single-molecule localization super-resolution microscopy: Deeper and faster. *Microscopy and Microanalysis*, *18*, 1419–1429.

Jahn, K. A., Barton, D. A., Kobayashi, K., Ratinac, K. R., Overall, R. L., & Braet, F. (2012). Correlative microscopy: Providing new understanding in the biomedical and plant sciences. *Micron*, *43*, 565–582.

Jahn, K. A., Barton, D. A., Su, Y., Riches, J., Kable, E. P. W., Soon, L. L., et al. (2009). Correlative fluorescence and transmission electron microscopy imaging of the actin cytoskeleton of whole-Mount (breast) cancer cells. *Journal of Microscopy*, *235*, 282–292.

Jahn, K. A., & Braet, F. (2008). Monitoring membrane rafts in colorectal cancer cells by means of correlative fluorescence electron microscopy (CFEM). *Micron*, *39*, 1393–1397.

Kobayashi, K., Cheng, D., Huynh, M., Ratinac, K. R., Thordarson, P., & Braet, F. (2012). Imaging fluorescently labelled complexes by means of multidimensional correlative light and transmission electron microscopy: A practical approach. *Methods in Cell Biology*, *111*, 1–20.

Koster, A. J., & Klumperman, J. (2003). Electron microscopy in cell biology: Integrating structure and function. *Nature Reviews. Molecular Cell Biology*, (Suppl), S6–S10.

Kremer, J. R., Mastronarde, D. N., & McIntosh, J. R. (1996). Computer visualization of three-dimensional image data using IMOD. *Journal of Structural Biology*, *116*, 71–76.

LeicaMicrosystems. (2014). *Sample preparation for ground state depletion (GSD) super-resolution imaging, version 2.0 ed.*

Maji, S., & Bruchez, M. P. (2012). Inferring biological structures from super-resolution single molecule images using generative models. *PLoS One*, *7*, e36973.

Müller-Reichert, T., & Verkade, P. (2012). *Correlative light and electron microscopy* (1st ed.). Oxford: Elsevier.

Requejo-Isidro, J. (2013). Fluorescence nanoscopy. Methods and applications. *Journal of Chemical Biology*, *6*, 97–120.

Schermelleh, L., Heintzmann, R., & Leonhardt, H. (2010). A guide to super-resolution fluorescence microscopy. *Journal of Cell Biology*, *190*, 165–175.

Sedgewick, J. (2008). *Scientific imaging with photoshop: Methods, measurement, and output* (1st ed.). Berkeley: New Riders Press.

Sjollema, K. A., Schnell, U., Kuipers, J., Kalicharan, R., & Giepmans, B. N. (2012). Correlated light microscopy and electron microscopy. *Methods in Cell Biology*, *111*, 157–173.

Spiegelhalter, C., Laporte, J. F., & Schwab, Y. (2014). Correlative light and electron microscopy: From live cell dynamic to 3D ultrastructure. *Methods in Molecular Biology*, *1117*, 485–501.

Su, Y., Nykanen, M., Jahn, K. A., Whan, R., Cantrill, L., Soon, L. L., et al. (2010). Multidimensional correlative imaging of subcellular events: Combining the strengths of light and electron microscopy. *Biophysical Reviews*, *2*, 121–135.

CHAPTER

8 Correlative Cryo-Fluorescence and Cryo-Soft X-Ray Tomography of Adherent Cells at European Synchrotrons

Raffaella Carzaniga*[,1], Marie-Charlotte Domart*[,1], Elizabeth Duke[†], Lucy M. Collinson*

**Electron Microscopy Unit, London Research Institute, Cancer Research UK, London, United Kingdom*

[†]Diamond Light Source, Harwell Science and Innovation Campus, Didcot, Oxfordshire, United Kingdom

CHAPTER OUTLINE

[1]These authors contributed equally to the work.

Methods in Cell Biology, Volume 124, ISSN 0091-679X, http://dx.doi.org/10.1016/B978-0-12-801075-4.00008-2

Abstract

Cryo-soft X-ray tomography (cryo-SXT) is a synchrotron-hosted imaging technique used to analyze the ultrastructure of intact, cryo-prepared cells. Correlation of cryo-fluorescence microscopy and cryo-SXT can be used to localize fluorescent proteins to organelles preserved close to native state. Cryo-correlative light and X-ray microscopy (cryo-CLXM) is particularly useful for the study of organelles that are susceptible to chemical fixation artifacts during sample preparation for electron microscopy. In our recent work, we used cryo-CLXM to characterize GFP-LC3-positive early autophagosomes in nutrient-starved HEK293A cells (Duke et al., 2013). Cup-shaped omegasomes were found to form at "hot-spots" on the endoplasmic reticulum. Furthermore, cryo-SXT image stacks revealed the presence of large complex networks of tubulated mitochondria in the starved cells, which would be challenging to model at this scale and resolution using light or electron microscopy. In this chapter, we detail the cryo-CLXM workflow that we developed and optimized for studying adherent mammalian cells. We show examples of data collected at the three European synchrotrons that currently host cryo-SXT microscopes, and describe how raw cryo-SXT datasets are processed into tomoX stacks, modeled, and correlated with cryo-fluorescence data to identify structures of interest.

INTRODUCTION

Correlative light and electron microscopy (CLEM), in its many different forms, has answered biological questions that were previously intractable using either modality alone. As CLEM is adopted more widely, there has been a drive to improve the technique by streamlining sample preparation protocols to minimize artifacts and image cells as close to their living state as possible.

The combination of cryo-preparation and cryo-imaging maintains cells at the near-native state while protecting them from the microscope vacuum. Cryo-electron tomography (cryo-ET) has been used to image subcellular structure within thin peripheral regions of vitrified cells (~1 μm thick) (Berriman et al., 2009), and cryo-focused ion beam scanning electron microscopy has been used for serial imaging through small volumes (~5 μm thick) of vitrified cells and tissues (Schertel et al., 2013). The most recent advances in native-state imaging use X-rays rather than electrons in order to collect data from even larger volumes of vitrified material.

Cryo-soft X-ray tomography (cryo-SXT) (Carzaniga, Domart, Collinson, & Duke, 2013; Larabell & Nugent, 2010; Schneider et al., 2010) requires a source of X-rays at ~500 eV within the so-called "water window," which lies between the carbon *k*-edge (284 eV) and the oxygen *k*-edge (543 eV). This water window is critical because it exploits the natural absorption of X-rays by carbon at an energy where the sample is virtually transparent to oxygen-containing moieties (including ice). As a result, cell membranes are visible without the need for additional stains. Furthermore, entire cells can be imaged intact because soft X-rays have a penetration depth of ~10 μm through organic material. The trade-off for native-state ultrastructure is currently resolution, which is approximately an order of magnitude lower than cryo-ET at ~40 nm^3, depending on both the microscope optics and the quality of the sample preparation.

Cryo-SXT imaging of cells is inherently compatible with correlative workflows for several reasons. Fluorescent proteins are preserved during the freezing process and are also more resistant to photobleaching at cold temperatures (Schwartz, Sarbash, Ataullakhanov, McIntosh, & Nicastro, 2007). Light (brightfield and/or fluorescence) microscopy is an integral part of the cryo-SXT workflow, required to map the positions of cells on grids either pre- or postfreezing, to maximize X-ray data collection at the synchrotron beamline. Finally, there are no disruptive sample preparation steps (e.g., dehydration, ultrathin sectioning) between light microscopy and X-ray microscopy; so, the correlation accuracy of fluorescence and X-ray images is high. Thus, cryo-correlative light and X-ray microscopy (cryo-CLXM) is unique in providing nanoscale 3D information from entire cells at near-native state, enabling studies of cellular events that cannot be captured using light, electron, or X-ray microscopes alone.

8.1 RATIONALE

Light microscopy is optimal for imaging fluorescent proteins within live and fixed cells, tissues, and model organisms. Electron microscopy is optimal for imaging the ultrastructure of fixed cells, tissues, and model organisms at nanometer resolution. CLEM links the imaging modalities to give functional and structural information. However, some cellular structures are particularly susceptible to fixation artifacts and require cryo-preparation. These structures can be imaged in their entirety by cryo-SXT.

Cell biological questions that can be answered using cryo-SXT fall into several subcategories: those that benefit from the cell being fully hydrated (e.g., quantitative volumetric analysis of cells and organelles); those that require the avoidance of chemical treatments (e.g., for elemental analysis); and those that are adversely affected by sectioning (e.g., investigations of large complex 3D organelle networks). The application examples below are intended to give a flavor of the kind of biological study that would benefit from this novel imaging platform, though applications will undoubtedly expand as cell biologists adopt the technique.

Pioneering work from the Larabell group demonstrated the use of cryo-SXT for imaging intact yeast cells, which are optimally preserved by cryo-preparation. Volumetric measurements of fully hydrated organelles including nuclei, mitochondria, vacuoles, and lipid droplets have been made (Gu, Etkin, Le Gros, & Larabell, 2007; Larabell & Le Gros, 2004; Parkinson, McDermott, Etkin, Le Gros, & Larabell, 2008; Uchida et al., 2011) and used in a study of the effect of antifungal peptides on *Candida albicans* (Uchida et al., 2009).

Cryo-SXT data can be mined for compositional information, exploited by Hanssen et al. in studies of the malaria parasite *Plasmodium falciparum* within red blood cells. Iron-rich hemozoin crystals, deposited as the parasite matures, are easily distinguished within the carbon-rich cell (Hanssen et al., 2011, 2012). In other studies, nanoparticles made from titanium oxide (Ashcroft et al., 2008), silver-enhanced nanogold (Meyer-Ilse et al., 2001), and nanogold alone (Duke et al., 2013) have been used as tracers or immunolabels.

Complex organelle networks can be studied within intact mammalian cells because the soft X-rays penetrate through ~10 μm of organic material. Our cryo-CLXM study focused on autophagosomes (Duke et al., 2013), which are susceptible to chemical fixation and dehydration artifacts during sample preparation for TEM (Eskelinen, 2008; Eskelinen & Kovacs, 2011). A range of other organelles contribute to the formation of early autophagosomes including endosomes, the endoplasmic reticulum (ER), and mitochondria. Cryo-SXT highlights these interorganelle connections. Indeed, we were able to map multiple nanogold-labeled endosomes surrounding "hot-spots" of early autophagosome formation at the ER. Furthermore, in nutrient-starved cells, we were able to track mitochondrial networks over many microns (Duke et al., 2013). Mitochondrial tubulation is thought to protect the organelle from autophagosomal degradation (Gomes, Di Benedetto, & Scorrano, 2011; Rambold, Kostelecky, Elia, & Lippincott-Schwartz, 2011).

Currently, scientists who wish to use cryo-SXT must enter the world of the synchrotron light source. While the macromolecular crystallography community are regular users of synchrotrons, the cell biology community are generally less familiar with protocols for accessing and running experiments at these massive multiuser facilities. This chapter is thus intended to guide the scientist through the entire cryo-CLXM workflow, from conception to correlation (Box 8.1), covering sample preparation, microscope access, data collection, and data processing.

8.2 METHODS

8.2.1 CRYO-PREPARATION OF ADHERENT CELLS ON GRIDS

8.2.1.1 Principle

As with any imaging technique, cryo-SXT requires excellent sample preparation for successful data collection. The first steps in the protocol involve grid preparation, cell culture, fluorescence microscopy, plunge freezing, and grid storage.

For cryo-SXT of adherent cells, a substrate must be chosen that does not significantly absorb soft X-rays and can be blotted prior to freezing to leave a thin,

BOX 8.1 WORKFLOW FOR CORRELATIVE CRYO-FLUORESCENCE AND CRYO-SOFT X-RAY MICROSCOPY: FROM CONCEPTION TO CORRELATION

At the home laboratory

Live

- Apply for beamtime at the synchrotron
- Prepare or purchase holey carbon grids
- Grow cells on grids and transfect if required
- Image the cells by live fluorescence microscopy

Frozen

- Plunge freeze the grids in liquid ethane and store in liquid nitrogen
- Image the frozen grids by cryo-fluorescence microscopy, noting the position of cells of interest
- Ship the grids to the synchrotron in a dry nitrogen shipper

At the synchrotron

- Load grids into the microscope
- Trace back the position of cells of interest and image using the in-vacuum visible light microscope
- Acquire an X-ray mosaic image of the area of interest
- Determine the final field of view and perform a test rotation to establish the extreme tilt angles and increment steps
- Calculate dose and exposure time
- Run the tilt series

At the home laboratory

- Import the data into, e.g., IMOD software
- Align the tilt series (using fiducial markers)
- Perform BP and SIRT reconstructions
- Align the fluorescence and X-ray datasets
- Identify potential structures of interest
- Manually segment the structures and render into a 3D model
- Overlay fluorescence image with 3D model to back-check correlation

Carzaniga et al. (2013).

homogenous layer of liquid across the cell monolayer. The preferred support is a gold grid, which is inert and non-toxic to cells in culture, coated with a thin film of holey carbon. The grid design differs depending on the cryo-SXT microscope used. The microscopes at the ALBA and Diamond synchrotrons support conventional 3-mm TEM grids; the microscope at the HZB-BESSY II synchrotron currently accepts high-tilt grids (HZB-2) (Hagen et al., 2012) and will soon also accept 3-mm TEM grids. In both cases, reference patterns are used for orientation and to enable mapping of cell positions for correlative microscopy. An additional consideration is the spacing of the grid bars, which must be large enough to allow data collection at high tilts but not so large that the carbon film tears. The holey carbon film offers support to the cells while allowing liquid to pass through the film during blotting, and may be homemade or commercially produced. Homemade films can be produced in large quantities, but the hole size and pattern are less reproducible than the commercial alternatives (e.g., Quantifoil™ R2/2).

Cell density is a critical parameter for cryo-SXT. A confluency of ~50% at the time of freezing ensures that cells are sparse enough that they do not overlap at high tilt during cryo-SXT data collection. It should be noted that some cell lines do not grow well on carbon films and therefore require a higher cell density at plating; this will need to be optimized for each cell line. After plating and transfection, grids are imaged by fluorescence microscopy to map cells of interest. Grids are then transferred as quickly as possible to the plunge freezer. Fiducial markers are added immediately prior to freezing to prevent uptake into the cells by endocytosis or phagocytosis. Fiducial markers (usually 250-nm diameter gold particles or fluoronanospheres) are critical for alignment during tomographic reconstruction, with at least three particles required per field of view.

Manual or commercial vitrification devices are suitable for preparing cells for cryo-SXT (Dobro, Melanson, Jensen, & McDowall, 2010; Duke et al., 2013; Hagen et al., 2012), as long as the grid is only blotted from the back so as not to disturb the cells. Blotting is essential to create an ice layer thin enough for imaging, though the blotting process is rarely reproducible, in part due to differences in cell height and non-uniform distribution of cells across the film.

After freezing, good cryo-handling techniques are essential to ensure that the cells remain vitrified. A cryo-fluorescence screening step in the home lab is useful primarily to assess the quality of each freezing run. However, cryo-fluorescence mapping of all grids may be a requirement if traveling to a synchrotron that does not have a visible light microscope (VLM) with fluorescence capability inside the cryo-SXT microscope.

8.2.1.2 Protocol

Preparation of homemade holey carbon grids

1. Fifty milliliter of 0.5% (w/v) Formvar® in chloroform was made in a volumetric flask with a tight-fitting lid and stirred for 20 min at room temperature to dissolve. Twenty drops of a 1:1 glycerol:ddH_2O solution were added (Fig. 8.1A), and the mixture was shaken vigorously for 30 s to disperse the

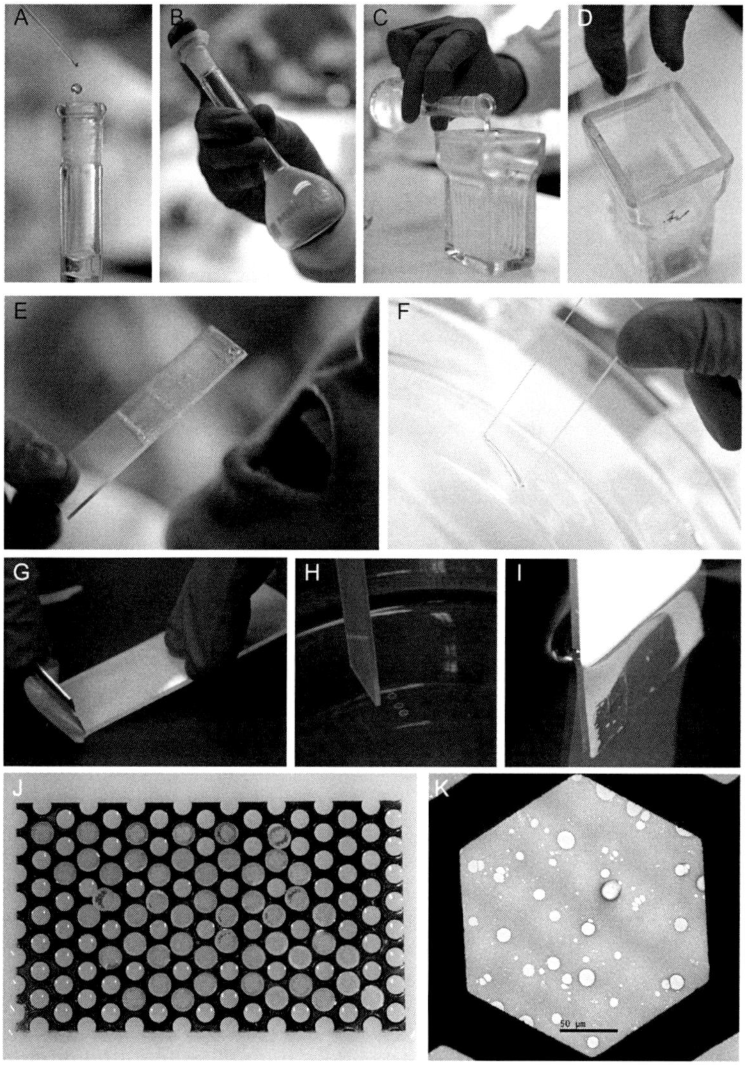

FIGURE 8.1

Preparation of homemade holey carbon grids. (A) Glycerol is added to Formvar®/chloroform in a volumetric flask. The mixture is shaken (B) and sonicated to produce glycerol microdroplets. The mixture is poured into a Coplin jar (C) and films are cast onto glass slides (D). The film is removed by "huffing" (E) and floating onto ddH_2O (F). A glass slide is coated with an adhesive label (G) and used to pick up grids placed onto the film (H and I). Grids are released from the slide and placed into a grid coating plate with Formvar® films facing down (J). The film is etched with methanol, coated with carbon, and the Formvar® removed using 1,2-dichloroethane to give carbon films with a suitable size and density of holes per grid square (K). Scale bar: 50 μm.

aqueous glycerol into the solvent, causing the solution to become milky (Fig. 8.1B). The mixture was then immediately sonicated for 60 s using a microprobe sonicator to reduce the glycerol microdroplet size (settings will depend on the type of sonicator used).

2. The solution was immediately transferred to a Coplin jar (Fig. 8.1C). A clean test slide was dipped into the solution and removed at constant speed (Fig. 8.1D). Several slides can be made from the solution, but this must be done quickly because the solution is unstable and the glycerol droplets slowly coalesce. Film-coated slides were allowed to air dry and were stored in a humid chamber (a metal box containing a dish of water).
3. A deep glass bowl was filled with ddH_2O. To remove the holey Formvar® films, the edges of the slide were scored with a clean razor blade, and the slide was "huffed" on each side (Fig. 8.1E).
4. The bottom edge of the slide was held on the meniscus of the water until the edge of the films began to detach from both sides of the slide. The slide was then pushed vertically downward as the films floated off onto the surface of the water (Fig. 8.1F). The color of the film should be silver (thickness around 50–60 nm). Several test grids were placed onto the film.
5. A glass slide was cleaned and covered with a shiny white address label, which was then trimmed closely around the slide using a razor blade (Fig. 8.1G). The bottom edge of the slide was placed at a short edge of the floating film, and the slide pushed vertically downward (Fig. 8.1H and I). The Formvar® film and grids should attach to the slide. The top of the film was submerged, and then the slide was removed vertically from the water at constant speed. The slide was air dried.
6. One of the test grids was checked in a transmission electron microscope (TEM) for appropriate size and distribution of the holes (10–20 holes per grid square on a 100-mesh grid). If the conditions were satisfactory, then more grids were coated using the remaining films; otherwise, the process was repeated from Step 1.
7. Grids were released from the slide and placed on a grid-coating plate with the Formvar® film facing down (Fig. 8.1J). The holder was placed onto No.4 Whatman paper in a glass Petri dish, and the filter paper was soaked with methanol to remove any glycerol and sharpen the edges of the holes. After 10 min, the holder was moved to dry filter paper and a grid was checked again in the TEM to evaluate the "etching" process.
8. The grids were then coated with 5–10 nm of carbon in a high-resolution ion beam coater, and the Formvar® film removed by incubation on 1,2-dichloroethane for 2 h (in a glass Petri dish as in Step 7). Finally, the grids were checked in the TEM (Fig. 8.1K).

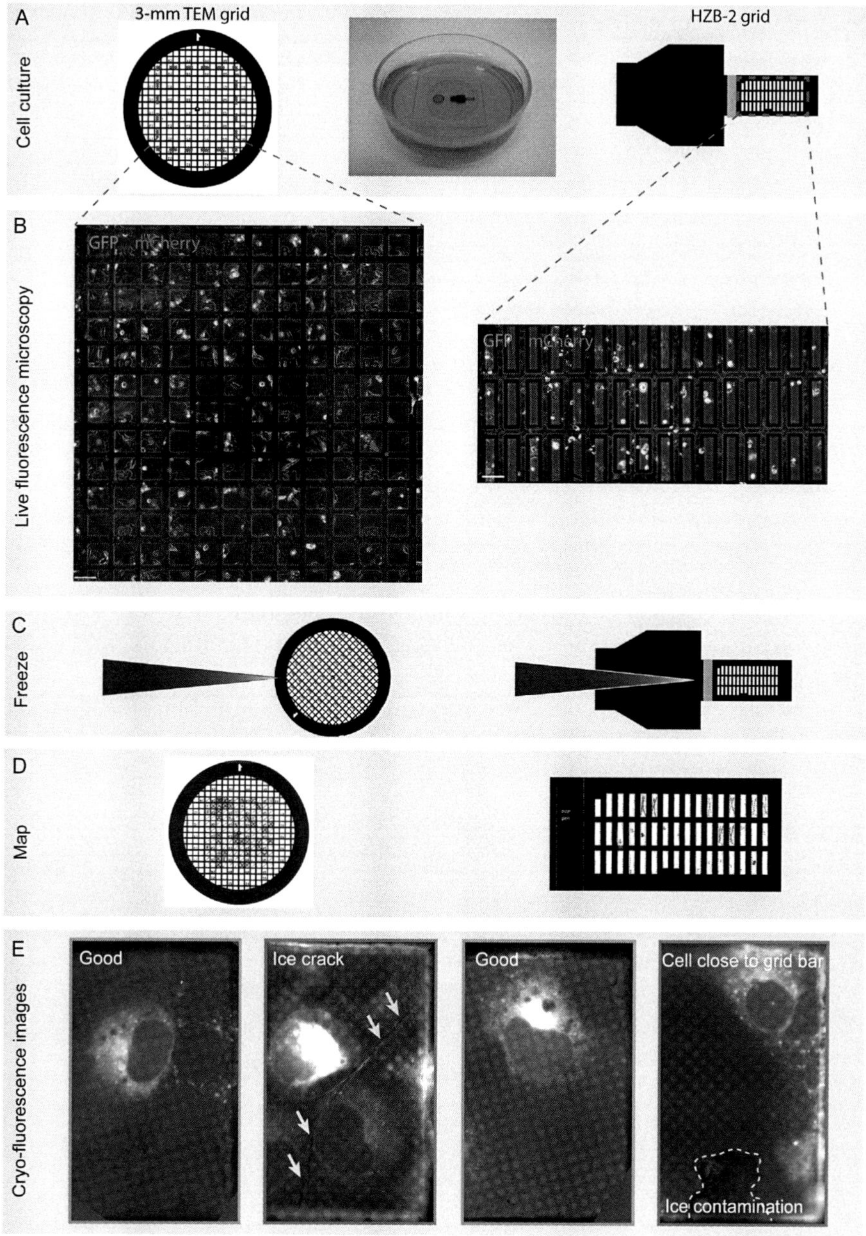

FIGURE 8.2 See legend on next page.

Growing adherent cells on grids

9. When using HZB-2 grids, a thin line of liquid blocker pen (PAP pen) was drawn on both sides of the grid, next to the observation region, using a cocktail stick (Hagen et al., 2012) (Fig. 8.2A; green line).
10. The grids were glow discharged with the carbon film facing upward in a glass Petri dish.
11. Grids were placed in cell culture dishes containing cell culture medium with the carbon film facing upward. Cells were added and allowed to settle for 24 h before transfection (if required).

Fluorescence microscopy

12. Prior to freezing, the grids can be imaged with an inverted (confocal) fluorescence microscope equipped with a temperature- and CO_2-controlled chamber for live cell imaging. To achieve optimal resolution, the grid should be inverted so that the cells are facing downward and placed in a glass-bottom dish (e.g., 35 mm MatTek dish) containing fresh cell culture medium (Fig. 8.2A). In this orientation, the carbon film does not block the light path to the cells, which would otherwise considerably reduce the fluorescence signal.
13. Phase contrast or differential interference contrast images were collected to assess the integrity of the carbon film and the cell confluency. For correlative work, fluorescence images were collected to permit quantification of transfection efficiency and fluorescent protein expression levels prior to freezing. A map of the grid was also collected using a tile scan, to aid in selection of cells of interest for cryo-fluorescence and/or cryo-SXT imaging (Fig. 8.2B).

FIGURE 8.2

Workflow for cryo-preparation of adherent cells on grids. (A) Diagrams representing the grids used for cryo-SXT: a 3-mm TEM finder grid used at ALBA and Diamond, and an HZB-2 grid used at HZB-BESSY II. The green line on the HZB-2 grid represents the position of the liquid blocker. Adherent cells are grown on the grids, which are imaged using confocal fluorescence microscopy immediately prior to freezing (B). The images shown here are composed of merged GFP (green), mCherry (red), and phase contrast images and represent a 1.7-μm thick confocal *z*-section acquired using a Ph2 20 × dry objective. The left image consists of 3 × 3 tile scan images of HEK-293A cells, the right of 3 × 1 tile scan images of COS-7 cells. (C) Diagram indicating the recommended positioning of the tweezers for plunge freezing. (D) Illustrative grid maps assembled after live fluorescence and cryo-fluorescence screening. The highlighted grid slots were selected by live fluorescence imaging for optimal cell confluency and by cryo-fluorescence for cells expressing the fluorescent markers/proteins of interest in regions of suitable ice (Carzaniga et al., 2013). (E) Areas with low cell confluency, good fluorescence levels, good positioning of the cell (away from the grid bars), and optimal ice thickness are selected (green frame). Areas with cracks in the ice (arrows) or ice contamination (dense crystals—dashed white line) or cells too close to the grid bars were discarded (red frame, right top corner). Scale bars: (B) 100 μm; (E) 10 μm. (See the color plate.)

Plunge-freezing grids

14. The stock solution (3.60×10^8 particles per mL) of 250-nm gold nanoparticles was resuspended and dispensed into 1-mL aliquots. The gold was allowed to settle, and immediately prior to use, 900 μL of the supernatant was removed from the aliquot.
15. The 3-mm TEM grids were picked up with the tweezers taking great care not to bend the grid (Fig. 8.2C), and the tweezers were then placed into the plunge freezer (held at 70% relative humidity and 37 °C). Immediately prior to freezing, 1.5 μL of 250-nm gold fiducial aliquot was added to the grid. The grids were then manually blotted (e.g., Vitrobot Mark IV; FEI Company) or automatically blotted (e.g., GP; Leica Microsystems) using Whatman® No.1 filter paper, and rapidly plunge frozen in liquid nitrogen-cooled liquid ethane (held at −174 °C).

Note: HZB-2 grids need further attention. The tweezer tips need to be positioned above the PAP pen line (Fig. 8.2C), and this area needs to be carefully blotted on both sides to avoid ice formation that would hinder insertion of the grid into the cryo-holder of the HZB-BESSY II microscope. One microliter of 250-nm gold fiducial was then added to the grid before blotting and plunge freezing (as described above).

Grid storage

16. The 3-mm TEM grids were transferred from the liquid ethane into cryo-grid boxes under liquid nitrogen and stored in a 50-mL centrifuge tube weighted with a metal nut in liquid nitrogen. HZB-2 grids were transferred to 2-mL cryo-tubes and stored in cryo-canes in liquid nitrogen. Note that it is wise to adopt a sensible naming convention with easy-to-read labels, as it is common to ship upward of 50 samples to the synchrotron.

8.2.2 CRYO-FLUORESCENCE MICROSCOPY (OPTIONAL STEP)

8.2.2.1 Principle

A cryo-fluorescence mapping step (Fig. 8.2D) can be added in the home laboratory to locate cells of interest on the grids, screen for artifacts and evaluate ice thickness (Fig. 8.2E). Although ice thickness cannot be accurately quantified in the light microscope, it can be estimated based on cell morphology (overblotting leads to thin ice and disrupted cell morphology) and image quality (underblotting leads to thick ice, causing scattering of light, leading to reduced fluorescence intensity and poor focus). Cryo-fluorescence screening is critical for correlative imaging if the cryo-SXT microscope does not have an in-vacuum fluorescence microscope (Fig. 8.3) and also helps to maximize throughput at beamlines that do have an in-vacuum VLM.

8.2.2.2 Protocol

1. Grids were transferred under liquid nitrogen to a standard cryo-cassette (3-mm TEM grids) (van Driel, Valentijn, Valentijn, Koning, & Koster, 2009) or a modified cryo-cassette (HZB-2 grids) inside a pre-cooled CMS-196 cryo-correlative stage (Linkam Scientific Instruments, UK). The cryo-cassette was

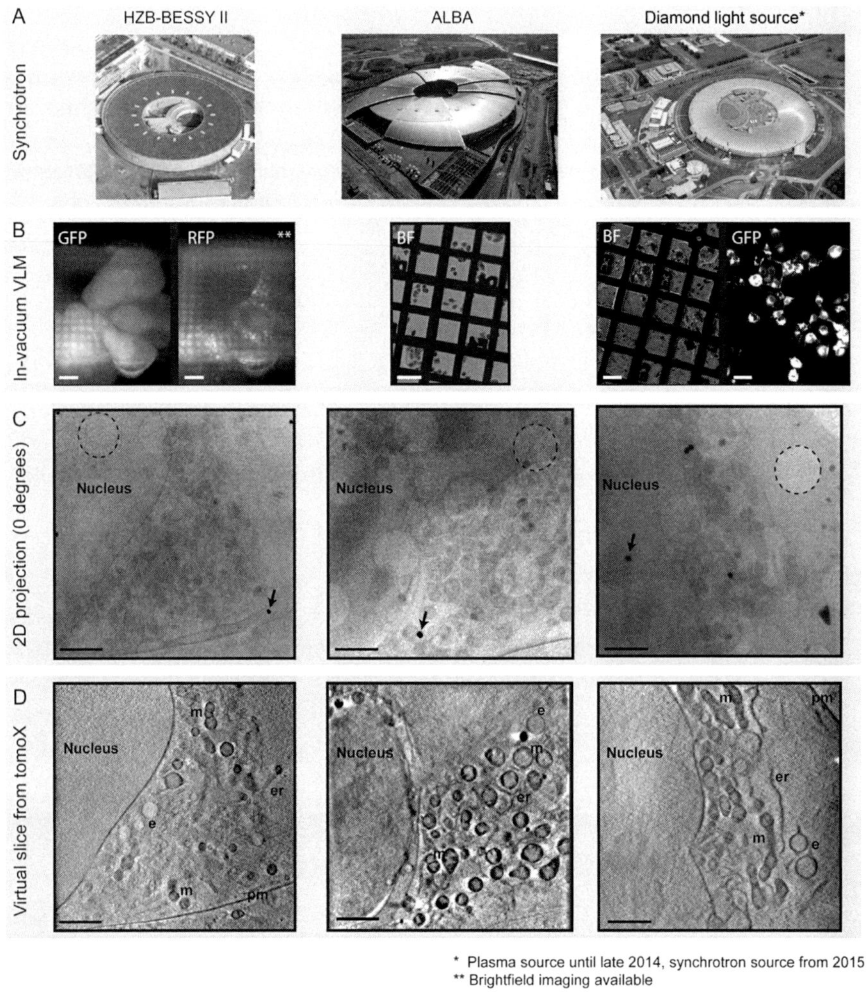

FIGURE 8.3

Illustrative data from adherent cells imaged at the European cryo-soft X-ray microscopes. (A) Cryo-SXT microscopes are hosted at the U41-TXM beamline at HZB-BESSY II in Germany (image courtesy of https://www.helmholtz-berlin.de/aktuell/pr/mediathek/bildarchiv/#1393587589561_4); the BL09 Mistral beamline at ALBA in Spain (image courtesy of ALBA), and the Diamond Light Source in the UK (image courtesy of Diamond Light Source). (B) The cryo-SXT microscopes have in-vacuum visible light microscope (VLM) for brightfield navigation (all) and imaging of fluorophores (BESSY, Diamond). (C) 2D projection images of cells acquired with soft X-rays at each of the synchrotrons. The dashed circle highlights Quantifoil™ holes and the arrows indicate the 250-nm gold fiducials. Cell features can already be recognized, including the nucleus and nuclear envelope. (D) Virtual slices from the processed tomoX stacks resemble TEM images even though no heavy metal stains were used during sample preparation. Organelles including the nucleus, mitochondria (M), endoplasmic reticulum (er), endosomes (e), and plasma membrane (pm) are visible. Scale bars: (B, left) 10 μm; (B, middle and right) 100 μm; (C and D) 2 μm.

placed onto the cooled viewing platform and the cryo-stage was mounted on a widefield epifluorescence light microscope (Axio Scope.A1; Zeiss, Cambridge) (Fig. 2 in Chapter 17).

2. Grids were imaged at −196.5 °C using a long working distance objective (EC Epiplan-Neofluar 100×/0.75) to locate suitable cells for cryo-SXT imaging (Fig. 8.2E, green boxes) and assess freezing artifacts including cracks in the ice (Fig. 8.2E, white arrows) and ice contamination (Fig. 8.2E, dashed white line). Positions of cells of interest were recorded and brightfield images were collected to show the carbon film and grid bars to aid in relocation of cells and overlay of images during correlation. Fluorescence images were recorded for correlative work.
3. After screening, grids were returned to liquid nitrogen storage.

8.2.3 CRYO-SOFT X-RAY TOMOGRAPHY

8.2.3.1 Principle

At the current time, the majority of biological soft X-ray microscopes use a synchrotron radiation source to supply the X-rays. However, with recent developments in the field, it is now possible to use alternatives such as nitrogen plasma sources to power the microscopes.

There are currently three cryo-soft X-ray microscopes operating at synchrotron facilities in Europe (Fig. 8.3): at the HZB-BESSY II electron storage ring in Berlin, Germany; at the ALBA Light Source in Barcelona, Spain; and at the Diamond Light Source in the UK, where the microscope is running on a nitrogen plasma source. Beamline construction at Diamond is underway, and the microscope running with a synchrotron source should become available for users in 2015. In the United States, the beamline XM2 at the Advanced Light Source, Berkeley, California, operated by the National Centre for X-ray Tomography (http://ncxt.lbl.gov/) provides facilities dedicated to cryo-soft X-ray tomography of biological cells. Elsewhere in the world, beamlines are at various stages of design and construction in order to take advantage of the increasing number of biological users keen to apply this new technique to answer their own biological questions. Each of these synchrotron-hosted cryo-soft X-ray microscopes is slightly different, though all are governed by the same laws of physics.

To gain access to the microscopes, regular calls are made for which anyone can submit a proposal. Each facility has their own mechanism, with proposals judged on scientific merit and beam time on the microscope awarded accordingly. It is always advisable to contact the local experts for advice on the application process. Access to the microscope is free at the point of use for academics who commit to publishing their results in peer-reviewed journals, and there are often EU funding streams available for travel and subsistence for the experiments. At the current time, BioStruct-X (http://www.biostruct-X.eu) provides this.

The practice developed for shipping samples to the synchrotron borrows heavily from techniques that are now standard practice in macromolecular crystallography, where synchrotron data collection on cryo-preserved samples is the norm. Samples are shipped to the synchrotron in a dry nitrogen shipper. These Dewars have an adsorbent matrix layer that, when precooled with liquid nitrogen, remains cold for an extended period (at least a week if the Dewar is not opened). As the Dewars are not pressurized, they are acceptable for transportation by most couriers.

It should be noted that each synchrotron X-ray microscope is different. The comments and suggestions given here relate primarily to the European facilities, which are all in-vacuum microscopes that accept samples grown on EM grids. At each facility, a significant amount of support is provided by the staff, from sample loading to acquiring data for the user.

8.2.3.2 Protocol

Transporting grids to the synchrotron

1. Pre-cool the dry shipper by filling it with liquid nitrogen until the central bore is full. Allow to settle for several hours and refill if necessary. Repeat until the level does not drop, indicating that the matrix is full. Just before shipping, pour any remaining liquid nitrogen from the central bore. Transfer the storage tubes into the dry shipper as quickly as possible to prevent thawing.
2. Place the dry shipper into its carry case and ship to the synchrotron. Note that each synchrotron has a preferred method of handling the arriving Dewars, and busy synchrotrons often have tens of Dewars a day passing through the Goods Handling Office, so correct labeling of the Dewar is critical. Check details with local facility staff.

Sample loading at the synchrotron

3. Pre-cool the sample stage and sample loading system with liquid nitrogen. Transfer the cryo-tube/grid box from the dry shipper and load it into the workstation under liquid nitrogen. It is very important when handling the grids that they do not get bent.
4. At HZB-BESSY II, the grid is loaded into a cryo-holder that is inserted directly into the microscope via a vacuum load lock mechanism. The interstitial space is then pumped manually by switching pumps on/off and manually opening valves, and the cryo-holder is inserted directly into the data collection position. At ALBA and Diamond, multiple grids are loaded into sample cartridges on a shuttle. The shuttle is then placed inside a special transfer chamber that has been precooled with liquid nitrogen and maintained at vacuum. The user is taken through this process via a graphical user interface on the workstation that controls the necessary pumps and valves. This transfer station is then loaded onto the microscope and another graphical user interface guides the user through the necessary vacuum steps that ultimately allow the shuttle to be inserted into the microscope chamber. The first grid to be imaged is then selected using the robot interface and positioned at the data collection position.

Navigation to the area of interest

5. Each of the cryo-SXT microscopes contains a VLM (Fig. 8.3B), which aids in the identification of cells suitable for data collection with X-rays. In the case of

BESSY and Diamond, the VLM is also set up to detect fluorescence. Use should also be made of the live fluorescence and cryo-fluorescence grid maps created prior to shipping the samples to the synchrotron (Fig. 8.2B and D).

6. Once a cell is identified, X-rays are used to set up the tilt series for data collection (Fig. 8.3C). A montage or tiled array of X-ray images can be acquired to image large fields of view, locate the cell, and determine imaging conditions. Key features to avoid are:
   a. Cells too close to a grid bar (Fig. 8.2E, right panel). When the cell is rotated, the grid bar comes into the field of view. At ALBA and Diamond, vertical grid bars are problematic; at BESSY, the horizontal bars should be avoided.
   b. Thick ice, leading to poor X-ray penetration and dark images.
   c. Cracked ice (Fig. 8.2E), which is unstable during data collection, probably due to heating effects.
   d. Overlapping cells, which interfere at high tilt.
   e. Absence of fiducial markers, which hinders image alignment.

Evaluation of dose and exposure; tilt series acquisition

7. To set up the tilt series, the extreme high-tilt angles must first be determined. Each microscope has physical limits on the rotation of the grid in order to avoid clashes between the grid/grid holder and the objective zone plate optic. Rotation of approximately $\pm 65^\circ$ may be physically possible, but the increased path length of the X-rays at high-tilt angles remains the main limiting factor. At high tilt, the increased path length results in fewer X-rays reaching the detector and dark images that lack contrast and therefore biological information. It is possible to partially offset the longer path length of the X-rays at higher tilts by adjusting the exposure time according to tilt angle. Thus, the decision on extreme tilt angle is made with consideration to balancing X-ray dose on the sample versus the amount of additional information obtainable.
8. The tilt series is collected as a series of "still" images with the cell rotated by a fixed amount between each image (usually 1° steps). As X-rays are damaging to the sample, the exposure time per image is selected in order to gain sufficient information in each image without damaging the sample during the course of data collection. It is therefore important to monitor dose effects by evaluating a zero degree image taken before and after the tilt series. Currently, damage is evaluated by comparing these two images "by eye." Susceptibility to damage varies from cell to cell, and may be related to cell morphology, chemistry, or vitrification quality.

8.2.4 DATA PROCESSING

8.2.4.1 Principle

The raw tilt series acquired at the synchrotron is processed to give a series of sequential images through the volume of the sample, following the same principles established for electron tomography (ET). Indeed, at the present time, reconstructions are performed using softwares developed for ET that take no account of the specific properties of X-rays (rather than electrons) and the thickness of the sample being imaged (approximately an order of magnitude thicker than for ET). Therefore, there is scope for developments in image processing that may deliver improved resolution and biological information in the final tomograms.

Nevertheless, programs such as IMOD (Kremer, Mastronarde, & McIntosh, 1996) using either Back Projection (BP) or Simultaneous Iterative Reconstruction Technique (SIRT) algorithms deliver impressive results. The protocol below is intended as a general guide to the reconstruction workflow (with consideration to specific cryo-SXT settings) rather than an exhaustive method. The user should refer to the IMOD home page (http://bio3d.colorado.edu/imod/), etomo tutorial (http://bio3d.colorado.edu/imod/doc/etomoTutorial.html), and the introduction to 3dmod (http://bio3d.colorado.edu/imod/doc/3dmodguide.html) for further information.

8.2.4.2 Protocol

Data reconstruction using IMOD

1. Open a new tomogram (.st file), select Single axis and specify the *Pixel size* (e.g., 16 nm), *Fiducial diameter* (e.g., 250 nm), and *Image Rotation* (currently 90° for HZB-BESSY II and 0° for ALBA and Diamond), as well as the starting angle (e.g., −55°) and increment (e.g., 1°). *View Raw Image Stack*, and then *Create Com Scripts.*
2. In the Pre-processing window: *Find X-rays, Create Fixed Stack and Use Fixed Stack → Done.*
3. In the Coarse Alignment window: *Calculate Cross-Correlation, Generate Coarse Aligned Stack* and *View Aligned Stack in 3dmod → Done* (Fig. 8.4A).
4. In the Fiducial Model Gen. window: *Seed Fiducial Model.* Using 3dmod, untick *Autocenter*, and in the image window select gold fiducials, pressing *N* for each new fiducial (Fig. 8.4B). Check that each fiducial remains within the field of view through the whole tilt series. Select at least three fiducials if possible, and press *S* to save the model. In etomo, press *Track Seed Model* and then *Fix Fiducial Model.* In the image window, riffle through the slices and correct any fiducials that have not been selected properly, and press *S* to save → *Done* (Fig. 8.4C).
5. In the Fine Alignment window: *Compute Alignment* and *View/Edit Fiducial Model.* Move points by residual as appropriate and repeat *Compute Alignment* process until no residuals remain to be examined. Try to reduce the Residual error mean below 0.5 → *Done* (Fig. 8.4D).
6. In the Tomogram Positioning window: Use the GPU if present for fast processing, set a Sample tomogram thickness of around 1200, check Use whole tomogram and use Binning 3, then *Create Whole Tomogram. Create Boundary Model*, and Edit-Image-Flip in the 3dmod window (Fig. 8.4E). In the image window, draw lines at the top and bottom of the cell at three different planes through the stack (giving six lines in total) (Fig. 8.4F). Press S to save model, and *Compute Z Shift & Pitch Angles* in etomo. *Create Final Alignment → Done.*

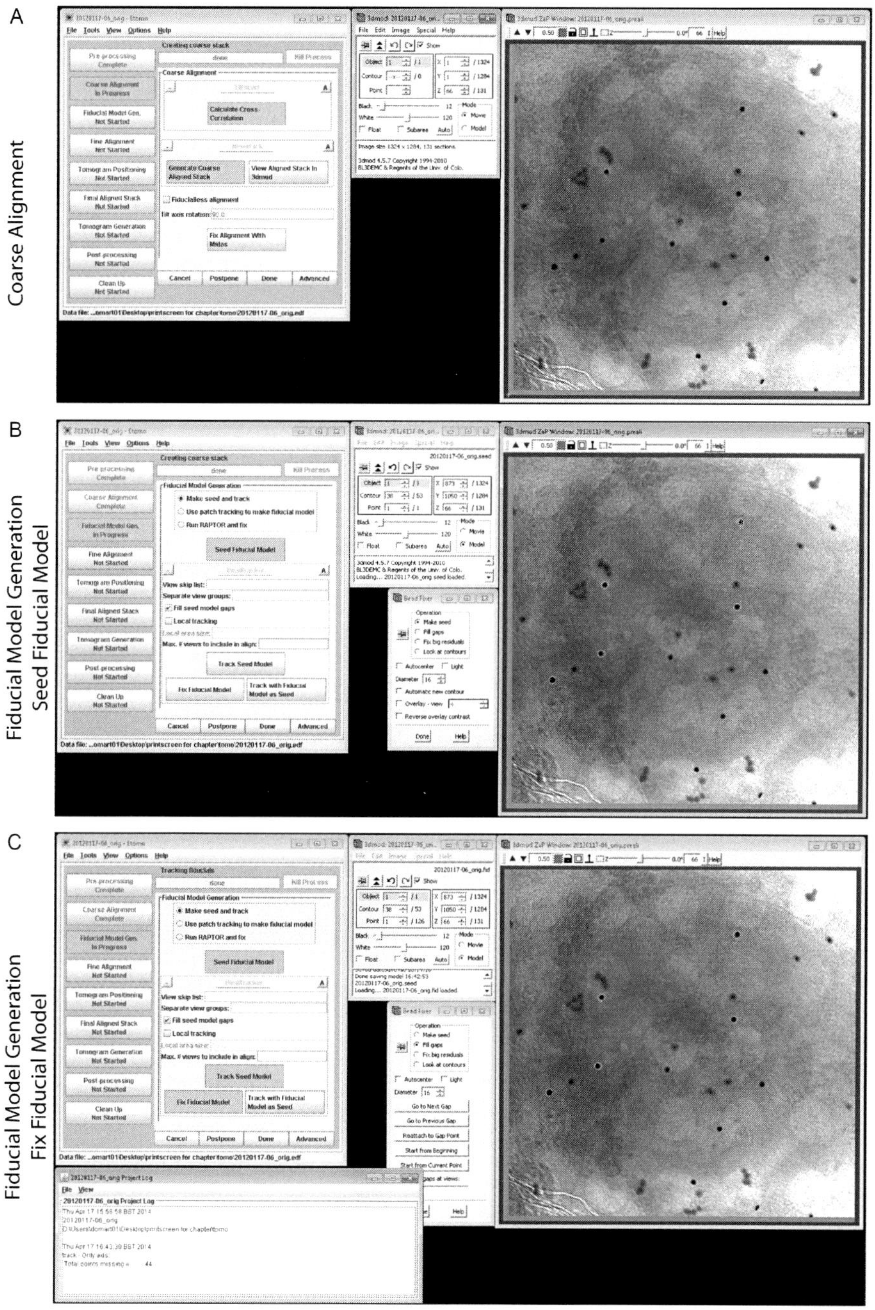

FIGURE 8.4—Cont'd.

D

Fine Alignment

E

Tomogram Positioning

F

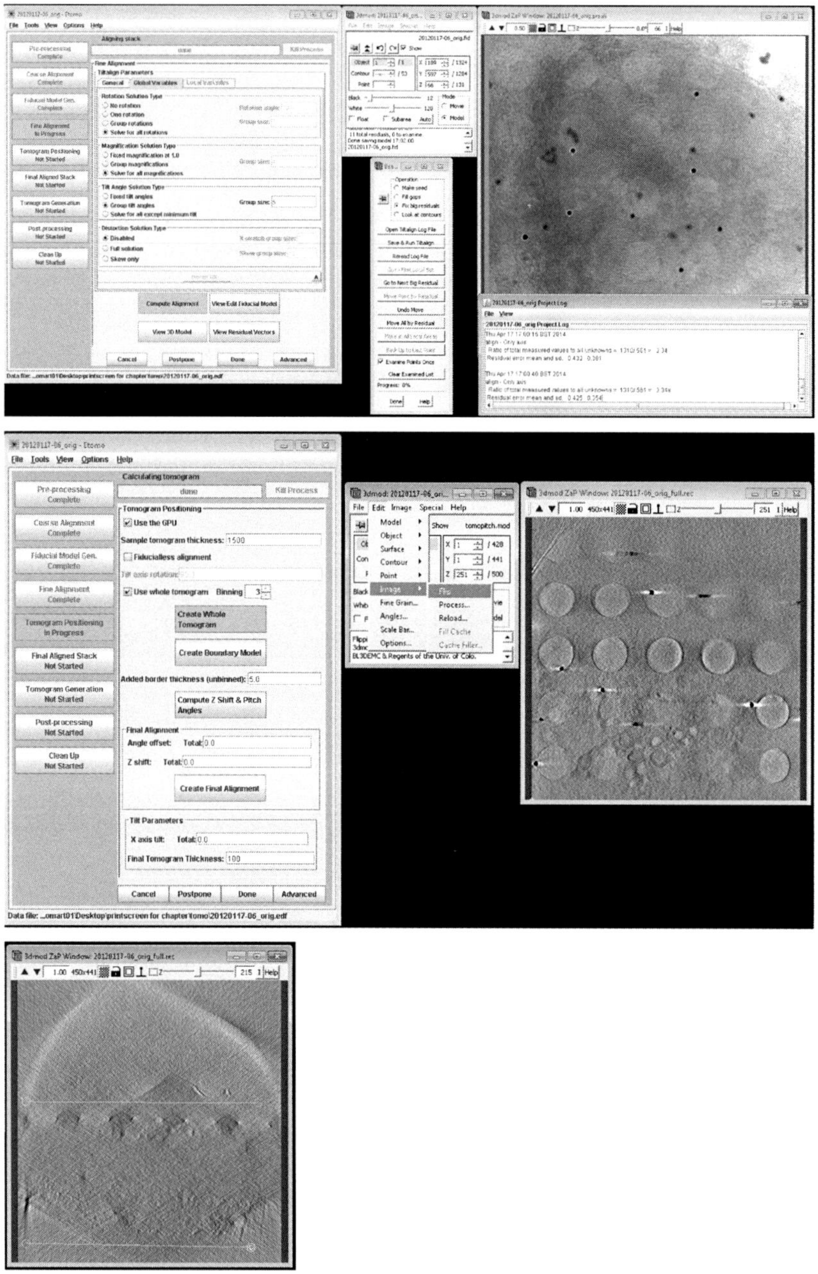

FIGURE 8.4 See legend on next page

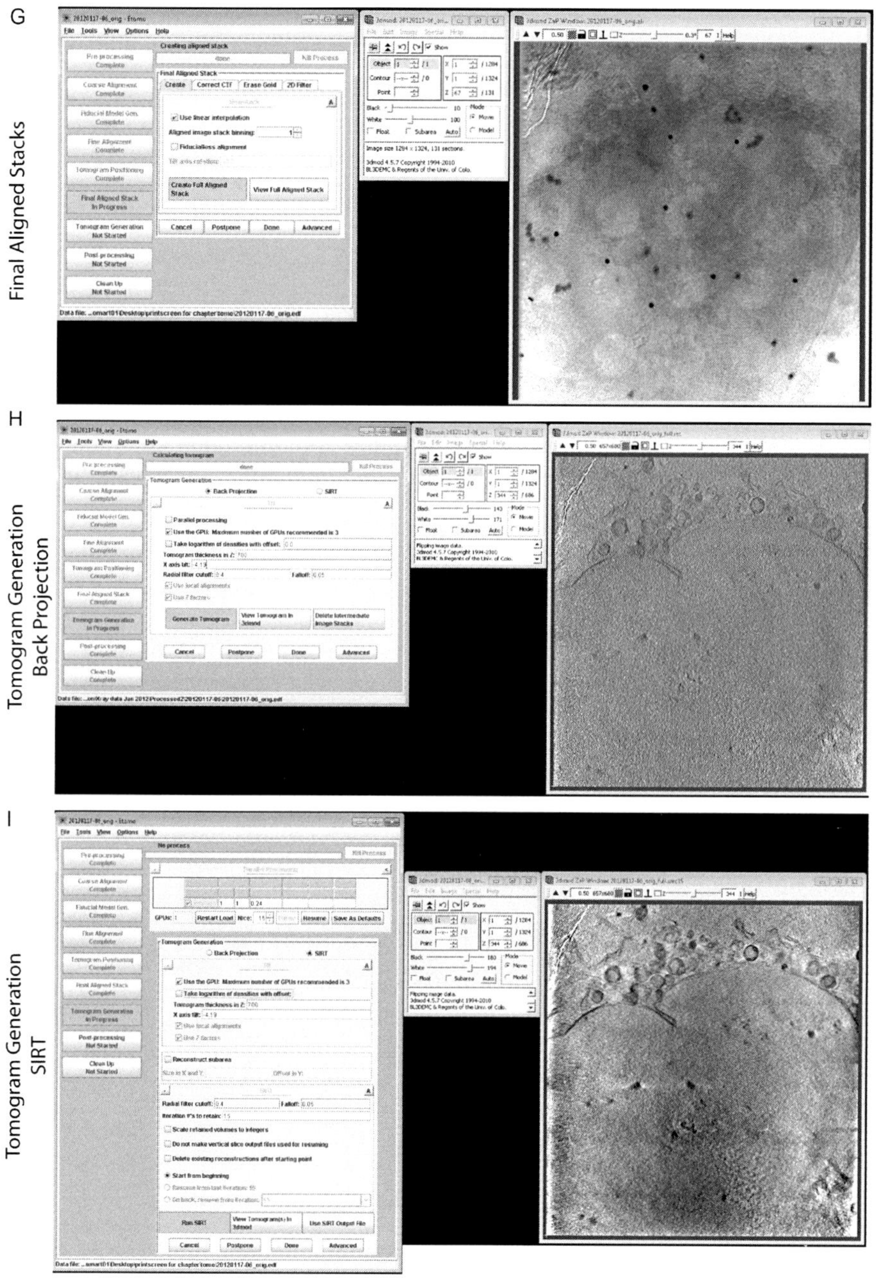

FIGURE 8.4

Data reconstruction using IMOD. Screenshots of the IMOD windows showing the key steps of data reconstruction. See text for details.

7. In the Final Aligned Stack window: Use linear interpolation with binning 1 (if the workstation is capable). *Create Full Aligned Stack* and *View Full Aligned Stack*, at which point the stack should flow smoothly as you riffle through the slices → *Done* (Fig. 8.4G).
8. In the Tomogram Generation window: Select Back Projection in the first instance. Use the GPU if present, uncheck Take logarithm of densities with offset, *Generate Tomogram* and *View Tomogram in 3dmod* → *Done* (Fig. 8.4H).
9. If the resulting tomogram is well aligned, perform a SIRT reconstruction as well (in the same window), using an iteration of 15 in the first instance, which will result in a slight loss of resolution but an increase in contrast so that the images resemble traditional TEM micrographs (Fig. 8.4I).
10. Organelles including the nucleus, mitochondria, ER, endosomes, and Golgi should be visible (Fig. 8.3D). In tomograms with optimal sample preparation, data collection, and reconstruction, nuclear pores and the two bilayers of the nuclear envelope will be visible where they run perpendicular to the imaging plane.
11. Artifacts are visible in tomograms of samples with suboptimal sample preparation, including cracks in the ice (Fig. 8.5A), ice contamination on the

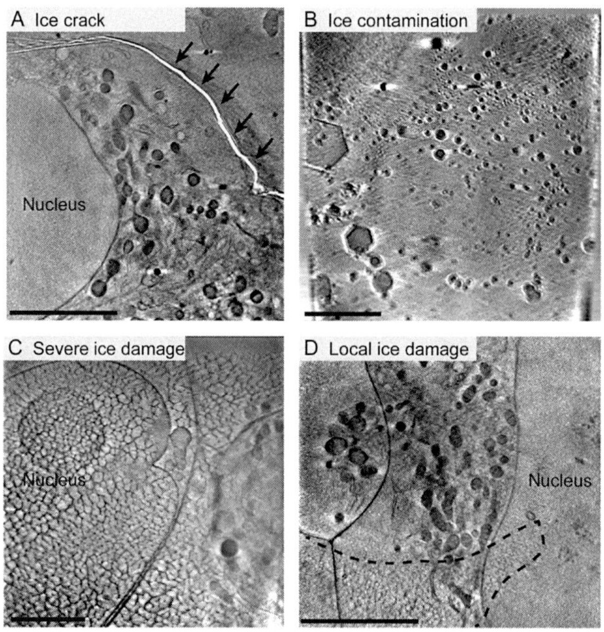

FIGURE 8.5

Virtual slices from tomoX stacks illustrating the appearance of common freezing artifacts. (A) A crack in the ice running along the edge of a cell (arrows). (B) Ice crystal contamination formed on the surface of the vitrified layer. (C and D) Where vitrification has failed, damage caused by hexagonal ice crystals is visible within the cells, giving rise to a "fish scale" or "frog spawn" artifact. In severe cases, this can disrupt entire cells (C) or may be limited to subregions of cells (D). Scale bars: 5 μm.

surfaces of the vitrified layer (Fig. 8.5B), and ice crystal damage within the sample leading to segregation patterns on a global (Fig. 8.5C) or local scale (Fig. 8.5D, dashed line).

Correlating fluorescence and X-ray images

12. Brightfield and fluorescence images from the cryo-correlative stage in the home lab (Fig. 8.6A) and the in-vacuum VLM at the HZB-BESSY II beamline (Fig. 8.6B) were compared with the cryo-SXT image stack. Features like the grid bars were used to correct differences in orientation. Using standard image processing software (e.g., Adobe Photoshop, ImageJ) the brightfield, red, and green channels were overlaid to locate colocalized GFP-LC3 and RFP-Atg9 (Fig. 8.6C). The combined image was then overlaid onto a movie of the tomoX stack and the transparency adjusted so that both datasets could be seen simultaneously. The tomoX plane containing the Quantifoil™ holes and the 250-nm gold fiducial particles was located by scrolling through the virtual slices. The holes, which can also be seen in the cryo-brightfield and cryo-fluorescence images, were used to further align the two datasets (Fig. 8.6C).
13. The fluorescent structures were then correlated to the X-ray structures by scrolling through the tomoX slices (Fig. 8.7). Landmarks that can be seen in the fluorescence images, such as the shape of the nucleus and plasma membrane, can be used to aid the search for the tomoX plane that corresponds to the fluorescent punctae. In the autophagy study, the GFP-LC3 spots colocalizing with RFP-Atg9 (indicating early autophagosomal structures) were cup-shaped omegasomes (arrowheads; Fig. 8.7C–E & modeled in green in Fig. 8.7F). These structures were connected to the ER (blue, Fig. 8.7F), which could be followed over many microns due to the volume of the tomoX stack. Endosomes (e) containing anti-transferrin-receptor gold particles were clustered around the omegasomes (asterisk; Fig. 8.7C–E & modeled in yellow in Fig. 8.7F).
14. Manual segmentation and 3D models of organelles were made using Amira software (FEI Company Visualisation Sciences Group, Eindhoven, The Netherlands) as described in our previous work (Bushby, Mariggi, Armer, & Collinson, 2012) (Fig. 8.7F). Overlay of the 3D model onto the original cryo-fluorescence image can aid in back-checking the correlation.

8.3 MATERIALS

8.3.1 EQUIPMENT

Personal protective equipment (including goggles, gloves, apron, laboratory coat)
Magnetic stirrer
Microprobe sonicator
High resolution ion beam coater (e.g., MED020 High Vacuum Metal Evaporation Coater; Leica Microsystems, Vienna, Austria)

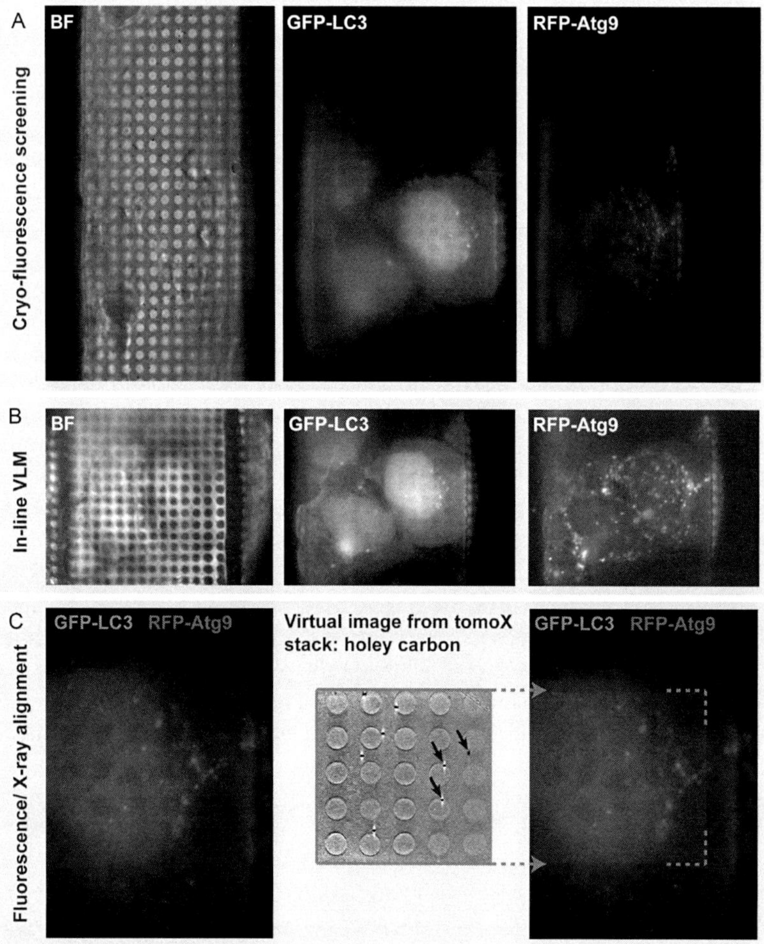

FIGURE 8.6

Alignment of cryo-fluorescence and cryo-SXT images for CLXM. Starved HEK293A cells expressing GFP-LC3 and RFP-Atg9 (Duke et al., 2013). (A) Images from the same cells acquired using the benchtop cryo-fluorescence microscope and (B) the in-vacuum visible light microscope at HZB-BESSY II. (C) The GFP and RFP cryo-fluorescence images are overlaid onto the cryo-brightfield image, which is then aligned with the virtual image of the Quantifoil™ holes from the tomoX stack. 250 nm gold fiducials can also be seen in all imaging modalities and are useful for alignment (arrows). (See the color plate.)

TEM (e.g., 120 kV Tecnai G2 Spirit TWIN, FEI, Eindhoven, the Netherlands)
Glow discharge system (e.g., SC7620; Quorum Technologies, West Sussex, UK)
CO_2-incubator for cell culture
Inverted (confocal) fluorescence microscope equipped with a temperature- and CO_2-controlled chamber for live cell imaging

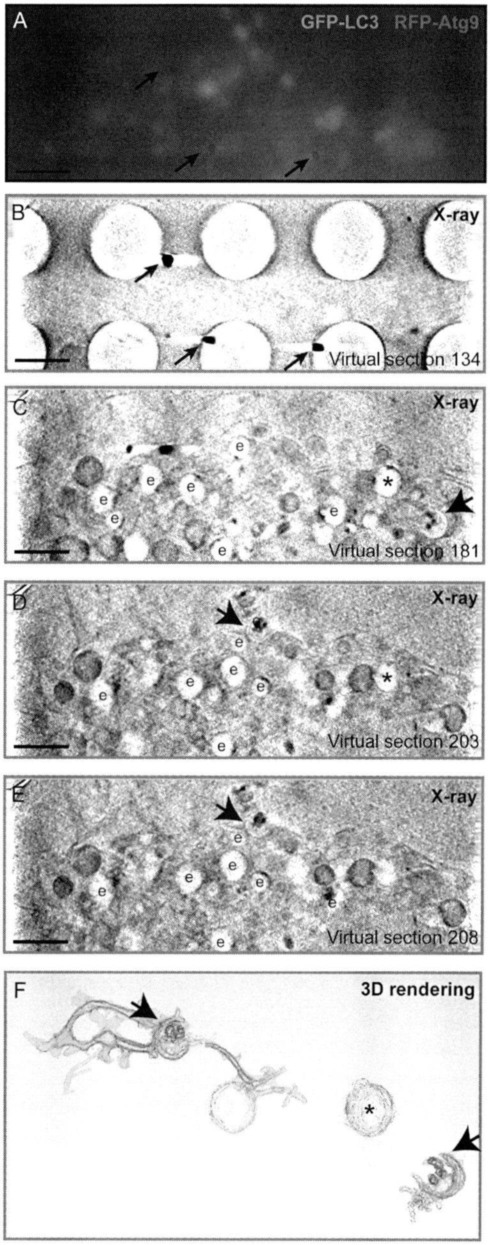

FIGURE 8.7 See legend on next page.

Commercial vitrification device (e.g., EM GP; Leica Microsystems, Vienna, Austria; Cryoplunge; Gatan Inc., Pleasanton; Vitrobot, FEI, Eindhoven, the Netherlands) or homemade vitrification device (Dobro et al., 2010)
Liquid nitrogen storage dewar
Cryo-light microscopy stage (e.g., CMS196 cryo-correlative stage; Linkam Scientific Instruments, Guildford, UK)
Widefield epifluorescence light microscope (e.g., Axio Scope.A1; Zeiss, Cambridge, UK)
Dry nitrogen shipper (e.g., CX-100, Taylor-Wharton, Minnetonka, USA)
Workstation for 3D reconstruction (with, e.g., 96 GB RAM, NVidia Quadro 6000 graphics card and 2 TB data storage)
Interactive pen display (e.g., Wacom Cintiq 21UX; Wacom Technology Corp., Vancouver)

8.3.2 MATERIALS AND CHEMICALS

G200F1 finder grids (Gilder Grids, Grantham, UK)
HZB-2 grids (Gilder Grids, Grantham, UK)
G200F1 and HZB-2 grids pre-coated with QUANTIFOIL® R2/2 (Quantifoil, Jena, Germany)
Formvar® (TAAB Laboratories Equipment Ltd., Aldermaston, UK)
Glass slides
Shiny white address labels
Single edge razor blades (TAAB Laboratories Equipment Ltd., Aldermaston, UK)
Glass Coplin staining jar (TAAB Laboratories Equipment Ltd., Aldermaston, UK)
Deep glass bowl (TAAB Laboratories Equipment Ltd., Aldermaston, UK)
Grid coating plates (AGG3653 Agar Scientific, Stansted, UK)
Whatman® paper No.1 and No.4 (Sigma-Aldrich)

FIGURE 8.7

Correlative cryo-fluorescence and cryo-SXT identifies forming autophagosomes. Starved HEK293A cell expressing GFP-LC3 and RFP-Atg9 from Fig. 8.6 (Duke et al., 2013). (A) The region of interest, containing GFP-LC3-positive autopagosomes in close proximity to RFP-Atg9, is cropped and rotated from the overlay in Fig. 8.6C. Note that the arrows in both figures indicate the same three gold fiducial particles. (B–E) Scrolling through the tomoX stack in this region from the Quantifoil™ layer (B) upward, structures are visible that correlate to the GFP-LC3 spots (C–E). The GFP-LC3 signal is localized to cup-shaped omegasomes that are forming from hotspots on the endoplasmic reticulum. (F) The omegasomes (green), endoplasmic reticulum (blue), and transferrin receptor gold-filled endosomes (yellow) were manually segmented through sequential tomoX slices, and rendered into a 3D model, which was then overlaid onto the cryo-fluorescence images to back-check the correlation. Scale bars: 2 μm. (See the color plate.)

Double distilled water (ddH_2O)
Chloroform (Sigma-Aldrich)
Acetone (Sigma-Aldrich)
Methanol (Sigma-Aldrich)
Glycerol (Sigma-Aldrich)
1,2-Dichloroethane (Sigma-Aldrich)
Wooden cocktail sticks
Metal grid holder (TAAB Laboratories Equipment Ltd., Aldermaston, UK)
Glass Petri dish (Agar Scientific, Stansted, UK)
Liquid blocker pen/PAP pen (Agar Scientific, Stansted, UK)
Glass bottom culture dishes (e.g., P35G-1.5-14-C, MatTek, Ashland, USA)
Gold nanoparticles 250 nm (EMGC250, BBI Group, Cardiff, UK)
Cryo-grid boxes (Agar Scientific, Stansted, UK)
15 and 50-mL centrifuge tubes
Metal nuts (internal diameter 15 mm)
Cryo-tubes and cryo-canes
Cryo-tweezers
Fine tweezers
Black cryo-pen
Liquid nitrogen
Ethane gas cylinder

OUTLOOK AND FUTURE DIRECTIONS

Correlative light and electron microscopy is now a field in its own right, encompassing a variety of methods for linking the two imaging modalities that are finding wide applications in biomedical research. Recent developments include the addition of new probes that can be detected using light and electrons; the use of super resolution light microscopes to improve localization precision; the use of volume EM techniques for faster image acquisition and larger samples; the integration of light and electron microscopes for faster, more accurate correlation; and the advent of semi-automated algorithms for faster image analysis. Our recent work (Duke et al., 2013) and this chapter highlight how correlative light and soft X-ray microscopy are now pushing the limits of 3D native-state imaging.

Our workflow for cryo-CLXM, allowing fluorescent proteins to be localized to organelles within intact fully hydrated cells, draws on workflows developed for CLEM. However, our modifications take account of the unique properties of soft X-rays in the water window, which enable (and require) cryo-preparation. Fortuitously, cryo-preparation has already been shown to preserve cells as close to native-state as possible for in-vacuum imaging. For cellular structures that are difficult to preserve and image intact using traditional EM and CLEM techniques, cryo-CLXM may provide the answer.

Cryo-SXT and cryo-CLXM are still in their infancy. Future improvements to the workflow and instrumentation may result in data at higher resolution and therefore ultimately additional biological information. Currently, there is active work taking place to improve the resolution of the objective zone plate which is one of the main drivers influencing the resolution of the final data. The highest resolution zone plate in routine use for biological samples is 25 nm. Ten nanomter zone plates have been developed but their use in 3D work is limited due to the restriction on sample rotation imposed by their very short focal length. The development of alternative cell-mounting technologies (e.g., capillaries that are routinely used on the Advanced Light Source-based microscope) for in-vacuum microscopes could have an impact here.

Clearly, to take full advantage of any improvements in microscope resolution, it is vital that there is total confidence in every stage of the sample preparation protocol. At present, most involved in this work would argue that we still have much to learn about every step. The technique of correlative light and X-ray microscopy is even younger than cryo-SXT, and so is a field wide open to those interested in novel approaches. The development of combined labels suitable for both X-rays and fluorescence would be an excellent place to start. Looking further along the technique pipeline, little attention has been paid to data analysis. At present, EM software is used for data analysis with scant regard taken of the differences between X-rays and electrons as they interact with the sample. Developments in this area are long overdue. However, perhaps the most critical factor now is a wider engagement of the biomedical research community in cellular imaging at synchrotrons, which will undoubtedly drive biomedical discovery.

ACKNOWLEDGMENTS

We thank Dr. Sharon Tooze and Dr. Minoo Razi for the samples, biological questions, and experimental work resulting in Figs. 8.6 and 8.7. We thank Prof. Clare Futter for the anti-TfnR-gold, Linkam Scientific Instruments for modifications to the cryo-fluorescence stage, Ela Kubiak-Goswami for the photographs in Fig. 8.1, and Dr. Fabienne Beuron for the holey carbon film protocol. We thank Dr. Peter Guttmann, Dr. Eva Pereiro, and all the HZB-BESSY II and ALBA beamline staff for their continued help and support. The research leading to these results has received funding from the European Community's Seventh Framework Programme (FP7/2007-2013) under BioStruct-X (Grant agreement N°283570). Dr. Raffaella Carzaniga, Dr. Marie-Charlotte Domart and Dr. Lucy M. Collinson are grateful for support from Cancer Research UK.

REFERENCES

Ashcroft, J. M., Gu, W., Zhang, T., Hughes, S. M., Hartman, K. B., Hofmann, C., et al. (2008). TiO2 nanoparticles as a soft X-ray molecular probe. *Chemical Communications (Cambridge, England)*, *21*, 2471–2473. http://dx.doi.org/10.1039/b801392f.

Berriman, J. A., Li, S., Hewlett, L. J., Wasilewski, S., Kiskin, F. N., Carter, T., et al. (2009). Structural organization of Weibel-Palade bodies revealed by cryo-EM of vitrified endothelial cells. *Proceedings of the National Academy of Sciences of the United States of America*, *106*(41), 17407–17412. http://dx.doi.org/10.1073/pnas.0902977106.

Bushby, A. J., Mariggi, G., Armer, H. E., & Collinson, L. M. (2012). Correlative light and volume electron microscopy: Using focused ion beam scanning electron microscopy to image transient events in model organisms. *Methods in Cell Biology*, *111*, 357–382. http://dx.doi.org/10.1016/b978-0-12-416026-2.00018-2.

Carzaniga, R., Domart, M. C., Collinson, L. M., & Duke, E. (2013). Cryo-soft X-ray tomography: A journey into the world of the native-state cell. *Protoplasma*. http://dx.doi.org/10.1007/s00709-013-0583-y.

Dobro, M. J., Melanson, L. A., Jensen, G. J., & McDowall, A. W. (2010). Plunge freezing for electron cryomicroscopy. *Methods in Enzymology*, *481*, 63–82. http://dx.doi.org/10.1016/s0076-6879(10)81003-1.

Duke, E. M. H., Razi, M., Weston, A., Guttmann, P., Werner, S., Henzler, K., et al. (2013). Imaging endosomes and autophagosomes in whole mammalian cells using correlative cryo-fluorescence and cryo-soft X-ray microscopy (cryo-CLXM). *Ultramicroscopy*, *143*, 77–87. http://dx.doi.org/10.1016/j.ultramic.2013.10.006.

Eskelinen, E. L. (2008). To be or not to be? Examples of incorrect identification of autophagic compartments in conventional transmission electron microscopy of mammalian cells. *Autophagy*, *4*(2), 257–260.

Eskelinen, E. L., & Kovacs, A. L. (2011). Double membranes vs. lipid bilayers, and their significance for correct identification of macroautophagic structures. *Autophagy*, *7*(9), 931–932.

Gomes, L. C., Di Benedetto, G., & Scorrano, L. (2011). During autophagy mitochondria elongate, are spared from degradation and sustain cell viability. *Nature Cell Biology*, *13*(5), 589–598. http://www.nature.com/ncb/journal/v13/n5/abs/ncb2220.html, supplementary-information.

Gu, W., Etkin, L. D., Le Gros, M. A., & Larabell, C. A. (2007). X-ray tomography of Schizosaccharomyces pombe. *Differentiation*, *75*(6), 529–535. http://dx.doi.org/10.1111/j.1432-0436.2007.00180.x.

Hagen, C., Guttmann, P., Klupp, B., Werner, S., Rehbein, S., Mettenleiter, T. C., et al. (2012). Correlative VIS-fluorescence and soft X-ray cryo-microscopy/tomography of adherent cells. *Journal of Structural Biology*, *177*(2), 193–201. http://dx.doi.org/10.1016/j.jsb.2011.12.012.

Hanssen, E., Knoechel, C., Dearnley, M., Dixon, M. W., Le Gros, M., Larabell, C., et al. (2012). Soft X-ray microscopy analysis of cell volume and hemoglobin content in erythrocytes infected with asexual and sexual stages of Plasmodium falciparum. *Journal of Structural Biology*, *177*(2), 224–232. http://dx.doi.org/10.1016/j.jsb.2011.09.003.

Hanssen, E., Knoechel, C., Klonis, N., Abu-Bakar, N., Deed, S., LeGros, M., et al. (2011). Cryo transmission X-ray imaging of the malaria parasite P. falciparum. *Journal of Structural Biology*, *173*(1), 161–168. http://dx.doi.org/10.1016/j.jsb.2010.08.013.

Kremer, J. R., Mastronarde, D. N., & McIntosh, J. R. (1996). Computer visualization of three-dimensional image data using IMOD. *Journal of Structural Biology*, *116*(1), 71–76. http://dx.doi.org/10.1006/jsbi.1996.0013.

Larabell, C. A., & Le Gros, M. A. (2004). X-ray tomography generates 3-D reconstructions of the yeast, saccharomyces cerevisiae, at 60-nm resolution. *Molecular Biology of the Cell*, *15*(3), 957–962. http://dx.doi.org/10.1091/mbc.E03-07-0522.

Larabell, C. A., & Nugent, K. A. (2010). Imaging cellular architecture with X-rays. *Current Opinion in Structural Biology*, *20*(5), 623–631. http://dx.doi.org/10.1016/j.sbi.2010.08.008.

Meyer-Ilse, W., Hamamoto, D., Nair, A., Lelievre, S. A., Denbeaux, G., Johnson, L., et al. (2001). High resolution protein localization using soft X-ray microscopy. *Journal of Microscopy*, *201*(Pt. 3), 395–403.

Parkinson, D. Y., McDermott, G., Etkin, L. D., Le Gros, M. A., & Larabell, C. A. (2008). Quantitative 3-D imaging of eukaryotic cells using soft X-ray tomography. *Journal of Structural Biology*, *162*(3), 380–386. http://dx.doi.org/10.1016/j.jsb.2008.02.003.

Rambold, A. S., Kostelecky, B., Elia, N., & Lippincott-Schwartz, J. (2011). Tubular network formation protects mitochondria from autophagosomal degradation during nutrient starvation. *Proceedings of the National Academy of Sciences of the United States of America*, *108*(25), 10190–10195. http://dx.doi.org/10.1073/pnas.1107402108.

Schertel, A., Snaidero, N., Han, H. M., Ruhwedel, T., Laue, M., Grabenbauer, M., et al. (2013). Cryo FIB-SEM: Volume imaging of cellular ultrastructure in native frozen specimens. *Journal of Structural Biology*, *184*(2), 355–360. http://dx.doi.org/10.1016/j.jsb.2013.09.024.

Schneider, G., Guttmann, P., Heim, S., Rehbein, S., Mueller, F., Nagashima, K., et al. (2010). Three-dimensional cellular ultrastructure resolved by X-ray microscopy. *Nature Methods*, *7*(12), 985–987. http://dx.doi.org/10.1038/nmeth.1533.

Schwartz, C. L., Sarbash, V. I., Ataullakhanov, F. I., McIntosh, J. R., & Nicastro, D. (2007). Cryo-fluorescence microscopy facilitates correlations between light and cryo-electron microscopy and reduces the rate of photobleaching. *Journal of Microscopy*, *227*(Pt 2), 98–109. http://dx.doi.org/10.1111/j.1365-2818.2007.01794.x, pii: JMI1794.

Uchida, M., McDermott, G., Wetzler, M., Le Gros, M. A., Myllys, M., Knoechel, C., et al. (2009). Soft X-ray tomography of phenotypic switching and the cellular response to antifungal peptoids in Candida albicans. *Proceedings of the National Academy of Sciences of the United States of America*, *106*(46), 19375–19380. http://dx.doi.org/10.1073/pnas.0906145106.

Uchida, M., Sun, Y., McDermott, G., Knoechel, C., Le Gros, M. A., Parkinson, D., et al. (2011). Quantitative analysis of yeast internal architecture using soft X-ray tomography. *Yeast*, *28*(3), 227–236. http://dx.doi.org/10.1002/yea.1834.

van Driel, L. F., Valentijn, J. A., Valentijn, K. M., Koning, R. I., & Koster, A. J. (2009). Tools for correlative cryo-fluorescence microscopy and cryo-electron tomography applied to whole mitochondria in human endothelial cells. *European Journal of Cell Biology*, *88*(11), 669–684. http://dx.doi.org/10.1016/j.ejcb.2009.07.002.

CHAPTER

9

Critical Step-by-Step Approaches Toward Correlative Fluorescence/ Soft X-Ray Cryo-Microscopy of Adherent Mammalian Cells

Kyle C. Dent*,†, Christoph Hagen†, Kay Grünewald†

**Diamond Light Source, Harwell Science and Innovation Campus, Didcot, United Kingdom*

†Oxford Particle Imaging Centre, Division of Structural Biology, Wellcome Trust Centre for Human Genetics, University of Oxford, Oxford, United Kingdom

CHAPTER OUTLINE

Methods in Cell Biology, Volume 124, ISSN 0091-679X, http://dx.doi.org/10.1016/B978-0-12-801075-4.00009-4

Abstract

Soft X-ray cryo-microscopy/tomography with its extraordinary capability to map vitreous cells with high absorption contrast in their full three-dimensional extent, and at a resolution exceeding super-resolution fluorescence microscopy, is a valuable tool for integrative structural cell biology. Focusing on cell biological applications, its ongoing methodological development gained momentum by combining it with fluorescence cryo-microscopy, thus correlating highly resolved structural and specific information *in situ*. In this chapter, we provide a basic description of the techniques, as well as an overview of equipment and methods available to carry out correlative soft X-ray cryo-tomography experiments on frozen-hydrated cells grown on a planar support. Our aim here is to suggest ways that biologically representative data can be recorded to the highest possible resolution, while also keeping in mind the limitations of the technique during data acquisition and analysis. We have written from our perspective as electron cryo-microscopists/structural cell biologists who have experience using correlative fluorescence/cryoXM/T at synchrotron beamlines presently available for external users in Europe (HZB TXM at U41-FSGM, BESSY II, Berlin/Germany; Carl Zeiss TXMs at MISTRAL, ALBA, Barcelona/Spain, and B24, DLS, Oxfordshire, UK).

Abbreviations

ALBA	ALBA Synchrotron Light Facility, Barcelona/Spain
BESSY II	BESSY II electron storage ring, Berlin/Germany
cryoEM/T	electron cryo-microscopy/tomography
cryoFM	fluorescence cryo-microscopy
cryoXM/T	soft X-ray cryo-microscopy/tomography
DLS	Diamond Light Source, Oxfordshire/UK
DOF	depth of focus
EM	electron microscopy
FOV	field of view
HZB	Helmholtz-Zentrum Berlin
NA	numerical aperture
ROI	region of interest
SNR	signal-to-noise ratio
TXM	transmission soft X-ray microscope
ZP	zone plate

INTRODUCTION

Soft X-ray cryo-microscopy/tomography (cryoXM/T) is able to reveal ultrastructural information of intact biological objects at a 3D resolution approaching 30 nm (Carrascosa et al., 2009; Hagen et al., 2012; Parkinson, Knoechel, Yang, Larabell, & Le Gros, 2012; Schneider et al., 2010). When imaging using photons in the "water-window" wavelength range 4.37–2.29 nm, i.e., between the K-shell absorption edges of carbon and oxygen, 284–543 eV (Wolter, 1952), contrast arises from the differential absorption of X-rays by carbon as well as other heavier atoms, and oxygen (for a textbook, see Attwood, 2007). For unstained and hydrated biological specimens—consisting primarily of carbon, nitrogen, oxygen, and phosphate—carbon atoms, for example, will absorb photons an order of magnitude more strongly than oxygen will (for review, see Falcone et al., 2011; Schneider, Guttmann, Rehbein, Werner, & Follath, 2012). The implication is significant "naturally" occurring (elemental) image contrast for cryo-immobilized, near-native specimens. Most commonly the photon energy is selected to be 510–520 eV serving to maximize the photon penetration depth, allowing hydrated specimens up to ~20 μm in thickness to be imaged as a function of soft X-ray transmission (Weiner et al., 2013). In contrast to cryoEM/T (electron cryo-microscopy/tomography) limited to a sample thickness of 0.5–1 μm (Lučić, Förster, & Baumeister, 2005), cryoXM/T is able to image unsectioned (intact) specimens, revealing carbon dense features, such as cellular membranes, but also heavy metal clumps or aggregates (e.g., colloidal gold particles), and is consequently readily applicable as an *in situ* method for high-resolution imaging of eukaryotic cells that have been cryo-immobilized to minimize Brownian motion and cellular/intracellular movement, as well as to protect from radiation damage.

By applying approaches for tomographic data collection and 3D reconstruction, intact cells can be studied in three dimensions. Indeed, in both conception and implementation, cryoXM/T has similarities to cryoEM/T, but it is very important to keep in mind that fundamental differences in image formation exist (Bertilson, von Hofsten, Hertz, & Vogt, 2011; Oton et al., 2012; Rehbein, Guttmann, Werner, & Schneider, 2012). Contrast in the X-ray absorption image is heavily influenced by the chemical nature of the target, and consequently, the view of the cell provided by cryoXM is quite unlike that of those achieved by cryoEM; one example being nucleoli, which are generally not easily identifiable in cryoEM/T of (unstained) vitreous sections, but extremely prominent in X-ray tomogram slices (cf. Fig. 3A in Hagen & Grünewald, 2008; Fig. 5a in Hagen et al., 2012). Loosely speaking, however, in quantifying attainable resolution of cryoXT (30–50 nm), specimen penetration depth, as well as areas imaged (10–15 μm), estimates for these can be calculated by scaling the corresponding values for cryoET by an order of magnitude. The implication of course is that a large amount of information of relevance to questions pertaining to cellular mechanism can be recorded relatively quickly. It may be argued then that cryoXT is a method for high-throughput structural cell biology, filling the resolution gap between diffraction-limited fluorescence microscopy and cryoET. The high contrast of 2D images, as well as reconstructed tomograms (3D), typically reveals the distribution of cellular organelles as distinct compartments of clear and characteristic morphology. Because the technique relies on absorption contrast (rather than phase contrast as in cryoET), it has been argued that different organelles can be identified not only by their morphology but also quantitatively by linear absorption coefficient (Uchida et al., 2011). One can expect prominent features in tomograms to include lipid bodies, nucleoli, the nuclear envelope, mitochondria, endoplasmic reticula, vesicles, vacuoles, as well as the cell boundary. In principle, the entire "catalog" of cellular membrane-bound organelles is accessible to the technique (cf. Müller et al., 2012). From the study of published work, one will find that the resolution in reconstructed 3D tomograms is typically high enough to reveal the perinuclear space (Hummel et al., 2012), as well as cristae of mitochondria and the endoplasmic reticulum (Duke et al., 2014). In some cases, even components of the cytoskeleton were visible (Schneider et al., 2010).

CryoXM/T has been used to image algae, bacteria, yeast, viruses, as well as mammalian cells, prepared adherently or in suspension (Carrascosa et al., 2009; Chichón et al., 2012; Larabell & Le Gros, 2004; Thieme, Schneider, & Knöchel, 2003; Weiss et al., 2000). Following the development of modern transmission soft X-ray microscopes (TXMs), pioneered at the University of Göttingen/Germany (for historical review, see Kirz, Jacobsen, & Howells, 1995), the U41-FSGM beamline of the BESSY II electron storage ring in Berlin/Germany and XM-2 beamline at the Advance Light Source (Berkeley, California) have recently furthered efforts to apply correlative fluorescence cryo-microscopy (cryoFM) in this context (Cinquin et al., 2014; Le Gros, McDermott, Uchida, Knoechel, & Larabell, 2009; Schneider, Guttmann, et al., 2012; Schneider, Heim, Guttmann, Rehbein, & Niemann, 2006). The original workflows for cryoXM relied on specimens prepared within thin-walled glass capillaries (recently described, for instance, by Parkinson et al., 2013)—a different approach to that advocated in this chapter—however, this development nevertheless motivated an increasing drive to improve the accessibility of the technique. The beamline U41-FSGM at BESSY II subsequently made important advancements by applying a high-efficiency soft X-ray illumination system, a spherical plane-grating monochromator coupled to a single-bounce reflecting condenser capillary (Guttmann et al., 2009), which allowed a larger space for the specimen, and rectangular specimen grid format, which permitted adherent cells to be studied in planar (or flat) geometry; these requiring more room for tilt experiments (Schneider et al., 2010). In addition, their design included an in-column fluorescence microscope for high-numerical aperture (NA) fluorescence imaging to be carried out on specimens without need of an offline cryo-stage (Guttmann et al., 2013). Thus, the experimental but fully functional and user-friendly endstation at U41-FSGM, HZB TXM, has served the community with a stable external user access application scheme since spring 2008 (Guttmann et al., 2013) and has inspired the design and manufacture of a commercially available TXM by Carl Zeiss X-ray microscopy (formerly known as Xradia, Pleasonton, CA), subsequently leading to the construction of cryoXM/T beamlines at ALBA and DLS. The properties of currently operating beamlines for correlative cryoXM/T are summarized in the literature (Carzaniga, Domart, Collinson, & Duke, 2014; Guttmann et al., 2013; Pereiro, Nicolas, Ferrer, & Howells, 2009).

Full-field TXMs are characterized by their use of zone plate (ZP) objective lenses. These are microscopic lenses composed of concentric metal rings that diffract X-rays to form a magnified image of the specimen (Werner, Rehbein, Guttmann, Heim, & Schneider, 2010). These components operate with relatively low efficiencies (~15%), and consequently synchrotron radiation sources are currently favored for high-resolution cryoXM because the X-ray fluxes achieved at these beamlines considerably reduce image acquisition times; a 1–8 s exposure time per image is typical for the kind of work described here. Other X-ray sources have also been applied effectively. These are typically laser plasma sources; however, exposure times are then on the order of minutes (Hertz et al., 2012). Standalone "table-top" sources of this kind are under active development and will certainly play an important role in making cryoXM more generally available in the future (Martz et al., 2012). Presently, tomography experiments of the kind useful for correlative biological/biomedical studies demand the use of synchrotron-based X-ray sources as each tilt series typically requires that 90–180 images are recorded per tilt series (recorded over 10–40 min, with 5–20 min of actual radiation exposure). Furthermore, for any particular study, analysis of a number (30–50) of tomograms will be required for any particular system/research question, implying that multiple synchrotron visits may be required. For the correlative workflows described here, a 12-h imaging session may allow up to 16 tilt series to be recorded (F. J. Chichón, personal communication), this time being consumed by not only image acquisition and readout but also manual searches to find regions of interest (ROIs) and setup tilt series collections.

Cell biology research projects frequently take advantage of fluorescent labeling in order to localize specific macromolecular components within cells (Dedecker, De Schryver, & Hofkens, 2013). Furthermore, as is well appreciated, fluorescence microscopy alone does not provide structural information about the remainder of the cell, and as result correlative light-electron cryo-microscopy methods have been developed to provide complementary context information (Zhang, 2013). However, owing to the complexities of specimen preparation for cryoEM, as well as image correlation, these "cryo-correlative" methods can be particularly challenging and require a considerable investment to be mastered. Given this fact, it is clear that many (nonspecialist) groups may benefit from a more easily implemented method that might provide an overview of the near-native ultrastructural changes associated with a specific biological event of interest. Appropriately designed experiments are able to investigate the roles fulfilled by specific (fluorescently labeled) proteins in the context of subcellular organization and compartmentalization (i.e., behavior revealed at ultrastructural resolutions of ~30–75 nm).

Previous studies have shown, when applied in correlation with other imaging modes, that cryoXM can be applied to questions relating to the mechanistic basis of organelle formation and function, i.e., questions benefiting most from larger fields of view. For this reason, cryoXM is of significant benefit when studying the cell nucleus when compared to a technique such as cryoEM. The general approach is to image as many snapshots of a process as possible (i.e., to observe the same process at different stages and in multiple cells using both cryoFM and cryoXM/T), and then to apply methods for quantitative modeling of the data (i.e., segmentation and statistics) so as to apply the experimental data to hypothesis formulation and testing. The application of correlative cryoFM/XM is still an emerging technique; however, important examples of how this technique may be applied are those already reported with regard to autophagosome formation (Duke et al., 2014), viral influence over the nucleoplasmic reticulum (Hagen et al., 2012), as well as apoptosis (Hagen et al., 2014). In these cases, the thickness of the cellular volume imaged allows a more complete view of the cellular environment. However, it must be reemphasized that the technique is still in its infancy, and the future is likely to reveal important (possibly serendipitous) benefits of the unique image formation and contrast mechanisms at work in the X-ray microscope.

Owing to various experimental constraints, as well as fortuitous benefits, the most appropriate manner of pursuing this sort of insight is to work with cryogenically immobilized specimens that are thereby in their best possible, near-native preservation state (Bleck et al., 2010; Weinhausen et al., 2014), also between visible light/X-ray photon imaging experiments, and additionally are maintained in a state resistant to forms of radiation damage that may be responsible for either structural alteration or bleaching of fluorophores (Le Gros et al., 2009; Moerner & Orrit, 1999; Sartori et al., 2007; Schwartz, Sarbash, Ataullakhanov, McIntosh, & Nicastro, 2007). Often overlooked, cryo-immobilization is central to high-resolution near-native studies, as Brownian motion would otherwise blur features of interest. While the approaches described here have great potential for cell biology studies owing to their broad applicability and sensitivities, the hardware and software involved will still see significant improvements over the coming decade(s). Despite this, the value of correlative cryoXM in the context of integrative structural biology initiatives (Fig. 9.1) has already become clear (Chichón et al., 2012; Duke et al., 2014; Hagen et al., 2012; Zeev-Ben-Mordehai, Hagen, & Grünewald, 2014).

9.1 RATIONALE

This chapter provides an overview of equipment and methods available to carry out correlative cryoFM/XT experiments on frozen-hydrated cells grown on a planar support. Our aim is to suggest ways of recording biologically representative data at the highest resolution possible, while also mentioning some of the possible limitations of the technique. We are not attempting to provide precise technical or physical descriptions of the processes involved at each step, but we have prepared this guide as (structural) cell biologists for cell biologists who are new to cryo-microscopy, and we hope to discuss important aspects pertaining to how specimens can be prepared in the home laboratory for the synchrotron/beamline visit. We have written from our

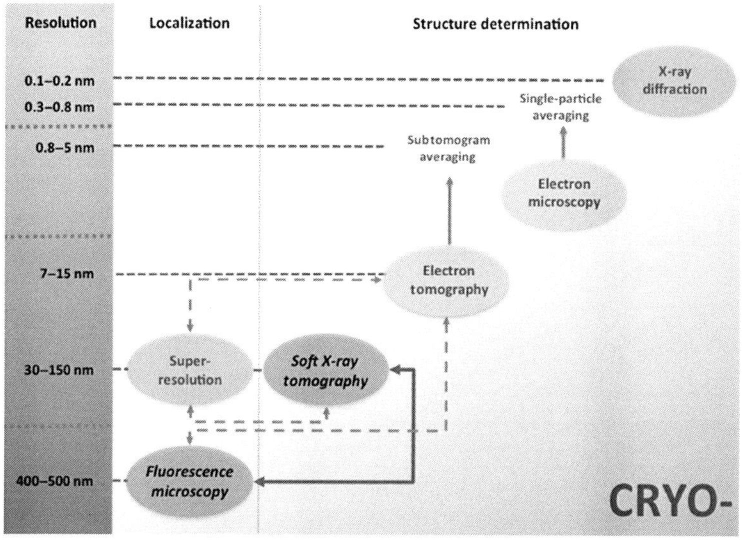

FIGURE 9.1

Schematic layout of integrative structural biological methods that benefit from rapid cryo-immobilization and native preservation of the specimen. Soft X-ray microscopy (cryoXM/T) in this context fills the "resolution gap" between fluorescence microscopy (cryoFM) and electron cryo-tomography (cryoET).

perspective, i.e., as electron cryo-microscopists who have experience using cryoXM at synchrotron beamlines presently available in Europe (HZB TXM at U41-FSGM, BESSY II, Berlin/Germany; Carl Zeiss TXMs at MISTRAL, ALBA, Barcelona/Spain and B24, DLS, Oxfordshire, UK).

Once an appropriate model system/cell line has been chosen, as well as a strategy for noninvasive fluorescent labeling established, correlative cryoFM/XT experiments proceed through four principal stages (Fig. 9.2). Here we provide only basic information regarding strategies for fluorescence labeling, and rather focus on cryo-image acquisition as well as the subsequent data-processing steps.

9.2 CRITICAL DISCUSSION OF THE CORRELATIVE CRYOFM/XT WORKFLOW

Because specimen preparation for cryoXM essentially involves two primary steps: cell culture on a microscope compatible support followed by immobilization by rapid cryogenic freezing (vitrification mostly by plunge freezing), cryoXT can provide a large amount of structural information in a comparably short amount of time—avoiding the investment required for preparation for cryoEM (e.g., cryo-electron microscopy of vitreous sections, CEMOVIS; Dubochet, 2012). However, despite

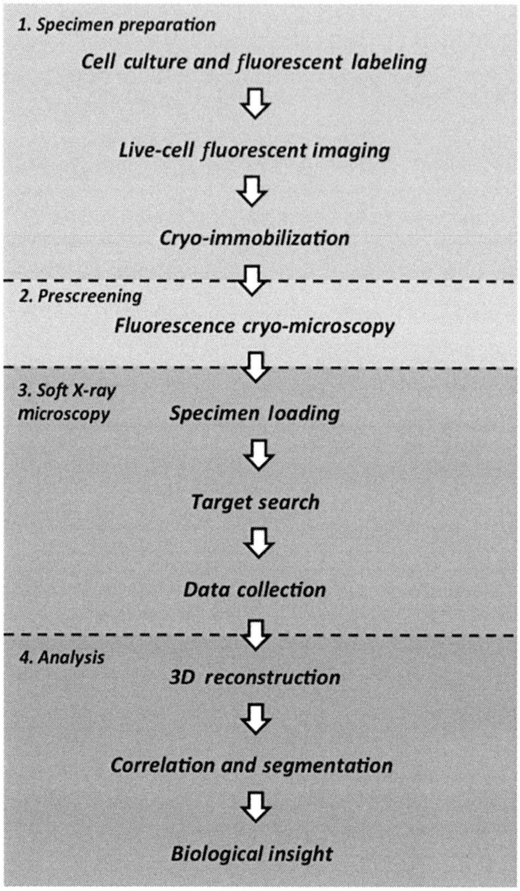

FIGURE 9.2

General workflow for correlative soft X-ray cryo-microscopy of adherent mammalian cells. After preparation of cryo-immobilized specimens, biological events of interest must be identified by high-resolution cryoFM, and following this, cryoXM/T is used to provide a detailed view of cell structure in the specific regions of interest. Finally, correlation of images allows biological interpretation of the experimental data. Each step is of fundamental importance, as the quality of a study can be limited by weakness in any of these areas.

these conveniences, there is still a "learning curve" involved with respect to adequate specimen preparation as well as ensuring that the biological events of interest are accessible for 3D reconstruction and fluorescence correlation. Fundamental to successful execution of this approach is also the fluorescent-labeling strategy used to identify the biological events of interest—although this should be developed without input of cryoXM until the later stages.

To apply the correlative approach described here, an application for beam time must be submitted to one or more of the available facilities. Important considerations prior to such an application may be:

- Is information about cell ultrastructural changes important for a full understanding of the mechanism under study?
- Can the cells be prepared in an adherent (planar) format, and how thick is the specimen expected to be over a typical tomographic tilt range?
- Has a suitable (noninvasive) system for fluorescent labeling been established?
- Is there access to equipment for rapid cryo-immobilization in the home laboratory, and is this facility nearby to where the cells are being cultured?
- What size are the cellular features under study?

This guide assumes that cryoXM is being explored as a means of extending a pre-established fluorescence imaging project, and it is likely that if an adherent mammalian cell line is already in use, this should be accessible to cryoXM/T. Most cell lines of this kind vary in thickness from 3 to 8 μm, but keep in mind that in the case of the planar specimen geometry, the apparent thickness of a cell will increase with the tilt angle (e.g., a 5-μm-thick cell will be 10 μm thick at 60°). With respect to features of interest, most studies focus on features at least 100 nm in size, with highest contrast being revealed for carbon or electron dense structures. Protein components of the cell (e.g., cytoskeletal structures) are usually much harder (although not impossible) to resolve as these are less carbon dense than are phospholipid bilayers. For studies where a metal nanoparticle is being identified (and possibly used as a specific labeling agent), these structures are most visible when they occur in clusters (Duke et al., 2014).

If the experimental system falls within these parameters, specimens will be prepared for transportation to the synchrotron storage ring, where imaging will be carried out at the beamline's endstation. The endstation is the X-ray microscope that uses the synchrotron, and primary beamline optics, as a source of finely "tuned" (collimated and monochromatic) illuminating X-rays. Specimen loading and imaging will be facilitated by synchrotron support staff, and while this is in progress, datasets can be analyzed (and reconstructed) by the user. This approach provides valuable feedback allowing refinement of the data-collection strategy. The sections which follow attempt to elaborate on these considerations by providing information about imaging strategies (choice of ZP, magnification, and data-acquisition parameters) that should support the collection of informative experimental data; this is in consideration of the properties of the specimen at hand.

It should be kept in mind that specimens prepared using the approaches described here are generally compatible with all three major bioimaging modalities (fluorescence, electron, X-ray). However, our focus is on correlative light/fluorescence/X-ray modalities which are least restrictive with respect to specimen thickness and can conceivably be used to study many cellular processes *in situ* as long as protein/organelle-labeling strategies have been well thought out and implemented. Specimens prepared for cryoXM are in many cases also amenable to "thin-edge" electron imaging (Lučić, Rigort, & Baumeister, 2013; Zeev-Ben-Mordehai et al., 2014), depending on the cell type used. For thicker regions (i.e., the cell body of typical cell lines, nucleus), electron transparent lamellae may be prepared by dual-beam (focused ion beam-scanning electron microscopy) micromachining—an approach under active development that will see widespread application in the future (Hsieh, Schmelzer, Kishchenko, Wagenknecht, & Marko, 2014; Rigort et al., 2012). The intention here would be to leverage the experimental results from cryoXM to inform the identification of separate but similar ROIs for electron imaging. CryoET would allow for higher (7–15 nm) resolution imaging, with the additional possibility of subtomogram averaging to investigate structural properties of macromolecular complexes in the environments of interest.

9.2.1 SAMPLE PREPARATION

Sample preparation is a fundamental aspect of every successful imaging experiment. The outcome of this phase is the preparation of hydrated (near-native) biological cells, maintained at temperatures below devitrification ($\leq$138 K; Dubochet, 2007) that also meet well-defined specifications with respect to thickness and ice quality (Fig. 9.3). Our aim is to immobilize the cells in as thin a layer of ice as reasonably possible while seeking to preserve their physiological state immediately prior to plunge freezing. It is for this reason that cells must not be disturbed by direct

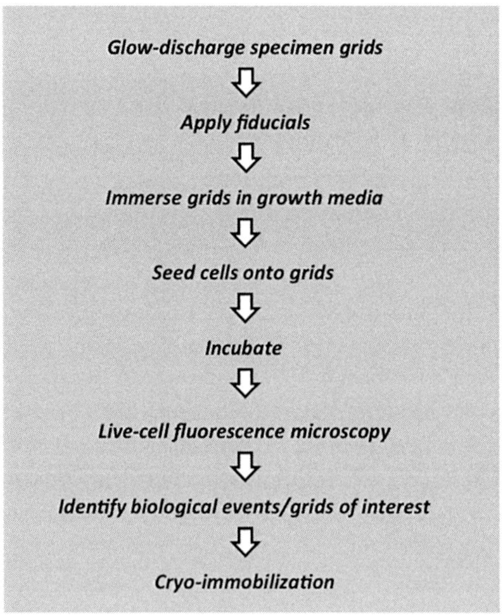

FIGURE 9.3

Workflow for preparation of cryo-immobilized adherent cells with fluorescently labeled features.

handling and should not experience significant changes in temperature or environmental composition (buffer/growth media conditions) immediately prior to cryo-immobilization. Careful planning ahead is, therefore, also needed for labeling target structures with fluorescent markers; however, in situations such as these, possible changes to cells should be characterized by live-cell fluorescence imaging.

9.2.1.1 Sample holder/grid format

Our workflows focus on correlative cryo-microscopy of adherent cells prepared on a planar support mesh. We argue here that in the context of biomedical imaging requirements, there are important reasons for choosing this approach over the approach of detachment and suspension of cells in cylindrical capillaries. The planar specimen format is most often applied in the context of live/fixed-cell fluorescence microscopy (thus appealing to the widest possible audience), but also has specific benefits for broader image correlation strategies in general (described in brief later). Furthermore, of particular importance to the overall validity of experiment design, the planar specimen geometry allows cells to be grown and prepared in a manner that does not alter their structural state. Based on our own experience, cells are profoundly sensitive to their environments and are able to undergo dramatic structural reorganization of the cytoskeleton within seconds/minutes if stressed by environmental changes. We argue that procedures requiring cell detachment must be seen as severe stressors, and it is our opinion that it is simply not possible to detach cells and cryo-immobilize them within sufficiently small time frames. While the planar geometry has implications for imaging strategy, with subsequent introduction of imaging artifacts such as the "missing wedge" (for review, see Lučić et al., 2005), these artifacts are well understood, and interpretation of data is carried out with these in mind. Possible cell changes as a result of an irrelevant stressor, on the other hand, are more difficult to characterize and may therefore strain the validity of experimental results—negating the benefits of cryo-immobilization as a means of "near-native" high-resolution imaging. The use of glass capillaries for cryoXT originally emerged owing to severe space constraints at the specimen loading position. These limitations have been overcome by the use of different condenser optics implying that a more widely applicable format should now take precedence for studies, at least of the kind described here with adherent cells (Schneider et al., 2010).

As of mid-2014, cryoXM beamlines at ALBA, DLS, and BESSY II will be compatible with the circular 3.05-mm electron microscopy (EM) support grid format. In order to increase the tomographic tilt range, beamline U41-FSGM of the BESSY II electron storage ring in Berlin/Germany has previously also designed their own rectangular formats (IFR-1, HZB-2; Guttmann et al., 2013), which consist of a large base for grid handling and mounting. The different grid formats each have their own benefits and drawbacks. Circular grids are more difficult to orient in the specimen holder and may be somewhat more difficult to handle without bending. The BESSY II design provides a large area for grid handling and mounting; however, these grids must nevertheless also be handled with care in order to preserve their planarity, and demand some extra preparation steps to achieve proper blotting during plunge freezing (Hagen et al., 2012). The BESSY II design has the benefit that the rectangular meshes are always aligned perpendicular to the tilt axis and, therefore, grid bars are unlikely to obstruct the view of the cell during tilting. In terms of availability from suppliers (including the perforated carbon coats), and concerning the stage design, as well as for possible correlation with cryoFM, and cryoEM, the circular 3.05 mm format could be argued to be the most convenient at the present time.

3.05-mm EM grids can be purchased in a variety of formats and coated with regular perforated carbon by Quantifoil (Jena, Germany) or Protochips (Raleigh, NC; Ermantraut, Wohlfart, & Tichelaar, 1998; Quispe et al., 2007). Alternatively, these suppliers may offer precoated finder grids that can simply be ordered directly. We have typically chosen 100- or 200-square mesh grids (grid codes: G100F1, G200F1, G200F2, G200F4 available from Gilder Grids, Lincolnshire, UK), with the important consideration for tilt experiments being the square size, which influences the likelihood that a particular cell can be viewed over an entire single-axis tilt geometry (±60–65°). For image correlation, any finder grid design can be successfully applied; however, certain designs have "per square" identification schemes, which can speed searches by eliminating ambiguity over the field of view (FOV) associated with typical high-resolution wide-field fluorescence microscopes. When culturing mammalian cells over periods of 18–36 h it is important that gold (Au) grids are used, as these are nontoxic (inert) and are proven to be compatible with cell culture. The perforated carbon support most commonly used is the "2/2" format that consists of regular array of holes 2 μm in diameter, separated from each other by a distance of 2 μm. Keep in mind that different cell lines may, however, benefit from different types on carbon support. Directly before use, specimen support grids must first be rendered hydrophilic by glow-discharge, a method available in most EM facilities.

9.2.1.2 Overview of strategies for fluorescent labeling of cellular features

Fluorescent markers may be introduced in a number of ways. Clearly, for a truly noninvasive correlative strategy, cells should not be permeabilized to allow access of fluorescent dyes or labeled antibodies into the cell and/or subcompartments, and consequently methods for protein labeling are limited to the expression of fluorescently labeled protein (i.e., following transfection or infection with a genetically modified virus). Multiple proteins may be labeled with different tags exciting and fluorescently in different channels (i.e., different colors), and biological events identified in areas where these colocalize. Furthermore, there are organelle-specific (as well as nonspecific) fluorescent dyes that can be incubated with cells to label and facilitate identification of organelles (e.g., MitoTracker; Chazotte, 2011; Wombacher & Cornish, 2011). In near future, fluorescent nanoparticles with functionalized surfaces might develop into useful tools for highly specific labeling of entire compartments or of suborganellar target structures (Huang, Leshuk, & Gu, 2011; Xu et al., 2013).

9.2.1.3 Alignment and correlation markers

For soft X-ray tilt-series alignment, the most commonly used approach is to include colloidal gold fiducials (typically 200–250 nm in diameter) that remain visible over the full angular range of the tilt series—despite the depth of focus (DOF) limitations of the TXM. Some specimens may include inherently dense particles, e.g., lipid bodies or large viruses (Carrascosa et al., 2009), which can also serve this purpose; however, by adding colloidal gold one ensures that the data can be readily aligned—to the highest precision—using popular and freely available software packages such as IMOD (Kremer, Mastronarde, & McIntosh, 1996; Mastronarde, 2008) and BSOFT (Heymann, Cardone, Winkler, & Steven, 2008). Colloidal gold (250 nm) can be purchased, for instance, from BBI Solutions (Cardiff, UK).

Achieving an adequate distribution of alignment particles across the carbon support mesh can be challenging, as the concentration of gold particles may be found to vary with each aliquot and, if applied to the cells immediately prior to rapid immobilization, may settle unevenly across the cell during blotting. It is for this reason that preparing grids with fiducials prior to cell seeding may be the best approach. Using this approach, grids can be preevaluated (prior to cell seeding) by transmission EM or possibly even high-NA light/fluorescence microscopy to ensure that gold is adequately distributed (Duke et al., 2014). Grids can be glow-discharged, and an aliquot of gold suspension applied to the carbon side and allowed to dry.

For correlation all images must include dense/bright features that allow them to be aligned to a common coordinate system. Ideally, one answer should be found for both requirements, a possibility being to include multimodal markers that would allow the localization/correlation to be determined to high precision, as it has been demonstrated recently for cryoFM/EM (Schellenberger et al., 2014; Schorb & Briggs, 2014). FluoSpheres tested for that purpose in cryoXT were applicable only in thin samples (<5 μm; Hagen et al., 2012). However, recently, a combination of both high soft X-ray contrast and fluorescence emission was demonstrated by multi-functionalized gold particles, and these properties were used to facilitate correlative cryoFM/XT imaging of nuclear structures (Hagen et al., 2014).

9.2.1.4 Seeding, incubation, and live-cell imaging

Immediately following glow-discharge and application of markers, the grids should be immersed in cell growth media and hydrated with occasional agitation to allow any air bubbles to disperse. At this stage, other support coating protocols can also be applied to increase cell adherence in later cultivation steps (fibronectin, poly-lysine). Following this, grids can be transferred into cell growth chambers. Grids can be placed on coverslips in either 6-well or standard Petri dishes; however, we use microscope slide growth chambers (μ-slide 2 × 9 well, Ibidi GmbH, Munich, Germany) to facilitate either phase contrast or fluorescence microscopy monitoring of the cell culture prior to cryo-immobilization (Hagen et al., 2012). Use of slide chambers insures that no grid handling is required at this stage. Depending on the cell type, as well as imaging requirements, the appropriate cell confluence (and cell seeding densities) will vary. For most projects, seeding to a confluence of 20% with an incubation time of 24 h will yield an average cell confluence of approximately 40%. Cell confluencies of 50% are typically not exceeded for most projects, although certainly, cryoXM can tolerate higher densities than will cryoET.

Once cells have adhered to the support grid, live-cell fluorescent imaging allows the dynamics of labeled molecules to be observed and, thus, allows cells to be monitored to the point that an event of interest takes place. Specimens can then be rapidly frozen to trap the cell in a state of interest.

9.2.1.5 Cryo-immobilization

Cryo-immobilization is carried out by blotting the rear of the specimen grid (to avoid contacting the cells) for a defined period of time with Whatman No. 1 filter paper and then rapidly plunging the specimen into a liquid nitrogen-cooled liquid preparation of either ethane or ethane/propane mix (Tivol, Briegel, & Jensen, 2008). Blotting shortly before plunging ensures that excess liquid/medium is removed so that an ice layer of minimal thickness is formed while still surround the cells. Plunging into hydrocarbon cryogen, it is essential to achieve cooling rates of at least 10,000 °C— such that the water layer is vitrified without formation of crystalline ice. It has been suggested that this approach is able to achieve acceptable levels of vitrification for specimens up to 10 μm thick (Hurbain & Sachse, 2011). Plunge freezing may be achieved using a manual (gravity-based) plunger, although commercial plungers that allow single-side blotting are also available (e.g., Leica EM GP, Leica Microsystems, Wetzlar, Germany).

Michael Elbaum and coworkers have developed a method for high-pressure freezing of cells grown on planar supports between two metal shims (Weiner et al., 2013). While plunge freezing may not guarantee vitrification throughout the entire cell (in all instances), high-pressure freezing is expected to achieve complete vitrification for even the thickest cells. This approach may be considered somewhat more challenging, also requiring more specialized equipment. At this time, most studies (including our own) have been carried out by plunge freezing, with some exceptions (Kapishnikov et al., 2012).

It is important to note that the ice thickness will most probably vary across the specimen grid and even within particular squares of the mesh. Cell density within the squares will influence how readily growth medium is able to be blotted through the holes of the carbon support, and hence blotting times must be chosen with this in mind. We have found that blotting times of 2–3 s will typically yield adequately thick ice for cells at a confluency of 50%. It is necessary to attempt to ensure that each grid is handled consistently, such that ice thicknesses are similar from grid to grid. CryoXM is somewhat forgiving with respect to specimen ice thickness, as a relatively wide range of blotting conditions are still likely to yield useful images; however, for those new to cryo-microscopy, grids are typically either excessively or insufficiently blotted. For high-resolution tomographic experiments, the thinnest specimens will yield the most isotropically resolved datasets as contrast within each image will decrease with increasing specimen thickness (as the specimen is tilted). The preference for thinner ice should not be taken too far, however, as it is important not to dehydrate the cells, and this may also negatively influence the tolerance of the specimen to the "heat load" of soft X-ray beam. As a result, some thought should be invested in considering the thickness of ice layer one is attempting to prepare. It is this aspect of specimen preparation that is perhaps the hardest to learn master at first, because direct feedback on ice thickness requires a visit to the synchrotron. Fluorescence microscopy combined with phase contrast light microscopy can, however, also reveal important information about the ice thickness in relevant regions (see Section 9.2.2). Users who have access to electron microscopes capable of cryo-microscopy may also consider prescreening the grids to assess ice thicknesses in regions surrounding cells (care should be taken not to bleach the fluorophores with the electron beam, and ROIs should be avoided; Lučić et al., 2007).

9.2.2 CORRELATIVE CRYO VISIBLE LIGHT/FM

Visible light microscopy whether by bright-field, phase contrast, differential interference contrast, or epi-fluorescence is the first line of cryogenic specimen grid screening and biological imaging (Fig. 9.4). By imaging immobilized specimens mounted in an appropriate cryo-stage, we are able to assess ice thickness as well as confirm that the cells exhibiting biological events of interest are still intact. Most importantly, images recorded during this step are fundamental (i) for fast localization of target structures in the TXM and (ii) for biological and scientific interpretation of the soft X-ray data. Consequently, there is a strong motivation to use a high-resolution/high-NA objective lens to identify the events of interest with adequate resolution, sensitivity, and precision.

9.2.2.1 External setup

Commercially available standalone cryo-stages for use with upright and inverted light microscopes are available or under development from/by Linkam Scientific Instruments (Guildford, UK), INSTEC Inc. (Boulder, CO), and FEI Company (Eindhoven, The Netherlands). Linkam cryo-stages allow cryogenic imaging using $100\times$, 0.75 NA lenses; however, recently a modified version of the stage achieved a NA of 0.95 with a $50\times$ lens operating at a working distance of only 300 μm (Schorb & Briggs, 2014). Linkam cryo-stages are able to mount specimen grids of standard 3.05 mm EM, HZB-2 U41, or FEI autogrid formats and are said to be widely compatible with a range of upright microscopes. At present, the stage does not support X/Y motorization; however, this is under development. FEI Company is developing their own product based on the Cryostage^2 (second-generation cryo-stage, MPI of Biochemistry, Martinsried) that has similar capabilities but for inverted microscopes. We use a Cryostage^2 in our laboratory that has been modified to accommodate HZB-2 grids; however, the standard 3.05-mm grid format is also supported (Hagen et al., 2014; Rigort et al., 2010). The Cryostage^2 is placed on a DC 120×100 motorized stage (Märzhäuser, Wetzlar, Germany) and grids are imaged using a Carl Zeiss Axio Observer.Z1 inverted microscope. At present, our strategy for semi-automated grid scanning is to record mosaics at $20\times$, over four different channels to identify ROIs, followed by high-resolution imaging using a $63\times$, NA 0.75 lens (working distance: 1.7 mm). Using the Zeiss AxioVision or ZEN software, mosaic images (grid scans) can be collected, across different channels, and overlayed

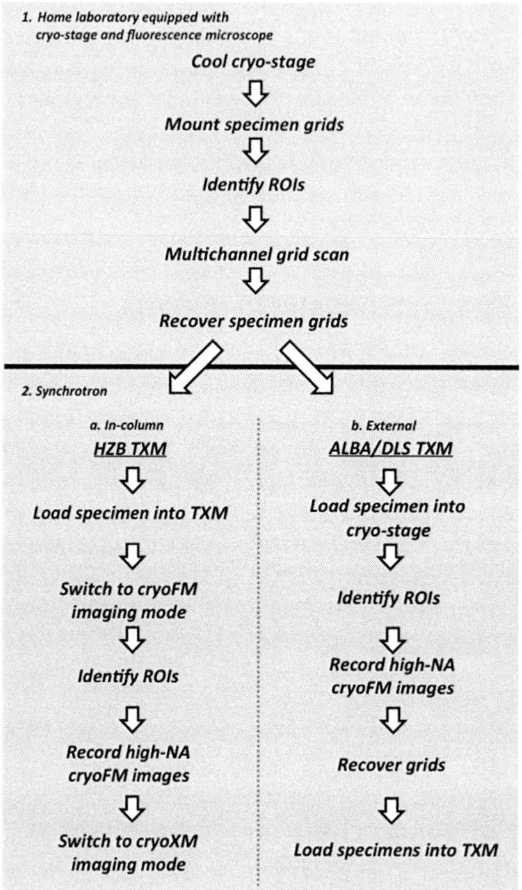

FIGURE 9.4

Alternative workflows for external or in-column cryoFM acquisition of high-resolution (high-NA) fluorescence images. The principal factors determining which workflow is followed are the availability of a cryo-stage in the home laboratory, as well as which synchrotron is to be visited.

for analysis later. The MISTRAL beamline at ALBA currently offers a Linkam cryo-stage to their users.

Potential pitfalls at this stage of the investigation are possible contamination of the specimen by ice crystals as well as condensation from the surrounding atmosphere. Because grids must be handled and mounted into the stage, there is also opportunity for specimen damage either by warming or by mechanical force. Clearly, the cryo-stage must be designed to minimize these dangers, while also maximizing the spatial resolution achieved in the images. As already mentioned it is desirable to use high-NA lenses for imaging—these typically operating at very small working distances. At present, these cryo-stages are operated using "dry" lenses that are most often maintained at room temperature. That the objective must not be exposed to an unstable thermal gradient and that the nearby specimen must not warm are the primary challenges associated with high-NA cryoFM imaging—these serving to constraint the resolution of diffraction-limited cryoFM imaging at this time.

9.2.2.2 In-column setup

In-column fluorescent microscopes have the benefit that once the specimen is loaded, no additional handling is required. The challenge for the manufacturer of the TXM is devising a system that allows a high-NA lens to fit-in among the soft X-ray optics which also operate with working distances of a few millimeters (or less). A detailed description of the in-column reflected light/epi-fluorescence microscope of the HZB TXM can be found in Schneider, Guttmann, et al. (2012). In principle, there the capillary condenser mirror is retracted, and a 100 × NA 0.75 objective comes in place. Thus, the sample can be imaged by light and fluorescence microscopy (three channels altogether), at a relatively high resolution for an in-column approach.

The Carl Zeiss TXM available at DLS offers an integrated epi-fluorescence microscope with a 20 × (0.42 NA, DOF ± 1.6 μm, working distance: 20 mm) objective lens. For either bright-field, epi-fluorescence, or reflective imaging, the specimen must be rotated to −60° and illuminated by either a white or a monochromatic LED source. Users are able to select one of the three Nikon "filter cubes" providing the excitation and emission filters for fluorescence imaging. Images are recorded on a CCD camera using exposure times of 1–3 s, providing a FOV of ~325 μm. Images are focused by moving the sample along the *Z* (optical) axis, and this may be automated by "Focus Function" within the Carl Zeiss XMCONTROLLER software for image acquisition. The alignment of the visible light objective is set to match that of the ZP, serving to facilitate a quick transition between imaging modes once the fluorescence images have allowed a ROI to be identified.

9.2.3 SOFT X-RAY CRYO-MICROSCOPY

9.2.3.1 Overall layout of the TXMs

The components of TXMs are housed within an evacuated chamber (typically maintained at pressure of $\leq 10^{-7}$ torr) to exclude molecules that would interfere with image formation by interacting with the soft X-ray beam. The specimen, as well as each optical component, are each mounted on motorized stages that can be used to adjust their individual positions. The specimen is maintained at cryogenic temperatures at all times during imaging allowing the hydrated specimen to survive the low pressure of the imaging chamber, and also protecting it against radiation damage. An evacuated chamber is also essential to keep the cryogenic specimen from accumulating contaminants over time. In the context of the HZB TXM microscope, this is achieved using a Gatan 630/914 cryo-holders which are cooled externally by a liquid nitrogen dewar. In the case of the Zeiss design, the samples are loaded into the microscope on small "cartridges." In these microscopes, the sample stage, cryo-shield, and robotic cartridge "gripper" are cooled by conduction via a liquid nitrogen dewar mounted on the side of the microscope. This dewar is kept filled by an automated refill system, allowing specimens to remain cold within the microscope for long periods without user intervention.

Soft X-ray microscopes house two primary optical components. These are a condenser (reflective capillary or Fresnel ZP) that focuses the X-rays to illuminate the specimen and a Fresnel ZP objective lens that forms of magnified image of the specimen onto a CCD detector. The capillary is a precisely shaped ellipsoid that reflects and focuses the incident beam to a finely formed spot (typically 1–2 μm in diameter). This spot is typically much smaller than that of the FOV, and hence the illumination must be oscillated to achieve full coverage on the CCD camera. The ZP is a circular diffraction grating, consisting a number of concentric opaque (strongly X-ray absorbing metal) rings that typically number in the hundreds. The ZP is overlayed on a silicon nitride membrane usually ~100 nm thick. The precisely shaped metal rings serve to form a series of transparent zones of decreasing width closer to the edge of the ZP. The resolution limit of the microscope is in large part determined by the width of the outermost zone. Both 40 and 25 nm ZP are routinely available at BESSY II/ALBA/DLS. These differ not only in resolution limit but also in DOF, which are ~2.6 and ~1 μm, respectively (Oton et al., 2012; Schneider, Guttmann, et al., 2012). As a fast exchange of ZP is not (yet) feasible, the imaging magnification can be adjusted only in a limited range (factor of ~2), by setting the distance of the detector from the ZP. The working distance of the ZP is ~2 mm, and particular care must be taken when making adjustments to *X*/*Y*/*Z* parameters to prevent mechanical damage to these delicate physical components. The Rayleigh resolution (or "information limit") of the optical system is influenced by the properties of the incident radiation (principally whether incoherent or partially coherent), as well as the outermost zone width of the ZP. It is important to note that for unstained biological specimens consisting of relatively weakly absorbing low-atomic number elements, this information limit may not be guaranteed in the images or tomograms, as it is the operator's responsibility to ensure that images are recorded at adequate signal-to-noise ratios (SNRs) while preserving the structural integrity of the specimen by avoiding radiation damage.

9.2.3.2 Sample mounting

Sample mounting involves loading the cryogenically preserved specimen into either a holder or a cartridge that is compatible with the microscope goniometer. The key aspects of this process are obviously not to damage the specimen grid, but most importantly not to warm the specimen to the point that ice devitrification occurs (>138 K). Although ice contamination is not as critical for cryoXM as for cryoEM, it should be avoided. Furthermore, a method of controlling the orientation of the grid (that it is kept upright and not flipped upside down), and ideally that the mesh should be oriented at right angles (for circular grids) to the tilt axis, should be applied. Cryogenic grid handling should be practiced in advance of beamline visits, and ideally, at the start of a visit, one should practice using spare grids so that grids can be loaded without unexpected snags or delays. Grids are most commonly handled with finely shaped self-clamping tweezers. Before touching a specimen grid, the tweezers should be precooled by immersing in liquid nitrogen. A good indicator of the tweezers temperature is cessation of nitrogen bubbling, after this point, holding the tweezers for another 2–3 s should bring them to a reliable temperature. Once a grid has been grabbed, the movement must be both quick and decided, so that the grid is moved and returned to contact with a metal surface at cryogenic temperatures. Upon releasing the grid, the tweezers should be recooled prior to making another adjustment of the grids position.

Sample mounting for the HZB TXM at BESSY II was described in Schneider et al. (2010) and Hagen et al. (2012). For the handling of the fragile IFR-1/HZB-2 grids, especially under cryo-conditions in spatial constrained loading cryo-stations of the Gatan grid-tilting cryo-holders used here, use of "cover slip forceps" with a 45°-bended tips are exceptionally helpful (#11251-33, Fine Science Tools GmbH, Germany; cf. Fig. 2B in Hagen et al., 2014).

The Carl Zeiss full-field TXM (MISTRAL—ALBA, B24—DLS, Product code: UltraTXM-L200c/Xradia 825 Synchrotron) features a cartridge/shuttle mounting system which allows four specimens to be loaded into the microscope at a time, and then by automated robot, mounted one by one into the imaging position. A difference between the BESSY II and Carl Zeiss design is that the cells should be mounted facing the detector in the Zeiss microscopes, whereas in the BESSY II, these should face the soft X-ray source (condenser capillary). Zeiss cartridges are designed to interface with the shuttle in a specific orientation (marked with a flat edge), and great care must be taken to ensure that cartridges are loaded into the shuttle in the correct orientation. Zeiss cartridges have been designed to accept either the BESSY II grid design (IFR-1, HZB-2) or standard 3.05-mm EM grids. In future, a separate design will also be available for the FEI Autogrid (Eva Pereiro, personal communication).

When working with the Carl Zeiss cryo-transfer station, standard 3.05-mm EM grids are loaded into cartridges that use a similar grid mounting mechanism as the Gatan 914 (high-tilt) cryo-holder; a standard tomography holder used for side-entry electron microscopes. The specimen cartridge is placed vertically into the "loading block," rotated 90° into the grid-loading position, and the cartridge springs are opened to allow the specimen to be loaded. If ice contamination on the grid becomes visible to the eye, grids may be washed with a gentle stream of liquid nitrogen to dislodge ice crystals that may have attached to the grid during transfer. This procedure may produce variable results depending on the nature of ice contamination, and great caution must be taken not to cause ice cracks, or to detach the immobilized cells from the support; the appropriateness of this step will depend on cell type. Liquid nitrogen should be used sparingly and added to the transfer station precisely, to keep the level in the loading chamber within a narrow range. Nitrogen levels too high will lead to condensation and ice formation in unwanted areas, potentially leading to grid contamination. Additionally, turbulence from liquid nitrogen bubbling will complicate specimen loading. A nitrogen level too low, however, risks compromising the temperature of the specimens as they have to remain in the transfer station while up to three additional cartridges are loaded.

The shuttle (with loaded cartridges) is manipulated on a rail system using a retractable arm on the Carl Zeiss transfer chamber, which allows the shuttle to move from the loading station (cryo-workstation), while under vacuum and at cryogenic temperatures, into the microscope without difficulty. Cryo-transfer involves various steps of transfer chamber evacuation as well as venting to atmospheric pressure with nitrogen gas. These sequences are automated using vacuum PLC display screens that clearly prompt the user to carry out various actions. The robot utilizes a gripper system that is compatible with both the base of cartridges and the lids of the specimen cartridge covers. Users can expect automated cartridge load/unload times of ~4 min, procedures which involve removing a specimen cover, moving the cover to another position in the specimen shuttle, and subsequently lifting the specimen out of the shuttle and lowering into the imaging position. At the time of the load, the X-ray optics are automatically withdrawn and then reextended into the previous position automatically.

9.2.3.3 Search

In general, visible light imaging of TXM-loaded specimens allows quality and ice thickness to be assessed prior to data collection. While this will have been confirmed during previous steps, changes to the specimen may have occurred during mounting. Once a ROI is centered using the in-column light microscope, one can switch to soft X-ray imaging to evaluate the region for possible data collection. However, searching for ROIs without having the opportunity to change magnification quickly can consume large portion of the allocated beam time, and consequently developing a strategy to efficiently move to ROIs may greatly improve the productivity of a synchrotron visit. For this reason, it is advisable to prescreen grids by offline cryoFM prior to the visit. Mapping the grid either onto a paper reference sheet or carrying out an automated "grid scan" and saving a digital montage are possible options. ROIs are of course fundamentally determined by the presence of the biological signal of interest (e.g., in many cases fluorescence colocalization); however, not all ROIs are suitable for cryoXM/T data collection owing to variable ice thickness/quality as well as availability of sufficient concentrations of fiducials. Searching using both visible and soft X-ray light should be employed to identify and exclude regions as relevant. However, it should be kept in mind that even low doses of soft X-ray irradiation are efficient in quenching the fluorescence of different fluorophores (Hagen et al., 2012, 2014). Thus, in practice, fluorescence imaging should precede exposure to soft X-rays. Once loaded into the microscope, multiple locations can be identified using the in-column light cryo-microscope, and specimen stage coordinates (X, Y, and Z, with description) noted down for screening in soft X-ray mode. Switching between modes requires 5–10 min, and every attempt to save time should be made.

The in-column fluorescence cryo-microscope of HZB TXM offers the benefit that high-resolution cryoFM images can be recorded immediately prior to soft

X-ray imaging. For users who do not have access to a high magnification/NA fluorescence microscope in their home lab, this functionality has numerous benefits. The Carl Zeiss TXM available at MISTRAL and B24 features low-magnification objective lens (20 ×, NA 0.42). When images recorded using this system are compared and used in conjunction with an offline higher NA cryo-stage, searching to identify a ROI is greatly facilitated. When searching for an area of interest using the Zeiss TXM, the best approach is imaging using simultaneous bright-field and fluorescence microscopy (two channels), as this allows one to differentiate between cells that have grown over the carbon support versus those which are on the grid bars, while also allowing identification of the fluorescent signal of interest. The bright-field (white) LED can then be turned off once a suitable cell has been identified, and fluorescence imaging used to refine ones position and focusing for soft X-ray imaging with reasonable precision.

Initial soft X-ray imaging allows the final confirmation that the ice in a particular region is vitreous. Ice damage is clearly visible as a dark network of crystallization-induced aggregation/segregation artifacts extending through regions of the specimen (Hagen et al., 2012; Weiner et al., 2013), which obscure biological details in the image and also suggest that cell is heavily damaged in these regions. Devitrification/crystalline ice may extend throughout the entire specimen (suggesting a problem in loading—grid warming) or be limited to only thick regions (suggesting an inadequacy in rapid freezing method or protocol).

When working in soft X-ray imaging mode, the Zeiss TXMs are able to record soft X-ray mosaics that allow larger FOVs to be inspected as a single tiled image. Mosaics are recorded by moving the specimen on the *X*- and *Y*-axis at a step size equal to the FOV on the CCD camera. The user has control over the exposure time, as well as the camera binning (bins 1–8), which means that images can be recorded quickly and at restricted radiation doses. For a single square of the grid mesh, for example, mosaics consisting of 10 × 10 images can be recorded at binning factors of 2 or 4 using 1-s exposure times, per tile. Such 2D soft X-ray mosaics may be exported as single images after reference correction and can also be correlated with fluorescence microscopy images (covering FOV of 75–150 μm or more) in order to study localization of signal of interest across entire cells.

9.2.3.4 Tilt-series acquisition

The aim of the following sections is to describe the basic workflow for setting up single-axis tilt-series data collection (Fig. 9.5). The fundamental principle involved in recording a tilt series is to fractionate a total soft X-ray dose over a number of exposures, each with a defined change in specimen orientation (the increment). The exposure and tilt parameters used here will heavily influence the properties and signal content of the individual images, as well as the final tomogram.

Each TXM design, whether Carl Zeiss or HZB TXM, is controlled by different operating software (also interfacing with unique hardware). As a result, the precise steps of setting up a collection will vary depending on which facility is visited. With respect to the tilt-series data collection, the following sections aims to inform what

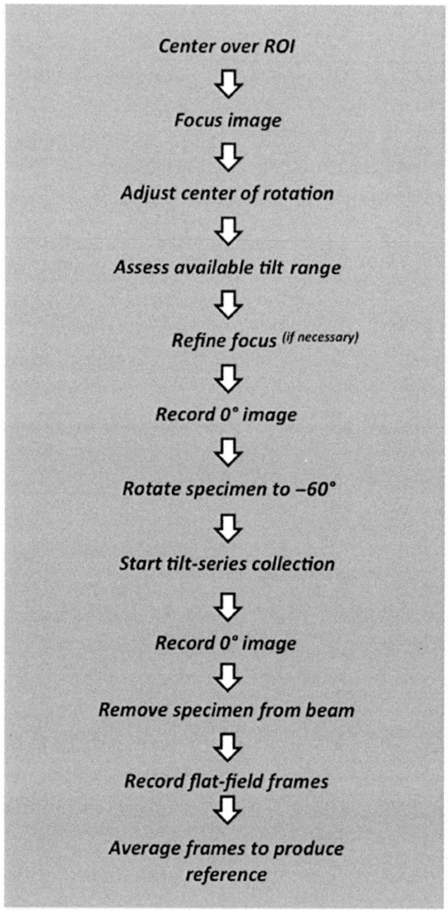

FIGURE 9.5

Workflow for cryoXT tilt-series acquisition.

options are available with respect to imaging parameters. However, it is important that plans are discussed with a beamline scientist in advance of the experiment.

9.2.3.4.1 Fresnel ZP and imaging magnification

A beamline visit will be for the use of either a 25- or a 40-nm ZP. At present, swapping ZPs requires that the microscope is vented to atmosphere and opened so that the ZPs can be carefully exchanged—a time-consuming process, especially for realignment after the exchange. It is for this reason that beamlines typically offer different periods of year for use of each ZP. Because DOF is limiting in soft X-ray experiments, the choice of ZP objective will influence the ease with which complete tilt series are recorded (and analyzed later on). In summary, data collection using a 25-nm ZP is often more challenging than that using a 40 nm. There are two primary reasons for this: first the DOF is more restricted and second 25-nm ZP is usually used with higher imaging magnifications (e.g., 2000 ×) and, as a result, the FOV is also reduced (influencing search times for example) and the beam intensity needed to achieve a similar SNR per image as compared to lower magnifications is increased (and, thus, radiation exposure). For these reasons, it is preferable to choose to start-out using the 40-nm ZP and then plan a second visit to take advantage of the higher resolution ZP if necessary.

Once a ZP has been chosen, the next step is to choose the imaging magnification. The soft X-ray micrographs (images) will be characterized by a pixel size or sampling. The resolution limit of information encoded in a digital image is twice that of the sampling size (Shannon–Nyquist theorem). The image sampling (or object pixel size) will depend on the physical size of the CCD camera pixel (13 μm—PIXIS-XO 1024B, 20 μm—PIXIS-XO 1300B, Princeton Instruments) as well as the magnification. Most commonly this will be on the order ~10–16 nm/pixel. According to the image processing theory, this value will directly limit the resolution of images and tomograms (with corresponding Nyquist resolutions of 20–32 nm). A change of magnification may be used to adjust the specimen sampling and can be expected to take about 5 min if this only involves moving the camera (Hagen et al., 2012). Also it should be kept in mind that choice of magnification will influence exposure time, as image contrast will typically be reduced at higher magnifications, implying that higher X-ray doses are required to achieve similar image quality.

9.2.3.4.2 Focusing

The first step once over a ROI is to focus the image. Focusing is a central aspect of restricted (low-) dose imaging protocols in cryoEM/T, where focusing (prior to exposure) occurs in a region adjacent to the ROI. In the case of cryoXT, where relatively large areas are imaged, the imperfections in planarity of the specimen often become significant (resulting in an apparent specimen *Z* shift >1 μm). This means that methods involving focusing in adjacent regions are prone to fail. In addition to this, for thick cells we often have to attempt to focus within a particular plane of the cell. As a result, we are forced to focus precisely on the area we are planning to image and must accept that some radiation damage is likely to take place prior to data collection. As a result, focusing steps should use minimal exposure times, as well as intermediate camera binning so as to achieve the desired result with minimum radiation dose.

NB: The Zeiss "Focus Function" allows a series of images to be recorded over a specified specimen *Z* range. We suggest initially recording 11 images, over 100 μm, and then a further 11 images over a range of 10 μm. By inspecting the image series in conjunction with the associated numerical score, one is able to choose the "best" specimen *Z* position.

9.2.3.4.3 Angular range and center-of-rotation

Focusing is important for other reasons. When appropriately aligned, the imaging plane of the ZP should coincide with the plane of the goniometer tilt axis. This physical reference point, referred to as the center-of-rotation, or in cryoEM as the eucentric height, is essential for tilt-series collection because it ensures that the ROI remains centered in the FOV over the full angular range. In addition, it should be expected that the tilt axis is aligned to pass through the center of the FOV on the CCD camera. Misalignment of these components will be apparent by large shifts in image features as the specimen is rotated. The easiest way to confirm that the center-of-rotation is aligned is to track the position of fiducials in short-exposure images as the specimen is tilted. In certain circumstances, the position of the ZP may need to be refined and should only be carried out in the presence of user support.

Assessing the available angular range for a particular ROI is important. Not all cells on a specimen grid will be accessible to tilt-series data collection, because at higher angles, the specimen may either become too thick or be obscured by a grid bar. Experience will help to predict the behavior of an area under tilting; however as a general rule, the closer a cell is to a grid bar (parallel to the tilt axis), the more likely it is that the bar will interfere with collection. For an adequate 3D reconstruction, an angular range of at least 90° (±45°) must be covered, although ±60–65° range should be considered standard for planar specimens—implying that a dataset consisting of at least 81–121 images is necessary (assuming 1° increment between images). The tilt range used will determine the size of the missing wedge in the 3D tomogram, as well as the extent of blurring along the direction of the beam in the reconstruction. A reliable way of testing if the ROI is visible over a particular tilt range may be to record short-exposure images over the desired tilt range with a coarse angular increment (10–20°). Tilting to extremes should then show a minimal specimen shift, and this step will confirm whether the range is obstructed by grid bar. Any adjustments to the center-of-rotation will require small corresponding focus adjustments of the ZP objective.

9.2.3.4.4 Tilt-series parameter choice and dataset collection

After an area has been chosen for imaging, parameters for tomographic data collection have to be specified. The precise details of a data-collection "recipe" will vary according to the behavior of the specimen under the X-ray beam (i.e., its sensitivity to radiation damage) as well as its thickness. The data-collection geometry, i.e., angular range (e.g., ±65°) and increment (typically 1°) will influence final tomogram resolution, predicted theoretically according to the Crowther criterion (Lučić et al., 2005). However, the SNR of individual frames (influenced by the exposure time used for each tilted image) will also play an equally important role. For these reasons, careful selection of exposure time and number of images recorded is important. At the start of a beamline visit, an option may be to find an "uninteresting" cell that has similar properties to other ROIs, and use this as a trial. This will allow radiation damage to be assessed (discussed later) and also allow typical exposure times (and corresponding CCD saturations) to be evaluated. Exposure times can also be chosen based on 0° images used for focusing (applicable for all ROIs).

As mentioned previously, according to the principles of dose fractionation, while individual frames may suffer from low SNRs, the final tomogram, essentially a synthesis of all the data, will have improved contrast (SNR and resolution). Soft X-ray datasets are often noted for their high contrast in comparison to cryoET (Le Gros et al., 2012). While this is essentially true, it is important to keep in mind that contrast will vary as a function of spatial frequency (resolution) and also differ depending on magnification. In principle, the more photons detected per pixel, the better the SNR of the images; however, this comes at the cost of radiation-induced specimen damage (particularly costly in the context of low-efficiency ZP-based imaging).

The dose of soft X-ray beam on the specimen can be evaluated by reading the current on a photodiode inserted into the path of the condenser illumination. This current reading can be used to calculate the photon flux over a particular exposure time and allows the beam to be adjusted by opening or closing the exit slits of the beamlines' frontend (primary optics) or the monochromator slit (HZB TXM). Because the illumination spot is oscillated to cover the detector, longer exposure times are preferred so as to achieve an even illumination. Owing to the significant heat load placed on the specimen, as well as the extremely low thermal conductivity of water (in comparison to the gold grid), it may be advisable to adjust data-collection times to allow for heat dissipation. Using HZB TXM, 2–2.6 s exposures were applied over a total data-collection time of 35 min, with no reported evidence of observable radiation damage (Carrascosa et al., 2009). We have used exposure times of 4 s with data-collection schemes over similar duration and observed radiation damage principally starting on the beam-facing surface of the sample (Hagen et al., 2012). Furthermore, at these exposure times, we have found that we can exploit the full dynamic range of the detector (at least with the 40 nm ZP), ensuring that the performance of the detector is optimal. Our experience is that, owing to variable ice thicknesses, radiation sensitivity of the specimen varies greatly, and that in some cases one may find that adjusting the exposure time for each frame is necessary. Based on performance measurements on the soft X-ray detectors (in particular the PIXAS-XO), we have seen that maintaining an average pixel saturation of at least 20% is appropriate, as reducing the exposure time to achieve less than this would be to use the detector outside of its linear range, compromising the quality of the image, and introducing unnecessary noise into the data.

9.2.3.4.5 Alternative data-collection strategies

Previous correlative soft X-ray studies have generally collected data over a 120° range (±60°), with a 1° step. This approach has yielded excellent data; however, the community may consider exploring alternatives, e.g., the Saxton scheme (Saxton, Baumeister, & Hahn, 1984), particularly for thicker specimens. By applying the Crowther criterion, we can predict that by reducing the angular step to 0.5° or even 0.25°, we would expect an improvement in resolution in the reconstructed 3D dataset (more closely matching the 2D information limit). Certainly, the principles of dose fractionation imply that this can be achieved in some situations without increasing the total exposure time (radiation dose), e.g., if exposure time were halved in each frame (for example), while doubling the number of images recorded. Our experiments have shown that the tilt series recorded at similar total exposures show equivalent SNRs when tomograms are evaluated by the NLOO-2D/3D methods (see

Section 9.2.4.3). Recording more images is also beneficial for tomogram voxel statistics. Unlike single-particle cryoEM averaging approaches, for cryoXT, the only opportunity to overcome noise by averaging is during computational back projection.

Contrast in images, as already stated, arises from the absorption of X-rays by the specimen. For planar specimens, as the specimen is tilted it increases in thickness, and as a result, absorption will vary dramatically across the tilt series. For similar reasons, some cryoEM studies have used variable exposure time collection schemes (for instance, 1/cos scheme; Cope, Heumann, & Hoenger, 2011) that increase exposure so as to preserve pixel counts (and consequently SNRs) at high-tilt angles. Ultimately this ensures that the resolution of resultant tomogram is as isotropic as possible.

Last but not least, a dual tilt-axis data-collection scheme may be worth considering as a means of alleviating "missing wedge" artifacts for planar specimens (Mastronarde, 1997). Whereas the HZB TXM may not have sufficient space to accommodate a standard cryoET dual tilt-axis holder, corresponding modifications to the cartridges of the Carl Zeiss TXMs should be feasible.

9.2.3.4.6 Assessing radiation damage

During data collection, it is important to be aware of changes to the specimen as a result of exposure to the beam. While severe radiation damage will become obviously apparent as "bubbling," it is a common practice to record a "before" and a "after" untilted (0°) image. This has the benefit of revealing pronounced (low-resolution) structural changes within the specimen, which if observed, calls for a reduction in total exposure time for future datasets of similar specimen regions. It must be considered, however, that when operating within a dose-fractionated exposure regime, individual frames may not reveal radiation damage for higher resolution (smaller) features (owing to poor SNRs in the individual images). Averaging a series of 10 images, for example, may be more informative in some cases. That being said, it should be kept in mind that the point of data collection is to provide datasets that reveal as detailed view of the specimen as possible (in 3D). For this to be achieved reliably, some degree of radiation must be deemed acceptable, as damage in later images of tilt series should not contribute significantly to the appearance of the final tomogram.

We have observed that a specimen's sensitivity to the incident soft X-ray beam will vary according to the carbon/nitrogen density of a particular region; behavior not unexpected for absorption contrast imaging. Dense protein aggregates or lipid bodies are seen to undergo changes such as mass loss more quickly than other areas. A particularly important damage process seems to arise from the accumulated "heat load" of the exposure series. The details of this apparent localized warming are presently poorly understood, but we have observed that thin regions with minimal surrounding water are particularly prone to this kind of damage.

When attempting to track and manage radiation damage, it is best to take note of both the flux of the incident radiation onto the specimen, as well as the statistics for detector pixel saturations for each dataset. The transmissivity for a particular region of a cell is also useful and can be estimated by comparing mean intensities for flat-fields with those of experimental images. Transmissivities will typically vary from 15% to 40%, depending on the cell thickness.

9.2.4 DATA ANALYSIS

9.2.4.1 Data normalization and flat-field correction

Owing to optical design, the illumination profile (or "shape" of the beam) of TXMs is not typically even. Flat fielding is important to overcome these properties of the illumination, and subsequent normalization of pixel intensities allows images recorded at different tilt angles (differing thickness) to be viewed on a similar scale. Flat-fields should be recorded after each dataset. With the specimen removed, 10 successive images are recorded at the exposure time used for the 0° experimental image. These are then averaged to overcome noise to form a reference image. The reference can then be dived into each experimental image, thereby overcoming the shape of the beam to produce an absorption corrected image suitable for tomographic reconstruction.

9.2.4.2 Volume reconstruction

Presently, X-ray transmission images are treated as 2D projections of the specimen, and owing to the complexity of the cell interior, 3D reconstruction by back projection is necessary for unambiguous analysis of fine structural features that would otherwise remain superimposed (Grünewald, Medalia, Gross, Steven, & Baumeister, 2003; Le Gros et al., 2012).

Software packages originally written for 3D reconstruction of EM data (e.g., IMOD, BSOFT) and XMIPP (de la Rosa-Trevín et al., 2013) can be applied to soft X-ray data without significant changes. Detailed tutorials for the use of these packages are available online and may include example (EM) datasets and tutorials. The IMOD package (Boulder, CO) is probably the easiest to use and includes a graphic user interface (Etomo) that keeps track of progress through each reconstruction step. Herein, we will cover some of the basic steps in an IMOD reconstruction (Fig. 9.6).

Reference corrected datasets are exported in MRC format (MRC-LMB, Cambridge, UK). In addition to these files, necessary starting parameters for IMOD reconstruction are object pixel size (e.g., 16 nm) as well as the angular position of the tilt axis. For the Carl Zeiss microscopes, this is 0°, while for the HZB TXM, this value is 90°. The first step prior to coarse alignment of the dataset is the removal of hot pixels using the CCDERASER program of IMOD. Our experience is that this works well for soft X-ray data using default parameters.

Tilt-series alignment serves to align images to a common coordinate system and is of fundamental importance. Fiducial-less methods (e.g., the ALIGNATOR software from the Frangakis group for alignment of tilt series of thin-vitreous sections; Castaño-Díez, Scheffer, Al-Amoudi, & Frangakis, 2010) or patch tracking as implemented in IMOD has been applied to soft X-ray data in certain cases (Hummel et al., 2012; Parkinson et al., 2012), although for thick specimens, especially these are

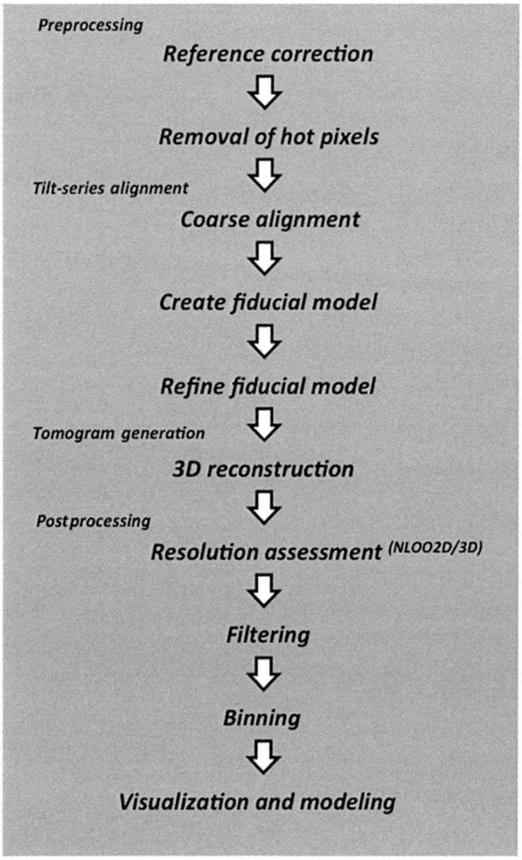

FIGURE 9.6

Computational workflow for 3D reconstruction of tilt series.

prone to failure at high tilt angles (distortions due to DOF, and also low SNR in high-tilt images). At present, fiducial-based alignment using a software package such as IMOD is the most easily and reliably applied. Fiducial models can be created by hand (manually); however, automated methods for bead selection and tracking (e.g., as implemented in the RAPTOR software package as part of the Etomo GUI of IMOD; Amat et al., 2008) can be successfully applied but require at least 10 fiducials to be present in the FOV. For the time being, manually supervised fiducial alignment remains the most reliable option for high-quality alignment.

Once a tilt series has been aligned, 3D reconstruction is typically achieved using one of the two methods: weighted back projection or the simultaneous iterative reconstruction (SIRT; also implemented in IMOD). We have observed that SIRT can provide important contrast improvements for particularly noisy datasets. SIRT may produce reconstructions of improved visual quality, a property related to the removal of noise after each reconstruction iteration. Our experience has been that choice of how many iterations to carry out is somewhat dataset dependent (and is also subjective) and will need to be assessed by trial and error.

9.2.4.3 Resolution assessment and postprocessing

As described previously, the strength of the signal in "water-window" (absorption contrast) soft X-ray imaging is heavily influenced by the atomic composition of the specimen. It is for this reason that the Rayleigh (or information) limit of the microscope is only achieved under precise imaging conditions when imaging thick biological specimens (consisting of low-atomic number elements). Simply put, a resolution test using a metal test pattern (while likely to reveal the imaging performance of the microscope) will not provide a reliable estimate for bioimaging.

For this reason, it is important to apply quantitative methods for resolution assessment that can be used to support interpretation of the experimental data. The quality of any study will in part be determined by our understanding of a particular dataset's limitations, and we must keep in mind that while a particular microscope may be capable of imaging at a certain resolution, not all images/tilt-series will be recorded with equal properties. In particular, we advise that tomograms should be filtered to remove noise prior to segmentation, as interpretation and modeling of noisy data can only be marred by subjectivity.

Features that influence a datasets quality include, but are not limited to, drifted images, or low SNRs (depending on specimen thickness and exposure time). These factors have directly quantifiable effects and serve to limit resolution and contrast. While the resolution limit of the microscope is primarily influenced by the outermost spacing of the ZP, the detector itself also has performance limitations that may vary depending on use. Given all of these factors, how then should we go about quantifying the quality (and hence resolution) of a particular dataset?

Resolution estimation in this context is a complex topic that has seen considerable development for cryoEM data—this owing to the presence of relatively high levels of noise. While soft X-ray datasets are characterized by high contrast, this contrast varies over the spatial frequency range of each image just as it does in cryoEM. One reason for this is the contribution of noise by the detector, a property referred to as the detective quantum efficiency. At low resolutions (e.g., features greater than 100 nm), the SNR may be relatively high (responsible for giving the cryoXT images there impressive appearance); however, for features smaller than 100 nm (for example), noise may increasingly make the identification of features difficult. In short, noise has the property of obscuring important biological details in the tomogram, and it is for this reason that we argue here that once 3D reconstruction has been carried out it is advisable that resolution assessment is carried out using a method such as that used for 2D or 3D noise-compensated leave-one-out estimation NLOO-2D/3D (Cardone, Grünewald, & Steven, 2005). This method was adapted from the Fourier shell correlation method applied in cryoEM, where averaging of multiple "single-particles" is applied (in 2D or 3D datasets), but is formulated specifically for the situation where 3D reconstruction is calculated from a dose-fractionated tilt series. Essentially, the NLOO method allows noise to be quantified within a particular dataset, and consequently the resolution to be assessed in accordance with this.

The ELECTRA package (containing amongst others the programs NLOO-2D, NLOO-3D) is maintained and made available on request from Giovanni Cardone (University of California, San Diego) and has been conveniently integrated into either IMOD or BSOFT reconstruction workflows. NLOO-2D, by plotting resolution as a function of tilt angle, is particularly useful in that it allows one to spot weaknesses in the dataset, such as drifted (blurry) images, as well as whether specimen thickness or radiation damage is compromising the reconstruction's overall resolution in specific images. One benefit of this process is that severely problematic images can then be excluded from the reconstruction. We have found the most reliable estimate of cryoXT resolution to be that corresponding to an NLOO-3D value of 0.3 (Hagen et al., 2012), as proposed for cryoEM data in Cardone et al. (2005). For soft X-ray data, the even/odd method is likely to underestimate the true resolution of the dataset. It must be kept in mind that these procedures are probably the most computationally intensive steps in the experimental workflow described in this chapter.

9.2.4.4 Visualization and quantitative measurements

3D tomograms can be visualized using a number of different software packages. For those using IMOD, the 3DMOD program offers the most convenience as it can be accessed directly from the Etomo GUI (Fig. 9.7). Within programs such as 3DMOD, tomogram *XY* slices (1 voxel thick—typically along the *Z*-axis) are typically inspected. Certain programs (e.g., 3DMOD) also include the option of increasing the "slab" thickness, and we have found that visualizing slabs 100–200 nm thick can be informative, providing a view more similar to plastic section-EM.

Segmentation involves actively defining boundaries within the tomogram, thus allowing quantitative measurements regarding organelle volumes (for example) to be made (Hanssen et al., 2012). An added benefit of this process is that the entire depth of a tomogram can be visualized with various regions colored to assist communication of results. Programs for segmentation include Amira (FEI, Eindhoven, The Netherlands), IMOD (3DMOD), as well as UCSF Chimera (Yang et al., 2012).

Segmentation also has an important role to play for cryoFM correlation. Using presently available cryo-stages, images from cryoFM can only be recorded in 2D. Because fluorescence localization is, therefore, only established in the *XY* plane, correlation of the datasets can be complicated by the uncertainty along the *Z*-axis. Future developments may allow for confocal cryo-microscopy to partially overcome this problem; however, segmentation according to density thresholds represents a way of 'reducing the dimensionality' of cryoXT data such that two sets of images (representing features over full axial range) are directly comparable.

9.2.4.5 Correlation of specimen images

Clearly, the "correlative approach" requires that images recorded using two or more different modalities are combined so that a composite image can be interpreted in consideration of a specific biological question (Fig. 9.7). At present, image

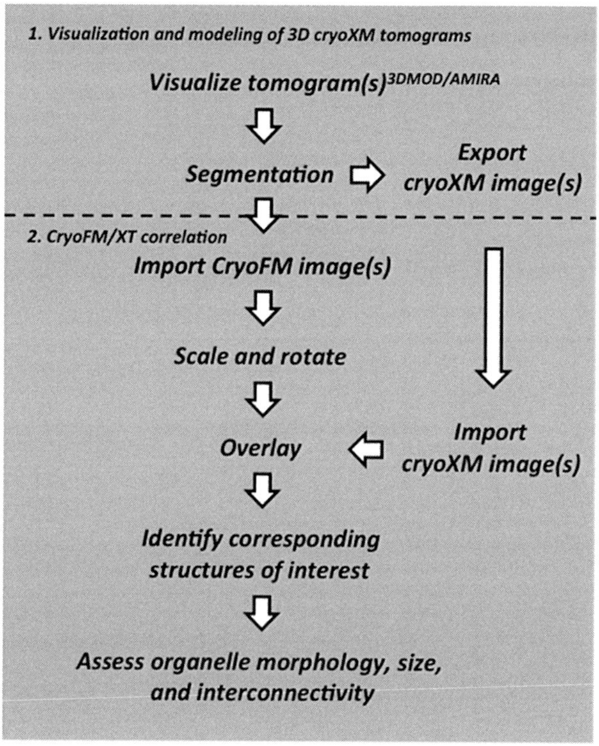

FIGURE 9.7

Workflow for correlation of cryo-microscopy images.

correlation of cryoXT (~30–100 nm resolution) with diffraction-limited cryoFM (~450 nm resolution) is considerably less challenging than for correlative EM and does not require the accuracy afforded by precision alignment of fluorescent beads (Schellenberger et al., 2014; Schorb & Briggs, 2014). Nevertheless, mutually visible fluorescent markers can (or should) be included; however, any shared features (i.e., organelles revealed by specific dyes) can also be used. In our work, we have applied 200 nm FluoSpheres in the hope that these could serve as markers for tilt-series alignment (Hagen et al., 2012). These particles are not dense enough to remain visible over a full tilt range (for thick specimens), but these do provide a strong fluorescent signal that can be used for quantitative correlation procedures (if appropriate) and therefore help to eliminate any ambiguity that may arise during downstream processing.

Composite images can be created using standard packages for image processing. CryoFM images can be scaled and oriented, and then subsequently overlaid over the soft X-ray images. Most groups have used Adobe Photoshop to carry out this process (Duke et al., 2014; Hagen et al., 2012); however, other packages such as ImageJ may also be considered (Schneider, Rasband, & Eliceiri, 2012).

CONCLUSION AND PROSPECTS

CryoXM is an emerging technique for structural cell biology that is becoming increasingly accessible to a broader community. The technique offers a rapid cryogenic approach to "near-native" high-resolution imaging, while also offering the potential of high-throughput visualization of cellular organelles in relatively large volumes. The unique contrast mechanisms at work in cryoXM provide a distinct view of the hydrated cell that is more sensitive to the chemical composition than other imaging methods. For this reason, cryoXM offers benefits when imaging even thin specimens, especially when applied in conjunction with EM. Certainly, the technique also holds important promise for integrative approaches to structural biology that seek to unify the mechanistic and structural changes of macromolecules with the greater (wholistic) perspective of the living cell.

As a scientist hoping to apply cryoXM, the primary limitation at present is securing beam time, and then subsequently ensuring that this time translates into valuable experimental data. Access to a nearby epi-fluorescence microscope with cryo-stage that will allow grids to be prescreened and ice thickness to be evaluated prior to a synchrotron visit is key. While such an approach cannot guarantee success at the synchrotron, future developments may see synchrotrons offering laboratory facilities that allow users to culture and prepare their specimens immediately prior to, or during beamline visits (E. M. H. Duke, personal communication). Furthermore, it is important that datasets are evaluated by 3D reconstruction during the beam time itself and we highly recommend that (at least initially) some time is dedicated to exploring different data-collection parameters in consideration of the behavior of the specimen.

The ability of correlative cryoFM/XM to inform our understanding of cellular mechanism is currently limited by the resolution of images recorded using cryoFM. While future development of high-NA cryo-immersion objectives will lead to some improvements in the context of wide-field diffraction-limited imaging, the full potential of the correlative approaches described here will only be achieved by fluorescent imaging at 100 nm resolution or less (below the diffraction limit of visible light). As a result, there are efforts to apply super-resolution methods (e.g., STORM, PALM, and SIM) to cryogenic specimens (Chang et al., 2014; Kaufmann, Hagen, & Grünewald, 2014; Kaufmann, Schellenberger, et al., 2014). These methods, also relying on the development of cryo-immersion objectives, will allow protein localization to areas within organelles with good sensitivity, allowing cellular mechanisms to be elucidated in further detail.

Soft X-ray microscopy has a number of limitations, most notably the DOF of ZPs that do not match the penetration depth associated with ~520 eV photons. Phase contrast cryoXM/T can, in principle, be carried out by moving to higher energies (2.1–2.5 keV) and inserting a Zernike phase ring after the ZP. This offers the potential to overcome the DOF limitation and will also allowing thicker specimens to be imaged (Schneider, Guttmann, et al., 2012). However, it remains to be seen whether sufficient contrast will be able to reveal structural information of a useful resolution and without severe radiation damage, and it is likely that detector developments—similar in essence to those recently seen in cryoEM (Ruskin, Yu, & Grigorieff, 2013)—will be required for the method to realize its full potential.

In this chapter, we have advocated the use of the planar specimen format. While the missing wedge is an undesirable consequence of the data-collection geometry associated with specimens of this kind, for adequately recorded data, the missing wedge will (in most cases) not significantly impact on the interpretability of the data. That being said, it is possible that the application of dual-beam micromachining, for instance to high-pressure frozen samples (Hayles et al., 2010), even in biopsies (Vanhecke et al., 2003), may allow soft X-ray tilt-series geometries to routinely be extended to $\pm 80°$, which will significantly reduce image artifacts associated with specimens prepared in this way.

Finally, the soft X-ray imaging field is rapidly developing, with new methods and understanding emerging continually. While this chapter will certainly not be the last word regarding the use of correlative cryoFM/XM, we are confident that cryoXM will be adopted as another important method for the study of the biological mechanisms of eukaryotic cells.

ACKNOWLEDGMENTS

We thank Gerd Schneider (Helmholtz-Zentrum Berlin für Materialien und Energie GmbH), Eva Pereiro (ALBA Synchrotron Light Source, Barcelona), Elizabeth M. H. Duke (Diamond Light Source, Oxfordshire), and their beamline teams for support and hospitality. We acknowledge the TXM teams at Carl Zeiss X-ray microscopy and Energetiq for ongoing support. We thank Jürgen Plitzko and the correlative team at the MPI of Biochemistry for sharing experience and supporting the Cryostage^2 hardware. We thank Cathy Whittle for help in the lab. This work was supported by grants from the Human Frontier Science Program (RGY0079/2009-C to K. G.), a Wellcome Trust Senior Research Fellowship to K. G. (090895/Z/09/Z), a Wellcome Trust JIF award (060208/Z/00/Z), a Wellcome Trust equipment grant (093305/Z/10/Z) to the Oxford Particle Imaging Centre, and the Wellcome Trust core award 090532/Z/09/Z to the Wellcome Trust Centre for Human Genetics. We acknowledge the Helmholtz-Zentrum Berlin—electron storage ring BESSY II for provision of synchrotron radiation at beamline U41-FSGM. The research leading to these results/experiences has received funding from the European Community's Seventh Framework Programme (FP7/2007–2013) under BioStruct-X (grant agreement no. 283570 and no. 226716).

REFERENCES

Amat, F., Moussavi, F., Comolli, L. R., Elidan, G., Downing, K. H., & Horowitz, M. (2008). Markov random field based automatic image alignment for electron tomography. *Journal of Structural Biology*, *161*, 260–275.

Attwood, D. (2007). *Soft X-rays and extreme ultraviolet radiation: Principles and applications*. New York: Cambridge University Press.

Bertilson, M., von Hofsten, O., Hertz, H. M., & Vogt, U. (2011). Numerical model for tomographic image formation in transmission X-ray microscopy. *Optics Express*, *19*, 11578–11583.

Bleck, C. K. E., Merz, A., Gutierrez, M. G., Walther, P., Dubochet, J., Zuber, B., et al. (2010). Comparison of different methods for thin section EM analysis of *Mycobacterium smegmatis*. *Journal of Microscopy*, *237*, 23–38.

Cardone, G., Grünewald, K., & Steven, A. C. (2005). A resolution criterion for electron tomography based on cross-validation. *Journal of Structural Biology*, *151*, 117–129.

Carrascosa, J. L., Chichón, F. J., Pereiro, E., Rodriguez, M. J., Fernandez, J. J., Esteban, M., et al. (2009). Cryo-X-ray tomography of vaccinia virus membranes and inner compartments. *Journal of Structural Biology*, *168*, 234–239.

Carzaniga, R., Domart, M. C., Collinson, L. M., & Duke, E. (2014). Cryo-soft X-ray tomography: A journey into the world of the native-state cell. *Protoplasma*, *251*, 449–458.

Castaño-Díez, D., Scheffer, M., Al-Amoudi, A., & Frangakis, A. S. (2010). Alignator: A GPU powered software package for robust fiducial-less alignment of cryo tilt-series. *Journal of Structural Biology*, *170*, 117–126.

Chang, Y.-W., Chen, S., Tocheva, E. I., Treuner-Lange, A., Löbach, S., Søgaard-Andersen, L., et al. (2014). Correlated cryogenic photoactivated localization microscopy and cryo-electron tomography. *Nature Methods*, *11*, 737–739.

Chazotte, B. (2011). Labeling mitochondria with MitoTracker dyes. *Cold Spring Harbor Protocols*, *2011*, 990–992.

Chichón, F. J., Rodriguez, M. J., Pereiro, E., Chiappi, M., Perdiguero, B., Guttmann, P., et al. (2012). Cryo X-ray nano-tomography of vaccinia virus infected cells. *Journal of Structural Biology*, *177*, 202–211.

Cinquin, B. P., Do, M., McDermott, G., Walters, A. D., Myllys, M., Smith, E. A., et al. (2014). Putting molecules in their place. *Journal of Cellular Biochemistry*, *115*, 209–216.

Cope, J., Heumann, J., & Hoenger, A. (2011). Cryo-electron tomography for structural characterization of macromolecular complexes. *Current Protocols in Protein Science*, *Chapter 17*, Unit17.13.

Dedecker, P., De Schryver, F. C., & Hofkens, J. (2013). Fluorescent proteins: Shine on, you crazy diamond. *Journal of the American Chemical Society*, *135*, 2387–2402.

de la Rosa-Trevín, J. M., Oton, J., Marabini, R., Zaldivar, A., Vargas, J., Carazo, J. M., et al. (2013). Xmipp 3.0: An improved software suite for image processing in electron microscopy. *Journal of Structural Biology*, *184*, 321–328.

Dubochet, J. (2007). The physics of rapid cooling and its implications for cryoimmobilization of cells. In J. R. McIntosh (Ed.), *Cellular electron microscopy* (pp. 7–21). San Diego: Elsevier Academic Press.

Dubochet, J. (2012). Cryo-EM - the first thirty years. *Journal of Microscopy*, *245*, 221–224.

Duke, E. M. H., Razi, M., Weston, A., Guttmann, P., Werner, S., Henzler, K., et al. (2014). Imaging endosomes and autophagosomes in whole mammalian cells using correlative cryo-fluorescence and cryo-soft X-ray microscopy (cryo-CLXM). *Ultramicroscopy*, *143*, 77–87.

Ermantraut, E., Wohlfart, K., & Tichelaar, W. (1998). Perforated support foils with predefined hole size, shape and arrangement. *Ultramicroscopy*, *74*, 75–81.

Falcone, R., Jacobsen, C., Kirz, J., Marchesini, S., Shapiro, D., & Spence, J. (2011). New directions in X-ray microscopy. *Contemporary Physics*, *52*, 293–318.

Grünewald, K., Medalia, O., Gross, A., Steven, A. C., & Baumeister, W. (2003). Prospects of electron cryotomography to visualize macromolecular complexes inside cellular compartments: Implications of crowding. *Biophysical Chemistry*, *100*, 577–591.

Guttmann, P., Rehbein, S., Werner, S., Henzler, K., Tarek, B., & Schneider, G. (2013). Nanoscale spectroscopy and tomography with the HZB X-ray microscope: Applications in materials and life sciences. *Journal of Physics Conference Series*, *463*, 012032.

Guttmann, P., Zeng, X., Feser, M., Heim, S., Yun, W., & Schneider, G. (2009). Ellipsoidal capillary as condenser for the BESSY full-field X-ray microscope. *Journal of Physics Conference Series*, *186*, 012064.

Hagen, C., & Grünewald, K. (2008). Microcarriers for high-pressure freezing and cryosectioning of adherent cells. *Journal of Microscopy*, *230*, 288–296.

Hagen, C., Guttmann, P., Klupp, B., Werner, S., Rehbein, S., Mettenleiter, T. C., et al. (2012). Correlative VIS-fluorescence and soft X-ray cryo-microscopy/tomography of adherent cells. *Journal of Structural Biology*, *177*, 193–201.

Hagen, C., Werner, S., Carregal-Romero, S., Malhas, A. N., Klupp, B. G., Guttmann, P., et al. (2014). Multimodal nanoparticles as alignment and correlation markers in fluorescence/soft X-ray cryo-microscopy/tomography of nucleoplasmic reticulum and apoptosis in mammalian cells. *Ultramicroscopy*, *146*, 46–54.

Hanssen, E., Knoechel, C., Dearnley, M., Dixon, M. W. A., Le Gros, M., Larabell, C., et al. (2012). Soft X-ray microscopy analysis of cell volume and hemoglobin content in erythrocytes infected with asexual and sexual stages of *Plasmodium falciparum*. *Journal of Structural Biology*, *177*, 224–232.

Hayles, M. F., de Winter, D. A. M., Schneijdenberg, C., Meeldijk, J. D., Luecken, U., Persoon, H., et al. (2010). The making of frozen-hydrated, vitreous lamellas from cells for cryo-electron microscopy. *Journal of Structural Biology*, *172*, 180–190.

Hertz, H. M., von Hofsten, O., Bertilson, M., Vogt, U., Holmberg, A., Reinspach, J., et al. (2012). Laboratory cryo soft X-ray microscopy. *Journal of Structural Biology*, *177*, 267–272.

Heymann, J. B., Cardone, G., Winkler, D. C., & Steven, A. C. (2008). Computational resources for cryo-electron tomography in BSOFT. *Journal of Structural Biology*, *161*, 232–242.

Hsieh, C., Schmelzer, T., Kishchenko, G., Wagenknecht, T., & Marko, M. (2014). Practical workflow for cryo focused-ion-beam milling of tissues and cells for cryo-TEM tomography. *Journal of Structural Biology*, *185*, 32–41.

Huang, J. G., Leshuk, T., & Gu, F. X. (2011). Emerging nanomaterials for targeting subcellular organelles. *Nano Today*, *6*, 478–492.

Hummel, E., Guttmann, P., Werner, S., Tarek, B., Schneider, G., Kunz, M., et al. (2012). 3D ultrastructural organization of whole *Chlamydomonas reinhardtii* cells studied by nanoscale soft X-ray tomography. *PLoS One*, *7*, e53293.

Hurbain, I., & Sachse, M. (2011). The future is cold: Cryo-preparation methods for transmission electron microscopy of cells. *Biology of the Cell*, *103*, 405–420.

Kapishnikov, S., Weiner, A., Shimoni, E., Guttmann, P., Schneider, G., Dahan-Pasternak, N., et al. (2012). Oriented nucleation of hemozoin at the digestive vacuole membrane in *Plasmodium falciparum*. *Proceedings of the National Academy of Sciences of the United States of America*, *109*, 11188–11193.

Kaufmann, R., Hagen, C., & Grünewald, K. (2014). Fluorescence cryo-microscopy: Current challenges and prospects. *Current Opinion in Chemical Biology*, *20*, 86–91.

Kaufmann, R., Schellenberger, P., Seiradake, E., Dobbie, I., Jones, E. Y., Davis, I., et al. (2014). Super-resolution microscopy using standard fluorescent proteins in intact cells under cryo-conditions. *Nano Letters*, *14*, 4171–4175.

Kirz, J., Jacobsen, C., & Howells, M. (1995). Soft X-ray microscopes and their biological applications. *Quarterly Reviews of Biophysics*, *28*, 33–130.

Kremer, J. R., Mastronarde, D. N., & McIntosh, J. R. (1996). Computer visualization of three-dimensional image data using IMOD. *Journal of Structural Biology*, *116*, 71–76.

Larabell, C. A., & Le Gros, M. A. (2004). X-ray tomography generates 3-D reconstructions of the yeast, *Saccharomyces cerevisiae*, at 60-nm resolution. *Molecular Biology of the Cell*, *15*, 957–962.

Le Gros, M. A., Knoechel, C. G., Uchida, M., Parkinson, D. Y., McDermott, G., & Larabell, C. A. (2012). Visualizing sub-cellular organization using soft X-ray tomography. In E. H. Egelman (Ed.), *Comprehensive biophysics* (pp. 90–110). Amsterdam: Elsevier.

Le Gros, M. A., McDermott, G., Uchida, M., Knoechel, C. G., & Larabell, C. A. (2009). High-aperture cryogenic light microscopy. *Journal of Microscopy*, *235*, 1–8.

Lučić, V., Förster, F., & Baumeister, W. (2005). Structural studies by electron tomography: From cells to molecules. *Annual Review of Biochemistry*, *74*, 833–865.

Lučić, V., Kossel, A. H., Yang, T., Bonhoeffer, T., Baumeister, W., & Sartori, A. (2007). Multiscale imaging of neurons grown in culture: From light microscopy to cryo-electron tomography. *Journal of Structural Biology*, *160*, 146–156.

Lučić, V., Rigort, A., & Baumeister, W. (2013). Cryo-electron tomography: The challenge of doing structural biology in situ. *Journal of Cell Biology*, *202*, 407–419.

Martz, D. H., Selin, M., von Hofsten, O., Fogelqvist, E., Holmberg, A., Vogt, U., et al. (2012). High average brightness water window source for short-exposure cryomicroscopy. *Optics Letters*, *37*, 4425–4427.

Mastronarde, D. N. (1997). Dual-axis tomography: An approach with alignment methods that preserve resolution. *Journal of Structural Biology*, *120*, 343–352.

Mastronarde, D. N. (2008). Correction for non-perpendicularity of beam and tilt axis in tomographic reconstructions with the IMOD package. *Journal of Microscopy*, *230*, 212–217.

Moerner, W. E., & Orrit, M. (1999). Illuminating single molecules in condensed matter. *Science*, *283*, 1670–1676.

Müller, W. G., Heymann, J. B., Nagashima, K., Guttmann, P., Werner, S., Rehbein, S., et al. (2012). Towards an atlas of mammalian cell ultrastructure by cryo soft X-ray tomography. *Journal of Structural Biology*, *177*, 179–192.

Oton, J., Sorzano, C. O. S., Pereiro, E., Cuenca-Alba, J., Navarro, R., Carazo, J. M., et al. (2012). Image formation in cellular X-ray microscopy. *Journal of Structural Biology*, *178*, 29–37.

Parkinson, D. Y., Epperly, L. R., McDermott, G., Le Gros, M. A., Boudreau, R. M., & Larabell, C. A. (2013). Nanoimaging cells using soft X-ray tomography. In A. A. Sousa & M. J. Kruhlak (Eds.), *Nanoimaging* (pp. 457–481). New York: Springer Science+Business Media, LLC.

Parkinson, D. Y., Knoechel, C., Yang, C., Larabell, C. A., & Le Gros, M. A. (2012). Automatic alignment and reconstruction of images for soft X-ray tomography. *Journal of Structural Biology*, *177*, 259–266.

Pereiro, E., Nicolas, J., Ferrer, S., & Howells, M. R. (2009). A soft X-ray beamline for transmission X-ray microscopy at ALBA. *Journal of Synchrotron Radiation*, *16*, 505–512.

Quispe, J., Damiano, J., Mick, S. E., Nackashi, D. P., Fellmann, D., Ajero, T. G., et al. (2007). An improved holey carbon film for cryo-electron microscopy. *Microscopy and Microanalysis*, *13*, 365–371.

Rehbein, S., Guttmann, P., Werner, S., & Schneider, G. (2012). Characterization of the resolving power and contrast transfer function of a transmission X-ray microscope with partially coherent illumination. *Optics Express*, *20*, 5830–5839.

Rigort, A., Bäuerlein, F. J. B., Leis, A., Gruska, M., Hoffmann, C., Laugks, T., et al. (2010). Micromachining tools and correlative approaches for cellular cryo-electron tomography. *Journal of Structural Biology*, *172*, 169–179.

Rigort, A., Bäuerlein, F. J. B., Villa, E., Eibauer, M., Laugks, T., Baumeister, W., et al. (2012). Focused ion beam micromachining of eukaryotic cells for cryoelectron tomography. *Proceedings of the National Academy of Sciences of the United States of America*, *109*, 4449–4454.

Ruskin, R. S., Yu, Z. H., & Grigorieff, N. (2013). Quantitative characterization of electron detectors for transmission electron microscopy. *Journal of Structural Biology*, *184*, 385–393.

Sartori, A., Gatz, R., Beck, F., Rigort, A., Baumeister, W., & Plitzko, J. M. (2007). Correlative microscopy: Bridging the gap between fluorescence light microscopy and cryo-electron tomography. *Journal of Structural Biology*, *160*, 135–145.

Saxton, W. O., Baumeister, W., & Hahn, M. (1984). Three-dimensional reconstruction of imperfect two-dimensional crystals. *Ultramicroscopy*, *13*, 57–70.

Schellenberger, P., Kaufmann, R., Siebert, C. A., Hagen, C., Wodrich, H., & Grünewald, K. (2014). High-precision correlative fluorescence and electron cryo microscopy using two independent alignment markers. *Ultramicroscopy*, *143*, 41–51.

Schneider, G., Guttmann, P., Heim, S., Rehbein, S., Mueller, F., Nagashima, K., et al. (2010). Three-dimensional cellular ultrastructure resolved by X-ray microscopy. *Nature Methods*, *7*, 985–987.

Schneider, G., Guttmann, P., Rehbein, S., Werner, S., & Follath, R. (2012). Cryo X-ray microscope with flat sample geometry for correlative fluorescence and nanoscale tomographic imaging. *Journal of Structural Biology*, *177*, 212–223.

Schneider, G., Heim, S., Guttmann, P., Rehbein, S., & Niemann, B. (2006). Novel X-ray microscopes for 3-D and fs-imaging at BESSY. *IPAP Conference Series*, *7*, 349–352.

Schneider, C. A., Rasband, W. S., & Eliceiri, K. W. (2012). NIH Image to ImageJ: 25 years of image analysis. *Nature Methods*, *9*, 671–675.

Schorb, M., & Briggs, J. A. G. (2014). Correlated cryo-fluorescence and cryo-electron microscopy with high spatial precision and improved sensitivity. *Ultramicroscopy*, *143*, 24–32.

Schwartz, C. L., Sarbash, V. I., Ataullakhanov, F. I., McIntosh, J. R., & Nicastro, D. (2007). Cryo-fluorescence microscopy facilitates correlations between light and cryo-electron microscopy and reduces the rate of photobleaching. *Journal of Microscopy*, *227*, 98–109.

Thieme, J., Schneider, G., & Knöchel, C. (2003). X-ray tomography of a microhabitat of bacteria and other soil colloids with sub-100 nm resolution. *Micron*, *34*, 339–344.

Tivol, W. F., Briegel, A., & Jensen, G. J. (2008). An improved cryogen for plunge freezing. *Microscopy and Microanalysis*, *14*, 375–379.

Uchida, M., Sun, Y., McDermott, G., Knoechel, C., Le Gros, M. A., Parkinson, D., et al. (2011). Quantitative analysis of yeast internal architecture using soft X-ray tomography. *Yeast*, *28*, 227–236.

Vanhecke, D., Graber, W., Herrmann, G., Al-Amoudi, A., Eggli, P., & Studer, D. (2003). A rapid microbiopsy system to improve the preservation of biological samples prior to high-pressure freezing. *Journal of Microscopy*, *212*, 3–12.

Weiner, A., Kapishnikov, S., Shimoni, E., Cordes, S., Guttmann, P., Schneider, G., et al. (2013). Vitrification of thick samples for soft X-ray cryo-tomography by high pressure freezing. *Journal of Structural Biology*, *181*, 77–81.

Weinhausen, B., Saldanha, O., Wilke, R. N., Dammann, C., Priebe, M., Burghammer, M., et al. (2014). Scanning X-ray nanodiffraction on living eukaryotic cells in microfluidic environments. *Physical Review Letters*, *112*, 088102.

Weiss, D., Schneider, G., Niemann, B., Guttmann, P., Rudolph, D., & Schmahl, G. (2000). Computed tomography of cryogenic biological specimens based on X-ray microscopic images. *Ultramicroscopy*, *84*, 185–197.

Werner, S., Rehbein, S., Guttmann, P., Heim, S., & Schneider, G. (2010). Towards high diffraction efficiency zone plates for X-ray microscopy. *Microelectronic Engineering*, *87*, 1557–1560.

Wolter, H. (1952). Spiegelsysteme streifenden Einfalls als abbildende Optiken für Röntgenstrahlen. *Annalen der Physik*, *10*, 94–114.

Wombacher, R., & Cornish, V. W. (2011). Chemical tags: Applications in live cell fluorescence imaging. *Journal of Biophotonics*, *4*, 391–402.

Xu, J. M., Chang, J. S., Yan, Q., Dertinger, T., Bruchez, M. P., & Weiss, S. (2013). Labeling cytosolic targets in live cells with blinking probes. *Journal of Physical Chemistry Letters*, *4*, 2138–2146.

Yang, Z., Lasker, K., Schneidman-Duhovny, D., Webb, B., Huang, C. C., Pettersen, E. F., et al. (2012). UCSF chimera, MODELLER, and IMP: An integrated modeling system. *Journal of Structural Biology*, *179*, 269–278.

Zeev-Ben-Mordehai, T., Hagen, C., & Grünewald, K. (2014). A cool hybrid approach to the herpesvirus 'life' cycle. *Current Opinion in Virology*, *5*, 42–49.

Zhang, P. J. (2013). Correlative cryo-electron tomography and optical microscopy of cells. *Current Opinion in Structural Biology*, *23*, 763–770.

CHAPTER

Correlative Cryo-Fluorescence Light Microscopy and Cryo-Electron Tomography of *Streptomyces*

10

Roman I. Koning*, Katherine Celler[†], Joost Willemse[†], Erik Bos*, Gilles P. van Wezel[†], Abraham J. Koster*

**Department of Molecular Cell Biology, Section Electron Microscopy, Leiden University Medical Center, Leiden, The Netherlands*

[†]Molecular Biotechnology, Institute of Biology Leiden, Leiden University, PO Box 9505, Leiden, The Netherlands

CHAPTER OUTLINE

Methods in Cell Biology, Volume 124, ISSN 0091-679X, http://dx.doi.org/10.1016/B978-0-12-801075-4.00010-0

Abstract

Light microscopy and electron microscopy are complementary techniques that in a correlative approach enable identification and targeting of fluorescently labeled structures *in situ* for three-dimensional imaging at nanometer resolution. Correlative imaging allows electron microscopic images to be positioned in a broader temporal and spatial context. We employed cryo-correlative light and electron microscopy (cryo-CLEM), combining cryo-fluorescence light microscopy and cryo-electron tomography, on vitrified *Streptomyces* bacteria to study cell division. Streptomycetes are mycelial bacteria that grow as long hyphae and reproduce via sporulation. On solid media, *Streptomyces* subsequently form distinct aerial mycelia where cell division leads to the formation of unigenomic spores which separate and disperse to form new colonies. In liquid media, only vegetative hyphae are present divided by noncell separating crosswalls. Their multicellular life style makes them exciting model systems for the study of bacterial development and cell division. Complex intracellular structures have been visualized with transmission electron microscopy. Here, we describe the methods for cryo-CLEM that we applied for studying *Streptomyces*. These methods include cell growth, fluorescent labeling, cryo-fixation by vitrification, cryo-light microscopy using a Linkam cryo-stage, image overlay and relocation, cryo-electron tomography using a Titan Krios, and tomographic reconstruction. Additionally, methods for segmentation, volume rendering, and visualization of the correlative data are described.

INTRODUCTION

Correlative light and electron microscopy (CLEM) is a combination of imaging techniques in which the same object is visualized using both light microscopy (LM) and electron microscopy (EM). CLEM is employed if a specific biological question cannot be answered by LM or EM alone and cellular morphology as well as functional, dynamic, or positional information of cellular structures is needed. In CLEM on biological samples, fluorescence microscopy (FM) is generally employed to target sparse, transient, or specifically labeled (protein) structures for structural imaging using EM.

The specific advantage of EM is that cell morphology and specific macromolecular structures can be imaged in three dimensions at nanometer scale resolution ($\sim 2\times 10^{-9}$ m). EM provides images of fixed, static structures with a maximum thickness of several hundred nanometers and a field of view of several micrometers, often much smaller than one cell. Larger volumes can be imaged by serial sectioning and the field of view can be enhanced by combining images into montages (Faas et al., 2012). The advantage of LM is that large areas can be dynamically imaged using a variety of genetically encoded or chemical fluorescent labels. Imaging resolution is, however, two orders of magnitude lower ($\sim 1\times 10^{-7}$ m) and does not provide detailed information about labeled or unlabeled structures.

The main difficulty of CLEM is related to compatibility of specimen preparation techniques. Standard (chemical) fixation techniques for LM generally do not optimally preserve the ultrastructural for EM. Conventional EM specimen techniques, which include dehydration, plastic embedding, and heavy metal staining, negatively affect the usability of fluorescent dyes, though useful protocols are emerging (Karreman et al., 2012; Kukulski et al., 2011). These disadvantages are not encountered when cryo-fixation (either plunge freezing or high-pressure freezing) is used in combination with cryo-electron microscopy (cryo-EM) (Le Gros, McDermott, Uchida, Knoechel, & Larabell, 2009; Schwartz, Sarbash, Ataullakhanov, Mcintosh, & Nicastro, 2007; van Driel, Valentijn, Valentijn, Koning, & Koster, 2009). Vitrification is a cryo-fixation process in which the cooling rate is so high (>100.000 °C/s for plunge freezing) that water does not crystallize, but instead adopts a glass-like structure that perfectly preserves biological samples (Dubochet et al., 1988).

Vitrification is essential for cryo-CLEM on cells. First, whole bacteria and parts of cells that are thinner than a few microns can be excellently preserved in a (close to) native state, allowing unperturbed and detailed three-dimensional high-resolution cryo-imaging of subcellular structures (Koning, 2010; Lucic, Leis, & Baumeister, 2008; Pilhofer, Ladinsky, McDowall, & Jensen, 2010). Second, most fluorescent dyes and labels that are employed for live cells seem to retain their fluorescence after vitrification while being kept at cryo-conditions. Vitrified cells can be imaged using both (phase contrast) cryo-EM, without using additional contrasting agents, and fluorescence with bright field LM. Samples that are thinner than half a micron can be imaged whole, without specimen thinning (by cryo-sectioning or focused ion beam milling) (Marko, Hsieh, Schalek, Frank, & Mannella, 2007; Rigort et al., 2010). *Streptomyces* are 500–800 nm in thickness and are therefore suitable for visualization by whole-cell cryo-EM.

The major disadvantage of cryo-EM is that prolonged exposure to the electron beam increasingly damages the sample, as electrons are interacting strongly with vitrified biological samples. Therefore, searching must be performed fast, using a limited electron dose rate and at low magnification, resulting in noisy images and obscuring potential structures of interest. In addition, techniques to tag and identify specific proteins or molecules by labeling inside cells with electron dense labels are currently very limited (Diestra, Fontana, Guichard, Marco, & Risco, 2009; Risco et al., 2012; Wang, Mercogliano, & Lowe, 2011). Consequently, it is challenging to identify and target specific subcellular structures other than those which can be identified by their morphology, like membranes, the cytoskeleton, or ribosomes. FM offers advantages to cryo-EM with respect to labeling and targeting specific structures. Macromolecular structures inside cells can be fluorescently labeled using genetic fusion proteins like eGFP or using site-specific chemical labels. This allows searching relatively large areas for locations with labeled structures, while minimizing imaging using electrons.

Cryo-CLEM on thin samples is a relatively fast technique compared to many other CLEM techniques, as lengthy specimen preparation procedures for freeze substitution, preparation of markers, and sectioning are not needed. Fluorescent labeling, vitrification, cryo-fluorescence imaging transfer, and relocation in the transmission electron microscope (TEM) followed by cryo-ET can be carried out within 1 day. Numerous positions on one EM grid can be targeted in the light microscope after which tens of tomograms can be recorded in batch in a few days.

In conclusion, cryo-CLEM, combining FM and cryo-electron tomography, surmounts the limitations of each individual technique with respect to resolution, scale, dynamics, and labeling. Cryo-CLEM allows live imaging of infrequent and/or transient events and targeting of labeled structures inside cells using LM, providing a broader historical and spatial context to high-resolution structural 3D maps obtained using EM.

Cryo-CLEM on whole cells is facilitated by their natural thickness. With an average diameter of a few microns, bacteria are generally one or two orders of magnitude smaller than eukaryotic cells. Some bacteria (including *Streptomyces*) even fall within the thickness limit for tomography (~500 nm), making them particularly suitable for whole-cell cryo-CLEM without the need for sectioning (Morris & Jensen, 2008). In addition, culturing and genetic manipulation is more straightforward in prokaryotic organisms, which provides an additional advantage. Additionally, correlation of LM and EM images of fluorescently labeled bacteria on an EM grid is easier, since the pattern in LM is comparable to the electron dense pattern in cryo-EM. This is even more apparent for filamentous bacteria like *Streptomyces*, which form specific thread-like patterns (Celler, Picioreanu, van Loosdrecht, & van Wezel, 2012). These multicellular filamentous bacteria are commonly found in the soil (Claessen, Rozen, Kuipers, Sogaard-Andersen, & van Wezel, 2014). Their complex life cycle starts with the germination of a spore, which grows out by apical extension and branching to form a dense network of hyphae that is referred to as vegetative (or substrate) mycelium (Chater & Losick, 1997). Upon nutrient depletion, this mycelium is autolytically degraded to produce the nutrients for the next phase of the life cycle, whereby erected sporogenic hyphae are produced that form the aerial mycelium. This stage of the *Streptomyces* life cycle coincides with antibiotic production (Bibb, 2005; van Wezel & McDowall, 2011). The aerial hyphae differentiate into chains of spores involving a spectacular process of coordinated cell division and DNA segregation and DNA segregation and condensation, whereby up to 100 uninucleoid spores are produced (Schwedock, McCormick, Angert, Nodwell, & Losick, 1997).

Although the Streptomycetes are commercially widely used as producers of natural products (in particular antibiotics) and industrial enzymes (van Wezel, McKenzie, & Nodwell, 2009), there is still much to be learned about the molecular and structural basis of growth, morphogenesis, and division. During the *Streptomyces* life cycle, two distinct forms of cell division occur (McCormick, 2009). During sporulation-specific cell division, a thick spore wall consisting of peptidoglycan is synthesized and remodeled during the metamorphosis of unicellular prespore compartments into mature spores (Jakimowicz & van Wezel, 2012). In vegetative hyphae, crosswalls form at irregular intervals of roughly 5–10 μm (depending on the specific strain) to delimit multigenomic compartments. These compartments usually contain an irregular number of chromosomes and replication machinery in different stages of assembly and disassembly, sometimes within the same compartment. Unlike aerial septa, crosswalls do not constrict the cell and do not result in cell fission, effectively constituting a completely different form of cell division. Many of the divisome components which are required for septal peptidoglycan synthesis in canonical cell division such as found in *Bacillus subtilis* and *Escherichia coli* (Errington, Daniel, & Scheffers, 2003), and during sporulation-specific cell division in *Streptomyces*, are not required for cross-wall formation (Mistry, Del Sol, Wright, Findlay, & Dyson, 2008). These crosswalls delimit the syncytial cells, but do not constrict, and do not lead to cell fission. Live imaging of vegetative hyphae has demonstrated that membranes, enriched in phosphatidylglycerol and cardiolipin, dynamically localize during growth, ultimately localizing to the sites of cross-wall formation. In order to study these membranes in more detail, we developed cryo-CLEM, combining cryo-electron tomographic structural studies with functional LM assays.

10.1 RATIONALE

We performed cryo-CLEM on *Streptomyces* (i) to investigate the localization of DNA and lipids vegetative hyphae, (ii) to develop a workflow for correlative fluorescence LM and cryo-electron tomography of vitrified whole cells, and (iii) to develop a workflow in which all correlative data, including (fluorescent) LM, cryo-EM, cryo-ET, and surface-rendered data, are visualized in a conceptual manner.

Streptomyces are suitable for the development of a cellular cryo-CLEM workflow for several reasons. (1) *Streptomyces* are thin enough (500–800 nm) for whole-cell cryo-fixation by plunge freezing, which ensures excellent preservation of structural detail. (2) The three-dimensional structure can be determined using cryo-ET along consecutive parts of whole, unsectioned, vegetative mycelium of *Streptomyces*, when liquid culture samples are deposited sparsely on an EM grid and vitrified. (3) Position relocation from LM to EM and correlation of recorded images can be performed without the use of additional markers or finder grids, because the mycelium has a unique and distinct morphology that are similar in both imaging modalities.

Cryo-EM is useful for CLEM research since vitrification preserves fluorescence of many chemical dyes and fluorescent proteins, which makes LM at cryogenic temperatures possible. Another advantage is that the cryo-CLEM technique described here is fast and can be performed within a day.

10.2 METHODS

Cryo-CLEM is described here in six general steps comprising two specimen preparation steps, two imaging steps, and two image processing steps (Fig. 10.1), which comprise the following practical sequence: cell culturing, fluorescent labeling and fiducial marker addition, cryo-fixation, cryo-LM, image overlay, relocation, cryo-EM, cryo-electron tomography, tomographic reconstruction, segmentation and volume rendering, and visualization of correlative data.

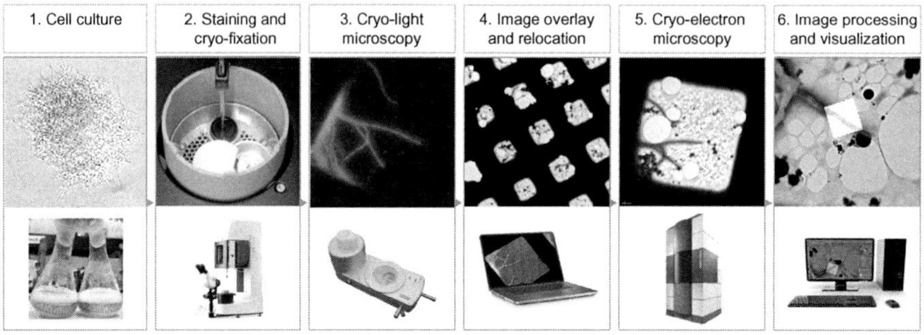

FIGURE 10.1

Schematic overview of the workflow used for the investigation of *Streptomyces* bacteria using cryo-CLEM. The complete workflow includes six steps. (1) Cells are cultured and fluorescently stained. (2) These cells are applied to an EM support grid, blotted, and vitrified using a plunge-freezing device. (3) The vitrified fluorescently stained cells on the EM grid are transferred to a cryo-stage and imaged at cryogenic conditions by cryo-FM. (4) The specimen is transferred into a TEM and the positions at which the cryo-FM images were acquired are visually relocated using the morphology of the *Streptomyces* mycelium. (5) Cryo-EM projection images at different magnifications are acquired for correlation and overlay of images. Cryo-ET is performed at targeted positions. (6) Visualization was performed by overlaying cryo-LM, cryo-EM, and 3D cryo-ET data using Photoshop, Amira, and Premiere. (See the color plate.)

10.2.1 CELL CULTURING

For plunge freezing, *Streptomyces* bacteria were grown in liquid culture, where they form a branched network of hyphae otherwise known as a vegetative mycelium.

1. *Germinating spores.* Spores 20 μL of *Streptomyces albus* J1074 containing roughly 10^7–10^9 spores/mL were added to 20 mL 2 × YT medium (Kieser, Bibb, et al., 2000) containing 10% sucrose.
2. *Growing cells.* The spore suspension was grown for 10–12 h at 30 °C at 300 rpm stirring speed.

10.2.2 FLUORESCENT LABELING AND FIDUCIAL MARKER ADDITION

Immediately prior to vitrification, cell cultures were washed, and dyes for fluorescence LM and fiducial nanogold for electron tomography were added.

1. *Washing cell culture.* 1 mL of cell culture was pipetted into an Eppendorf and 0.5 mL PBS buffer was added to wash the culture and remove sucrose. After mixing by inverting the Eppendorf, the culture was spun down gently at 1000 rpm, and 1 mL of supernatant was removed. The washing was repeated twice; in each case, 1 mL of PBS was added.
2. *Adding fluorescent dyes.* Depending on the structures to be labeled, different fluorescent dyes were added. Note that stains that were used are routinely used for

live microscopy to minimize secondary effects on morphological and functional integrity. FM5-95 was used to stain anionic phospholipids, including phosphatidylglycerol. 10-Nonyl acridine orange (NAO) was used to preferentially stain cardiolipin. DAPI was used to stain DNA. Note that the dyes were chosen such that the excitation/emission spectra allowed dual labeling and combination of FM5-95 and DAPI, or NAO and DAPI. All dyes were obtained from Molecular Probes. Roughly 1–5 μL of dye was added to a 40 μL sample, depending on dye intensity.

3. *Adding fiducial gold*. Fiducial markers were added to the samples to aid tracking and alignment during tilt series acquisition and tomographic reconstruction. Final dilution varied with the concentration of the Au particles and the desired density of the fiducials on the grid. In general, a good fiducial density was obtained when to a 40 μL sample, 1 μL of 15 nm protein A-stabilized Au particles were added.

10.2.3 CRYO-FIXATION

1. *Selecting EM grid and support film*. In CLEM, finder grids and relocation markers are often used to target positions between LM and EM modes. *Streptomyces* vegetative mycelium formed a unique pattern when placed on an EM grid. This pattern was similar when imaged by transparent cryo-LM and low-magnification cryo-EM conditions. Therefore, targeted positions were easily recognized and correlated, and it was not necessary to use finder grids for relocation. In practice, a variety of different specimen support grids were used. In order to maintain support film stability and allow tilting for tomography, 150–300-mesh-sized grids were used. As specimen support film, generally Quantifoil or lacey carbon films were used. Though in cryo-EM on adherent cells inert gold grid were necessary to allow cell growth, in this case copper grids were used since the bacteria were only in contact with the grids for a limited amount of time (<1 min) prior to vitrification.
2. *Glow-discharging the EM grid*. To render the EM grids hydrophilic, grids were glow-discharged before use. They were placed onto a metal mesh with the carbon side facing up. Grids were glow-discharged for 1–2 min at ~20 mA in 2×10^{-1} mbar air. Grids were used within an hour after glow-discharging.
3. *Preparing the cryo-fixation device*. The plunge-freezing device (Leica EM GP) was switched on, the water container for the humidifier was filled with 60 mL distilled water, and filter paper was placed (Whatman no 4. or Schleicher and Schuell no. 595 or 597) into position inside the environmental chamber. The filter paper counter was reset to zero before starting. A cryo-storage box for vitrified grids was marked for identification of the sample throughout the experiment and positioned into the transfer container. The liquid nitrogen Dewar was filled with (1.5 L) liquid nitrogen. When the ethane container reached liquid nitrogen temperature, the (black) secondary cryogen container was filled with ethane or ethane/propane (Tivol, Briegel, & Jensen, 2008) using the cryogen liquefier.

4. *Setting vitrification parameters*. Using the control panel, the environmental chamber temperature was set to 22 °C and the humidity (Hr) to 95%. The blotting time was set to 1–2 s, blotting once from one side, with no waiting time after blotting. The liquid ethane temperature was set to −181 °C when liquid ethane was used, or to −191 °C when a liquid ethane/propane mixture was used.
5. *Sample vitrification*. The sample (3 μL) was applied to the carbon side of the glow-discharged grid. This was done outside the environmental chamber, since cell growth and blotting was performed at room temperature. The grid was clamped with a forceps such that the sample was blotted from the carbon facing side. The Leica EM GP rod was set in the "Load forceps" position and the tweezers were mounted on the forceps interlock. The automatic vitrification process was activated, initiating lowering of the environmental chamber, grid blotting, and plunge freezing.
6. *Storage and transfer of vitrified grid*. After plunging, the vitrified grid was transferred from the liquid ethane into a cryo-grid box for transfer and storage. At this point, if desired, more grids were vitrified (repeat step 5). The cryo-grid box was stored under liquid nitrogen in 50-mL polystyrene tubes (Greiner, The Netherlands), which were placed in liquid nitrogen Dewars (Cryotech, The Netherlands). When the grids were directly used for cryo-LM, up to three grids were positioned in a Linkam transfer cassette that could be held in a transfer box (Fig. 10.2C).

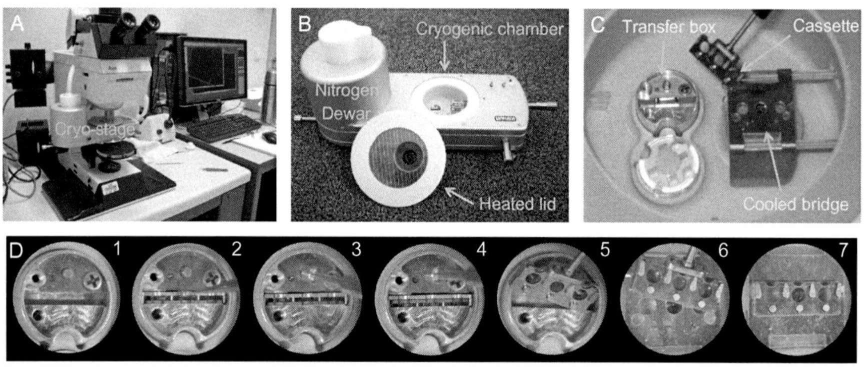

FIGURE 10.2

The Linkam correlative CMS196 cryo-LM stage. The cryo-LM setup consists of an upright light microscope under which the cryo-stage is mounted (A). The cryo-stage has a nitrogen Dewar and a cryogenic chamber for the samples, which is covered by a heated lid with a depressed opening in which the lens is positioned (B). In the cryogenic chamber, vitrified grids can be transferred from a cryo-grid box to the transfer box, which is used to transfer grids into a cassette onto the cooled bridge for imaging (C). The transfer box (D1) holds the cassette in which a EM grid is placed (D2) and a mechanism for closing the cassette (D3) so the EM gird is magnetically clamped inside (D4). Using a magnetic tool, the cassette can be picked up (D5) and transferred—at liquid cryogenic conditions—to the cooled bridge (D6 and C) at which it can be imaged (D7). (See the color plate.)

More detailed descriptions and protocols for using the Leica EM GP or vitrobot plungers can be found elsewhere (Iancu et al., 2006; Resch, Brandstetter, Königsmaier, Urban, & Pickl-Herk, 2011).

10.2.4 CRYO-LM

The Linkam correlative cryo-stage. For cryo-LM, a Linkam CMS196 correlative microscopy cryo-stage was used (Linkam Scientific Instruments, UK). This stage was specifically designed for light microscopic imaging of cryogenically cooled samples that are prepared on EM grids. The cryo-stage is suitable for upright light microscopes (Fig. 10.2A), in both fluorescence and bright field mode. The cryo-stage is equipped with its own liquid nitrogen Dewar that is coupled to a filling system, which automatically regulates the nitrogen levels in the cryogenic chamber (Fig. 10.2B). This chamber could be kept at cryogenic temperatures for about 30 min without refilling the nitrogen Dewar. The cryogenic chamber has a heated lid, which prevents frost contamination, with a depressed circular opening in which the lens is positioned (Fig. 10.2B). Manual sample movement in the horizontal plane was achieved by rotation of two screws on the cryo-stage, while focusing was performed by changing the height of the microscope stage on which it was mounted.

The cryo-stage is optimized for handling EM grids at cryogenic conditions while minimizing ice contamination from air and specimen warming. The stage includes a specific grid transfer system that comprises a cassette and a transfer box (Fig. 10.2C and D). The cassette holds up to three grids and is made up of two metal sheets with three holes that are clamped together with small magnets (Fig. 10.2D, 2–7). The transfer box holds one cassette for transfer, but also contains a mechanism to open and close the two magnetically clamped sheets of the cassette, allowing vitrified grids to be loaded into the cassette (Fig. 10.2C and D, 1–5). The cassette, with vitrified grids, can be swiftly moved with a special magnetic rod from the transfer box to a cooled metal bridge (Fig. 10.2C and D, 5–6). There the cassette is, again magnetically, clamped to the cooled bridge where it can be imaged (Fig. 10.2B and C). The Linkam cryo-stage has designated positions for two cryo boxes to facilitate transfer of grids that are stored in standard cryo boxes into the Linkam transfer box.

1. *Preparing the light microscope.* The light microscope and the lamp for bright field microscopy were activated. The postmagnification lens was set to 1.0 × and the 100 × objective lens (Fluortar 100 × dry objective lens with a numerical aperture (NA) of 0.71 and a working distance of 4.7 mm was selected). The camera control and image acquisition computer and the Leica AF Lite software were started.
2. *Preparing the cryo-stage.* The cryo-stage was switched on, the Dewar was filled with liquid nitrogen, and the automated filling system was activated. The system was ready for use within a few minutes, when stable at liquid nitrogen temperature.
3. *Transferring the EM grid.* The cryo-grid storage box and the cassette in the transfer box were placed into position in the working space of the cryo-stage. EM

grids were generally stored in liquid nitrogen after vitrification. The grids were transferred from the storage box to the cassette in the transfer box, which was subsequently transferred with the magnetic transfer tool to the cooled bridge for imaging (Fig. 10.2D).

4. *Mounting the cryo-stage*. The cryo-stage was transferred onto the light microscope, fastened onto the microscope stage with the screws, and the desired lens was selected. The stage was elevated to focus the grid.
5. *Cryo-LM imaging*. The appropriate filter was selected for scanning the grid in fluorescent mode for interesting positions. For targeting spots with excessive lipids, the fluorescent signal of FM5-95 was used. For targeting positions that were devoid of DNA, DAPI was used. Bright field images were always recorded alongside the fluorescent images. This image was necessary as an intermediate for correlating FM with EM images. Targeted positions were recorded preferentially near the center of the grid to improve relocation in TEM, because of the limited stage movement (2×2 mm) and potential off-center positioning of the grid in the TEM. Images were named and stored such that it enabled easy tracking of the positions on the grid later in the TEM and when making the overlays.
6. *Transferring grids and data*. After imaging, the stage was dismounted from the microscope, the cassette transferred back into the cryo-grid box, and the box placed in a liquid-nitrogen-cooled vessel for transfer to the TEM. Recorded images were copied on to a laptop to make overlay images and to relocate the imaged positions in the TEM.

10.2.5 IMAGE OVERLAY

The cryo-LM images were used for three purposes: (i) relocation in TEM, (ii) annotation of recorded cryo-ET positions, and (iii) as basis for overlaying the cryo-LM, cryo-EM, and surface-rendered data. For relocation in the TEM, the cryo-LM images were first aligned to each other and overlaid in an Adobe Photoshop file, using layers.

1. *Loading the images into Photoshop*. The cryo-LM images that were recorded with different optical filters were imported into separate layers in Photoshop: *File–Scripts–Load files into stack–Browse*... and select images.
2. *Aligning the layers*. The layers could be automatically aligned during import into Photoshop by checking the box before "*Attempt to Automatically Align source files*" in the previous step. Dependent on the image content, this might not result in correctly aligned images. If the alignment did not work, a single layer was selected and aligned manually. Correct alignment of the layers was compared either by changing the opacity of the top layer or by alternating the visibility of the layer.
3. *Coloring the images*. If the fluorescent images were not colored, colors were assigned according to the fluorescent emission wavelength (e.g., red, green, or blue). The desired layer was selected (*Window–Layer*) and all other layers were

made invisible. The channel panel was opened (*Window–Channel*) and the green and blue channels were shift-selected for a red color, red and blue channels for a green color, and red and green channels for a blue color. The selected layers were filled with black (*Edit–Fill* or Shift+F5) using the options: Contents: Use: Black; Blending Mode: Normal; Opacity: 100%.

4. *Adjusting brightness and contrast*. To optimize visualization, the brightness and contrast of all layers were adjusted. This was done in a reversible way using adjustment layers. The desired layer was selected (*Window–Layer*), and selected layer and all other layers were made invisible. An adjustment layer was inserted (*Layer–New adjustment layer–Brightness/Contrast*) using the options: Color: none; Mode: Normal; Opacity 100%. The adjustment layer appeared above the selected layer and acted on the first visible layer beneath it. To keep the adjustment layer acting on the desired cryo-LM image, there were several options. Both layers were linked: select both layers and select *Link layers*; or both layers were grouped: select both layers and *Make new group from layers*. "Linear dodge (Add)" was selected to overlay different color on a low-magnification LM or EM background. The *brightness/contrast* icon in the adjustment layer (in the properties tab) was selected, legacy mode was selected, and brightness and contrast were changed using the sliders.
5. *Save the image*. The layered image was saved as a Photoshop document or layered tiff document. As filename the increasing CLEM experiment number were used, including, e.g., CLEM-1401.

The aligned cryo-LM images were transferred to a laptop that was able to open the layered images at the TEM. The transparent LM images were used to navigate to the correct grid square using the overall morphology of the mycelium. The fluorescence LM images and additional visible features on the EM grid such as grid bars, holes in the support film, and ice contamination were used to navigate to the desired spot within the mesh.

10.2.6 CRYO-EM

Cryo-EM was performed on a Titan Krios electron microscope (FEI, Eindhoven, The Netherlands). The use of the Titan Krios had several advantages for cryo-CLEM on *Streptomyces*. First, the use of 300 keV electrons in combination with zero-loss energy filtering enabled cryo-ET of the 500–800 nm thick *Streptomyces*. Higher-energy electrons have longer mean-free paths (the average distance between two scattering events), which reduced inelastic scattering. Zero-loss energy filtered imaging additionally filtered out the inelastically scattered electrons, thereby increasing the signal-to-noise ratio in the acquired images. This was especially important while imaging thick samples using tomography, since tilting increases the effective specimen thickness (twofold at 60°). Furthermore, the Titan Krios TEM system allowed high-throughput data collection of cryo-electron tomograms and enabled imaging of many positions on one (or multiple) EM grid(s). Effectively, this increased the yield of data collection 5- to 10-fold compared to imaging on earlier types of microscopes (e.g., Tecnai F20) that have side entry holder for only one grid.

1. *EM at lowest magnification*. The EM grid of choice was loaded into the electron microscope and visualized in the electron microscope at low magnification (typically 20 × to 100 ×). The beam and image shifts between LM, M, and SA magnification modes on the TEM were accurately aligned to keep the same feature centered in all magnifications.
2. *Showing the cryo-LM images*. The aligned cryo-LM image from the grid that was present in the TEM was opened using Photoshop on a laptop and the bright field image was displayed.
3. *Relocating the grid square*. Grid squares matching the LM imaging positions were relocated by eye and the stage positions were stored. When finder grids were used, the letters served as navigation. When nonfinder grids were used, relocation was guided by visual comparison of the *Streptomyces* morphology and the arrangement of holes in the carbon film. Sometimes the TEM projections and LM image were mirrored because the grid was flipped. This was corrected by flipping the LM image (Photoshop: *Image–Image Rotation–Flip Canvas Horizontal/Vertical*).
4. *Defining tomogram positions*. The stored grid square positions were revisited and imaged at low magnifications, typically at 560 × (comprising one complete grid square) and 2250 × (for accurate determination of tomogram positions). Stage positions for tomograms were stored and annotated in the cryo-LM Photoshop documents. After tomographic data collection, additional projection images at 8700 × or 15,000 × were recorded for visualization and correlation of LM and EM data.

10.2.7 CRYO-ELECTRON TOMOGRAPHY

Tilt series were recorded using Xplore3D software (FEI Company) using the low-dose system and batch tomography. The stored positions for recording tomogram were revisited during set up of the batch tomography. Images were recorded with a 2k × 2k CCD camera mounted on a postcolumn energy filter (Gatan Quantum SE 963, Gatan GmbH, Germany) in zero-loss mode using a slit width of 20 or 30 eV. Tomographic tilt series were recorded in two halves starting from 0° to respective negative and positive maximum tilt angles. A linear tilt scheme was used with tilt angle increments of 3° or 2° and maximum angles of 51° or 60°, respectively. Depending on the chosen tilt angle increment, images were recorded with an exposure time of 0.25 or 0.5 s. The total electron dose that was used to acquire the full tilt series was in the order of 200–300 e/Å^2. This dose was experimentally determined and kept the final images below the point of visual deterioration. The exposure time was increasing linearly with a factor of 1.6 between 0° and 60°, while the dose was distributed over the whole series. Magnifications varied between 8.700 × and 11,500 ×, resulting in pixel sizes between 1.8 and 1.4 nm. The defocus ranged between −15 and −20 μm. Specimen shift correction during data collection was measured by the "tracking after" option on filtered images (long wavelength cutoff at 100 ± 3 nm and 2 ± 0.5 nm low wavelength cutoff, with a taper filter of 8 pixels and the Sobel filter and remove X-ray options on). At the highest tilt angles, defocus was checked every two tilt angles, while at the lower tilt angles, defocus was not checked. Total acquisition time ranged between 20 and 55 min, depending on acquisition parameters; in an overnight, batch up to 35 tomograms could be recorded.

10.2.8 TOMOGRAPHIC RECONSTRUCTION

Image processing was performed using the IMOD software suite (Kremer, Mastronarde, & McIntosh, 1996). Here, we briefly describe the steps taken and parameters chosen for the reconstruction of cryo-electron tomograms of *Streptomyces* taken on the Titan Krios described above. A more detailed description on tomographic reconstruction can be found elsewhere (Koning & Koster, 2013).

1. *Preparing tilt series.* The recorded electron tomography tilt series and accompanying files were copied to a separate folder. The .mrc extension of the tilt series was renamed to .st.
2. *Opening tilt series.* Etomo was started and the tilt series was opened. "*Single axis*" and "*Single frame*" were selected in the "*Data type*." Next, "*Scan header*" was selected and the value for the diameter of the used fiducials was added in the appropriate box. "*View Raw Image Stack*" was selected and the numbers of bad images (e.g., images with partly blocked beam, blurred images, low contrast images, or badly focused images) were selected in the "*Exclude views*" box. "*Create Com Scripts*" was clicked.
3. *Preprocessing tomogram.* Hot pixels were removed via one detection round of CCD eraser function using default settings: Peak criterion: 10; Difference criterion: 8; Maximum radius: 2.6. If high-intensity streaks were present in the images, several rounds of the CCD eraser were used or the parameters were changed to fully remove the hot pixels.
4. *Prealigning tilt series.* A first, course alignment of the tilt series was performed by cross-correlation using the default parameters. In some cases, the course alignment did not work correctly, e.g., because parts of the images were obscured by thick ice, resulting in abrupt image shifts. In this case, the alignment was improved by changing the default values of the cross-correlation parameters, by trimming pixels, or by manual alignment of the tilt series using MIDAS.
5. *Generation fiducial model.* Gold fiducials of 15 nm were used for the fine alignment of the tilt series using the "*Make seed and track option.*" In our case, patch tracking was less efficient because of limited contrast in the tilt series images of the *Streptomyces*. Up to 10 fiducials positioned as close as possible to the aligned object were initially selected and tracked along the whole tilt series. In subsequent rounds, more fiducials were added. Alternatively, the RAPTOR option within IMOD was used to automatically generate an initial fiducial model.

6. *Fine align tilt series.* Fine alignment was performed starting with the following parameters: "Threshold for residual report" 3; "Relative to all neighbouring views"; "Do not sort fiducials into 2 surfaces"; "Enable local alignments" off; "Solve for all rotations"; "Fixed magnification at 1.0"; "Group tilt angles"; "Distortion Solution type" disabled. Tracked beads with big residuals were manually centered to the beads. Bead positions in low signal-to-noise positions and aggregated beads were removed. After one or two rounds of refinement, residual report was decreased in steps to 1, and "Solve for all magnifications" and "Solve for all except minimum" tilt angles were activated. "Final global mean residual" was below 1.
7. *Positioning tomogram.* Since the specimens were thick whole *Streptomyces* cells (and not sections), "Create boundary model" was not critical or necessary to minimize the reconstructed volume or straighten the sample in the volume. "Sample tomogram thickness" was set to 1200; "Use whole tomogram" and "Binning" 3 were used to "Create a whole tomogram" and determine the thickness of the final tomogram.
8. *Fine align stack.* Fine alignment of the stack was done using "Binning" 1 and visually inspected for quality. "CTF correction," "Gold erasing," and "2D filtering" were not usually performed.
9. *Generating tomogram.* Tomograms were generated using "Back Projection" using default values. Alternatively, SIRT reconstructions were calculated using 5–10 iterations. SIRT reconstructions generated tomograms with more contrast for large features.
10. *Postprocess tomogram.* The tomograms were postprocessed by "Swap y and z dimensions"; "Conversion to bytes" and standard values were used for "Scaling from sections." To reduce the tomogram file size for loading and rendering in Amira, sometimes volume trimming and "Squeeze vol" (using an isotropic reduction factor of 2) were performed.
11. *Filter tomogram.* Noise reduction and contrast enhancement of (weighted back-projected) tomograms were performed using nonlinear anisotropic diffusion in etomo (*File–New Nonlinear Anisotropic Diffusion*). To determine suitable parameter settings for this algorithm, a typical single Z slice was selected from the file and the effect of k values and iterations were evaluated for contrast enhancement by visual inspection. Typical values used were between 20 and 200 for k and between 10 and 20 for the number of iterations.

10.2.9 SEGMENTATION AND VOLUME RENDERING

The main objective for performing segmentation and surface rendering was to extract and visualize specific structures inside tomograms. Surface rendering provides better impression of structures in 3D and makes the information in a tomogram accessible for a broader audience. Segmentation, surface rendering, and three-dimensional visualization were performed using Amira. Segmentation was performed by rough manual masking of structures, followed by fine-tuning by histogram thresholding of highest density voxels and subsequent smoothing by a combination of small islands removal and iterative growing/shrinking of selected voxels (erosion/dilation). Image segmentation was performed on nonlinear anisotropic diffusion-filtered or SIRT-reconstructed datasets only. Nevertheless, manual segmentation was necessary because of the high noise levels in the tomograms.

1. *Loading tomogram*. The tomogram was loaded into AMIRA using "*File–open data* (or Ctrl+O)" or "Open data..." in the pool. An orthoslice was attached to the data: select data and select "Orthoslice" in the pool (or right-click data and select orthoslice). In the properties window, the orientation of the slice, the contrast, and the slice number could be varied.
2. *Rough segmentation*. Specific structures of interest were manually selected using the AMIRA segmentation editor. In the editor, a new label was created, and new materials were created for each structure to be segmented. First, the complete *Streptomyces* cell was outlined and selected by interpolating several circles or ovals that were positioned along the length of the cell in end-on views every 10th to 50th slice using the brush tool. Other structures, like the cytoplasm or vesicles, were segmented in a similar fashion and added to the separate materials. Note that in the data area, the visibility in 2D slices and 3D volume can be switched on and off and a material can be selected or locked, to prevent further changes.
3. *Fine segmentation*. Fine segmentation was performed by selecting the voxels with highest density from the roughly segmented structures using a histogram. In "Display and Masking," the "Masking: enable" and "Masking: show" buttons were ticked. Selected voxels became marked in purple. "Threshold" with options "All slices" and "Current Material Only" on was selected. The sliders of the histogram window were changed such that lowest values up to the desired gray value were included. The existing material with the current selection was replaced.
4. *Smoothing selection*. The segmented voxels after histogram selection made up a noisy and discontinuous selection that did not result in a usable surface rendering. The segmentation was improved by removing small islands ("Segmentation–Remove Islands" using "Mode: 3D volume" increasing the size from 15 to a desired value until most small islands are removed), smoothing of labels ("Segmentation–Smooth labels" selecting "Mode: 3D volume" and size 3 and using several rounds of smoothing to get desired effect), and subsequent growing and shrinking of the selection in 3D.
5. *Make surface rendering*. A surface-rendered view of the segmented data was visualized in the object pool. (A "SurfaceGen" module was applied to the selected data object ending with ".labels." A surface view module was applied to the data object ending with ".surf" to surface render the segmented volume.) Note that different materials could be switched on and off and surfaces could be made transparent. For a mix of transparent and solid views of different materials, a duplicate of the "SurfaceView" was necessary.

6. *Make animation*. The surface-rendered objects were used to produce an animation that was later integrated with the LM and 2D EM data. Only a short overview of the procedure is given here since the choice for animations varies with the dataset and individual preferences. Movie sequences were made using the "DemoMaker" module ("Create–Animation/Demo–DemoMaker" and select "DemoDirector" to visualize the timeline). Multiple consecutive events were added to make the complete movie ("New event. . ." to add the first sequence. For example, for making a tomographic slice sequence, expand "Orthoslice" and select "Slice Number." In the "Properties Window," select the start and end values of both slice number and time). Rotating objects were created using "Create–Animation/Demo–Object rotate." Once the object rotate module was activated, it could be selected in the "DemoMaker" module. The "MovieMaker" module was used to make a movie from the sequence.

10.2.10 VISUALIZATION OF CORRELATIVE DATA

Our goal for the visualization was to convincingly correlate the LM, EM, and surface-rendered tomographic data into a continuous sequence. In our experience, before starting to visualize the correlation, it was convenient to group all data necessary for a single correlation into a single computer folder assigned with a CLEM experiment number. One CLEM experiment typically contains cryo-LM data from one grid square position and cryo-EM 2D overview images from the square, multiple higher-magnification cryo-EM images with accompanying cryo-electron tomograms and surface-rendered tomograms.

1. *Opening the overlay image*. The previously made file containing the aligned LM channels in a layered .psd file was opened in Adobe Photoshop.
2. *Adding EM overview image*. In order to place the complete EM overview image onto the LM image, the canvas size was increased by 200%. The low-magnification EM overview of a grid square was placed in a separate layer ("File–place"). This layer was made partly transparent (Set layer opacity to 50%) in order to view the EM and LM images simultaneously. The EM overview image was mirrored (if necessary), shifted, and rotated ("Edit–transform" or "Shift + T") such that all visible features (holey carbon and *Streptomyces*) optimally fitted the transparent LM and FM images.
3. *Adding high-magnification EM images*. The image size was increased ("Edit–Image size" (or alt + ctrl + I), selecting "Percent" in "Pixel dimensions" or "Document size" and setting these to 200%, 400%, or 800%) to place higher-magnification overview images without resolution loss. In order to prevent excessive document file size, unnecessary parts of the LM image were cropped. The medium-magnification EM overview were placed in a separate layer and fitted as was described above. For correct scaling and placement, the pixel size and microscope coordinates of the images were used. Tomographic slices were added in the same way as done for high-magnification images. Alignment of the tomographic slices over high-magnification EM images was aided by the

edges of holes in the carbon and fiducial gold markers. The resulting overlay .psd file served as basis for a movie including the surface-rendered data that can be made with Adobe Premiere.

4. *Making movies of correlative data.* Movie sequences were produced using Adobe Premiere. All correlative data were integrated to produce a movie that zoomed in from large overview LM images to the high-resolution EM and surface-rendered data. The image size of the Photoshop file was reduced to 10–25% of the original before they were imported into Adobe Premiere, because images were much larger than the resolution of HD movies. The Photoshop document was imported into a new Premiere project, retaining the layers and alignments ("*File–Import*," select Import as "*Individual layers*"). A new sequence was opened and the images were dragged into different video layers in the sequence. The image scale was adapted to fit the resolution of the project (*Select image* and in *video effects* "Motion–*Scale*" to 10%). Movies were produced from the correlative data by seamlessly zooming in from the LM images via the overview EM images, to the high-magnification EM images, by changing the scaling and transparencies. In the end, the 3D tomographic and surface-rendered data that were made in Amira were seamlessly integrated.

10.3 MATERIALS AND INSTRUMENTATION

a. *Cryo-LM*
   - Linkam CMS196 cryo-stage (Linkam, UK).
   - Leitz DMRB upright light microscope (Leica, Wetzlar, Germany) with an adapted microscope stage to hold the CMS196 and an adapted condenser lens (Linkam) for transparent light imaging. The microscope was equipped with a Hi Plan 10 × objective lens with a NA of 0.22 and a HCXPL Fluortar 100 × dry objective lens with an NA of 0.71 and a working distance of 4.7 mm. The microscope was equipped with a Leica TRITC filter with 515–560 nm excitation and 590 nm long-pass detection filter, a Leica UVA filter with excitation from 340 to 380 nm, and a 425 nm long-pass detection filter. Images were recorded using a 100 × dry objective with a working distance of 4.7 mm and a NA of 0.75. Images were recorded with a Leica DFC350FX CCD camera using Leica Application Suite Advanced Fluorescence Lite software (LAS-AF-Lite 2.6.0, Leica, Austria).

b. *Cryo-EM*
   - Gatan Digital Micrograph (Gatan, München, Germany).
   - Xplore3D, version 4 (FEI Company, Eindhoven, The Netherlands).
   - Titan Krios TEM (Krios 1, NeCEN) (FEI, Eindhoven, The Netherlands), equipped with Quantum SE/963 energy filter (Gatan, München, Germany).

c. *Image processing*
   - IMOD version 4.5 (Boulder, CO, USA).
   - Amira version 5.4 (FEI Company, Eindhoven, The Netherlands).
   - Photoshop CS6 and Premiere CS6 (Adobe, CA, USA).

10.4 DISCUSSION AND OUTLOOK

The presented method describes a two-step cryo-CLEM approach in which vitrified samples are imaged using cryo-fluorescence LM, followed by cryo-electron tomography. The acquired images and reconstructed data, including transparent and fluorescence LM, EM projection images, three-dimensional electron tomograms, and surface-rendered data, have different scales, dimensions, and modalities and are correlated and integrated into overlay images (Fig. 10.3) and a continuously scaling image sequence.

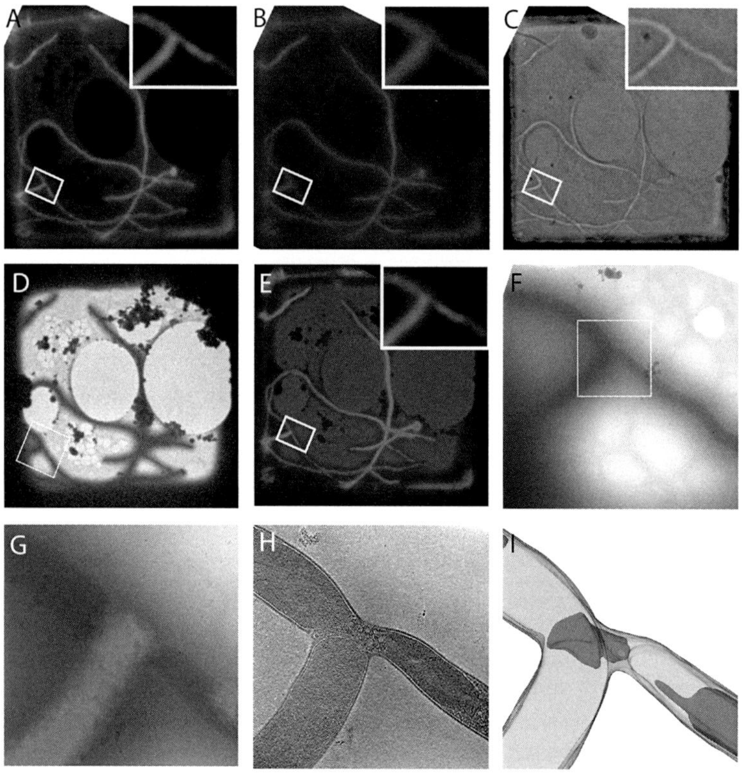

FIGURE 10.3

Cryo-CLEM of *Streptomyces*. The cryo-fluorescence microscopy using FM5-95 stain for membranes (A) and DAPI stain for DNA in (B) are used to determine target positions for cryo-ET. The insets in A and B show (denoted by the white rectangle in A, B, C, and E) a region of interest with membrane and DNA staining. The cryo-bright field microscopy image (C) and the cryo-electron microscopy projection overview (D) are used to aid correlation of the fluorescent images (A and B) with higher-magnification cryo-EM images (F, G) and electron tomogram slices (H) and surface-rendered tomogram (I, cell wall in blue and membranes in red). (See the color plate.)

This two-step cryo-CLEM approach, using a cryo-light microscopic stage in combination with a cryo-TEM, has been applied before, using different cryo-stages and TEMs (Carlson & Evans, 2011; Sartori et al., 2007; Schwartz et al., 2007; van Driel et al., 2009). The currently used setup with the improved specimen handling and transfer capabilities of the Linkam CMS196 freezing stage in combination with automated data collection using a FEI Titan Krios electron microscope was advantageous compared to our earlier experiments (van Driel et al., 2009). It improved the handling of cryogenically cooled specimens, the quality of specimens, the success rate of the experiments, and the amount of cryo-electron tomograms, resulting in tens of useful correlations within a few weeks of data acquisition.

The use of *Streptomyces* for the development of the here described cryo-CLEM method had several advantages. In earlier cryo-CLEM investigations, subcellular structures—microtubules, mitochondria, or viruses—were targeted in eukaryotic cells (Gruska, Medalia, Baumeister, & Leis, 2008; Jun et al., 2011; Sartori et al., 2007; Schwartz et al., 2007; van Driel et al., 2009). Finder grids were employed to navigate to the correct EM grid square, while visual clues were used for targeting specific subcellular structures. Recently, the addition of electron dense and fluorescent beads to the specimen was shown to enable high-precision alignment of LM and EM images (Fukuda et al., 2014; Schellenberger et al., 2014; Schorb & Briggs, 2014). *Streptomyces* are multicellular bacteria which grow in the form of a branching mycelium. When cryo-fixed on an EM grid, their distinct pattern allowed visual relocation of targets from the light to the electron microscope and reasonably accurate overlay of LM and EM images based solely on the *Streptomyces* hyphal morphology, without the use of finder grids.

One major disadvantage of the two-step cryo-CLEM method is that the extra cryo-LM imaging step increases the amount of handling and transfers of cryogenically cooled specimens needed as opposed to cryo-EM on its own. These vitrified EM grids are very sensitive to handling and ice contamination from humid air and accidental temperature fluctuations. Any added handling step decreases the success rate of an experiment. A solution is the recently developed integrated LM and EM systems (Faas et al., 2013; Iijima et al., 2014; www.fei.com/correlative-microscopy), in which no additional transfer step is necessary since LM is performed within the TEM. Additionally, these integrated systems allow fast and easy relocation of targeted structures and enable automatic overlay of 2D LM and EM images. Disadvantages are that the LM and EM microscope types are fixed, the resolution of the LM data in current light and electron microscopes is compromised by the relatively low NA of the lens, and the TEM extraction voltage (80–120 keV) is not optimal for cryo-ET on thick cells. In fact, most parts of any whole cell are too thick for cryo-ET imaging without being thinned to less than 300 nm. Cryo-sectioning of cells (Al Amoudi et al., 2004) and cryo-FIB methods (Marko, Hsieh, Moberlychan, Mannella, & Frank, 2006; Rigort et al., 2010) can be used to thin cells for imaging by cryo-ET, and these methods are being explored for integration with cryo-CLEM methods (Faas et al., 2013; Fukuda et al., 2014).

An important additional development in the field of CLEM is combining LM of live cells with EM, adding the temporal component of tagged structures to the static high-resolution cellular morphology. Currently used CLEM methods combine live imaging with subsequent fixation of live cells, either chemically (Mironov, Polishchuk, & Beznoussenko, 2008), high-pressure freezing (Verkade, 2008), or plunge freezing (Jun et al., 2011) with electron tomographic imaging. The recent integration of a light microscope into a plunge-freezing device reduces the amount of additional specimen transfer steps and sets the time resolution between live imaging and vitrification to a few seconds (Koning et al., 2014).

A key future improvement for cryo-CLEM applications is to bridge the resolution gap between LM (few hundred nanometers) and cryo-TEM (few nanometers) to improve the localization accuracy of fluorescently tagged structures. The current localization accuracy using markers is in the order of tens of nanometers (Schellenberger et al., 2014; Schorb & Briggs, 2014), which is still one order larger than the resolution that can be achieved by cryo-ET. The use of super-resolution LM techniques might enable precise localization of individual GFP-tagged proteins.

With respect to the visual integration of data, a single platform for visualizing LM and EM projection images with live imaging sequences, electron tomography, and surface-rendered data is desirable. While many software programs perform well in correlating LM and EM data (Adobe Photoshop), visualizing different types of data (Amira), and making movie sequences from these (Adobe Premiere), we have found no program that is tailored to performing both the correlation and the visualization of multidimensional data which varies several orders in size.

We think that the described cryo-CLEM methods that we developed here for Streptomyces are applicable with only minor adaptions—e.g., using reference grids for relocation, markers for correlation, different fluorescent labels, different freezing stages, and electron microscopes—for many other biological specimens.

ACKNOWLEDGMENTS

We would like to thank Marjon Mourik for critical reading of the manuscript. This research has been executed with support of Cyttron II (LSH framework FES0908), NeCEN, The Netherlands Centre for Electron Nanoscopy, Leiden, The Netherlands, and NWO, The Netherlands Organisation for Scientific Research and is partly financed by the European Regional Development Fund of the European Commission.

REFERENCES

Al Amoudi, A., Chang, J. J., Leforestier, A., McDowall, A., Salamin, L. M., Norlen, L. P., et al. (2004). Cryo-electron microscopy of vitreous sections. *The EMBO Journal*, *18*, 3583–3588.

Bibb, M. J. (2005). Regulation of secondary metabolism in streptomycetes. *Current Opinion in Microbiology*, *2*, 208–215.

Carlson, D. B., & Evans, J. E. (2011). Low-cost cryo-light microscopy stage fabrication for correlated light/electron microscopy. *Journal of Visualized Experiments*, (52).

Celler, K., Picioreanu, C., van Loosdrecht, M. C., & van Wezel, G. P. (2012). Structured morphological modeling as a framework for rational strain design of Streptomyces species. *Antonie van Leeuwenhoek*, *3*, 409–423.

Chater, K. F., & Losick, R. (1997). Mycelial life style of Streptomyces coelicolor A3(2) and it's relatives. Oxford University Press, 149–182.

Claessen, D., Rozen, D. E., Kuipers, O. P., Sogaard-Andersen, L., & van Wezel, G. P. (2014). Bacterial solutions to multicellularity: A tale of biofilms, filaments and fruiting bodies. *Nature Reviews. Microbiology*, *2*, 115–124.

Diestra, E., Fontana, J., Guichard, P., Marco, S., & Risco, C. (2009). Visualization of proteins in intact cells with a clonable tag for electron microscopy. *Journal of Structural Biology*, *3*, 157–168.

Dubochet, J., Adrian, M., Chang, J. J., Homo, J. C., Lepault, J., McDowall, A. W., et al. (1988). Cryo-electron microscopy of vitrified specimens. *Quarterly Reviews of Biophysics*, *2*, 129–228.

Errington, J., Daniel, R. A., & Scheffers, D. J. (2003). Cytokinesis in bacteria. *Microbiology and Molecular Biology Reviews*, *1*, 52–65, table of contents.

Faas, F. G., Avramut, M. C., van den Berg, B. M., Mommaas, A. M., Koster, A. J., & Ravelli, R. B. (2012). Virtual nanoscopy: Generation of ultra-large high resolution electron microscopy maps. *The Journal of Cell Biology*, *3*, 457–469.

Faas, F., Bárcena, M., Agronskaia, A., Gerritsen, H., Moscicka, K., Diebolder, C., et al. (2013). Localization of fluorescently labeled structures in frozen-hydrated samples using integrated light electron microscopy. *Journal of Structural Biology*, *3*, 283–290.

Fukuda, Y., Schrod, N., Schaffer, M., Feng, L. R., Baumeister, W., & Lucic, V. (2014). Coordinate transformation based cryo-correlative methods for electron tomography and focused ion beam milling. *Ultramicroscopy*, *143*, 15–23.

Gruska, M., Medalia, O., Baumeister, W., & Leis, A. (2008). Electron tomography of vitreous sections from cultured mammalian cells. *Journal of Structural Biology*, *3*, 384–392.

Iancu, C. V., Tivol, W. F., Schooler, J. B., Dias, D. P., Henderson, G. P., Murphy, G. E., et al. (2006). Electron cryotomography sample preparation using the Vitrobot. *Nature Protocols*, *6*, 2813–2819.

Iijima, H., Fukuda, Y., Arai, Y., Terakawa, S., Yamamoto, N., & Nagayama, K. (2014). Hybrid fluorescence and electron cryo-microscopy for simultaneous electron and photon imaging. *Journal of Structural Biology*, *1*, 107–115.

Jakimowicz, D., & van Wezel, G. P. (2012). Cell division and DNA segregation in Streptomyces: How to build a septum in the middle of nowhere? *Molecular Microbiology*, *3*, 393–404.

Jun, S., Ke, D., Debiec, K., Zhao, G., Meng, X., Ambrose, Z., et al. (2011). Direct visualization of HIV-1 with correlative live-cell microscopy and cryo-electron tomography. *Structure*, *11*, 1573–1581.

Karreman, M. A., Agronskaia, A. V., van Donselaar, E. G., Vocking, K., Fereidouni, F., Humbel, B. M., et al. (2012). Optimizing immuno-labeling for correlative fluorescence and electron microscopy on a single specimen. *Journal of Structural Biology*, *2*, 382–386.

Kieser, T., Bibb, M. J., Buttner, M. J., Chater, K. F., & Hopwood, D. A. (2000). *Practical Streptomyces genetics*. U.K., John Innes Foundation: Norwich.

Koning, R. I. (2010). Cryo-electron tomography of cellular microtubules. *Methods in Cell Biology*, *97*, 455–473.

Koning, R. I., Faas, F. G., Boonekamp, M., de Visser, B., Janse, J., Wiegant, J. C., et al. (2014). MAVIS: An integrated system for live microscopy and vitrification. *Ultramicroscopy*, *143*, 67–76.

Koning, R. I., & Koster, A. J. (2013). Cellular nanoimaging by cryo electron tomography. *Methods in Molecular Biology*, *950*, 227–251.

Kremer, J. R., Mastronarde, D. N., & McIntosh, J. R. (1996). Computer visualization of three-dimensional image data using IMOD. *Journal of Structural Biology*, *1*, 71–76.

Kukulski, W., Schorb, M., Welsch, S., Picco, A., Kaksonen, M., & Briggs, J. A. (2011). Correlated fluorescence and 3D electron microscopy with high sensitivity and spatial precision. *The Journal of Cell Biology*, *1*, 111–119.

Le Gros, M. A., McDermott, G., Uchida, M., Knoechel, C. G., & Larabell, C. A. (2009). High-aperture cryogenic light microscopy. *Journal of Microscopy*, *1*, 1–8.

Lucic, V., Leis, A., & Baumeister, W. (2008). Cryo-electron tomography of cells: Connecting structure and function. *Histochemistry and Cell Biology*, *2*, 185–196.

Marko, M., Hsieh, C., Moberlychan, W., Mannella, C. A., & Frank, J. (2006). Focused ion beam milling of vitreous water: Prospects for an alternative to cryo-ultramicrotomy of frozen-hydrated biological samples. *Journal of Microscopy*, *222*(Pt. 1), 42–47.

Marko, M., Hsieh, C., Schalek, R., Frank, J., & Mannella, C. (2007). Focused-ion-beam thinning of frozen-hydrated biological specimens for cryo-electron microscopy. *Nature Methods*, *3*, 215–217.

McCormick, J. R. (2009). Cell division is dispensable but not irrelevant in Streptomyces. *Current Opinion in Microbiology*, *6*, 689–698.

Mironov, A. A., Polishchuk, R. S., & Beznoussenko, G. V. (2008). Combined video fluorescence and 3D electron microscopy. *Methods in Cell Biology*, *88*, 83–95.

Mistry, B. V., Del Sol, R., Wright, C., Findlay, K., & Dyson, P. (2008). FtsW is a dispensable cell division protein required for Z-ring stabilization during sporulation septation in Streptomyces coelicolor. *Journal of Bacteriology*, *16*, 5555–5566.

Morris, D. M., & Jensen, G. J. (2008). Toward a biomechanical understanding of whole bacterial cells. *Annual Review of Biochemistry*, *77*, 583–613.

Pilhofer, M., Ladinsky, M. S., McDowall, A. W., & Jensen, G. J. (2010). Bacterial TEM: New insights from cryo-microscopy. *Methods in Cell Biology*, *96*, 21–45.

Resch, G. P., Brandstetter, M., Königsmaier, L., Urban, E., & Pickl-Herk, A. M. (2011). Immersion freezing of suspended particles and cells for cryo-electron microscopy. *Cold Spring Harbor Protocols*, *7*, 803–814.

Rigort, A., Bauerlein, F. J., Leis, A., Gruska, M., Hoffmann, C., Laugks, T., et al. (2010). Micromachining tools and correlative approaches for cellular cryo-electron tomography. *Journal of Structural Biology*, *2*, 169–179.

Risco, C., Sanmartin-Conesa, E., Tzeng, W. P., Frey, T. K., Seybold, V., & de Groot, R. J. (2012). Specific, sensitive, high-resolution detection of protein molecules in eukaryotic cells using metal-tagging transmission electron microscopy. *Structure*, *5*, 759–766.

Sartori, A., Gatz, R., Beck, F., Rigort, A., Baumeister, W., & Plitzko, J. M. (2007). Correlative microscopy: Bridging the gap between fluorescence light microscopy and cryo-electron tomography. *Journal of Structural Biology*, *2*, 135–145.

Schellenberger, P., Kaufmann, R., Siebert, C. A., Hagen, C., Wodrich, H., & Grunewald, K. (2014). High-precision correlative fluorescence and electron cryo microscopy using two independent alignment markers. *Ultramicroscopy*, *143*, 41–51.

Schorb, M., & Briggs, J. A. (2014). Correlated cryo-fluorescence and cryo-electron microscopy with high spatial precision and improved sensitivity. *Ultramicroscopy*, *143*, 24–32.

Schwartz, C. L., Sarbash, V. I., Ataullakhanov, F. I., Mcintosh, J. R., & Nicastro, D. (2007). Cryo-fluorescence microscopy facilitates correlations between light and cryo-electron microscopy and reduces the rate of photobleaching. *Journal of Microscopy*, *2*, 98–109.

Schwedock, J., McCormick, J. R., Angert, E. R., Nodwell, J. R., & Losick, R. (1997). Assembly of the cell division protein FtsZ into ladder-like structures in the aerial hyphae of Streptomyces coelicolor. *Molecular Microbiology*, *5*, 847–858.

Tivol, W. F., Briegel, A., & Jensen, G. J. (2008). An improved cryogen for plunge freezing. *Microscopy and Microanalysis*, *5*, 375–379.

van Driel, L. F., Valentijn, J. A., Valentijn, K. M., Koning, R. I., & Koster, A. J. (2009). Tools for correlative cryo-fluorescence microscopy and cryo-electron tomography applied to whole mitochondria in human endothelial cells. *European Journal of Cell Biology*, *11*, 669–684.

van Wezel, G. P., & McDowall, K. J. (2011). The regulation of the secondary metabolism of Streptomyces: New links and experimental advances. *Natural Product Reports: A Journal of Current Developments in Bioorganic Chemistry*, *7*, 1311–1333.

van Wezel, G. P., McKenzie, N. L., & Nodwell, J. R. (2009). Chapter 5. Applying the genetics of secondary metabolism in model actinomycetes to the discovery of new antibiotics. *Methods in Enzymology*, *458*, 117–141.

Verkade, P. (2008). Moving EM: The Rapid Transfer System as a new tool for correlative light and electron microscopy and high throughput for high-pressure freezing. *Journal of Microscopy*, *230*(Pt. 2), 317–328.

Wang, Q., Mercogliano, C. P., & Lowe, J. (2011). A ferritin-based label for cellular electron cryotomography. *Structure*, *2*, 147–154.

CHAPTER

Immuno Correlative Light and Electron Microscopy on Tokuyasu Cryosections

11

Viola M.J. Oorschot*, Tamar E. Sztal†, Robert J. Bryson-Richardson†, Georg Ramm*,‡

**Monash Micro Imaging, Monash University, Melbourne, Victoria, Australia*

†School of Biological Sciences, Faculty of Sciences, Monash University, Melbourne, Victoria, Australia

‡Department of Biochemistry and Molecular Biology, Monash University, Melbourne, Victoria, Australia

CHAPTER OUTLINE

Methods in Cell Biology, Volume 124, ISSN 0091-679X, http://dx.doi.org/10.1016/B978-0-12-801075-4.00011-2

Abstract

Finding a rare structure by electron microscopy is the equivalent of finding a "needle in a haystack." Correlative light- and immunoelectron microscopy (CLEM) on Tokuyasu cryosections is a sophisticated technique to address this challenge. Hereby, fluorescently labeled structures of interest are identified in an overview image by light microscopy and subsequently traced in electron microscopy. While the direct transfer and imaging of the same sections from optical to electron microscopy enables straightforward correlation, the sample preparation is crucial and technically demanding. We provide a detailed guide outlining the critical steps for sample embedding, cryosectioning, immunolabeling, and imaging. In the example provided, we use CLEM to trace aggregates formed in a zebrafish myopathy model expressing enhanced green fluorescent protein (eGFP) tagged actin. In our case, only a few muscle fibers express eGFP-actin with a subset of fibers containing aggregates. By fluorescence microscopy, we are able to identify the aggregates in the zebrafish tissue, and we subsequently, use immunoelectron microscopy to image the same structures at high resolution. The CLEM method described here using Tokuyasu cryosections can be applied to a large range of samples including small organisms, tissue samples, and cells.

INTRODUCTION

The high spatial resolution of electron microscopy can become a disadvantage when trying to locate small and rare structures in a larger specimen and the search for specific structures may become very time consuming. Often only few observations are feasible and there is little scope for statistical analysis. In addition, when using transiently expressed proteins or suppression of protein expression in a subset of cells, it is mostly impossible to distinguish the affected cell population by EM. Rare structures or regions within a specimen can nevertheless be easily identified when fluorescently labeled for light microscopy (Hassink et al., 2012). Immunocorrelative light and electron microscopy (immuno-CLEM) on Tokuyasu cryosections allows for the immunofluorescence (IF) to be performed on the same sample as the subsequent EM (Cortese et al., 2012, 2013; Vicidomini et al., 2010). Hereby, IF can provide a map to guide through the sample, to identify transiently transfected cells, and to identify rare structures of interest. Furthermore, immuno-CLEM allows for the further characterization of the rare structures by performing additional immuno-EM-labeling steps.

Tokuyasu embedding and cryosectioning (Slot & Geuze, 2007; Tokuyasu, 1973) is the method of choice for postembedding immunolabeling in EM due to the superior preservation of the antigenicity of epitopes. Tokuyasu cryosections can be labeled for both IF and immuno-EM (Geuze, Slot, van der Ley, & Scheffer, 1981; Tokuyasu, 1980). Probes that are visible in both IF and EM would be ideal for CLEM. Unfortunately, although some very promising probes are being developed, some limitations apply. Such labels include secondary probes that label primary antibodies directed against endogenous proteins or genetically encoded probes. For instance, quantum dots coupled to secondary antibodies are visible in both EM and IF (Giepmans, Deerinck, Smarr, Jones, & Ellisman, 2005). However, due to their size they have limited penetration into the sample, which will reduce the labeling efficiency for IF. In addition, when quantum dots are used on stained immuno EM sections, they are less visible than immunogold particles. Fluorescent-nanogold particles show better labeling efficiency, but require silver or gold enhancement for EM (Takizawa, Suzuki & Robinson, 1998). Another alternative is the use of genetic tags such as ReAsH, miniSOG, or GFP-APEX in combination with DAB cytochemistry. These have been used successfully in a number of cases (Gaietta et al., 2002; Martell et al., 2012; Shu et al., 2011) and generally tend to work well if expressed on the luminal side of intracellular compartments or structures that are densely packed such as cytoskeletal elements, but only allow for single labeling in EM and they work poorly in combination with the Tokuyasu technique.

The method described here is a compromise, but works well and is reliable, while not reducing the labeling efficiency in IF. It uses separate labels for IF and EM with sequential primary antibody labeling followed by secondary antibodies coupled to fluorophores and Protein-A-Gold particles. In addition, creating a tree-like structure using a bridging antibody provides extra enhancement of the fluorescent signal and also on the level of EM. An example is the use of biotinylated primary antibodies in combination with an intermediary anti-biotin antibody. It should be noted that more antigens are fluorescently labeled than gold labeled due to the low penetration of larger gold particles into the chemically cross-linked matrix of the tissue section.

Although in some cases alternating sections for IF microscopy and EM could be used, using a same section approach has clear advantages. The correlation is straightforward as distortions and accidental folding of the section during cutting and transfer are the same for IF and EM. In addition, small structures might only be present in a few serial sections. Using fluorescent probes as well as gold particles coupled to the same primary antibody ensures that exactly the same structure is traced from light to electron microscopy.

For the crude correlation of IF and EM data, orientation marks on the EM grid and the form and shape of the section can be used. We find that staining the nuclei with Hoechst provides good fiducial markers for a finer correlation of the data. More sophisticated techniques for the correlation step have been developed that allow for correlation in the light microscopy subdiffraction range (Kukulski et al., 2011).

11.1 RATIONALE

We are interested in the causes and cell biology of muscle disease and utilize the advantages of the zebrafish model system to investigate this. The use of fluorescent markers, such as eGFP, is widespread in zebrafish research, allowing *in vivo* imaging of tagged proteins and visualization of the onset and progression of pathology in our models. However, analysis of pathology in patient samples is routinely carried out through histology and EM. In order to unambiguously determine if the eGFP-labeled structures detected by fluorescence microscopy correspond to features identified in EM, it is necessary to have a combined approach allowing both the fluorescence and ultrastructure to be examined within the same specimen.

11.2 METHODS

11.2.1 ZEBRAFISH

1. Generation of mosaic expression in zebrafish

   Zebrafish embryos are injected at the 1-cell stage with a DNA construct to drive expression of fluorescent proteins. The injected fish are maintained in embryo water using standard methods as previously described (Westerfield, 1993).
2. Zebrafish selection

   Injected embryos are screened for the presence of the fluorescent label using a stereo-fluorescent microscope. Embryos positive for fluorescence are selected for fixation (Fig. 11.1).

11.2.2 FIXATION

Chemical fixation of the biological sample is crucial for the preservation of the ultrastructure and antigenicity of a specimen. The most frequently used fixatives for immunoelectron microscopy are paraformaldehyde and glutaraldehyde. The choice, combination, and concentrations can be critical to the outcome of the immunolabeling and morphology. Strong fixatives such as glutaraldehyde result in better ultrastructure preservation, but can decrease the antibody sensitivity. Fixation is also pH-dependent and strong buffers should be used where possible as it can result in minor improvements to the ultrastructure. An alternative is cryofixation by HPF followed by freeze substitution and rehydration (Van Donselaar, Posthuma, Zeuschner, Humbel, & Slot, 2007) or frozen hydrated sectioning with the section fixation method (Karreman, van Donselaar, Gerritsen, Verrips, & Verkleij, 2011). The specimen, antibody, and scientific question should be considered as key factors when choosing the fixation method.

Fixation was carried out as described (Slot & Geuze, 2007).

1. Fixation protocol

   Zebrafish are fixed by adding 4% paraformaldehyde and 0.4% glutaraldehyde in 0.1 M buffer to an equal volume of zebrafish water for 10 min, followed by

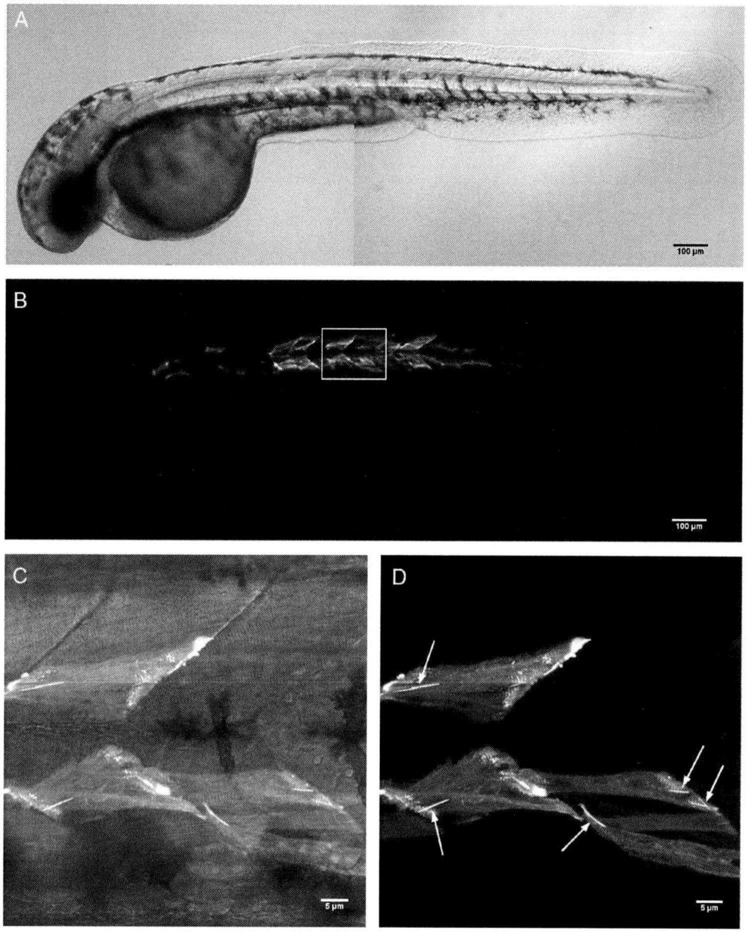

FIGURE 11.1

Actin-mutant ZF injected with GFP-actin. Overview of a 5× magnified fish (A and B). 20× magnification of the box in (B) showing GFP-positive muscle fibers (C and D), aggregates clearly visible (arrows). (See the color plate).

postfixation in 2% formaldehyde and 0.2% glutaraldehyde in 0.1 M buffer without fishwater and fixed overnight at 4 °C.

Phosphate buffer (PB), pH 7.4, or PHEM buffer, pH 7.2, was used, and good morphology was observed with both buffers.

2. Sample storage

Zebrafish are stored in 1% paraformaldehyde in phosphate buffered saline (PBS) at 4 °C until further processing.

11.2.3 SAMPLE PREPARATION

To give support to the tissue and facilitate the cryosectioning, the biological specimen needs to be embedded in a gelatin matrix. The embedding procedure can also be used to orientate the sample prior to freezing. This step might be crucial if targeting a specific area within the specimen and can easily be monitored. Unfortunately, gelatin will shrink upon sucrose infusion and this will cause irregular shaped blocks, which have a tendency to detach from the pins after freezing. Here, the flat embedding technique (Oorschot, de Wit, Annaert, & Klumperman, 2002; Van Rijnsoever, Oorschot, & Klumperman, 2008) is beneficial, whereby specimens, such as zebrafish, are flat embedded on a Petri dish between two gelatin layers. The blocks of gelatin-embedded specimens are only removed after sucrose infusion (to prevent ice crystal formation during freezing) and keep their straight shape. In addition, the optional fixation of solidified gelatin can decrease shrinkage and facilitate the cutting process.

1. Tokuyasu embedding

    Fixed tissue is washed with 0.15% glycine in PBS and infused in 5% gelatin in PB at 37 °C for 60 min followed by infusion of 10% gelatin in PB at 37 °C for 60 min (Fig. 11.2A).

    A Petri dish is filled with a thin layer of 37 °C gelatin, which is solidified at room temperature for 60 min. A second layer of gelatin prewarmed to 37 °C is then added to the Petri dish and the zebrafish is placed in this layer. The Petri dish with the zebrafish and gelatin is cooled down to 4 °C for 60 min to allow the gelatin to solidify.

    Note: the flat embedding method minimizes the shrinkage of gelatin upon sucrose infiltration resulting in square fish-gelatin blocks with better attachment to the pin.
2. Gelatin fixation

    Optionally, the gelatin can be fixed in 0.5% paraformaldehyde for 30 min at 4 °C. Although this step is not crucial, it can facilitate cryosectioning. However, blocks will be more brittle and fragile to handle. After fixation, the gelatin with zebrafish is washed three times with PB. We have used both fixed and unfixed gelatin in the current study.
3. Sample freezing

    For cryoprotection, the solidified zebrafish-gelatin is infiltrated with 2.3 M sucrose in PB at 4 °C for 2 days (Fig. 11.2B).

    The gelatin around the region with the embedded zebrafish is removed by cutting the gelatin with a scalpel (Fig. 11.2C) and removing the large part of the gelatin as indicated in Fig. 11.2D. The gelatin is then loosened with a spatula (Fig. 11.2E) and the zebrafish-gelatin slab is lifted out of the Petri dish, cut into small blocks using a razor blade (Fig. 11.2F), and mounted on an aluminum pins (Fig. 11.2G). The excess of sucrose is removed with filter paper and the samples on pins are frozen in liquid nitrogen or in the chamber of the −100 C cryo-ultramicrotome.
4. Sample storage

    Pins with zebrafish-gelatin blocks were stored in liquid nitrogen until or after cutting.

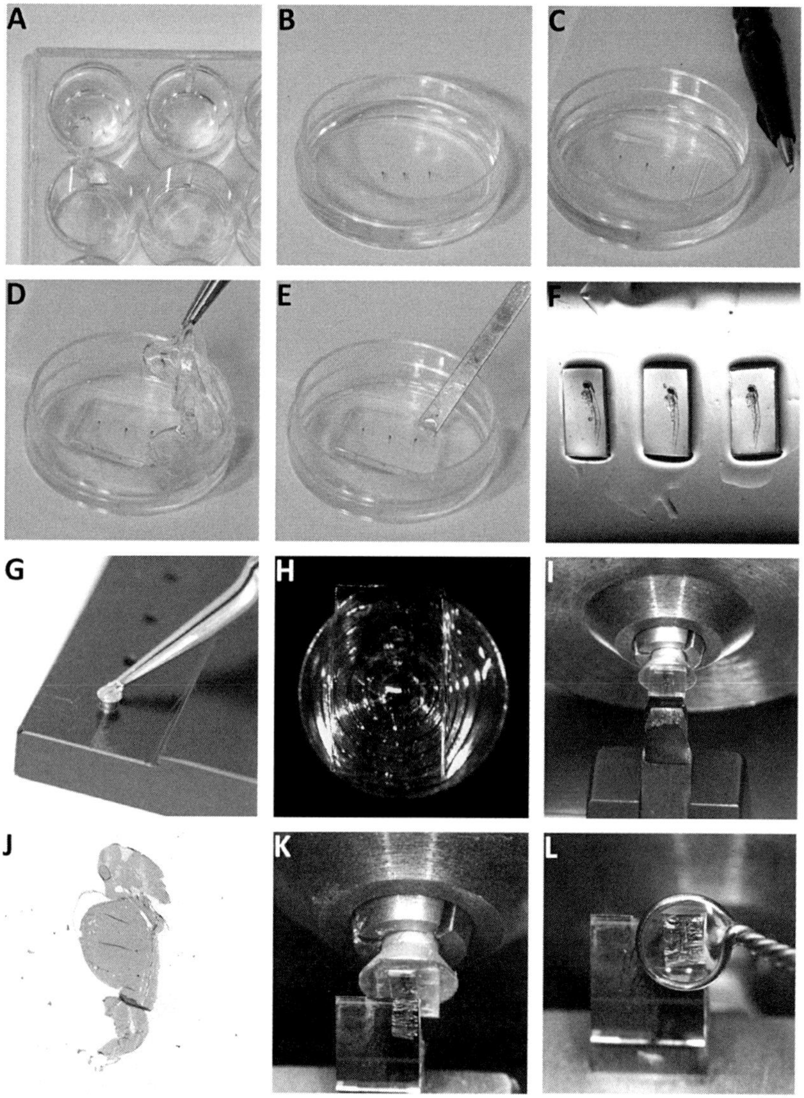

FIGURE 11.2

(A) Rinse ZF and infuse with gelatin at 37 °C. (B) Flat embed ZF in gelatin, solidify at 4 °C, and infuse with sucrose. (C) Cut around the ZF in the solidified gelatin. (D) Remove excess of gelatin. (E) Gently detach gelatin slab from the dish. (F) Cut ZF-gelatin in small blocks. (G) Mount ZF-gelatin block on a specimen pin and freeze in LN2. (H) Keep ZF in upright position during cutting. (I) Carefully trim the front-face and sides using a trimming knife. (J) Cut and stain a 300 nm section until the layer of interest is reached. (K) Switch to cryosectioning knife and cut ultrathin sections. (L) Pick up sections with MC/S, transfer to EM grids at RT.

Note: If no liquid nitrogen storage is available, zebrafish-gelatin block can be demounted from the pin, put back in cold 2.3 M sucrose, and stored at 4 °C. Prior to cutting, blocks have to be frozen again.

11.2.4 CRYOSECTIONING

Large organisms such as zebrafish embryos that are embedded in a large mass of gelatin are trickier to section. The principles of cryosectioning apply (Slot & Geuze, 2007) but additional tips and tricks might facilitate the process. When using fixed gelatin blocks, the cutting temperature should be set higher/warmer than that with nonfixed gelatin. The zebrafish should be placed in an upright position to minimize compression of the sections (Fig. 11.2H). Extreme care should be taken during trimming and a slow cutting speed is required during ultrathin sectioning.

1. Semithin sections

   The Leica UC7/FC7 cryo-ultramicrotome is allowed to cool down to −100 °C (−90 °C if using fixed gelatin), a Diatome cryotrim 20° knife is placed in the knifeholder. Place the pin in the specimen head with the zebrafish in an upright position to avoid compression during sectioning.

   The edges and front-face of the block are trimmed at a speed-setting of 100 mm/s and feed-forward of 200 nm, 100 nm in case of fixed gelatin (Fig. 11.2I).

   To locate the layer of interest, 100 nm sections are cut at speed of 3 mm/s, picked up with a mixture of 2% methylcellulose/2.3 M sucrose (1:1) placed on an object slide and stained with methylene blue/Azur II solution (Fig. 11.2J).
2. Ultrathin sections

   After the layer of interest is reached, the diamond cryotrim knife is replaced with a Diatome Cryo Immuno 35° knife and the cryotome chamber with the knife and sample cooled to −120 °C or −90 °C for fixed gelatin. Plain gelatin will give more compression than the sample itself (Fig. 11.2K); this will not cause problems for the ultrastructure. Serial ultrathin sections of 70 nm are cut at a speed of 0.8 mm/s and collected on 50-mesh cupper grids (with formvar film and carbon coat). Pickup of the sections is done with a mixture of 2% methylcellulose/2.3 M sucrose (1:1) (Fig. 11.2L).
3. Section storage

   Serial sections on grids can be stored at 4 °C in a humid-free box (Griffith & Posthuma, 2002).

11.2.5 IMMUNO-CLEM LABELING

The combined fluorescence- and immunolabeling was done as described (Vicidomini et al., 2008) with a few adaptations. The principle and details of immunolabeling are well described (Slot & Geuze, 2007). The critical step is the use of a bridging step to increase labeling efficiency.

1.	Place the grids with sections down on PBS at 37 °C	30 min (Fig. 11.3A)
2.	Rinse with PBS+0.15% glycine	4×2 min
3.	Incubate with PBS+1% bovine serum albumin (BSA)	5 min (Fig. 11.3B)
4.	Incubate with anti-GFP in PBS+1% BSA	45 min
5.	Rinse with PBS+0.1% BSA	4×2 min
6.	Incubate with bridging step in PBS+1% BSA	30 min
7.	Rinse with PBS+0.1% BSA	4×2 min
8.	Incubate with Goat anti-rabbit Alexa 488 in PBS+1% BSA	45 min
9.	Rinse with PBS+0.1% BSA	4×2 min
10.	Incubate with Protein-A-Gold 10 nm in PBS+1% BSA	20 min
11.	Rinse with PBS	2×2 min
12.	Rinse with distilled water	4×2 min
13.	Incubate with Hoechst 3342—1 μM in distilled water	20 min
14.	Rinse with distilled water	4×2 min

11.2.6 OPTICAL MICROSCOPY

Both wide-field and confocal microscopy can be used for imaging. We have tested a series of microscopes from all major brands and found all suitable for imaging Tokuyasu cryosections. However, if imaging weakly labeled epitopes, technical specifications will become more important such as the sensitivity of CCD cameras in wide-field microscopes, detectors in confocal microscopes, signal-to-noise levels, numerical aperture of the objective, and the wavelengths and efficiency of the fluorescent filters. As the grid with formvar and sections might be slightly bent or tilted, we recommend recording a z-stack of images to ensure focus throughout the section.

Before optical imaging, the grid is mounted in between a coverslip and a slide in the following way: after finishing the labeling, the grids are placed in a droplet of 50% glycerol in distilled water. The sections are facing up and the grid is covered completely by glycerol (Fig. 11.3C). Use fine tweezers to move the grid gently in the droplet to minimize air bubbles on the backside of the grid. Place the grid, sections facing down, with an adherent droplet of glycerol on a coverslip (Fig. 11.3D. Cover the coverslip/grid with an object slide, glycerol will spread out and is used as the mounting reagent (Fig. 11.3E and F). The grid is now ready for optical imaging.

Any fluorescent system can be used but in the examples presented we utilized a Leica AF6000LX microscope with a DFC 350FX camera with a 63×/1.3 Na glycerol objective. For better overview of the fish, the tiling option can be used. Since a formvar film on a grid can be wobbly, the z-stack setting is beneficial to get the entire height of the section in focus. For Hoechst, an excitation similar to DAPI (365 nm) is used. A bright-field image is obtained for recording the gridbar pattern.

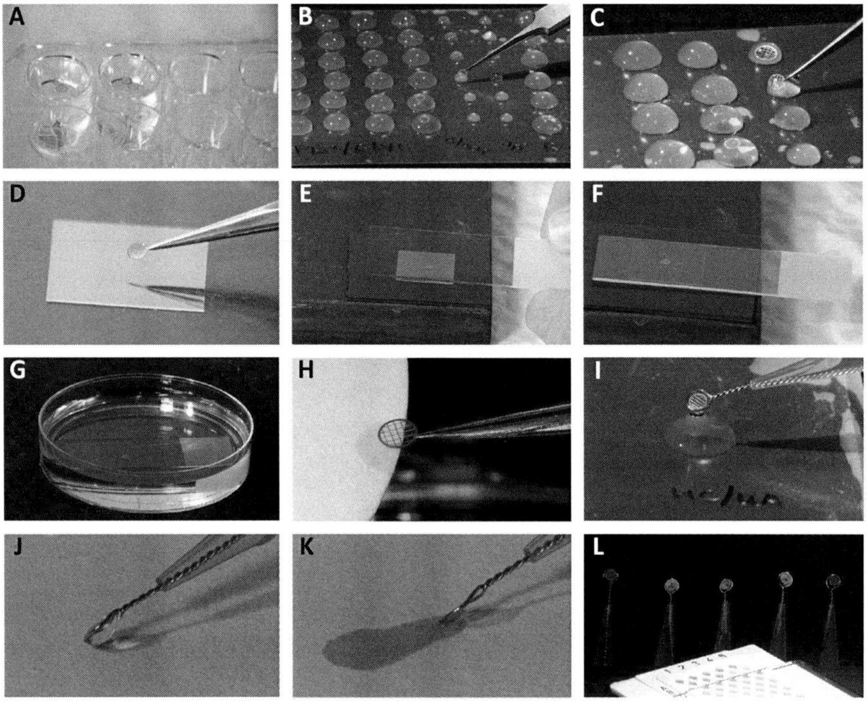

FIGURE 11.3

(A) Place grids with sections facing down on PBS at 37 °C to dissolve and rinse away gelatin. (B) During labeling, place grids on consecutive droplets on parafilm. (C) After labeling, dip grids with sections facing up in droplets of 50% glycerol. (D) Place grid with sections facing down with an adherent droplet of 50% glycerol on a coverslip. (E) Cover the grid with an object slide. (F) Grid mounted in between coverslip and object slide is ready for optical imaging. (G) After IF, place object slide in water, stir gently till the coverslip detaches, and collect grid for EM staining. (H) Carefully remove water from the backside of the grid with filter paper placed on the edge of the grid. (I) Stain sections with UAc and on a droplet MC/UA on ice, then place a loop underneath and lift the grid out. (J) Gently touch the loop with grid and adherent MC/UA on filter paper in an angle of 45°. (K) Remove the excess of methylcellulose by slowly dragging the loop over filter paper. (L) Air-dry grids and store in gridbox.

11.2.7 STAINING PROTOCOL FOR EM

1. Put the object slide/coverslip in a 10 cm dish with distilled water for atleast 5 min and unmount the coverslip from the object slide (Fig. 11.3G).
2. Carefully remove the EM grid and dry the backside of the EM grid by draining it sideways on filter paper (Fig. 11.3H). Be careful not to touch the surface or backside of the grid with the filter paper. Put the EM grid on a droplet of distilled water. If the backside is dry and the support film is intact, the grid should float on the droplet.

3. Rinse 6 × 2 min with distilled water.
4. Stain with uranyl oxalic acetate, pH 7.0, for 5 min.

   Note: We have not tested the relevance of this step in the ZF. UOA was introduced by Tokuyasu to stabilize membrane lipids and enhance final contrast. Although used in our study, this staining step is still controversial.
5. Rinse briefly with distilled water.
6. Stain with methylcellulose/uranylacetate (UA), pH 4.0, on ice for 5 min.
7. Place a wire loop underneath the grid and lift the grid from the MC/UA droplet (Fig. 11.3I). A large droplet of MC/UA is hanging in the loop; the grid is floating in it. Make sure the EM grid is located in the center of the loop. Gently touch the side of the loop to absorbing filter paper (Fig. 11.3J) and slowly drag along the filter paper until no more UA/MC comes off (Fig. 11.3K). The grid will be hanging in the loop and covered by a thin film of MC/UA. The thickness of the film is dependent on the speed of dragging the loop over the filter paper. The slower the speed, the thinner the film. This step requires some practice and testing to obtain the right thickness. It should take roughly 2–8 s to drain the EM grid. Air-dry the grid for 10 min and gently remove it from the loop by using fine tweezers. Store the grids in a dust-free gridbox (Fig. 11.3L).

11.2.8 ELECTRON MICROSCOPY AND CORRELATION

Standard TEM imaging at 80 kV is used and the images are digitally recorded. In our experimental design, additional fiducial markers apart from the Hoechst stain are not required. The actual correlation is largely done by eye. As a first step, a rough overlap between the IF data and the EM grid is based on the location of the gridbars and the relative position of the Tokuyasu sections. This narrows down the ROI to one grid-maze where the outline of the section is visible. The next step is to analyze the pattern of the Hoechst blue-stained IF nuclei. This correlation is done at a low EM magnification (4000 ×) at which the nuclei are clearly recognizable as large organelles (Fig. 11.4A). Higher magnification (> 12,000 ×) is used to analyze the antibody labeling with Protein-A-Gold and its correlation with the green fluorescent IF label (Fig. 11.4D and E).

A montage of the EM (Fig. 11.4A and C) and fluorescent confocal (Fig. 11.4A and B) images can be created using the TrakEM2 plugin for Fiji (Cardona et al., 2012; Saalfeld, Cardona, Hartenstein, & Tomancak, 2010). EM images are loaded into TrakEM2 and the workspace calibration set to that of the EM images. EM images are aligned with the montage function of TrakEM2 using the "elastic (nonlinear block correspondences)" mode. Block matching is performed using the default settings with a Rigid Transformation. Feature extraction and alignments are also performed using default settings with the expected transformation set to "Affine Transformation." The confocal image is subsequently imported and scaled to match the calibration of the workspace. The image is then rotated and positioned manually (Fig. 11.4A).

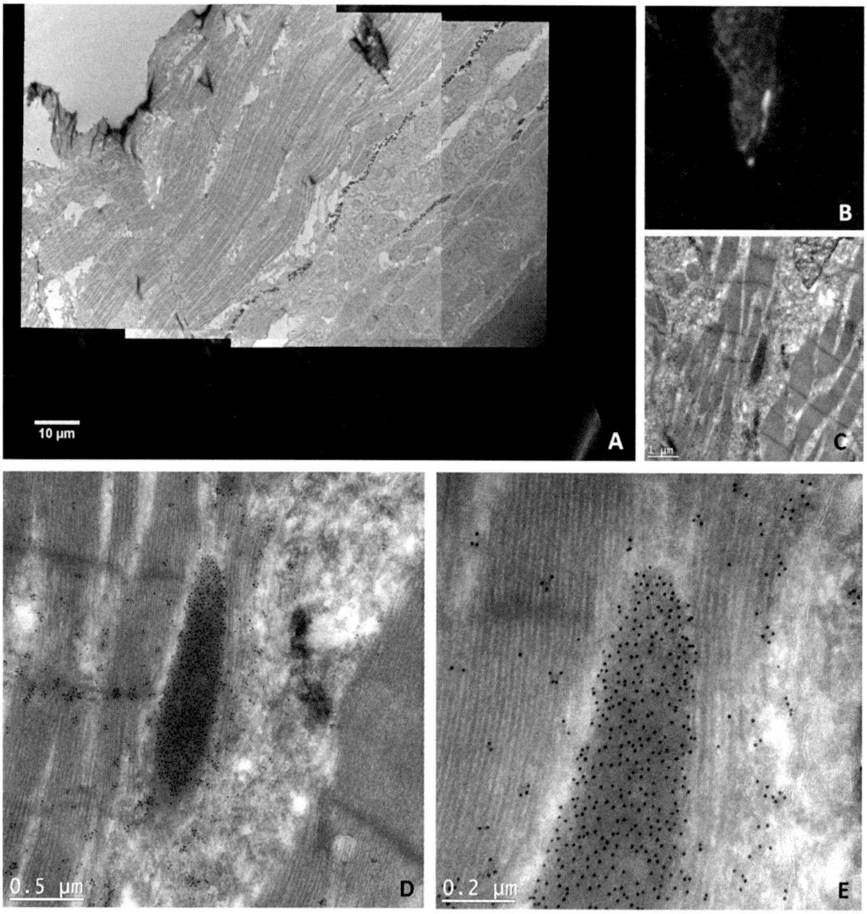

FIGURE 11.4

Overlay of the corresponding fluorescent and EM images showing a GFP-positive muscle fiber and the Hoechst-stained blue nuclei pattern (A). Higher magnification of (A) with the visible rod (B and C). Detailed EM pictures of the anti-GFP–PaGold 10 nm-labeled actin-rod (D and E). (See the color plate.)

11.3 MATERIALS

11.3.1 REAGENTS

Zebrafish culture

1.	Embryo water (E3) 5 mM NaCl 0.17 mM KCl 0.33 mM CaCl 0.33 mM $MgSO_4$ in H_2O	
2.	Tricaine methanesulfonate (pH 6.5–7.5) 3-amino benzoic acid ethylester at a final concentration of 0.16% in E3	Sigma-Aldrich A5040

Fixation

3.	0.2 M PB, pH 7.4	–
4.	0.1 M PBS, pH 7.4	–
5.	0.2 M PHEM buffer, pH 7.2	–
6.	PIPES	Merck 110220
7.	HEPES	Merck 110110
8.	EGTA	Sigma-Aldrich E0396
9.	16% Paraformaldehyde, EM grade	EMS 15710
10.	8% Glutaraldehyde, EM grade	Polysciences 00216A

Embedding

11.	10% Gelatin in 0.1 M FB	Sigma-Aldrich G1890
12.	2.3 M Sucrose (D+ saccharose)	Sigma-Aldrich S9378

Sectioning

13.	1.1% Formvar in chloroform	Merck 12164
14.	2% Methylcellulose 25 cp	Sigma-Aldrich M-6385
15.	2.3 M Sucrose (D+ saccharose)	Sigma-Aldrich S9378
16.	Methylene blue/Azur II	

Immuno-CLEM labeling

17.	1 M PBS, pH 7.4	–
18.	0.15% Glycine	Sigma-Aldrich G7403
19.	10% BSA	Sigma-Aldrich A9647
20.	Protein-A-Gold 10 nm	Department of Cell Biology, UMC Utrecht, Netherlands
21.	Hoechst	Invitrogen 33342

Fluorescent imaging

22.	50% Glycerol in distilled water	–

Electron microscopy contrast staining

23.	4% Uranyl acetate, pH 4	EMS 22400
24.	0.3 M Oxalic acid	Sigma-Aldrich O0376
25.	2% Uranyl oxalic acetate, pH 7	–
26.	2% Methylcellulose 25 cp	Sigma-Aldrich M-6385
27.	1.8% Methylcellulose-0.4% UA, pH 4	–

11.3.2 ANTIBODIES AND MARKERS

1.	Mouse anti-GFP 4C9	1:10	DSHB 4C9
2.	Rabbit anti-mouse IG	1:500	Rockland 610-4120
3.	Goat anti-GFP-biotin	1:300	Rockland 600-106-215
4.	Rabbit anti-biotin	1:10.000	Rockland 100-4198
5.	Goat anti-rabbit Alexa 488	1:300	Molecular probes-Life technologies
6.	Protein-A-Gold 10 nm	1:50	Department of Cell Biology, UMC Utrecht, Netherlands
7.	Hoechst	1 μM	Invitrogen 33342

11.4 INSTRUMENTATION

11.4.1 ZEBRAFISH SELECTION

1. Olympus SZX16 Research stereo microscope

11.4.2 SAMPLE PREPARATION

2. 37 °C stove
3. Dissecting microscope
4. Leica Ultracut UC7/FC7 cryotome
5. Leica EM Crion ionizer
6. Diatome Cryotrim 20° diamond knife
7. Diatome Cryo Immuno 35° diamond knife
8. 50-mesh cupper grids with 1% formvar film, glow discharged, and carbon evaporated
9. Aluminum pins EMS #70446
10. Standard Tokuyasu embedding tools like razorblades, tweezers, etc.

11.4.3 OPTICAL MICROSCOPES

Fluorescence imaging of fish

11. Zeiss LSM710
    5 ×/0.16 NA dry lens
    20 ×/1.0 NA dipping lens

Fluorescence imaging of immunolabeled sections

12. Leica AF6000LX
    Leica DFC 350FX Camera
    5 ×/0.15 NA dry lens (506001)
    63 ×/1.3 NA glycerol lens (50619)

Alternatives tested

13. Applied Precision DeltaVision deconvolution microscope
    CoolSNAP HQ2 1K × 1K camera (Roper Instruments)
    40 ×/1.35 NA oil lens, 60 ×/1.4 NA oil lens
14. Leica SP8 inverted laser scanning confocal microscope
    Leica HyD/PMT detector
    63 ×/1.2 NA water lens (506280)
15. Leica SP5 inverted laser scanning confocal microscope
    Leica PMT detector
    63 ×/1.2 NA water lens (506280)
16. Olympus Cell R system on a I × 81 widefield microscope
    Andor 885 iXon EMCCD 1K × 1K camera
    60 ×/1.2 NA lens

11.4.4 ELECTRON MICROSCOPES

EM Imaging of immunolabeled sections

17. Hitachi H7500 TEM at 80 KV, Gatan multiscan camera 1K × 1K

Alternatives

18. Any other electron microscope, preferable at 80 KV with a tiling/panel option.

11.5 IMAGE ANALYSIS SOFTWARE

11.5.1 FLUORESCENT IMAGES

Leica Application Suite Advanced Fluorescence Lite (LAS AF Lite)

11.5.2 ELECTRON MICROSCOPE IMAGES

Gatan Digital Micrograph
Fiji (Image J)

11.5.3 CORRELATION

Fiji (Image J), TrakEM2
Adobe Photoshop, Illustrator, InDesign

11.5.4 DATA STORAGE

My tardis (https://mytardis.org)
Dropbox (www.dropbox.com)

ACKNOWLEDGMENTS

We would like to thank Judy Gallaghan, Stephen Firth, and Alex Fulcher for technical assistance with optical microscopes. Keith Schulze for excellent software expertise. Ben Padman for photographical assistance and support.

Protocols for immuno-CLEM were developed by VO while working in the group of Prof. J. Klumperman at the Department of Cell Biology, UMC Utrecht, The Netherlands. Thanks to all the Dutch colleagues who contributed.

REFERENCES

Cardona, A., Saalfeld, S., Schindelin, J., Arganda-Carreras, I., Preibisch, S., Longair, M., et al. (2012). TrakEM2 software for neural circuit reconstruction. *PLoS One*, *7*(6), e38011.

Cortese, K., Vicidomini, G., Gagliani, M. C., Boccacci, P., Diaspro, A., & Tacchetti, C. (2012). 3D HDO-CLEM: Cellular compartment analysis by correlative light-electron microscopy on cryosections. *Methods in Cell Biology*, *111*, 95–115.

Cortese, K., Vicidomini, G., Gagliani, M. C., Boccacci, P., Diaspro, A., & Tacchetti, C. (2013). High data output method for 3-D correlative light-electron microscopy using ultra-thin cryosections. *Methods in Molecular Biology*, *950*, 417–437.

Gaietta, G., Deerinck, T. J., Adams, S. R., Bouwer, J., Tour, O., Laird, D. W., et al. (2002). Multicolor and electron microscopic imaging of connexin trafficking. *Science*, *296*(5567), 503–507.

Geuze, H. J., Slot, J. W., van der Ley, P. A., & Scheffer, R. C. (1981). Use of colloidal gold particles in double-labeling immuno electron microscopy of ultrathin frozen tissue sections. *The Journal of Cell Biology*, *89*, 653–665.

Giepmans, B. N., Deerinck, T. J., Smarr, B. L., Jones, Y. Z., & Ellisman, M. H. (2005). Correlated light and electron microscopic imaging of multiple endogenous proteins using Quantum dots. *Nature Methods*, *2*(10), 743–749.

Griffith, J. M., & Posthuma, G. (2002). A reliable and convenient method to store ultrathin thawed cryosections prior to immunolabeling. *Journal of Histochemistry and Cytochemistry*, *50*, 57–62.

Hassink, G., Slotman, J., Oorschot, V., van der Reijden, B. A., Monteferrario, D., Noordermeer, S. M., et al. (2012). Identification of the ubiquitin ligase Triad1 as a regulator of endosomal transport. *Biology Open*, *1*(6), 607–614.

Karreman, M. A., van Donselaar, E. G., Gerritsen, H. C., Verrips, C. T., & Verkleij, A. J. (2011). VIS2FIX: A high-speed fixation method for immuno-electron microscopy. *Traffic*, *12*(7), 806–814.

Kukulski, W., Schorb, M., Welsch, S., Picco, A., Kaksonen, M., & Briggs, J. A. G. (2011). Correlated fluorescence and 3D electron microscopy with high sensitivity and spatial precision. *The Journal of Cell Biology*, *192*(1), 111–119.

Martell, J. D., Deerinck, T. J., Sancak, Y., Poulos, T. L., Mootha, V. K., Sosinsky, G. E., et al. (2012). Engineered ascorbate peroxidase as a genetically encoded reporter for electron microscopy. *Nature Biotechnology*, *30*, 1143–1148.

Oorschot, V., de Wit, H., Annaert, W., & Klumperman, J. (2002). A novel flat-embedding method to prepare ultrathin cryosections from cultured cells in their in situ orientation. *The Journal of Histochemistry and Cytochemistry*, *50*(8), 1067–1080.

Saalfeld, S., Cardona, A., Hartenstein, V., & Tomancak, P. (2010). As-rigid-as-possible mosaicking and serial section registration of large ssTEM datasets. *Bioinformatics (Oxford, England)*, *26*(12), i57–i63. http://dx.doi.org/10.1093/bioinformatics/btq219.

Shu, X., Lev-Ram, V., Deerinck, T. J., Qi, Y., Ramko, E. B., Davidson, M. W., et al. (2011). A genetically encoded tag for correlated light and electron microscopy of intact cells, tissues, and organisms. *PLoS Biology*, *9*(4), e1001041.

Slot, J. W., & Geuze, H. J. (2007). Cryosectioning and immunolabeling. *Nature Protocols*, *2*(10), 2480–2491.

Takizawa, T., Suzuki, K., & Robinson, J. M. (1998). Correlative microscopy using FluoroNanogold on ultrathin cryosections. *Journal of Histochemistry and Cytochemistry*, *46*(10), 1097–1102.

Tokuyasu, K. T. (1973). A technique for ultracryotomy of cell suspensions and tissues. *The Journal of Cell Biology*, *57*, 551–565.

Tokuyasu, K. T. (1980). Immunocytochemistry of ultrathin frozen sections. *The Histochemical Journal*, *12*, 381–403.

Van Donselaar, E., Posthuma, G., Zeuschner, D., Humbel, B. M., & Slot, J. W. (2007). Immunogold labeling of cryosections from high-pressure frozen cells. *Traffic*, *8*(5), 471–485.

Van Rijnsoever, C., Oorschot, V., & Klumperman, J. (2008). Correlative light-electron microscopy (CLEM) combining live-cell imaging and immunolabeling of ultrathin cryosections. *Nature Methods*, *5*(11), 973–980.

Vicidomini, G., Gagliani, M. C., Canfora, M., Cortese, K., Frosi, F., Santangelo, C., et al. (2008). High data output and automated 3D correlative light-electron microscopy method. *Traffic*, *9*(11), 1828–1838.

Vicidomini, G., Gagliani, M. C., Cortese, K., Krieger, J., Buescher, P., Bianchini, P., et al. (2010). A novel approach for correlative light electron microscopy analysis. *Microscopy Research and Technique*, *73*(3), 215–224.

Westerfield, M. (1993). *The Zebrafish Book: A guide for the laboratory use of zebrafish (Brachydanio) rerio* (2nd ed.). Eugene, OR: University of Oregon Press.

CHAPTER

Step by Step Manipulation of the CryoCapsule with HPM High Pressure Freezers

12

Xavier Heiligenstein*,†, Ilse Hurbain*,†,‡, Cédric Delevoye*,†, Jean Salamero*,‡,§, Claude Antony¶, Graca Raposo*,†,‡

**Centre de Recherche, Institut Curie, Paris, France*

†Structure and Membrane Compartments, CNRS UMR144, Paris, France

‡Cell and Tissue Imaging Facility (PICT-IBiSA), CNRS UMR144, Paris, France

§Spatio-Temporal Modeling Imaging and Cellular Dynamics, CNRS UMR144, Paris, France

¶Department of Structural Biology and Genomics, Institut de Génétique et de Biologie Moléculaire et Cellulaire, INSERM, U964, CNRS UMR7104, Illkirch, France

CHAPTER OUTLINE

Methods in Cell Biology, Volume 124, ISSN 0091-679X, http://dx.doi.org/10.1016/B978-0-12-801075-4.00012-4

Abstract

The CryoCapsule is a tool dedicated to correlative light to electron microscopy experiments. Focused on simplifying the specimen manipulation throughout the entire workflow from live-cell imaging to freeze substitution following cryofixation by high pressure freezing, we introduce here a step by step procedure to use the CryoCapsule either with the high pressure freezing machines: HPM010 or the HPM100.

INTRODUCTION

Using a correlative light and electron microscopy (CLEM) approach to study dynamic events is a challenging investigative process, because there are several sample preparation steps with data collection required at different points in the process. In addition, the two modalities involve observing different elements of the sample where there is an order of magnitude difference in scale and resolution. Light microscopy (LM) selectively highlights one specific and dynamic protein population within a cell, regardless of the protein context, while electron microscopy (EM) shows the entire protein content and the cell ultrastructure at a nanometer resolution. One main challenge is to combine both data to conduct a complementary analysis. Ideally, imaging the specimen at each step in the process from the light to the electron microscope allows tracking the phenomenon of interest in detail, even in its finest dynamic and/or fragile structures.

CLEM has been extensively investigated and has raised a large number of technical challenges. These originate mostly from the profoundly different nature of sample preparation used in LM and EM.

To conduct transmission electron microscopy (TEM) analysis at room temperature, the sample must be (1) optimally preserved prior to (2) dehydration, (3) embedding, (4) thinning by sectioning, and (5) contrasting. To achieve the first step, two fundamentally different approaches are often used in life sciences: either a chemical fixation, based on proteins crosslinking by aldehydes, or a physical fixation, based on vitrification by rapidly decreasing the temperature (Erk, Nicolas, Caroff, & Lepault, 1998; Leforestier, Richter, Livolant, & Dubochet, 1996; McDowall et al., 1983).

The vitrification can be achieved by several means: plunge or slam freezing, double propane jet freezing, or high pressure freezing (HPF) (Moor, 1987). This last one is, to date, the only solution to vitrify specimens up to 300 μm (Moor, Bellin, Sandri, & Akert, 1980; Studer, Graber, Al-Amoudi, & Eggli, 2001). To achieve deep freezing, i.e., vitrification of biological sample, the pressure around the sample is raised to 2100 bars and cooled down to −196 °C within less than 10 ms. The high pressure lowers the water freezing point to −40 °C reducing ice crystal nucleation, especially the appearance of hexagonal ice that damages cell ultrastructure. After vitrification, specimens must be freeze-substituted and embedded in a resin to conduct CLEM observations at room temperature. The hardened block is then sliced and the sections are collected for TEM observation. Some applications require a second step of fluorescence imaging on the block or on the EM sections as already extensively described (Kukulski et al., 2011; Nixon et al., 2009).

CLEM is hence a very powerful technique but requires several sample preparation steps that are all equally challenging as they directly influence the success rate of the method. As a consequence, statistics around CLEM approaches are limited and method improvements will help to broaden the use of CLEM in biology.

As already discussed in other publications (Jahn et al., 2012), there are many methods for correlating fluorescent LM and EM information. In this chapter, we will introduce a new tool dedicated to CLEM experiments, called the CryoCapsule (Fig. 12.1), for use in the processing step from the sample preparation and conditioning to cryo-immobilization and embedding.

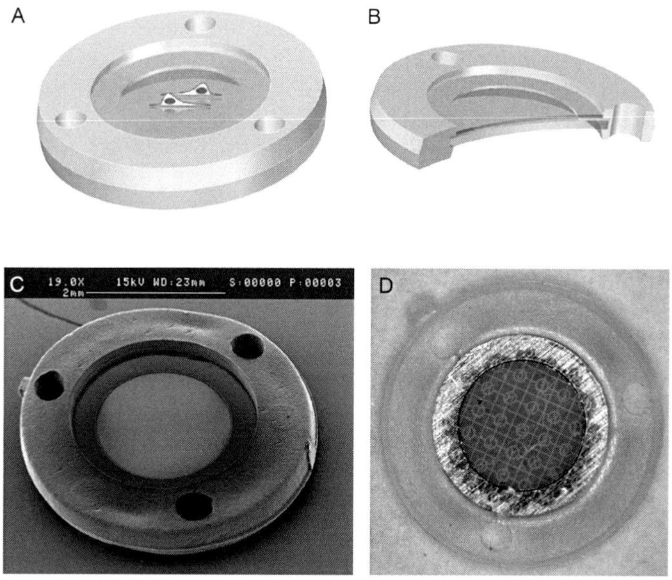

FIGURE 12.1

Views of the CryoCapsule. (A) Schematic representation of the CryoCapsule seeded with cells tagged with a fluorescent marker. (B) Schematic cross-section of the CryoCapsule closed by the covering sapphire disk, the 50-μm cavity preserves the cells during sample transfer and preparation. (C) Scanning electron micrograph of the CryoCapsule. (D) Overview picture of the CryoCapsule where the carbon landmarks are observable.

Figure adapted from Heiligenstein et al. (2014) with permission from John Wiley & Sons.

12.1 THE CRYOCAPSULE

12.1.1 WHAT IS IT MADE FOR?

The CryoCapsule is a simple tool developed to provide a transparent chamber with sapphire top and bottom where the specimen is placed, hence enabling LM observation and recording prior to vitrification by HPF. Therefore, the CryoCapsule can be compared to a small cell culture dish well suited for CLEM approaches. The CryoCapsule can be used in a variety of HPF devices including either the HPM010 (AbraFluid/BalTec, Switzerland), the Compact 01/02 (M. Wohlwend GmbH, Sennwald, Switzerland), or the HPM100 apparatus (Leica Mikrosystem GmbH, Vienna, Austria). Although mainly dedicated to CLEM approaches as described in this chapter, it can be used as a conventional sample carrier for HPF of bulk samples if cell physiology is affected by metal toxicity when aluminum or copper are used (Han, Huebinger, & Grabenbauer, 2012).

12.1.2 TECHNICAL CHARACTERISTICS: COMPOSITION, DIMENSIONS, AND PROPERTIES

The CryoCapsule is composed of three elements. The substrate for the biological specimen is a sapphire disk of 3 mm in diameter by 50 μm in thickness. The support for the biological material is landmarked by carbon evaporation through a mask. These landmarks are used to locate the area of interest and retrieve it after the HPF-FS cycle. We calculated the cooling time to be 0.2 ms to go from 20 °C to −196 °C (Heiligenstein et al., 2014).

On the top of the substrate sapphire disk, a ring made of pure gold serves as a spacer to create a chamber where the specimen will be isolated. The outer diameter of the spacer ring is 3 mm in diameter and the diameter of the hole is 2 mm. The thickness of the spacer ring is 50 μm but we have designed and are preparing deeper spacer rings to have a deeper CryoCapsule, suitable to thicker specimens. The calculated cooling time of the 50 μm spacer ring is 0.02 ms (Heiligenstein et al., 2014).

These two elements are held together by another ring, made of polyethylene. The outer diameter of the ring is 4.5 mm and the inner diameter of the hole is 2.8 mm. The thickness of the polyethylene ring is 600 μm. This last element defines the total dimensions of the CryoCapsule. We calculated its cooling time to −196 °C in 167 ms. In HPMs, the pressure is maintained to 2100 bars and −196 °C for approximately 340 ms (according to build-in measurements sensors).

12.1.3 HOW DO WE USE IT?

Comparable to a transparent reservoir or a Petri dish, the transparent bottom sapphire disk of the CryoCapsule is used to support the biological material. At the moment, the sample needs to fit within the 2 mm inner diameter of the gold spacer ring. Various cultured cell lines from robust to very sensitive (i.e., HeLa cells and neurons) were successfully grown in the CryoCapsule for different lengths of time from a few days up to several weeks.

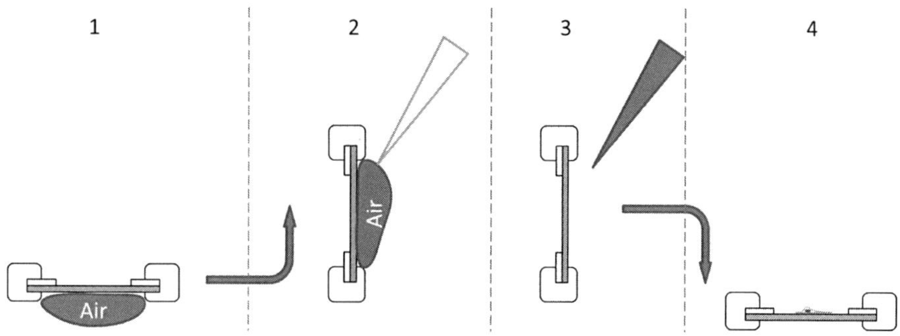

FIGURE 12.2

Removing an unwanted air bubble beneath the CryoCapsule. In some cases, an air bubble stays below the CryoCapsule during the cell seeding (1). Lift the CryoCapsule with tweezers (2) and gently remove the air bubble with a pipet tip (3). Deposit the CryoCapsule back in the Petri dish (4).

In practice, the CryoCapsules are baked at 60 °C for 48 h then sterilized in 70% ethanol (if required). The bakeout at 60 °C ensures that the carbon evaporated layer on the sapphire disk (required for the later transfer onto the resin at the sample recovery step (cf. Section 12.3.4)) hardens properly. The CryoCapsules placed into a Petri dish are kept and dried in sterile conditions. The next day, the cell culture is deposited onto the CryoCapsules and in the Petri dish to seed and grow the cells. When filled carefully, the CryoCapsules stick to the Petri dish and stay at the bottom. However, the CryoCapsules can sometimes start floating due to the presence of an air bubble below the substrate sapphire disk. The air bubble must be carefully removed with sterile tweezers and the CryoCapsule sunk back into the Petri dish (Fig. 12.2).

Once placed at the bottom, the CryoCapsule will no longer float as they are slightly denser than water (polyethylene 0.97 g/cm^3; sapphire density 3.98 g/cm^3; gold density 19.30 g/cm^3).

In the case of a tissue explant or biopsy, the sample can simply be deposited in its culture medium using a pipette.

Before imaging, a covering sapphire disk of 2.5 mm in diameter and 50 μm thick is placed in the CryoCapsule onto the gold spacer ring. The isolating imaging chamber is now closed and the sample is ready for imaging then/or HPF (Fig. 12.3).

12.2 COMBINING THE CRYOCAPSULE WITH HPF/HPMS

Currently, three major types of high pressure freezers exist on the market: the HPM010 from AbraFluid (formerly produced by BalTec, or its evolution, the Compact 01/02 from Wohlwend), the HPM100 and the EMPACT (1 and 2) from Leica MicroSystem. The ergonomics of each machine lead scientists to create tools better adapted to their specific question (Guizetti, Mäntler, Müller-Reichert, & Gerlich,

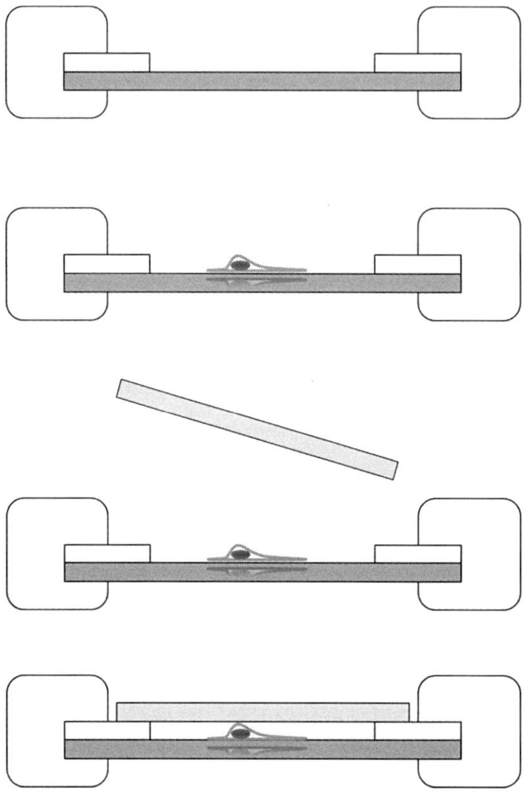

FIGURE 12.3

Closing the CryoCapsule for CLEM imaging. In the empty CryoCapsule (1), deposit a sample of interest (2), and enclose the specimen with the covering sapphire disk (3). The sample is ready for live-cell imaging (4).

2010; Hawes et al., 2007; Jiménez et al., 2010; Kolotuev, Schwab, & Labouesse, 2010; Mcdonald et al., 2010; Müller-Reichert, Hohenberg, O'Toole, & Mcdonald, 2003). As a single unit, the CryoCapsule allows similar handling on the HPM010 and the HPM100.

12.2.1 THE HPM010

The HPM010 is a manually operated machine. The sample is maintained in a clamp screwed into a large handle. We have designed a specific clamp to adapt the CryoCapsule to the HPM010.

Live-cell imaging was done on a plastic Petri dish with a drilled hole of 4 mm to ensure a direct light path and then quickly transferred into the clamp (Fig. 12.4).

The Supplementary Video S1 (http://dx.doi.org/10.1016/B978-0-12-801075-4.00012-4) illustrates the manipulation and shows how fast the loading can be

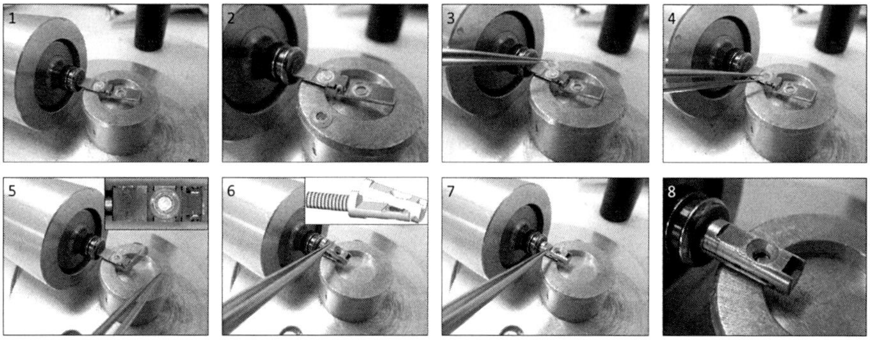

FIGURE 12.4

Loading the HPM010 holder. Open the HPM010 tip (1) and prepare the CryoCapsule containing the specimen of interest (2). Deposit the CryoCapsule in the cavity (3 and 4) and gently close the tip (4–6). Then screw the tip in the handle (8) and proceed with high pressure freezing. See also Supplementary Video S1 (http://dx.doi.org/10.1016/B978-0-12-801075-4.00012-4).

performed. After closing the clamp and screwing it into the handle, the sample can be rapidly high pressure frozen.

The CryoCapsule is then collected under liquid nitrogen for later processing. The clamp must be warmed up and dried before loading the next CryoCapsule.

12.2.2 THE HPM100

The HPM100 has a fundamentally different design therefore we also created a specific adaptor for this machine which we called the "plate." After deposing the covering sapphire disk in the CryoCapsule, the plate is placed above and clipped down. The CryoCapsule is maintained unless the plate is bent (cf. Section 12.3.1) (Fig. 12.5).

Once clipped, the CryoCapsule and plate can be loaded into an imaging chamber where humidity is preserved, thereby preventing rapid dehydration of the specimen during the live-cell imaging. Currently, we achieved this by simply drilling a 4 mm hole through a regular cell culture plastic Petri dish (4 cm in diameter) into which we added some wetted filter paper. The chamber is placed onto a cell culture microscope adaptor and imaged using a 40 × air, long working distance objective. In this condition, imaging up to 5 min at 37 °C was successfully conducted (Heiligenstein et al., 2014) (Fig. 12.6).

Once the event of interest is located, the sample is rapidly taken from the microscope and the plate is loaded into the HPM100 for HPF. This transfer step can be accomplished in less than 10 s for experienced users (Heiligenstein et al., 2014).

The plate is collected and stored as it is until freeze substitution (FS). Under liquid nitrogen, the plate is too rigid and cannot be bent, hence the CryoCapsule cannot be removed.

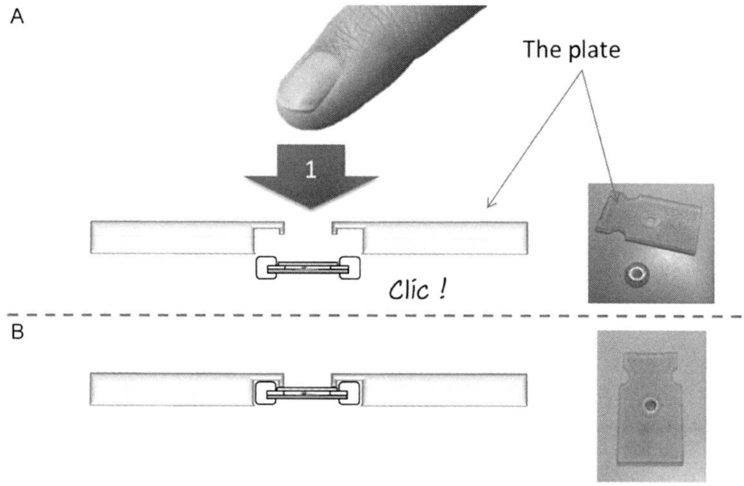

FIGURE 12.5

Loading the HPM100 plate. (A) The plate is placed above the CryoCapsule. (B) A gentle press on top of the plate will clip the CryoCapsule in.

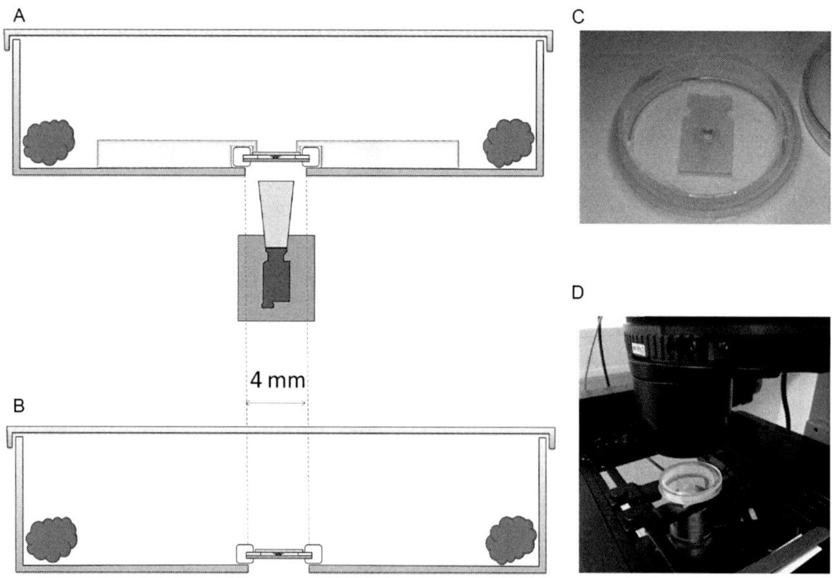

FIGURE 12.6

Live-cell imaging in a humid chamber. (A) Schematic representation of the CryoCapsule mounted in the plate and loaded in the plastic Petri dish. The bottom is perforated to allow direct fluorescence imaging. (B) Schematic representation of the individual CryoCapsule in the humid chamber prior to transfer in the HPM010 clamp. The hole for the light is 4 mm. (C) Picture of the CryoCapsule clipped in the plate and loaded in the humid chamber. (D) The humid chamber is loaded into the inverted microscope for live-cell imaging.

Note: If forced out, we experimentaly noticed that the covering sapphire disk detaches simultaneously and induces a freeze fracture effect on the cells. The risk of breaking the plate and the CryoCapsule is high so we do not recommend this procedure.

12.2.3 THE EMPACT

Although the CryoCapsule fits into the EMPACT-2 shuttles, the plastic ring around it is a thermo-insulator. We therefore expect the cooling process to be impaired by the plastic ring and hence the CryoCapsule not to be compatible with the EMPACT system for HPF. We are currently investigating the use of thermo-conductive polymers to render the CryoCapsule compatible.

12.3 AFTER HPF

12.3.1 COLLECTING AND OPENING THE CRYOCAPSULE

If you are using an HPM010, the CryoCapsule is free. We transfer it directly from the LN2 into the FS solution at −90 °C in an AFS-2 (Leica). If you are using a HPM100, the CryoCapsule is maintained in the plate. To collect it back, the plate is bent in the FS solution using a special pencil and the décapsuleur (Fig. 12.7). The procedure is as follows: sink the décapsuleur into the FS dish so it is covered by FS solution. Transfer the plate rapidly from the LN2 to the décapsuleur and align it in the compartment. Assert that the CryoCapsule faces down (Fig. 12.7E). Use the pencil to bend the middle plate to the bottom of the décapsuleur and press the piston to eject the CryoCapsule in the reservoir. Recycle the plate according to the reagents present in the FS solution (uranyl acetate, glutaraldehyde).

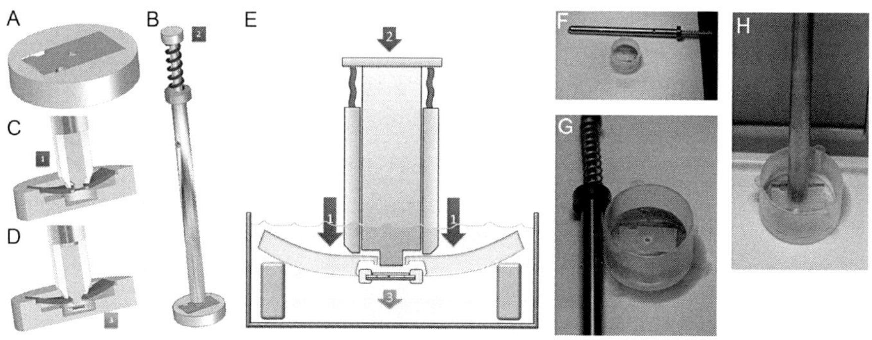

FIGURE 12.7

Use a décapsuleur to collect the CryoCapsule clipped in the plate when using a HPM100. (A) The CryoCapsule is transferred into the décapsuleur (F–G) while in the freeze substitution solution (E). (B, C, H) A piston presses on the middle plate to bend it (1) and (D) a secondary piston (2) expels the CryoCapsule (3) held in the bore of the décapsuleur (3) full of FS solution.

The CryoCapsule is then lifted from the side and the covering sapphire disk can be removed by a gentle shake. When using free-floating samples or unattached tissues, we recommend leaving the covering sapphire disk on the CryoCapsule until later steps of embedding, to prevent large movements and loss of correlation, as mentioned in Chapter 6.

12.3.2 THE FLOW-THROUGH CHAMBER

The CryoCapsule is only 4.5 mm in diameter, and fits perfectly into the AFS-1 (5 mm) or the AFS-2 (6 mm) flow-through chambers. So the FS solution exchanges can be done into all commercially available FS tools from Leica (Fig. 12.8).

12.3.3 FLAT OR EN-BLOC EMBEDDING

The 3-mm flat embedding tool from Leica is only 3 mm in diameter so as the Cryo-Capsule does not fit it and therefore the 6-mm flat embedding device is recommended.

12.3.4 REMOVING THE CRYOCAPSULE FROM THE RESIN BLOCK

Once the resin is polymerized, the CryoCapsule is at the bottom of the bloc. Gently trim the resin with a razor blade until you reach the CryoCapsule. If the plastic ring is nicely cleared, you can cut one side and unwarp it gently. The sapphire disk is

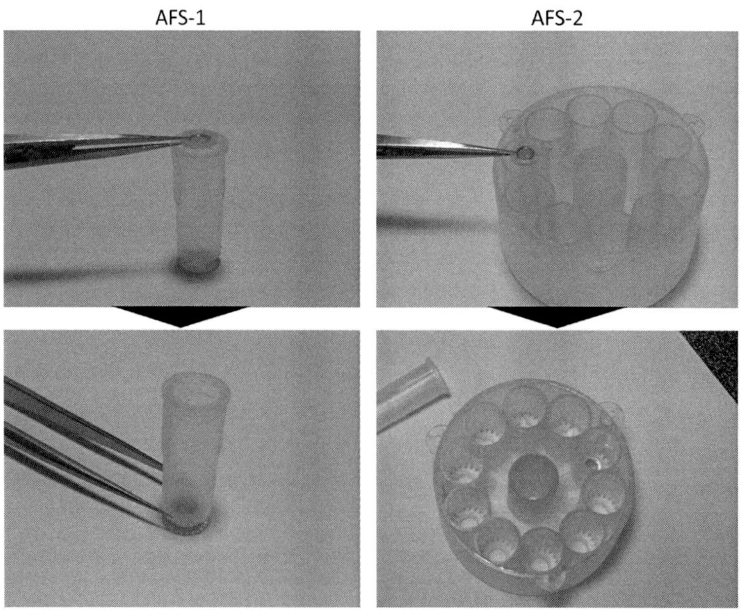

FIGURE 12.8

The CryoCapsule fits in the flow-through chambers from the Leica AFS-1 and -2 machines.

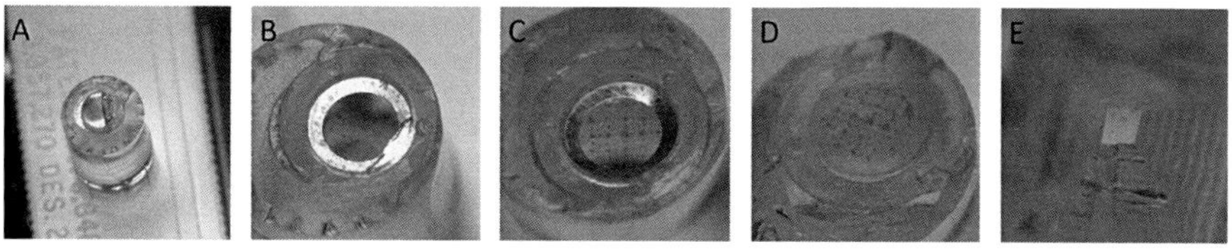

FIGURE 12.9

From the block to the trimmed specimen. (A) The full bloc is collected from the flow-through chamber. (B) The layer of resin below the CryoCapsule is cleared. (C) The resin surrounding the CryoCapsule is cleared and the plastic ring is gently removed. (D) Gentle mechanical removal of the sapphire disk and the gold spacer ring leaves the carbon landmark at the surface of the Lowicryl bloc. (E) Targeted trimming is done.

directly reachable on top of the sample and the letters are reversed. If the CryoCapsule was baked at 60 °C for at least 48 h before the experiment, sliding a very fine razor blade between the sapphire and the gold spacer ring will detach the sapphire while leaving the carbon landmark at the surface of the Lowicryl block. Attacking from various angles will help (Fig. 12.9).

Note: To detach the sapphire disk, a heat shock by rapid dip in LN2 may be required. However, a possible drawback is that the landmarks will not be transferred on hydrophilic resins such as Lowicryl. When using hydrophobic resins such as EPON resin, the carbon landmarks are easily transferred to the resin (both carbon and EPON are hydrophobic) but the fluorescence on block will be lost.

12.4 APPLICATION TO CLEM

12.4.1 CONSIDERATIONS ON OPTICAL PROPERTIES

The CryoCapsule is designed for CLEM approaches or to study pressure-sensitive samples (Heiligenstein et al., 2014).

In the latter case, optical properties are critical to ensure high-quality imaging that will be required to later relate with the EM structure. The design of the CryoCapsule is meant to allow direct imaging of the sample and reduce the refractive interfaces along the light path. The 50 μm substrate sapphire disk has a refractive index of $n \sim 1.67$, while most objectives are corrected for glass slides of 170 μm with a refractive index of $n \sim 1.52$.

12.4.2 MECHANICAL CONSIDERATION OF THE CRYOCAPSULE FOR LM

The technology used to produce the CryoCapsule allowed us to precisely overmold a brittle sapphire disk and a soft gold spacer ring neither breaking the first nor bending the second. However, in order to maintain the assembly together, a certain amount of polymer is required. It creates a distance of 0.23 mm from the bottom of the plastic ring to the bottom of the sapphire disk. Plus, the width of the cavity is 2.8 mm.

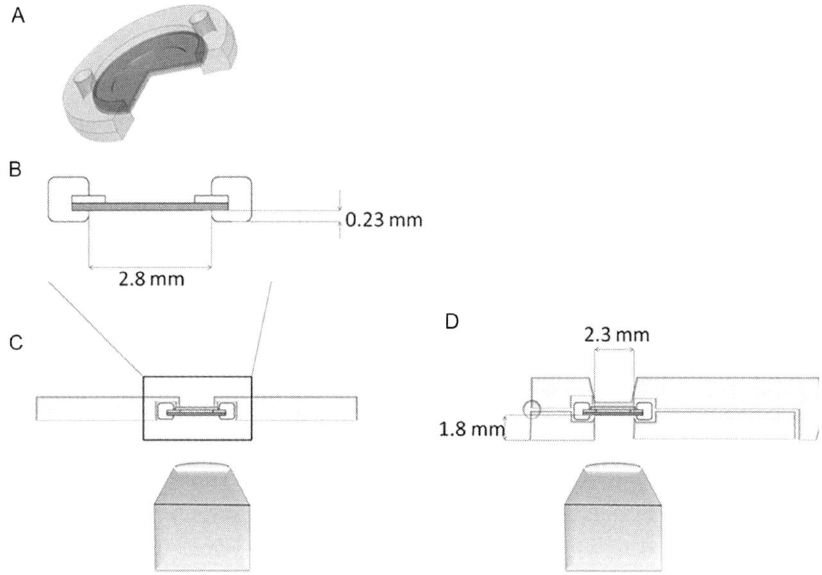

FIGURE 12.10

Live-cell imaging with long working distance objectives. (A) Schematic cross-section of the CryoCapsule, the distance of the sapphire disk from the surface of the plastic ring illustrates the necessity of using long working distance objectives. (B) Magnified view of the critical dimensions of the CryoCapsule that impose the objective selection. (C) Schematic representation of the CryoCapsule loaded into the middle plate adaptor with respect to the long working distance objective. (D) Schematic representation of the CryoCapsule loaded into the HPM010 clamp with respect to the long working distance objective.

Immersion objectives should be adequately selected so that their working distance could either fit inside the cavity or permit a 0.28 mm working distance.

Furthermore, the adaptors created to fit the CryoCapsule in the HPMs also present some particular constraints.

In the HPM100, the CryoCapsule is held by the sides only, allowing it to stand at the same level as the middle plate adaptor. This allows use of objectives with a working distance of 0.28 mm minimum.

In the HPM010, the CryoCapsule is enclosed in a larger metallic clamp and the working distance cannot be reduced to less than 1.8 mm. We therefore recommend loading the CryoCapsule after imaging, if the transfer time is not too constraining. However, we did not test it in real conditions (Fig. 12.10).

12.4.3 OBJECTIVES: AIR/WATER/OIL?

So far, to image our live specimens, we have only used long working distance air objectives. We used 40× air objectives corrected for 170 μm glasses. The refractive index of the sapphire was not specifically compensated. Often, the question of the need for immersion lenses was raised to improve the spatial resolution. We have not tested it yet. We have some concerns regarding the oil immersion objectives: the oil may be toxic for the cells and given the small shape of the CryoCapsule, oil can be easily spilled into the cell culture if not handled carefully. But most importantly, the immersion oil is thermo-resistant, thus the time required to blot it away reduces considerably the imaging transfer rate. Water immersion would then be a good compromise as long as the relevant objectives have long working distances (WD > 280 μm). We have not tested these options yet and cannot provide further feedbacks for the time being.

12.4.4 A CASE STUDY

To illustrate the results that can be obtained in a CLEM study we used M10-11G melanoma cells stably expressing YFP-langerin. These cells express a high level of fluorescence due to a protein stacking at the endoplasmic reticulum level (Heiligenstein et al., 2014; Lenormand et al., 2013). The high level of fluorescence allows identification of the cells at 40 × air. This cell is then tracked, and we show in this example a good ultrastructure preservation, illustrating the value of the CryoCapsule when conducting CLEM experiments (Fig. 12.11).

CONCLUSIONS

CLEM means adapting together two fundamentally different environments and sample preparation procedures that are used for fluorescence imaging and EM.

We designed first the CryoCapsule to simplify the manipulation steps from the light microscope to the high pressure freezer HPM010. The physical parameters of HPF imply working in very confined environment into which scientists have to fit their objects. Often, it results in assembling several small parts that tend to dissociate or have been specially created to fit the experimental parameters. The CryoCapsule associates all the elements that were previously separated into a single unit, ready for CLEM experiments. The simplification of the manipulation allows faster transfers and opens new ways toward super resolution CLEM.

More generally, the resolution limits are key elements in conducting informative CLEM studies. The CryoCapsule reduces the transfer time on both HPMs to less than 15 s (Heiligenstein et al., 2014). Hence, we achieved a good time resolution as discussed in Chapter 6.

A second key element is the spatial resolution. As mentioned in Section 12.4.3, the use of immersion objectives is important to improve resolution. However, the sapphire disk used in the CryoCapsule is 50 μm thick and has a refractive index of $n \sim 1.67$. Most objectives used in the cell biology community are designed to compensate 170 μm thick glass coverslips with a refractive index of $n \sim 1.52$. The difference results in a reduction of the spatial resolution and light detection. Spatial resolution is a combination of the refractive index, the sapphire thickness and the

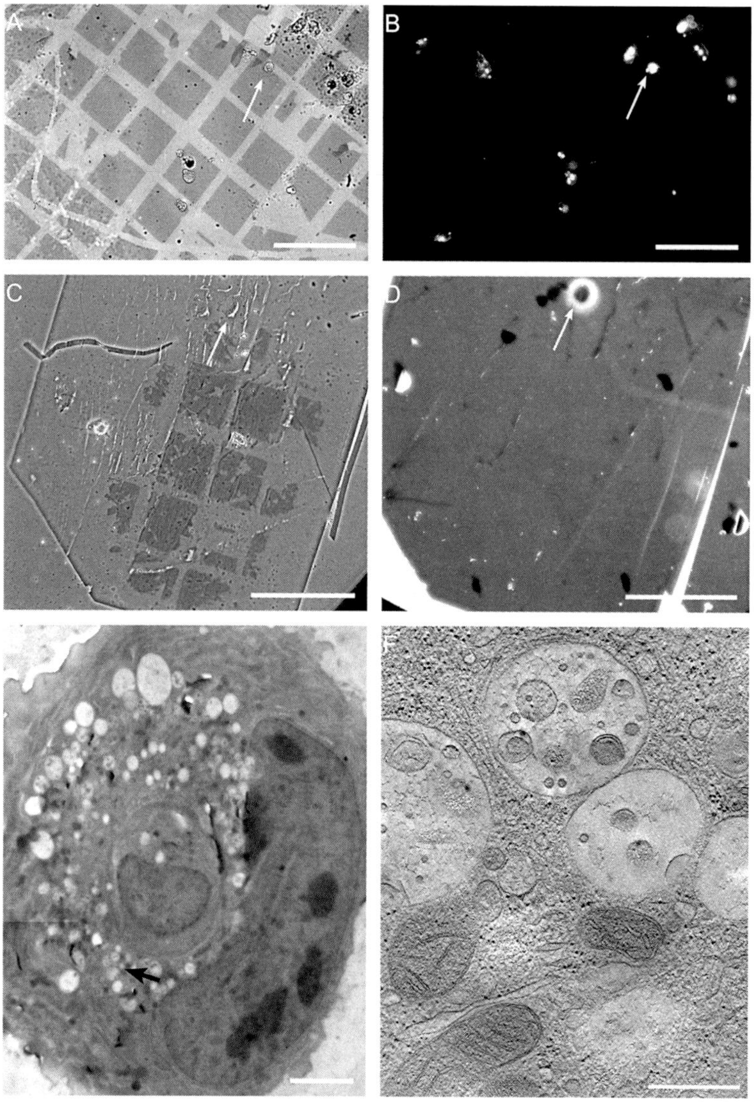

FIGURE 12.11

From live LM to EM. (A) Bright-field light microscopy at 40× air objective, we identify the cell of interest (white arrow; scale bar 50 μm). (B) The same cell is imaged by fluorescence light microscopy (white arrow; scale bar 50 μm). (C) The very first EM serial section shows the presence of the carbon landmarks, transferred to the embedding resin. The cell of interest is already identifiable (white arrow; scale bar 50 μm). (D) On the 10th serial thick section (400 nm), the cell of interest can be tracked (white arrow; scale bar 50 μm), the halo is due to high magnification imaging. (E) Low magnification stitched overview of the cell, the region of interest is localized (black arrow; scale bar 5 μm). (F) Electron tomography of the thick section (400 nm) shows good morphology preservation (scale bar 50 nm).

numerical aperture of the objective. Although we did not test it yet, we suggest using an objective (rather water or air) with adjustable thickness compensation, coating one CryoCapsule with fluorescent beads in the wavelength of interest and empirically testing the point spread function to find the most suitable parameters for your own setup.

Further developments will be introduced and tested on the CryoCapsule to broaden the application range in the coming years in order to facilitate CLEM studies.

REFERENCES

Erk, I., Nicolas, G., Caroff, A., & Lepault, J. (1998). Electron microscopy of frozen biological objects: A study using cryosectioning and cryosubstitution. *Journal of Microscopy*, *189*(Pt. 3), 236–248. http://www.ncbi.nlm.nih.gov/pubmed/9588022.

Guizetti, J., Mäntler, J., Müller-Reichert, T., & Gerlich, D. W. (2010). Correlative time-lapse imaging and electron microscopy to study abscission in HeLa cells. *Methods in Cell Biology*, *96*(10), 591–601. http://dx.doi.org/10.1016/S0091-679X(10)96024-X.

Han, H.-M., Huebinger, J., & Grabenbauer, M. (2012). Self-pressurized rapid freezing (SPRF) as a simple fixation method for cryo-electron microscopy of vitreous sections. *Journal of Structural Biology*, *178*(2), 84–87. http://dx.doi.org/10.1016/j.jsb.2012.04.001.

Hawes, P., Netherton, C. L., Mueller, M., Wileman, T., Monaghan, P., & Road, A. (2007). Rapid freeze-substitution preserves membranes in high-pressure frozen tissue culture cells. *Journal of Microscopy*, *226*(Pt. 2), 182–189. http://dx.doi.org/10.1111/j.1365-2818.2007.01767.x.

Heiligenstein, X., Heiligenstein, J., Delevoye, C., Hurbain, I., Bardin, S., Paul-Gilloteaux, P., et al. (2014). The CryoCapsule: Simplifying correlative light to electron microscopy. *Traffic*, *15*, 700–716. http://dx.doi.org/10.1111/tra.12164.

Jahn, K. A., Barton, D. A., Kobayashi, K., Ratinac, K. R., Overall, R. L., & Braet, F. (2012). Correlative microscopy: Providing new understanding in the biomedical and plant sciences. *Micron*, *43*(5), 565–582. http://dx.doi.org/10.1016/j.micron.2011.12.004.

Jiménez, N., Van Donselaar, E. G., De Winter, D. A. M., Vocking, K., Verkleij, A. J., Post, J. A., et al. (2010). Gridded Aclar: Preparation methods and use for correlative light and electron microscopy of cell monolayers, by TEM and FIB-SEM. *Journal of Microscopy*, *237*(2), 208–220. http://dx.doi.org/10.1111/j.1365-2818.2009.03329.x.

Kolotuev, I., Schwab, Y., & Labouesse, M. (2010). A precise and rapid mapping protocol for correlative light and electron microscopy of small invertebrate organisms. *Biology of the Cell/Under the Auspices of the European Cell Biology Organization*, *102*(2), 121–132. http://dx.doi.org/10.1042/BC20090096.

Kukulski, W., Schorb, M., Welsch, S., Picco, A., Kaksonen, M., & Briggs, J. A. G. (2011). Correlated fluorescence and 3D electron microscopy with high sensitivity and spatial precision. *The Journal of Cell Biology*, *192*(1), 111–119. http://dx.doi.org/10.1083/jcb.201009037.

Leforestier, A., Richter, K., Livolant, F., & Dubochet, J. (1996). Comparison of slam-freezing and high-pressure freezing effects on the DNA cholesteric liquid crystalline structure. *Journal of Microscopy*, *184*(Pt. 1), 4–13. http://www.ncbi.nlm.nih.gov/pubmed/8923755.

Lenormand, C., Spiegelhalter, C., Cinquin, B., Bardin, S., Bausinger, H., Angénieux, C., et al. (2013). Birbeck granule-like "organized smooth endoplasmic reticulum" resulting from the expression of a cytoplasmic YFP-tagged langerin. *PLoS One*, *8*(4), e60813. http://dx.doi.org/10.1371/journal.pone.0060813.

Mcdonald, K. L., Schwarz, H., Thomas, M., Müller-Reichert, T., Webb, R., Buser, C., et al. (2010). Tips and tricks for high-pressure freezing of model systems. *Methods in Cell Biology*, *96*, 671–693. http://dx.doi.org/10.1016/S0091-679X(10)96028-7.

McDowall, A. W., Chang, J. J., Freeman, R., Lepault, J., Walter, C. A., & Dubochet, J. (1983). Electron microscopy of frozen hydrated sections of vitreous ice and vitrified biological samples. *Journal of Microscopy*, *131*(Pt. 1), 1–9. http://www.ncbi.nlm.nih.gov/pubmed/6350598.

Moor, H. (1987). Theory and practice of high pressure freezing. In R. A. Steinbrecht & K. Zierold (Eds.), *Cryotechniques in biological electron microscopy* (pp. 175–191). Heidelberg: Springer, OL2393773M.

Moor, H., Bellin, G., Sandri, C., & Akert, K. (1980). The influence of high pressure freezing on mammalian nerve tissue. *Cell and Tissue Research*, *209*(2), 201–216. http://www.ncbi.nlm.nih.gov/pubmed/6994890.

Müller-Reichert, T., Hohenberg, H., O'Toole, E. T., & Mcdonald, K. L. (2003). Cryoimmobilization and three-dimensional visualization of C. elegans ultrastructure. *Journal of Microscopy*, *212*(Pt. 1), 71–80. http://www.ncbi.nlm.nih.gov/pubmed/14516364.

Nixon, S. J., Webb, R. I., Floetenmeyer, M., Schieber, N., Lo, H. P., & Parton, R. G. (2009). A single method for cryofixation and correlative light, electron microscopy and tomography of zebrafish embryos. *Traffic*, *10*(2), 131–136. http://dx.doi.org/10.1111/j.1600-0854.2008.00859.x.

Studer, D., Graber, W., Al-Amoudi, A., & Eggli, P. (2001). A new approach for cryofixation by high-pressure freezing. *Journal of Microscopy*, *203*(Pt. 3), 285–294. http://www.ncbi.nlm.nih.gov/pubmed/11555146.

CHAPTER

Live-Cell CLEM of Subcellular Targets: An Optimized Procedure for Polymer-Based Imaging Substrates

13

Benjamin S. Padman[*,†], **Georg Ramm**[*,†]

[*]*Department of Biochemistry and Molecular Biology, Monash University, Melbourne, Victoria, Australia*

[†]*Monash Micro Imaging, Monash University, Melbourne, Victoria, Australia*

CHAPTER OUTLINE

Methods in Cell Biology, Volume 124, ISSN 0091-679X, http://dx.doi.org/10.1016/B978-0-12-801075-4.00013-6

Abstract

Live-cell correlative light and electron microscopy permits the visualization of ultrastructure details associated with dynamic biological processes. On the optical level, fluorescence microscopy can be further combined with functional studies of intracellular processes and manipulation of biological samples using laser light. However, the major challenge is to relocate intracellular compartments in three dimensions after the sample has undergone an extensive EM sample preparation process. Here, we describe a detailed protocol for live-cell CLEM that provides easy guidance for 3D relocalization. Based on the use of the novel polymer film TOPAS as direct imaging substrate, we provide a setup that uses highly visible toner particles for tracking the region of interest in 2D and fiducial markers for the 3D relocation of intracellular structures. An example is given where a single mitochondria is targeted by laser micro-irradiation in live-cell fluorescence microscopy. After relocating the same structure in 3D in serial EM sections, the changes to the mitochondrial ultrastructure are observed by TEM. The method is suitable for correlation of live-cell microscopy of cells and can be performed using any inverted optical microscope.

INTRODUCTION

Correlative light and electron microscopy (CLEM) is a technical synergy of electron microscopy (EM) and light microscopy (LM). The powerful capability of live-cell CLEM and the spatiotemporal insight it can provide has the potential of driving major fundamental discoveries in cell biology. However, the technique remains technically challenging with only few laboratories performing live-cell CLEM routinely. There are three major technical challenges common to live-cell CLEM; live-cell imaging *in situ*, relocation of a target cell within the sample, and relocation of a subcellular region-of-interest within the target cell. In this chapter, we provide a technical outline of the strategies we have developed to address these three challenges.

Live-cell optical imaging usually requires that cells are cultured on an optically clear cell culture substrate. Glass coverslips are the most commonly used substrate for live-cell imaging, since most objective lenses are optimized to their thickness (160–180 μm) and refractive index (1.52–1.54). Alternatively, however, polymer films such as Aclar, Cultfoil, and TOPAS are increasingly being used instead of glass, as they are easier to handle and process for EM (Alpy et al., 2013; Gao et al., 2013; Guizetti, Mäntler, Müller-Reichert, & Gerlich, 2010; Kingsley & Cole, 1988; Lenormand et al., 2013; Müller-Reichert, Srayko, Hyman, O'Toole, & McDonald, 2007; Spiegelhalter et al., 2010), and are compatible with cryopreservation by high pressure freezing (Jimenez, Humbel, Van Donselaar, Verkleij, & Burger, 2006; Padman, Bach, & Ramm, 2014). While polymer films are usually imaged on top of a carrier glass coverslip, we have recently introduced a method to directly use polymer film as a substrate for optical imaging, thereby reducing the number of refractive interfaces between the sample and objective. We have also introduced the use of TOPAS as an alternative substrate, an optical grade polymer which has optical properties similar to glass (Padman et al., 2014).

Using a purpose designed film holder together with TOPAS eliminates the requirement for a supporting glass coverslip.

Relocating a target cell is typically achieved through the aid of a reference grid, which is used to navigate the sample. This can be achieved either by attaching a TEM finder grid to the sample (Verkade, 2008), depositing a layer of gold (Svitkina & Borisy, 1998) or carbon (McDonald, 2009) through a TEM finder grid, etching the sample surface with a needle or focussed ion beam (Jiménez et al., 2010), or using commercially available photo-etched glass coverslips (Polishchuk et al., 2000). Some of these grid preparation strategies have limited visibility, restrictions to the practical imaging area and/or reliance on specialized equipment. Here, we describe a procedure that produces high-contrast customizable reference grids consisting of toner particles. CLEM substrates are prepared using office equipment in our adapted version of the "toner transfer" technique (Abdelgawad & Wheeler, 2007; Gleskova, Wagner, & Shen, 1998).

Relocating a subcellular region in three dimensions is perhaps the most challenging task of live-cell CLEM. Common strategies rely on the 2D alignment of spatial reference points indicated by fiducial markers visible to both imaging modalities, or intrinsic morphological features of the sample imaged by phase contrast or differential interference contrast (DIC) microscopy (Hanson, Reilly, Lee, Janssen, & Phillips, 2010; Lucic et al., 2007; Razi & Tooze, 2009; Spiegelhalter et al., 2010; Wrede, Heller, Reitner, & Hoppert, 2008). Reliance on transillumination microscopy for spatial alignment is confounded by the fact that the resulting images are unrelated to the image contrast formed by TEM or fluorescence microscopy. Furthermore, 3D data acquired using phase contrast or DIC is more difficult to deconvolve and visualize than 3D fluorescence data (Arnison, Cogswell, Smith, Fekete, & Larkin, 2000; Conchello & Lichtman, 2005; Kam, 1998). Instead, we favor the use of fluorescence probes that stain biological structures that are also easily visible in EM images, such as the plasma membrane (CellMask) and the nucleus (DRAQ5). When imaged in 3D and corrected for artifacts by deconvolution, these stains provide the spatial markers required to relocate a subcellular target and correlate it between live-cell images and TEM data.

13.1 MATERIALS AND METHODS

The timing required for each of the six discrete stages is as follows. Substrate preparation and film holder assembly (A and B) together require approximately 1 day of labor, but the prepared polymer film can be stored for up to a year. Cell culture (C) requires 24–48 h depending on the conditions required for the experiment. Live-cell imaging and TEM sample preparation (D and E) can both be performed in 1 day, but requires 24–48 h for resin polymerisation (D and E can also be divided over 2 days provided the sample is osmicated on the first day). Ultramicrotomy, TEM imaging, and spatial alignment (F) will require at least 1 day, but this will largely depend on the experience of the ultramicrotomist. In our hands, the procedure (excluding A and B) requires 3–5 days to proceed from cell culture to correlation.

13.1.1 PREPARATION OF TONER-MARKED CLEM SUBSTRATES

The use of highly visible toner-based grids for CLEM is a central advantage of our procedure. The toner transfer technique enables the large-scale production of custom reference grids using commonly available equipment. In contrast to other methods for producing reference grids (Jiménez et al., 2010; McDonald, 2009), the toner-based grids are legible without the aid of a microscope (see Fig. 13.1E). Viewing at higher magnifications reveals the presence of individual toner particles which can serve as more precise extrinsic fiducial markers and which remain visible after resin embedding and trimming. The following protocol is similar to procedures used to create printed circuit boards (Abdelgawad & Wheeler, 2007; Gleskova et al., 1998). We also provide a troubleshooting guide to help the user in producing the best grids possible (see Table 13.1).

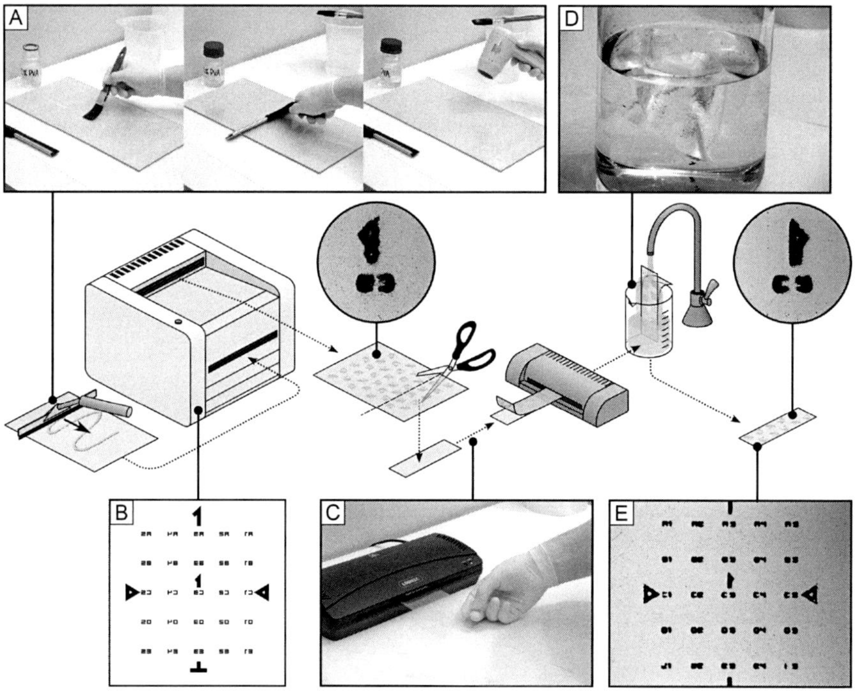

FIGURE 13.1

Preparation of toner grids on polymer-based CLEM substrates. (A–E) The toner transfer technique uses a water-soluble toner-release sheet to deposit toner on the CLEM substrate. (A) A toner-release sheet is prepared by drying multiple layers of 10% PVA on the surface of a laser printer projector transparency. (B) Grid patterns are designed using vector graphics software, then horizontally flipped before being printing on the PVA side of the toner-release sheet. (C) A strip of the printed toner-release is cut from the sheet and the PVA side is pressed against the CLEM substrate before feeding into a preheated laminator. (D) The substrate is separated from the toner-release using running water to dissolve the PVA or by soaking overnight. (E) The toner-marked CLEM substrate is now ready for use.

Table 13.1 Troubleshooting the toner transfer technique

Problem description	Caused by...	Remedy by cause
CLEM substrate becomes opaque or damaged during transfer	A, B	(A) *Transfer temperature too high* Exposure to high temperatures can damage the CLEM substrate. If an adjustable laminator is being used, reduce the lamination temperature or increase the roller speed of the laminator. If an alternative is required, a clothing iron pressed against a flat surface can be used in place of a laminator.
Transfer successful, but multiple toner images "shadows" are visible	B, C	(B) *Mechanical stresses during transfer* Excessive flexing or bending of the film during transfer can damage the CLEM substrate and alter the toner alignment, resulting in the transfer of multiple toner images in different locations. This can sometimes be resolved by taping the leading edge of the films to a thick piece of paper or cardstock before feeding into the laminator. If an alternative is required, a clothing iron pressed against a flat surface can be used in place of a laminator.
Toner remains on release sheet	C	(C) *Release sheet PVA layer too thin* Insufficient PVA prevents uniform adhesion between the substrate and release sheet during the transfer stage and inhibits the release of toner during the washing stage. Increase the number of PVA coatings when preparing the toner release or use a higher concentration of PVA to increase the layer thickness. Thinner than expected PVA layers are often caused by excessive or premature use of the squeegee.
Substrate and release sheet separate before the rinse stage	C, D, E	(D) *Transfer temperature too low* Toner cannot be transferred unless it is heated to above its melting point. If an adjustable laminator is being used, increase the lamination temperature or reduce the roller speed of the laminator. Partial transfers can often be remedied by increasing the number of times the film is passed through the laminator. If an alternative is required, a clothing iron pressed against a flat surface can be used in place of a laminator.
Transfer is inadequate, and toner does not remain on the release sheet after rinse stage	D, E, F	(E) *CLEM substrate insufficiently adherent* Toner will not be received by poorly adherent surfaces. Some substrates may require glow-discharge treatment, chemical modification, or slight surface abrasion to enhance the adhesion of toner.
Transfer initially successful, but toner progressively lost from substrate	E, F	(F) *Type of toner is inappropriate for transfer* The suitability of toner for heat-based transfer depends on its composition, which is generally considered a trade secret for most proprietary forms of toner. Trial-and-error is the only means of identifying products incompatible with toner transfer.

Materials and Reagents

Polyvinyl alcohol (PVA) powder (Sigma; #341584).
70% Ethanol.
A polymer-based CLEM substrate; Aclar (50.8 μm PCTFE; ProSciTech), Cultfoil (25 μm PTFE; Zeiss), or TOPAS 5013/6013/6015 (101.6 μm COC; TOPAS Advanced Polymers).
A4 (210 × 297 mm) overhead projector transparencies for laser printing.

Equipment

Hot-plate with magnetic stirrer.
Glow-discharge unit.
Silkscreen or window cleaning squeegee.
Laser printer (resolution above 600 dpi).
Office "pouch" laminator (heated).
Software for designing the grids (Inkscape).

Procedure

(1) Prepare a 10% PVA solution (at least 1 day before it is required).
  - **a.** Use a hot plate to boil 70 mL of water. Add a magnetic stirrer bar and remove from heat.
  - **b.** Slowly add 10 g PVA while stirring vigorously.
  - **c.** Gently heat the solution while stirring to dissolve the clumps of PVA (excessive heat will burn the PVA).
  - **d.** Dilute to 100 mL with water. The 10% PVA solution can be stored indefinitely.

(2) Prepare the toner-release sheet (see Fig. 13.1A).
  - **a.** Place a clean A4 overhead projector transparency on a flat bench-top or glass surface.
  - **b.** Use a paint brush to deposit a uniform coat of 10% PVA solution onto the projector transparency.
  - **c.** Partially dry the surface using the blow dryer for at least 15 s.
  - **d.** Draw the window wiper across the surface to flatten any bubbles and excess PVA. (Avoid gluing the sheet to the bench-top.)
  - **e.** Dry the surface using the blow dryer until the surface is no longer sticky to touch.
  - **f.** Repeat the steps (b–f) to deposit at least three layers of PVA.
  - **g.** Prior to printing, the sheet must be dried in a dust-free environment (overnight at room temperature or by heating at 60 °C for 1–2 h).

(3) Design the pattern or grid.
  - **a.** Open an A4 page template in Inkscape (or SVG software of choice).
  - **b.** Draw a circle (35 mm diameter, 0.25 pixel thickness dashed line, no fill).
  - **c.** Draw a second circle (16 mm diameter) and align centers of both circles.

   d. Design the grid pattern within the 16 mm circle. (See Note[1])
   e. Combine the designed components as an object group and horizontally flip the grouped object.
   f. Arrange the grid patterns on the A4 page template by using the "Create Tiled Clones. . ." function or by manually duplicating and aligning the grid patterns. (See Note[2])

(4) Print the design onto the dry PVA side of the A4 toner-release sheet.
   a. Determine the required page orientation before printing.
   b. Disable any printer settings relating to page rescaling, data compression, image rasterization, or economic toner use.
   c. Set the printer resolution to 600 dpi or higher.
   d. Print the design directly from the SVG software without converting to a different file format.

(5) Transfer the toner grids to the CLEM substrate.
   a. Cut a strip of the toner-release sheet containing the desired number of grids.
   b. Cut a strip of the CLEM substrate to the dimensions of the toner-release strip, and ensure that the toner-receiving surface is clean and dry.
   c. Expose the substrate to glow-discharge plasma for 30 s to enhance toner adhesion.
   d. Place the target surface of the CLEM substrate against the PVA/toner side of the toner release.
   e. Feed the combined films into a prewarmed laminator to melt the toner. Repeat this stage up as required (three to six times) to ensure complete transfer of the toner. Differences in thermal expansion coefficients may cause curling of the films with no ill effect.

(6) Separate the toner release from the CLEM substrate.
   a. Dissolve the PVA layer from the toner release by soaking the films in water on a mechanical rocker unit overnight, or until freely separated.
   b. Clean any poorly adherent toner particles from the CLEM substrate by gently wiping the surface with lens tissue paper and water. (See Note[3])

(7) The toner-marked CLEM substrates can be stored indefinitely in a clean, dust-free environment.
   a. If the results of toner transfer are unsatisfactory, please refer to the troubleshooting guide provided in Table 13.1.

---

[1]To increase the density of randomly scattered toner particles, set the fill property of the 16 mm circle to flat color, 0.5–7.5% opacity, black. Old or heavily used printers will produce randomly scattered toner particles regardless of the design.
[2]Leave space at the page margins and between groups of grid patterns to provide a border when cutting the sheet into strips.
[3]See "troubleshooting," and the toner grids can be removed from the substrate by firmly wiping the surface with lens tissue after softening the toner with 100% acetone.

13.1.2 ASSEMBLY OF THE POLYMER FILM HOLDER

We have designed a polymer film holder to optimize live-cell optical imaging through polymer films. The film holder is modeled on the dimensions of a 35 mm culture dish to be compatible with a wide range of inverted microscopes and produces a flat surface for imaging through a tensioning mechanism. The technical schematics for the film holder are available in the supplemental data of our previous publication (Padman et al., 2014).

Materials and Reagents

The polymer film holder.
Phillips head screw driver.
Low-power (8 w) or battery-operated soldering iron.

Procedure

(1) Ensure all components of the film holder are clean before use. (See Note[4])
  a. Clean all surfaces by wiping with 70% ethanol and inspect screw holes for possible blockages.

(2) Install the film between the main plate and base plate of the film holder (see Fig. 13.2).
  a. Cut a grid from the toner-marked CLEM substrate.
  b. Place the main plate on a flat surface and dampen the top surface with 70% ethanol.
  c. Place the CLEM substrate toner-side-down on top of the main plate and align the grid.
  d. Heat the soldering iron to the melting point of the CLEM substrate and pierce the film directly over one of the main plate screw holes.
  e. Place the base plate on top of the film and align one of the screw holes to the newly pierced hole.
  f. Install the screw in the aligned hole without completely tightening to allow final alignment of the components.
  g. Independently turn the film, main plate, and base plate until the grid is central and the screw holes are aligned.
  h. Ensuring that the film remains flat and aligned, use the soldering iron to pierce the film in the screw hole furthest from the first screw.
  i. Install the second screw. Repeat until all six screws are installed then trim any of the CLEM substrate protruding from the film holder.

(3) Install the tensioning mechanism of the film holder.
  a. Insert the tension cylinder into the tension plate. (See Note[5])
  b. Insert the tension cylinder of the tensioning mechanism into the center of the main plate.

---

[4]Proteinacious deposits on the film holder can be removed by cleaning with a weak ammonia solution in water. Objective lens immersion oil can be removed with acetone.

[5]Substrates needing minimal tensioning force may require the application of a silicone or vacuum grease to the tension cylinder to ensure a watertight seal.

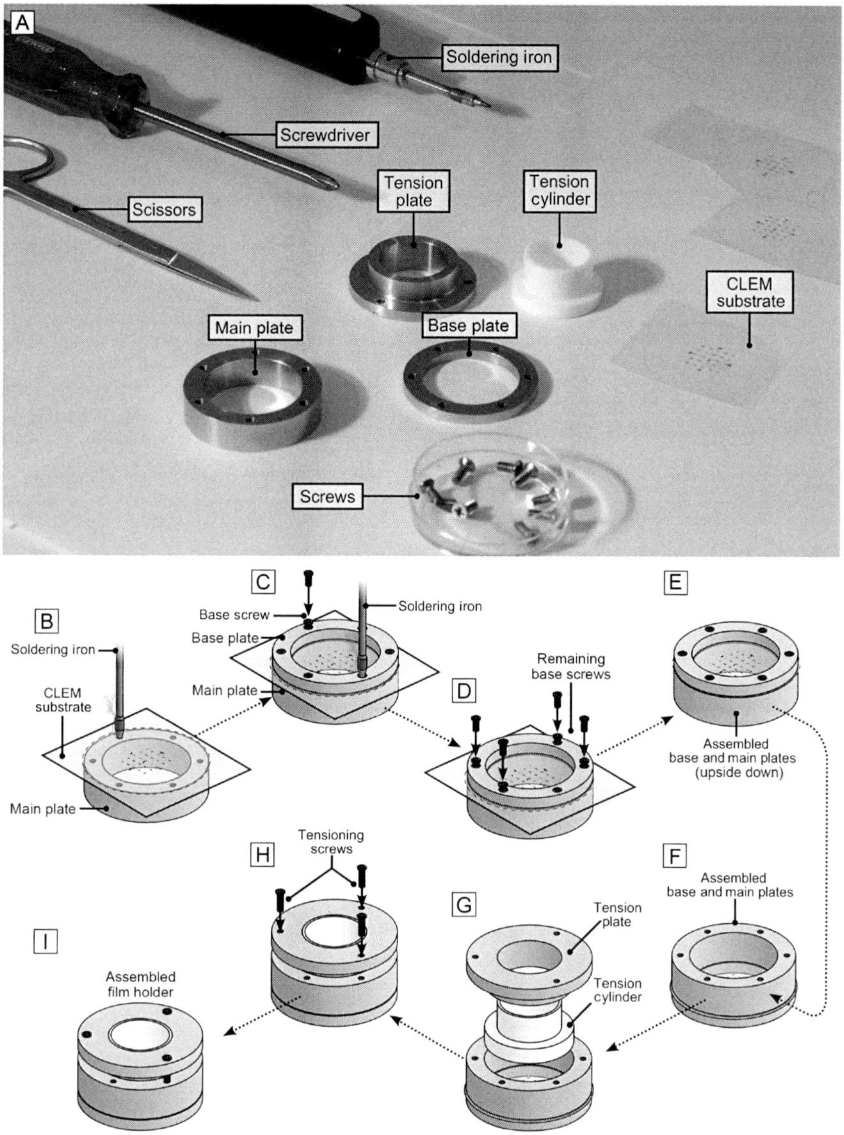

FIGURE 13.2

Assembly of the polymer film holder. (A) Polymer-based CLEM substrates are mounted into the film holder for cell culture and optical imaging. Assembly of the film holder requires a low-power soldering iron, screwdriver, pair of scissors, the CLEM substrate and the film holder (main plate, base plate, tension plate, tension cylinder, and assembly screws). (B–I) To assemble the film holder, (B) the CLEM substrate is placed upside-down on the main plate and a soldering iron is used to melt a hole through the CLEM substrate. (C) The base plate is loosely fastened to the main plate using a single screw through the melted hole. Once the base plate, substrate, and main plate are sufficiently aligned, the soldering iron can be used to melt the opposing screw hole. (D) After fastening the first two screws, the remaining four screws are installed similarly by melting each hole and installing the corresponding screw. (E) Once all six of the base screws have been installed, scissors are used to trim away any film protruding from the device. (F) The partially assembled device is then flipped over. (G) The tension cylinder and tension plate are inserted into the central cavity of the main plate. (H) The tensioning screws are installed through the three corresponding screw holes. (I) The polymer film is flattened by adjusting the top three tensioning screws and the assembled film holder is sterilized with 80% ethanol and UV light for cell culture.

c. Align the screw holes of the tension plate with any three screw holes of the main plate and loosely install the screws.
d. Slowly tighten the three screws until a watertight seal is formed between the substrate and the tension cylinder.
e. Independently tighten or loosen the screws to adjust the tension and flatness of the CLEM film.

(4) Prepare the film holder for tissue culture.
a. Reclean the assembled film holder with 70% ethanol then UV sterilize for 10 min. (See Note[6])
b. Use the lid from a 35 mm culture dish to cover the film holder for tissue culture. (See Note[7])

13.1.3 CELL CULTIVATION AND EXPERIMENT PREPARATION

The materials and procedures required will depend on the specific experimental objectives. We have successfully used the following procedure to investigate a variety of cell lines (3T3-L1, HeLa, MEF) cultured in DMEM + 10% FCS under a range of conditions (Padman, Bach, Lucarelli, Prescott, & Ramm, 2013; Padman et al., 2014).

Materials and Reagents

Culture media without phenol-red.
PBS (or other physiologically relevant buffer).

Procedure

(1) Prepare cells for the experiment.
a. Any transfection of cells should occur at least 24–48 h prior to optical imaging. (See Note[8])

(2) Prepare the assembled polymer film holder.
a. If the polymer film holder was not sterilized recently, resterilize with 70% ethanol and UV.
b. Pipette 1 mL of fresh culture media into the film holder and wait for 10–15 min.
c. If leaks are visible, tighten the tensioning screws until watertight.
d. Replace the culture media and prewarm the film holder in the incubator.

(3) Detach cells from tissue culture dish and seed the cells into the polymer film holder.
a. The polymer film holder supports culture volumes up to 2 mL.

(4) Culture the cells for at least 24 h prior to imaging.

---

[6]The film holder assembly can be treated with glow-discharge plasma to perform preliminary sterilization. This will also enhance cell adhesion for some CLEM substrates.
[7]The bottom imaging surface can be protected with a 35 mm culture dish lid.
[8]Cells can be successfully transfected while growing in the film holder, which has a culture surface area of 2.0 cm^2.

13.1.4 LIVE-CELL IMAGING AND PRIMARY FIXATION

Our procedure for live-cell CLEM involves two distinct stages of imaging. First, the living sample is imaged as under standard conditions for live-cell confocal microscopy until a target has been identified for correlation. The live-cell imaging stage is terminated during the acquisition of time-series data, when the primary fixative is pipetted directly into the film holder. The second stage of imaging begins with the acquisition of the high-resolution 3D optical data required for 3D relocation of the correlation target during ultramicrotomy. Overview images are then acquired of the surrounding region for relocation of the target cell during sample processing. These three data sets contain all the information required to relocate a subcellular target during ultramicrotomy.

Materials and Reagents

CellMask Deep-Red (1000 × stock solution).

(Optional) DRAQ5.

Primary fixative solution (1 mL of 8% paraformaldehyde (EM grade) solution in 0.1 M phosphate buffer (PB)).

Equipment

An inverted confocal microscope with the following characteristics and capabilities:

- Live-cell imaging chamber (CO_2 regulator, heated stage with 35 mm dish insert).
- Motorized axial focusing system, capable of z-stack acquisition and z-position measurement.
- Water immersion objective lens (60 ×; 1.0–1.2 NA; PlanApo).
- 10–20 × Dry/multi-immersion objective lens.
- Transmitted light detector.
- Standard multichannel imaging capability (two-channel minimum).
- Far-red imaging capability (633 nm HeNe laser line or equivalent diode laser excitation; electronic detector with Cy5 filter set or equivalent).

Procedure

(1) Stain the sample 1 h prior to optical imaging.

   a. Prepare a staining solution by diluting 1–2 μL of CellMask Deep-Red in 1000 μL of prewarmed PBS. (See Note[9])

   b. Aspirate the culture media from the sample and replace with 1000 μL of the staining solution.

   c. Incubate the cells with the staining solution for 15–30 min at 37 °C, 5% CO_2.

   d. Aspirate the staining solution and replace with 1000 μL of the phenol-free culture media.

[9]Any dyes that require "loading" should be included in the CellMask Deep-Red staining solution.

   e. Incubate the cells with the phenol-free culture media for 30–45 min before imaging.
   f. Prewarm the primary fixative solution in a 37 °C water bath or incubator.

(2) Begin the live-cell confocal imaging experiment.
   a. Select a cell of interest and begin the experiment.
   b. Focus on the basal surface of the cell in the far-red channel.
   c. Record the z-position to provide the first estimate of the "zero-depth" (see Fig. 13.3).
   d. Identify a correlation target and begin imaging in the relevant channels.
   e. Record the z-position to provide the first estimate of the "target plane."

(3) Primary fixation during time-series imaging.
   a. Once a correlation target has been selected, retrieve the prewarmed primary fixative.

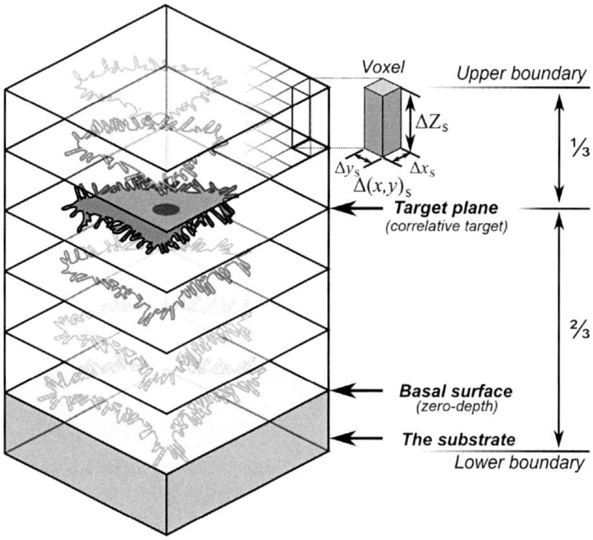

FIGURE 13.3

Guidelines for acquiring 3D optical data of intrinsic cellular features. Relocation of a "target plane" during ultramicrotomy requires the acquisition of 3D optical data optimized for deconvolution. The optical sectioning boundaries are determined by the position of the "target plane" relative to the "basal surface." To compensate for depth-dependent asymmetry in the imaging point spread function, the distance between the "upper boundary" and "target plane" should occupy ⅓ of the total z-stack volume. The remaining z-stack volume (⅔) should span the distance between the "target plane" and the "lower boundary," which should be located slightly below (~1 μm) below the "basal surface" of the cell. The voxel (volumetric pixel) size is determined by the Nyquist limits for the shortest excitation wavelength being used. The z-sampling interval (ΔZ_S) should be shorter than double the axial Nyquist length (see Note 11), and the lateral sampling resolution (pixel size; $\Delta(x,y)_S$) should be as close to the lateral Nyquist limit as possible. (See Note 10.)

b. Set up the time-series imaging parameters. Use the highest possible frame-rate with the lowest possible laser power. The pixel size (aka. lateral sampling; $\Delta(x,y)_s$) should be smaller than two to three times the lateral Nyquist limit. (See Note[10])
c. Begin the time series when the sample is ready for fixation.
d. If the target moves out of the focal plane during the time series, adjust the focus and record the new z-position.
e. When ready, directly pipette 1000 μL of the primary fixative into the film holder. Do not knock the sample or aim directly at the objective lens. The mass of the fixative will immediately move the sample out of focus.
f. End the time series and wait 15 min for the sample to restabilize. Do not move the sample.

(4) Postfixation z-stack for subcellular spatial alignment (See Fig. 13.3).
a. Refocus on the basal surface of the cell in the far-red channel. Record the z-position.
b. Calculate the new approximate position of the correlative target using the previously recorded values for the "zero-depth" and "target plane."
c. Confirm the position of the correlative target.
d. Define the z-stack imaging parameters as shown in Fig. 13.3. The boundaries of the z-stack are determined by the position of the "target plane" relative to the "basal surface." The voxel (volumetric pixel) size is determined by the axial and lateral Nyquist limits for the shortest excitation wavelength. (See Notes 10 and[11])
e. Acquire a multichannel z-stack using the lowest possible laser power to avoid photobleaching the sample.

(5) Postfixation image montages for sample navigation and cell relocation.
a. Imaging in the Deep-Red and transmitted light channels, refocus on the basal surface of the cell and acquire an image.
b. Increase the size of the scan field to the maximum allowable size and acquire an image in both channels. Nearby cells and toner particles should now be evident. (See Note[12])
c. Identify the nearest alphanumeric grid point, which should be visible to the unaided eye when looking at the sample.

[10]The lateral Nyquist length for confocal imaging is given by $\Delta(x,y)_{\mathrm{Nyquist}} = \lambda_{ex}/8n\sin(\theta)$, where $\theta = \sin^{-1}(\mathrm{NA}/n)$, NA is the numerical aperture, λ_{ex} is the excitation wavelength, and n is the refractive index of the immersion media.

[11]The axial Nyquist length for confocal imaging is given by $\Delta Z_{\mathrm{Nyquist}} = \lambda_{ex}/4n(1-\cos(\theta))$, where $\theta = \sin^{-1}(\mathrm{NA}/n)$, NA is the numerical aperture, λ_{ex} is the excitation wavelength, and n is the refractive index of the immersion media.

[12]If the z-stack used an extremely small scan field (high image zoom), collect additional images at intermediate scan field sizes to aid alignment with the subsequent images.

d. Imaging in both channels, acquire a montage of images surrounding the target cell position, then continue the montage to the nearest alphanumeric grid point.
e. Change to the low-magnification objective lens (10 × to 20 ×) and acquire an overview image.

13.1.5 TEM SPECIMEN PREPARATION

Our processing procedure uses the same protocol as our routine TEM processing procedure; secondary fixation with glutaraldehyde and osmium; En bloc staining with uranyl acetate (UA); dehydration with ethanol and acetone; and infiltration with Procure–Araldite resin for embedding. The protocol can be used for bench-top processing at room temperature or for microwave-assisted processing. Given that "routine protocols" vary between laboratories, the protocol for EM specimen preparation is subject to user preference.

Reagents

0.1 M PB.
Water (ultrapure).
2.5% Glutaraldehyde (GA) in 0.1 M PB.
1% Osmium tetroxide (OsO_4).
2% UA.
Ethanol.
Acetone.
Procure–Araldite resin (Procure-812, Araldite-502, DDSA, BDMA).

Equipment

Phillips head screw driver.
Low-power (8 W) or battery-operated soldering iron.
#3 Scalpel handle with #15 blade (or equivalent).
A polypropylene container or equivalent for processing. (See Note[13])
BEEM capsule (or equivalent).

Procedure

(1) Begin specimen preparation 1 h after primary fixation.
  a. Wash the sample 3 × for 5 min in 0.1 M PB (pH 7.4).
  b. Postfix the sample with 2.5% GA in 0.1 M PB for 1 h.
  c. Wash the sample 3 × for 5 min in 0.1 M PB (pH 7.4).
(2) Retrieve the region of interest for processing.
  a. Place the film holder in a separate dish or container to collect buffer spillage.
  b. Use the sample navigation images to identify the alphanumeric grid point closest to the target region.

[13]We usually cut a 70 mL polypropylene sample jar in half with a utility knife and use the bottom half for processing.

   c. Use a #15 scalpel to cut two parallel incisions either side of the target region. The incisions should be at least 1–2 mm away from the target region and as long as possible.
   d. Perform two more incisions to retrieve a strip of film approximately 2–4 mm wide and 10 mm long.
   e. Transfer the strip of film to the polypropylene container and use the soldering iron to weld two points of the film to the bottom surface. The welding points should be as far from the region of interest as possible. (See Note[14])
   f. Multiple pieces of film can be processed in the same container.

(3) Proceed with EM processing
   a. Wash the sample 3 × for 5 min with PB (pH 7.4).
   b. Postfix with 1% OsO_4 (1 h).
   c. Wash the sample 3 × for 5 min with water.
   d. En bloc stain with 2% UA (1 h).
   e. Wash the sample 3 × for 5 min with water.
   f. Dehydrate by graded series of ethanol (50%, 70%, 90%, 100%, and 100%; 10 min each) then acetone (100%, 100%; 15 min each).
   g. Infiltrate with Procure–Araldite resin by graded series of resin in acetone (25%, 50%, 75%, 100%, 100%; 45 min each).

(4) Retrieve the sample for embedding.
   a. Fill a BEEM capsule with Procure–Araldite resin. Ensure that the resin does not contain bubbles.
   b. Use a scalpel to cut the strip of film away from the welded points.
   c. Place the film upside-down on top of the resin-filled BEEM capsule.
   d. Move the film laterally until the target region is approximately the centered on the BEEM capsule.
   e. Polymerize the resin for 24–48 h at 60 °C.

13.1.6 ULTRAMICROTOMY, TEM IMAGING, AND SPATIAL ALIGNMENT

The procedure for relocating a subcellular target relies on optical images of intrinsic and extrinsic fiducial markers acquired during confocal microscopy. Using a multiscale system of fiducial markers provides sufficient information for an ultramicrotomist to hone in on the subcellular target across a range of magnifications, with enough redundancy to continue an experiment in the presence of unforeseen consequences (such as losing part of the sample). The ultramicrotomy skill requirement for our procedure is minimal as it does not rely on serial sectioning or the use of slot grids. That being said, an experienced ultramicrotomist will consistently relocate the "target plane" faster and more accurately than a novice. There are three hierarchical stages required for relocation of the correlative target. First, the thickness of

[14]The soldering iron temperature should be slightly above the melting point of the both plastics. Excessively high temperatures will cut through both plastics without welding them.

each ultrathin section is documented to provide the ultramicrotomist with the first estimate of section z-positions. Next, preliminary TEM imaging is then used to determine the accuracy of the initial estimates and correct them as required. Finally, imaging artifacts associated with electron and optical imaging are corrected before direct correlation between TEM and optical microscopy.

Materials and Reagents

TEM grids (grid style is subject to user preference).
4% UA grid stain.
Reynold's lead citrate (PbCit) grid stain.

Equipment

Standard ultramicrotome.
Trimming diamond or glass knives.
Ultramicrotomy knife (diamond or perfect glass).
Standard TEM for routine imaging of biological samples.

Software

Software for overlaying images (GIMP).
Software for 3D deconvolution (Huygens).
Software for EM distortion correction (ImageJ/Fiji).
Software for Montage stitching (GIMP/ImageJ/Fiji).

Procedure

(1) Planning the resin block trimming borders.
  - **a.** Align and combine the image montage into a composite image using your preferred software package.
  - **b.** Align and overlay the montage onto the low-magnification overview image.
  - **c.** Annotate the target cell and draw the intended borders of the final TEM sections. Avoid regions of thickly deposited toner as it may vary in hardness.
  - **d.** Horizontally flip the annotated image, then print it for reference as a "trimming plan."

(2) Trimming the resin block. (See Note[15])
  - **a.** Observe the resin block under a dissection microscope and manually trim excess resin and film without encroaching on the target region.
  - **b.** Use the printed trimming plan to identify the sample orientation and the approximate position of the target cell.
  - **c.** Load the resin block into a microtome so that it matches the orientation of the resin block plan.
  - **d.** Using a trimming diamond or glass knife, cut thick sections (~500 nm) directly into the polymer film until approximately half of the film thickness

[15]Some polymer films such as Aclar and Cultfoil can be peeled from the resin prior to trimming; however, this is not necessary as the film provides physical protection during trimming. Peeling the film can occasionally extract toner from the resin block, removing a key source of fiducial markers.

has been removed, which can be determined using the specimen feed counter found on most ultramicrotomes. Record this value for later use during ultramicrotomy. (See Note[16])

e. The sample features and toner should now be sufficiently visible to implement the trimming plan. (See Note[17]) Trim the preplanned borders of the resin block mesa on the microtome using a trimming diamond or glass knife.
f. Trim one corner of the mesa so that section orientation can be easily identified.
g. Use a dissection microscope to collect a final reference image of the resin block prior to sectioning. Aligning this image with the original "trimming plan" will enable identification of the target cell based on the position of the mesa boundaries, rather than extrinsic fiducials alone.

(3) Identification of the "target plane" for ultramicrotomy.

a. Deconvolve the optical z-stack data with Huygens 3D deconvolution or your preferred software package. The parameters for deconvolution will depend on the parameters used during optical imaging and the number of iterations preferred by the user.
b. The position of the basal surface and correlative target should now be determined with greater accuracy. Navigate the deconvolved z-stack and identify the basal surface. If this z-position differs from the estimates recorded during optical imaging, redefine the "zero-depth" as the z-position identified from the deconvolved data.
c. Relocate the correlation target in the deconvolved z-stack using the estimated "target plane" z-position recorded during optical imaging. Redefine the z-position of the "target plane" if required.
d. Calculate the physical distance between the "target plane" and "zero-depth." If the correlation target occupies multiple optical sections in the deconvolved data, calculate the physical distance between "zero-depth" and the first optical section containing the correlation target.
e. This calculated value is your "target depth" for ultramicrotomy.

(4) Ultramicrotomy

a. View the resin block from each side to ensure that the remaining film is uniform in thickness. Determine the approximate thickness of the remaining film by subtracting the specimen feed value (recorded during trimming) from the known polymer film thickness.
b. Cut thick sections (~500 nm) through most of the remaining film. Leaving approximately 1–2 μm of film on the mesa surface.

---

[16]The value provided by an ultramicrotome specimen feed counter should always be considered to be an approximation. Err on the side of caution.

[17]If the position of the trimming cuts cannot be confidently identified while viewing in the microtome, return the resin block to the dissection microscope and indicate the planned cut locations with needle scratches on the side of the block.

c. Replace the trimming knife with a glass or diamond knife suitable for ultrathin sectioning.
d. Cut ultrathin sections (70–80 nm) through the remaining film until the first evidence of resin. This can be determined by the visible change in section color (caused by the differing material birefringence), or by flattening the sections with heat or organic vapor (resin will expand and flatten significantly faster than the polymer substrate).
e. The "zero-depth" section is the first resin section containing the target region. The apparent thickness of all subsequent sections must now be recorded on the basis of their interference color (Peachey, 1958; Sakai, 1980), even if the section was lost due to mishandling.
f. Before proceeding, calculate or estimate the minimum number of sections required (divide the distance from "zero-depth" to "target depth" by the intended section thickness).
g. Proceed with ultramicrotomy while recording the section thicknesses. Maintain the order of the sections by ensuring that they form a ribbon. Serial-sectioning is desirable, but recording the section thickness is a higher priority.
h. Stop the microtome when the required number of sections have been cut (usually 5–10 sections more than the minimum number of sections required). The relative order of each section should now be directly relatable to the recorded list.
i. Flatten the sections using your preferred method (heat/organic vapor). Annotate the thickness list with the new apparent section thicknesses (will typically be around 90% of the original value).
j. Split ribbons of sections into retrievable groups of consecutive sections (three to five sections) and mount each section group in the center of a TEM grid. (See Note[18])
k. Circle the collected sections on the recorded list of thicknesses and document which grid they are mounted on. Each section should now be referred to by a unique identifier (i.e., Grid#, Section#).
l. Stain the grids with UA for 5 min and PbCit for 3 min.

(5) Preliminary TEM imaging.
a. Calculate the physical depth of each section using the cumulative section thicknesses after flattening. (See Note[19])

---

[18]The choice of TEM grid is subject to user preference and skill. Use of slot grids increases the risk of losing key sections, whereas meshed grids increase the risk of section occlusion. If meshed grids are preferred use hexagonally meshed grids, as square meshed grids occasionally align with the ribbon of sections to occlude all regions of interest.

[19]The difference between the preflattening and postflattening thickness can be considered the estimate error.

b. Load the grid closest to the "zero-depth" into the TEM.
c. Identify the first section on the grid and relocate the target cell (using the trimming plan prepared before ultramicrotomy). If there is no evidence of the cell (cloud of filapodia) at this location proceed to the next section. Repeat until the first evidence of the cell and appropriately offset the previous calculations of section depth.
d. Once the "zero-depth" grid has been identified, use the section thickness list to determine which grids contain sections at or near the "target plane" (previously identified from the deconvolved data).
e. Select a grid close to the center of the target volume and load the grid into the TEM.
f. Identify the section nearest to the "target plane" (using the section thickness list) and relocate the target cell (using the trimming plan). Confirm the identity of the cell by measuring its position relative to the section boundaries or other known features.
g. Acquire a low-magnification (1000–5000 ×) reference image of the cell. The cellular filapodia must be clearly visible.

(6) Confirmation of target depth.
a. Extract the "target plane" optical section from the far-red channel of the deconvolved z-stack for comparison with the recently acquired low-magnification TEM image.
b. Overlay the deconvolved optical section with the TEM image using GIMP or an equivalent software package. Manually rescale or rotate the data until the cell boundaries align. There are three possible outcomes from this alignment:
   (i) Correct cell, correct depth: The TEM data can be used to predict the position of filapodia in the optical data.
   (ii) Correct cell, incorrect depth: Cell borders align but there is no apparent relationship between the filapodia.
   (iii) Incorrect cell, depth unknown: Cell borders cannot be aligned. (Repeat the preliminary TEM imaging.)
c. If the initial depth estimates were incorrect, attempt to align the TEM data with an adjacent optical section. Offset the initial depth estimates by this distance then repeat preliminary imaging to confirm.

(7) TEM image montage for distortion correction and spatial alignment.
a. Once the section containing the correct cell at the "target plane" has been identified, begin acquiring TEM images for spatial alignment.
b. Collect a montage of images at a magnification of at least 10 kx, spanning the dimensions of the cell. The image montage will also serve as the calibration for distortion correction, which will require at least nine images overlapping each other by 50% horizontally and vertically in a 3 × 3 matrix.
c. Use the distortion correction plugin for FIJI to correct the EM lens distortion artifacts in the image montage (Kaynig, Fischer, & Buhmann, 2008; Kaynig, Fischer, Müller, & Buhmann, 2010).

d. Arrange and combine the distortion corrected images into a composite montage using GIMP (or your preferred equivalent).

(8) Final spatial alignment.

a. Load the distortion corrected TEM montage into GIMP (or your preferred equivalent) for spatial alignment.

b. Apply a Gaussian Blur (25–35 pixel radius), invert the color values, then apply automatic white balance to the image. It should now resemble the deconvolved far-red image.

c. Import the "target plane" optical section of the deconvolved far-red z-stack. (See Note[20])

d. Overlay both images and manually translate, rescale, and rotate the modified TEM montage to align filapodia visible in both data sets. Document the magnitude and type of every transformation operation for later use. (See Note[21])

e. Aligning the images will typically require iterative transformation; repeated adjustments until the correct alignment is achieved. Combine the documented transformation operations into cumulative values for translation, rescaling, and rotation (e.g., rescaling an image by 25%, 110% then 98% equates to a cumulative value of 26.95%).

f. Import the original distortion corrected TEM image montage and apply the cumulative transformation operations. Confirm that both montages are now aligned with the "target plane" from the deconvolved far-red optical data (adjust only if necessary).

g. The optical and TEM data are now aligned at the "target plane." This data should be stored digitally as a layered or multidimensional stack to allow the adjustment of image visibility without altering the spatial alignment. (We prefer to use the XCF file format for this purpose.) Do not flatten or merge the stack. Do not perform any further transformation operations on images in the stack.

(9) Correlation of the EM with a subcellular target identified during LM.

a. The stack generated during spatial alignment is the fundamental alignment for correlation, as it forms the basis of all subsequent alignment and correlation. To perform a correlation, all images must first be imported into the "alignment stack."

b. Using GIMP (or your preferred equivalent), import all deconvolved optical data into the "alignment stack" as individual layers, and organize them so their identities and physical positions are clearly indicated. Any transformation operations applied to the far-red image during the creation of the "alignment stack" must now be applied to all remaining optical data.

[20]Final spatial alignment will be aided by rescaling the optical data by 200–400% using bicubic interpolation. This is not advised if rescaling the "target plane" alters its alignment with any of the other optical images. Successful correlation relies on the assumption that all the optical data is intrinsically aligned with itself.

[21]The coordinates of filapodia can be used to calculate a 2D coordinate transformation matrix. This will still require manual intervention from the user.

(Do not proceed if the optical data cannot be aligned with itself or with the far-red deconvolved "target plane.")

c. By default, the optical data should now be laterally aligned with the TEM montage. Hide all optical layers except for the "target plane" to identify fluorescence from the correlative target. Adjust the opacity of this layer so that the TEM montage and fluorescence signal are both visible. This image provides a convenient indication of which regions need to be imaged at higher magnification in the TEM.
d. Print the image to use as a "map" during EM and annotate the regions that require closer inspection.
e. Load the appropriate grid into the TEM, identify the correct section, then begin imaging these regions of interest. Collect overview images of the cell at lower magnifications before proceeding to high magnification images (to prevent beam damage from being visible in the low-magnification images).
f. To perform a correlation, import a TEM image into the alignment stack. Hide all optical data and manually align a TEM image to the distortion corrected TEM montage. Document the rotation and rescaling values required for later use. Translation values will not be required.
g. Use the "Alpha Channel To Selection" function to create a selection mask with the same boundaries as the aligned image. (See Note[22])
h. Apply the "Grow selection" function (or equivalent) to this selection mask to expand its borders by 10–20 pixels, then copy this selected region from the layer containing the correlative target optical data at the "target plane."
i. Paste the copied optical data into a new image, then expand the "Canvas Size" to the dimensions of the original TEM image being correlated.
j. Use the documented transformation operation values (previously recorded during alignment of the image with the montage) to calculate the cumulative reverse transformation (e.g., rescaling the original image by 25%, 110% then 98% equates to a cumulative value of 26.95%, the reverse of which is 371.06%).
k. Apply the reverse transformations to the pasted fluorescence data. The optical data should now occupy an area slightly larger than the "Canvas Size" of the image. Ensure that the data is perfectly centered.
l. Import the original TEM image as a new layer and ensure it is perfectly centered. The TEM image and optical data are now aligned for correlation.
m. This procedure can also be used to correlate TEM data from adjacent ultrathin sections, provided they can be aligned to the original TEM montage (see, for example, Fig. 13.6D). The z-position of these sections (relative to the original montaged TEM section) should not differ by more than the axial Nyquist limit of the original optical data.

[22]Creating a path from this selection will simplify the annotation of images with an inset frame when making CLEM figures for publication, see Fig. 13.6A–C for example.

13.2 DEMONSTRATIVE EXAMPLE

Our laboratory currently investigates the dynamic mechanisms of mitophagy: the process by which eukaryotic cells use membrane structures to trap damaged mitochondria, and deliver them to the lysosome for autophagic degradation. Our CLEM procedure has proven to be a formidable asset for investigating mitophagy, enabling us to identify uncharacteristic mitochondria after extreme morphological disturbance (Padman et al., 2013), and to investigate the ultrastructural morphology of an individual mitochondrion after it was selectively damaged with laser light (Padman et al., 2014). This chapter will discuss the technical details from the latter of these two examples.

13.2.1 RATIONALE

High-intensity 488 nm laser light results in mitochondrial injury and irreversible depolarization (Aggarwal, Quintanilha, Cammack, & Packer, 1978; Kim & Lemasters, 2011). We sought to determine whether mitochondrial photodamage coincided with acute changes in mitochondrial ultrastructure. Mitochondrially targeted photoactivatable fluorescent proteins are a convenient way of identifying which mitochondria have been photoirradiated, but identifying depolarized mitochondria requires the use of potentiometric fluorophores like TMRM (Gerencser et al., 2012; Venable et al., 2013). Given that membrane potentials are not preserved by chemical fixation, live-cell imaging is the only valid way to identify the depolarized mitochondria for correlation.

13.2.2 EXPERIMENTAL CONDITIONS AND PARAMETERS

13.2.2.1 Substrate preparation

TOPAS 6013 film was marked with toner using the procedure outlined earlier. The toner-release sheet was prepared using PVA (Sigma; #341584), A4 (210 × 297 mm) overhead projector transparency (3M), nylon paint brush, and a window squeegee. The toner patterns were designed using Inkscape (Albert et al., 2013) and printed using a Xerox Phaser 6360 laser printer. The TOPAS film was prepared for toner deposition by exposure to argon glow-discharge plasma for 30 s and the toner was transferred by passing the TOPAS with toner release through a generic office pouch laminator (Lowell) five times. The film holder was assembled with as above, with a thin layer of vacuum grease under the tensioning cylinder to ensure a watertight seal.

13.2.2.2 Cell culture

HeLa cells were seeded into the film holder at an estimated confluency of 90% and cultured in DMEM+10% FCS media for 24 h. The cells were transfected with PAGFP-Mito (AddGene #23348) using a cationic lipid transfection reagent (Lipofectamine 2000; Life Technologies) in serum-free media, which was replaced with DMEM+10% FCS after 4 h.

13.2.2.3 Optical imaging and primary fixation (see Fig. 13.4)

The cells were stained with 500 nM TMRM, 2.5 μg/mL CellMask Deep-Red and 2.5 μM DRAQ5 in PBS for 30 min, then incubated with 1000 μL phenol-free culture media for 30 min. The film holder was transferred to the stage of an Olympus FluoView FV1000 inverted confocal microscope with a 60 × objective lens (PLAPO, NA

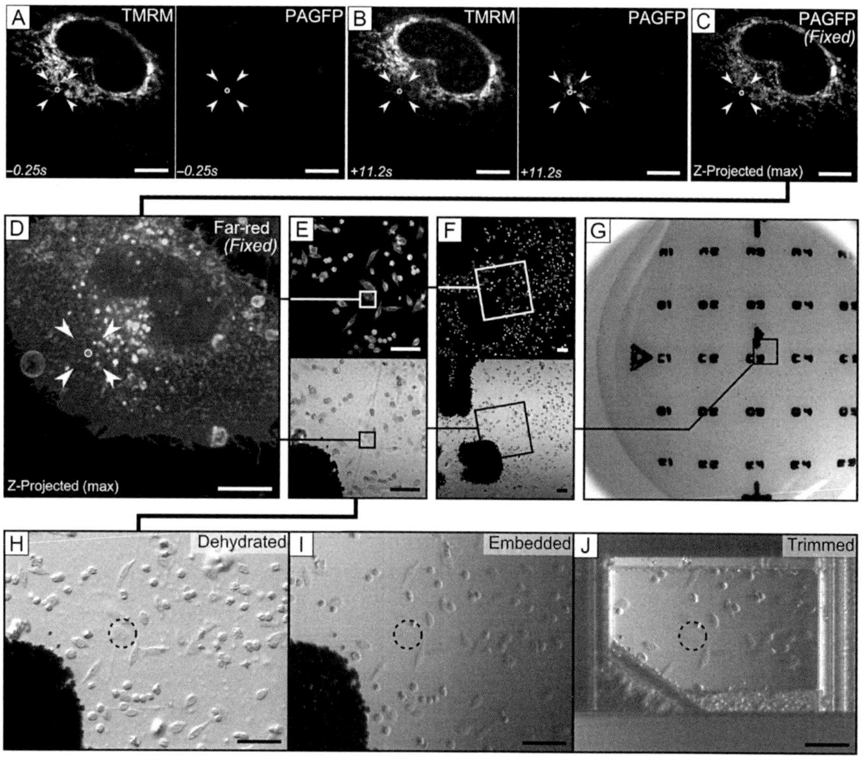

FIGURE 13.4

Live-cell CLEM of an individual mitochondrion after selective photodamage. (A) HeLa cells on toner-gridded TOPAS were transfected with PAGFP-Mito and stained with TMRM, CellMask Deep-Red, and DRAQ5 before live-cell imaging, and a 1.2 μm diameter region (arrowheads) within a cell was selected for photoirradiation during the acquisition of time-series data. (B) The primary fixative was added directly to the sample immediately after the 6th frame of the time series (11.2 s after photoirradiation). (C–D) The PAGFP was reactivated and the cell was optically sectioned in the (C) PAGFP and (D) far-red channels to provide the 3D optical data required for subcellular spatial alignment. (E–F) Transmitted light and far-red fluorescence images were acquired of the cell and its surroundings to determine the location of the cell. (G) After identifying the general location of the target cell the region of film was excised using a scalpel. (H) Welded to the bottom of a polypropylene container for sample processing. (I) After embedding the sample in Procure-Araldite. (J) Local toner particles were used to guide trimming in preparation for ultramicrotomy. The cell position is indicated by a dashed circle. (Images (I–J) are horizontally flipped. Scale bars: A–D = 10 μm; E, F, H, I, J = 100 μm.)

1.20; Olympus, UIS2). Upon identifying a target cell, the lateral sampling resolution was reduced to 2 × Nyquist limit of the PAGFP channel (103 nm/pixel) and the imaging parameters were adjusted for the acquisition of time-series data. A subcellular region of interest (1.2 μm in diameter) containing mitochondria was defined within the cell before beginning the time-series acquisition. Immediately after the first frame of the time-series, the subcellular region of interest was microirradiated for 250 ms using 405 nm and 488 nm lasers with postobjective powers of 287 and 967 μW, respectively. The acquisition of time-series data continued for 10 s until primary fixation with 1000 μL of 8% paraformaldehyde in 0.1 M PB.

After allowing the sample to stabilize for 10 min, the basal surface of the cell was relocated in the CellMask Deep-Red channel. The field of view was scanned with the 405 nm laser (645 μW for 500 ms) to reactivate the PAGFP-mito. The imaging parameters were adjusted for the acquisition of a z-stack in the CellMask/DRAQ5 and PAGFP channels, using a lateral sampling resolution of 0.5 × the Nyquist limit (34 nm/pixel) and an axial sampling resolution of 1.5 × the Nyquist limit (320 nm/section). The z-stack boundaries were 2.32 μm above and 1.28 μm below the estimated position of the basal surface, with the target plane approximately 0.64 μm above the basal surface. Transmitted light images acquired at low-magnification images revealed that the target cell was approximately 100 μm "north-east" of the central alphanumeric grid marker, reducing the need for a large navigation montage. The sample was removed from the microscope 45 min after primary fixation for EM processing.

13.2.2.4 EM specimen processing

The position of the target cell was identified in the processing laboratory using a Leica M125 dissection microscope with the IC80 HD camera (Leica) and a strip of film containing the target cell was cut directly from the film holder. The strip of film was heat welded to the bottom of a flat bottomed polypropylene sample jar for processing. The samples were processed using the procedure outlined earlier assisted by a BioWave Pro microwave system (Pelco) to reduce the time requirement of all stages after glutaraldehyde fixation. The samples were embedded in Procure–Araldite, which was polymerized at 60 °C for 48 h.

13.2.2.5 Ultramicrotomy and spatial alignment (see Fig. 13.5)

A trimming guide was drawn on printed copies of the low-magnification optical images and the resin block was loaded into a Leica Ultracut UCT ultramicrotome and for trimming with a glass knife. The TOPAS film was sectioned to a depth of 50 μm and the resin block was trimmed such that all borders were at least 100 μm from the target cell. The alphanumeric grid marker was removed due to the high density of toner (which often causes a region variable density or hardness). After trimming the resin block, a glass knife was used to section the remaining TOPAS film until the first sign of resin, which was considered to be the initial estimate of the zero-depth. Meanwhile, the optical z-stack data was deconvolved using Huygens Professional deconvolution software (Huygens professional 64 bit, version 4.2.1p8;

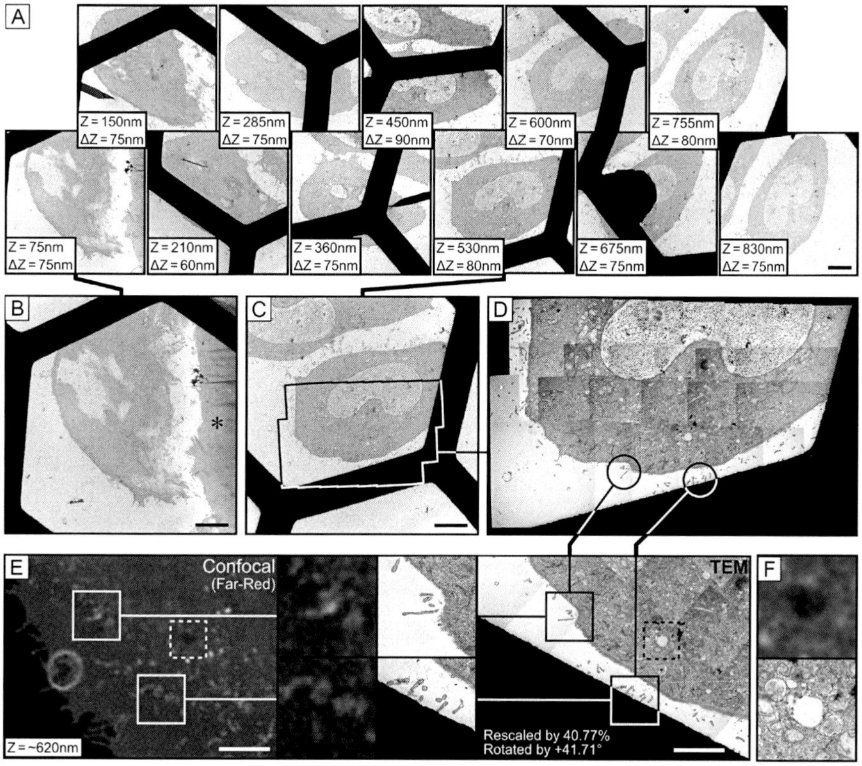

FIGURE 13.5

Ultramicrotomy, confirmation of target depth and spatial alignment. (A) The embedded sample was sectioned using a diamond knife. The thickness of each ultrathin section was recorded to determine its corresponding z-position within the sample. (B) The "zero-depth" section was identified to be the first section containing evidence of the cell within the target region. The section also contained a portion of the TOPAS substrate which provided further confirmation (indicated by the asterisk). (C) The deconvolved optical data was used to determine the z-position of the "target plane" (~640 nm) and a section near this position was selected for inspection. (D) A montage of images (25 frames; 10,000 × magnification) was acquired of the target section and then corrected for EM lens distortion. (E) Filapodia identified in the distortion corrected TEM montage were spatially aligned with equivalent structures in the deconvolved far-red "target plane" (the required transformation operations used for alignment are indicated). (F) Features visible in the optical data (dashed insets in E) were successfully used to predict the existence of features in the TEM data. (Scale bars: A–D = 10 μm; E = 5 μm.)

40 iterations; theoretical PSF; maximum likelihood estimation; signal-to-noise ratio: 20) to determine the depth of the target plane (~640 nm above basal surface). Ultrathin sections were cut using a diamond knife ($n = 32$; mean thickness: 74.8 nm ± 1.6 nm) until reaching an estimated cumulative distance of 2395 nm from

the zero-depth. The thickness of each section was recorded (as described earlier) and mounted on 300-mesh (Hex) copper grids. The grids were stained with UA and PbCit and imaged on a Hitachi H-7500 TEM at 80 kV using a Gatan 791 MultiScan side mount CCD camera and DigitalMicrograph (version 1.71.38) acquisition software.

TEM imaging revealed that the initial estimate of zero-depth was incorrect, as the cell was not detected until the 7th section (see Fig. 13.5A). The section depth estimates were offset by the cumulative value of the premature sections and the section depths were recalculated ($n=26$; mean thickness: 74.8 nm ± 2.2 nm; new distance

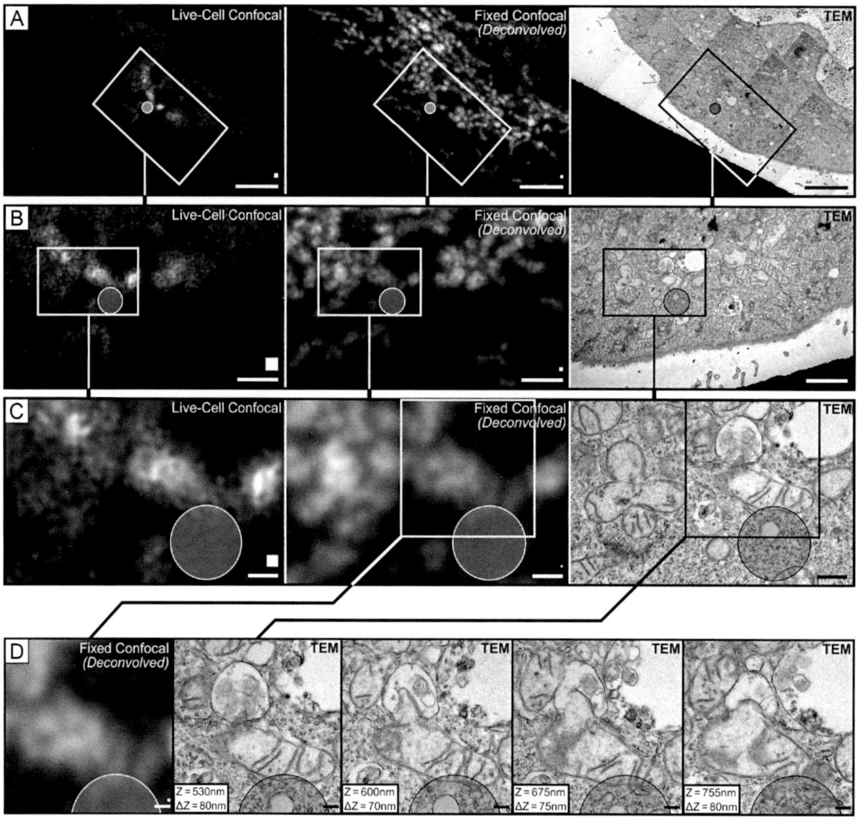

FIGURE 13.6

High-magnification images of the correlated subcellular region of interest. (A) The spatially aligned live-cell optical, fixed optical, and TEM data showing the photoirradiated region indicated by a circle, (B and C) with inset frames of the photoirradiated region at sequentially increasing magnification. (D) High-magnification insets of a mitochondrion within 1 μm of the photoirradiation region with TEM images acquired of adjacent sections. (Scale bars: A = 5 μm; B = 2 μm; C = 1 μm; D = 200 nm. The original pixel sizes of the optical data are indicated by squares above the scale bars: A = 10 × 10 pixel; B = 5 × 5 pixel; C, D = 1 × 1 pixel.)

from zero-depth: 1870 nm). The recalculated values indicated that the target z-plane (~640 nm from basal surface) was located within 11 sections of zero-depth (thicknesses: 75, 75, 60, 75, 75, 90, 80, 70, 75, 80, and 75 nm; cumulative value: 830 nm). The 7th of these 11 sections (80 nm thick; 530 nm from zero-depth) was confirmed to be at the target plane. A TEM montage was manually acquired (25 images at 10,000 × magnification), distortion corrected, overlayed, and aligned using the procedures outlined earlier. This alignment formed the basis of all subsequent alignments and correlations.

13.2.3 THE RESULTS OF CORRELATION

The resolution of the live-cell optical data was not sufficient for *direct* spatial alignment to the TEM data; however, this did not preclude *indirect* spatial alignment since the live-cell optical data was intrinsically aligned to the fixed-cell optical data (see Fig. 13.6). It was therefore a trivial process to identify an individually photodamaged mitochondrion for further inspection at higher magnifications in the TEM. These TEM images were aligned to the distortion corrected TEM montage, then used as a selection mask to extract the relevant region of fluorescence data. The PAGFP fluorescence signal from the fixed confocal data was found to extend beyond the apparent mitochondrial membranes, indicating that the aligned section contained only a portion of the mitochondrion. TEM images acquired of the mitochondrion in adjacent sections (thicknesses: 70, 75, and 80 nm; depth range: 600–755 nm) revealed an apparent interaction between the mitochondrion and an endosomal structure. While the physiological significance of this observation has yet to be determined, the experiment demonstrates the feasibility of our procedures for aligning subcellular structures between live-cell imaging and TEM of the same structure.

REFERENCES

Abdelgawad, M., & Wheeler, A. R. (2007). Rapid prototyping in copper substrates for digital microfluidics. *Advanced Materials*, *19*, 133–137. http://dx.doi.org/10.1002/adma.200601818.

Aggarwal, B. B., Quintanilha, A. T., Cammack, R., & Packer, L. (1978). Damage to mitochondrial electron transport and energy coupling by visible light. *Biochimica et Biophysica Acta*, *502*, 367–382. http://dx.doi.org/10.1016/0005-2728(78)90057-9.

Albert, M., Andler, J., Bah, T., Barbry-Blot, P., Barraud, J., Baxter, B., et al. (2013). Inkscape. www.inkscape.org.

Alpy, F., Rousseau, A., Schwab, Y., Legueux, F., Stoll, I., Wendling, C., et al. (2013). STARD3/STARD3NL and VAP make a novel molecular tether between late endosomes and the ER. *Journal of Cell Science*, *126*(Pt. 23), 5500–5512. http://dx.doi.org/10.1242/jcs.139295.

Arnison, M. R., Cogswell, C. J., Smith, N. I., Fekete, P. W., & Larkin, K. G. (2000). Using the Hilbert transform for 3D visualization of differential interference contrast microscope images. *Journal of Microscopy*, *199*, 79–84.

Conchello, J.-A., & Lichtman, J. W. (2005). Optical sectioning microscopy. *Nature Methods*, *2*, 920–931. http://dx.doi.org/10.1038/nmeth815.

Gao, Y., Berciu, C., Kuang, Y., Shi, J., Nicastro, D., & Xu, B. (2013). Probing nanoscale self-assembly of nonfluorescent small molecules inside live mammalian cells. *ACS Nano*, *7*, 9055–9063. http://dx.doi.org/10.1021/nn403664n.

Gerencser, A. A., Chinopoulos, C., Birket, M. J., Jastroch, M., Vitelli, C., Nicholls, D. G., et al. (2012). Quantitative measurement of mitochondrial membrane potential in cultured cells: Calcium-induced de- and hyperpolarization of neuronal mitochondria. *Journal of Physiology*, *590*, 2845–2871. http://dx.doi.org/10.1113/jphysiol.2012.228387.

Gleskova, H., Wagner, S., & Shen, D. (1998). Photoresist-free fabrication process for a-Si:H thin film transistors. *Journal of Non-Crystalline Solids*, *227–230*(Part 2), 1217–1220.

Guizetti, J., Mäntler, J., Müller-Reichert, T., & Gerlich, D. W. (2010). Correlative time-lapse imaging and electron microscopy to study abscission in HeLa cells. In T. Müller-Reichert (Ed.), *Methods in cell biology. Electron microscopy of model systems*, (pp. 591–601). Waltham, MA: Academic Press, Chapter 24.

Hanson, H. H., Reilly, J. E., Lee, R., Janssen, W. G., & Phillips, G. R. (2010). Streamlined embedding of cell monolayers on gridded glass-bottom imaging dishes for correlative light and electron microscopy. *Microscopy and Microanalysis*, *16*, 747–754. http://dx.doi.org/10.1017/S1431927610094092.

Jimenez, N., Humbel, B. M., Van Donselaar, E., Verkleij, A. J., & Burger, K. N. J. (2006). Aclar discs: A versatile substrate for routine high-pressure freezing of mammalian cell monolayers. *Journal of Microscopy*, *221*, 216–223. http://dx.doi.org/10.1111/j.1365-2818.2006.01558.x.

Jiménez, N., Van Donselaar, E. G., De Winter, D. A. M., Vocking, K., Verkleij, A. J., & Post, J. A. (2010). Gridded Aclar: Preparation methods and use for correlative light and electron microscopy of cell monolayers, by TEM and FIB-SEM. *Journal of Microscopy*, *237*, 208–220. http://dx.doi.org/10.1111/j.1365-2818.2009.03329.x.

Kam, Z. (1998). Microscopic differential interference contrast image processing by line integration (LID) and deconvolution. *Bioimaging*, *6*, 166–176. http://dx.doi.org/10.1002/1361-6374(199812)6:4<166::AID-BIO2>3.0.CO;2-Y.

Kaynig, V., Fischer, B., & Buhmann, J. M. (2008). Probabilistic image registration and anomaly detection by nonlinear warping. In *IEEE Conference on Computer Vision and Pattern Recognition. *Presented at the IEEE Conference on Computer Vision and Pattern Recognition* (pp. 1–8). http://dx.doi.org/10.1109/CVPR.2008.4587743.

Kaynig, V., Fischer, B., Müller, E., & Buhmann, J. M. (2010). Fully automatic stitching and distortion correction of transmission electron microscope images. *Journal of Structural Biology*, *171*, 163–173. http://dx.doi.org/10.1016/j.jsb.2010.04.012.

Kim, I., & Lemasters, J. J. (2011). Mitophagy selectively degrades individual damaged mitochondria after photoirradiation. *Antioxidants and Redox Signaling*, *14*, 1919–1928. http://dx.doi.org/10.1089/ars.2010.3768.

Kingsley, R. E., & Cole, N. L. (1988). Preparation of cultured mammalian cells for transmission and scanning electron microscopy using Aclar film. *Journal of Electron Microscopy Technique*, *10*, 77–85.

Lenormand, C., Spiegelhalter, C., Cinquin, B., Bardin, S., Bausinger, H., Angénieux, C., et al. (2013). Birbeck granule-like "organized smooth endoplasmic reticulum" resulting from the expression of a cytoplasmic YFP-tagged langerin. *PLoS One*, *8*, e60813. http://dx.doi.org/10.1371/journal.pone.0060813.

Lucic, V., Kossel, A. H., Yang, T., Bonhoeffer, T., Baumeister, W., & Sartori, A. (2007). Multiscale imaging of neurons grown in culture: From light microscopy to cryo-electron tomography. *Journal of Structural Biology*, *160*, 146–156. http://dx.doi.org/10.1016/j.jsb.2007.08.014.

McDonald, K. L. (2009). A review of high-pressure freezing preparation techniques for correlative light and electron microscopy of the same cells and tissues. *Journal of Microscopy*, *235*, 273–281. http://dx.doi.org/10.1111/j.1365-2818.2009.03218.x.

Müller-Reichert, T., Srayko, M., Hyman, A., O'Toole, E. T., & McDonald, K. (2007). Correlative light and electron microscopy of early Caenorhabditis elegans embryos in mitosis. In J. Richard McIntosh (Ed.), *Methods in cell biology. Cellular electron microscopy* (pp. 101–109). Waltham, MA: Academic Press.

Padman, B. S., Bach, M., Lucarelli, G., Prescott, M., & Ramm, G. (2013). The protonophore CCCP interferes with lysosomal degradation of autophagic cargo in yeast and mammalian cells. *Autophagy*, *9*, 1862–1875. http://dx.doi.org/10.4161/auto.26557.

Padman, B. S., Bach, M., & Ramm, G. (2014). An improved procedure for subcellular spatial alignment during live-cell CLEM. *PLoS One*, *9*, e95967. http://dx.doi.org/10.1371/journal.pone.0095967.

Peachey, L. D. (1958). Thin sections. I. A study of section thickness and physical distortion produced during microtomy. *Journal of Biophysical and Biochemical Cytology*, *4*, 233–242.

Polishchuk, R. S., Polishchuk, E. V., Marra, P., Alberti, S., Buccione, R., Luini, A., et al. (2000). Correlative light-electron microscopy reveals the tubular-saccular ultrastructure of carriers operating between Golgi apparatus and plasma membrane. *Journal of Cell Biology*, *148*, 45–58. http://dx.doi.org/10.1083/jcb.148.1.45.

Razi, M., & Tooze, S. A. (2009). Correlative light and electron microscopy. In Daniel J. Klionsky (Ed.), *Methods in enzymology. Autophagy in mammalian systems, Part B* (pp. 261–275). Waltham, MA: Academic Press, Chapter 17.

Sakai, T. (1980). Relation between thickness and interference colors of biological ultrathin section. *Journal of Electron Microscopy*, *29*, 369–375.

Spiegelhalter, C., Tosch, V., Hentsch, D., Koch, M., Kessler, P., Schwab, Y., et al. (2010). From dynamic live cell imaging to 3D ultrastructure: Novel integrated methods for high pressure freezing and correlative light-electron microscopy. *PLoS One*, *5*, e9014. http://dx.doi.org/10.1371/journal.pone.0009014.

Svitkina, T. M., & Borisy, G. G. (1998). Correlative light and electron microscopy of the cytoskeleton of cultured cells. In Richard B. Valee (Ed.), *Methods in enzymology. Molecular motors and the cytoskeleton. Part B* (pp. 570–592). Waltham, MA: Academic Press.

Venable, P. W., Taylor, T. G., Sciuto, K. J., Zhao, J., Shibayama, J., Warren, M., et al. (2013). Detection of mitochondrial depolarization/recovery during ischaemia-reperfusion using spectral properties of confocally recorded TMRM fluorescence. *Journal of Physiology*, *591*, 2781–2794. http://dx.doi.org/10.1113/jphysiol.2012.248153.

Verkade, P. (2008). Moving EM: The rapid transfer system as a new tool for correlative light and electron microscopy and high throughput for high-pressure freezing. *Journal of Microscopy*, *230*, 317–328. http://dx.doi.org/10.1111/j.1365-2818.2008.01989.x.

Wrede, C., Heller, C., Reitner, J., & Hoppert, M. (2008). Correlative light/electron microscopy for the investigation of microbial mats from Black sea cold seeps. *Journal of Microbiological Methods*, *73*, 85–91. http://dx.doi.org/10.1016/j.mimet.2008.02.020.

CHAPTER

14 Correlative Fluorescence and Scanning Transmission Electron Microscopy of Quantum Dot-Labeled Proteins on Whole Cells in Liquid

Diana B. Peckys*, Vera Bandmann†, Niels de Jonge*

**Innovative Electron Microscopy, INM-Leibniz Institute for New Materials, Saarbrücken, Germany*

†Biomineralization, INM-Leibniz Institute for New Materials, Saarbrücken, Germany

CHAPTER OUTLINE

Methods in Cell Biology, Volume 124, ISSN 0091-679X, http://dx.doi.org/10.1016/B978-0-12-801075-4.00014-8

Abstract

Correlative fluorescence microscopy combined with scanning transmission electron microscopy (STEM) of cells fully immersed in liquid is a new methodology with many application areas. Proteins, in live cells immobilized on microchips, are labeled with fluorescent quantum dot nanoparticles. In this protocol, the epidermal growth factor receptor (EGFR) is labeled. The cells are fixed after a selected labeling time, for example, 5 min as needed to form EGFR dimers. The microchip with cells is then imaged with fluorescence microscopy. Thereafter, STEM can be accomplished in two ways. The microchip with the labeled cells and one microchip with a spacer are assembled into a special microfluidic device and imaged with dedicated high-voltage STEM. Alternatively, thin edges of cells can be studied with environmental scanning electron microscopy with a STEM detector, by placing a microchip with cells in a cooled wet environment.

INTRODUCTION

The stoichiometric organization of protein complexes in the plasma membrane and their assembly into microdomains are responsible for many cellular functions. Direct visualization of this organization is challenging because it requires microscopic methods dealing with cells in their natural environment and providing a spatial resolution at the nanometer scale. Cellular imaging is mostly accomplished with light microscopy. But even with the improvement in resolution beyond the diffraction barrier achieved in recent years (Lippincott-Schwartz & Manley, 2009), it still lacks an order of magnitude to visualize the single constituents of protein complexes of dimensions around 5 nm. Standard transmission electron microscopy offers the necessary resolution but requires thin sections of fixed, dehydrated, and plastic-embedded samples, which do not represent the native environment of cells, or requires sections embedded in amorphous ice, which are difficult and time consuming to prepare (Hoenger & McIntosh, 2009; Kourkoutis, Plitzko, & Baumeister, 2012). To directly observe protein complex assemblies in their native environment, we recently developed a new approach in which whole cells in liquid can be analyzed with nanometer resolution. This microscopy method is named liquid scanning transmission electron microscopy (STEM) (de Jonge, Peckys, Kremers, & Piston, 2009; Peckys & de Jonge, 2014a). Sample dehydration and sectioning are avoided through the use of a special microfluidic chamber with electron transparent thin windows (Ring & de Jonge, 2010). This chamber is assembled from two silicon microchips supporting thin silicon nitride (SiN) membrane windows. Cells are grown in this chamber and can be imaged both with standard fluorescent microscopy and with STEM at 200-keV electron beam energy, whereby the images obtained with both microscopy modalities are correlated (Dukes, Peckys, & de Jonge, 2010). Alternatively, whole cells grown in the same manner on the thin supporting membrane of the microchip can be imaged with environmental scanning electron microscopy (ESEM) at 30 keV using a STEM detector (Peckys, Baudoin, Eder, Werner, & de Jonge, 2013). In this case, the microchip is not assembled into a microfluidic chamber, but one microchip is used and the cells are maintained in a thin layer of water at a temperature of around 3 °C in the low vacuum in the microscope chamber. The thin edges of the hydrated whole cells can then be studied with ESEM–STEM. To detect proteins in their native environment, they are labeled with nanoparticles. Nanoparticles with high atomic number (Z) such as gold provide the best contrast but various other nanoparticles can also be used. Of particular advantage for correlative fluorescence and electron microcopy is the labeling of proteins with quantum dots (QDs) (Gaietta et al., 2002). Compared to fluorescent dyes, QDs emit stronger fluorescence signals and are very resistant against bleaching. A cell with QD-labeled proteins can thus first be studied with fluorescence microscopy and then a selected region can be analyzed at high magnification with electron microscopy using the STEM detector.

Here, we describe a protocol for correlative fluorescence microscopy and STEM or ESEM–STEM of whole cells in liquid applied to the detection of plasma membrane localized epidermal growth factor (EGF) receptors (EGFRs) (Dukes et al., 2010). A subset of this protocol was also published elsewhere (Peckys, Dukes, & de Jonge, 2014). The EGFRs are labeled with EGF-conjugated QDs (Lidke et al., 2004). A microchip with cells is first imaged with fluorescence microscopy. Subsequently, the cells can be either enclosed in a microfluidic chamber for imaging with liquid STEM (Dukes et al., 2010), or the thin edges of the hydrated cells can be studied with ESEM–STEM. The fluorescence and liquid STEM or ESEM–STEM images are easily correlated via locating the corners of the window in the microchip. The choice of the electron microscopy modality for future experiments depends on the cellular region of interest. Liquid STEM provides nanometer resolution in the top micrometer of a liquid layer of around 5 μm. Cells in buffer solution and even live cells can be handled (Peckys & de Jonge, 2011, 2014a; Peckys, Mazur, Gould, & de Jonge, 2011). The labels can thus be studied anywhere in the cell as long as the QD labels are vertically located within about half a micrometer from the top supporting membrane. In case information about the thin cellular edges is sufficient and the transfer to pure water is allowed, then ESEM–STEM is probably the easiest method. The described protocols allow for the study of the spatial and stoichiometric distributions of selected proteins on a multiple of whole cells in liquid. The method is not much more time consuming than the standard methods used for fluorescence microscopy, so that statistics of the distribution over tens of cells can be readily obtained (Peckys et al., 2013; Peckys & de Jonge, 2014b).

14.1 MATERIALS (SEE NOTE 1 IN SECTION 14.3)

14.1.1 MICROFLUIDIC CHAMBER

1. Base microchip: the base microchip, manufactured by Protochips Inc., NC, USA, has dimensions of 2.00 × 2.60 × 0.30 mm (see Note 2 in Section 14.3). The SiN membrane window in the middle has dimensions of 50 μm × 200 μm × 50 nm (see Note 3 in Section 14.3) (Ring & de Jonge, 2010). The back surface of a base microchip is shown in Fig. 14.1A. The sides of the

FIGURE 14.1

Scanning electron microscopy (SEM) images of the microchips. The SEM images were recorded at 10 kV (using a S4700 Hitachi SEM). (A) Image of the backside of a microchip showing the silicon nitride (SiN) window through the opening etched in the silicon. (B) Image of the front side of the spacer microchip. The spacer structure is made of SU8 of 6 μm thickness. The SiN window is the dark shape in the middle.

Figure reproduced from Ring and de Jonge (2010) with permission from Cambridge University Press.

microchips are precision diced. The SiN side of the microchip is covered by a protective layer of standard photoresist material, for example, poly(methyl methacrylate) (see Note 4 in Section 14.3).

2. Spacer microchip: a microchip analogous to the base microchips but with the addition of a spacer layer of SU8 material patterned (see Note 5 in Section 14.3) via lithographic techniques (Fig. 14.1B).
3. Microspheres: polystyrene microspheres of 10 μm diameter in aqueous solution (from Structure Probe Inc., West Chester, PA, USA) (see Note 6 in Section 14.3).
4. Solution of 0.01% poly-L-lysine (PLL) in water (mol. wt 150,000–300,000, sterile-filtered, and cell culture tested).
5. Acetone, HPLC grade (see Note 7 in Section 14.3).
6. Ethanol, HPLC grade.
7. Pure water, HPLC grade.
8. Berkshire* Durx* 770 clean room wipers (from Berkshire Co., Great Barrington, MA, USA).
9. Tweezers with PTFE coating (see Note 8 in Section 14.3), one with flat, straight tips (Type EMS 2A) and one with fine and curved tips (Type EMS 7) (from Electron Microscopy Sciences, Hatfield, PA, USA).
10. 1.5-mL Microcentrifuge tubes.
11. Pipetters for volumes ranging between 0.1 and 2.5 μL and between 0.5 and 10 μL.
12. Two clean (see Note 9 in Section 14.3) glass beakers with a volume of around 200 mL.
13. Gel box with Gel X4 (clear plastic hinged, 50.8 × 50.8 × 6.4 mm, respectively, 2″ × 2″ × 1/4″) (from Ted Pella Inc., Redding, CA, USA).

14.1.2 CELL CULTURE

1. COS7 (African green monkey fibroblast) cells (from American Type Culture Collection (ATCC), Manassas, VA, USA).
2. Cell culture medium: Dulbecco's modified Eagle medium (DMEM) (from ATCC) supplemented with 10% certified fetal bovine serum (FBS), and additional L-glutamine (2 mM) (both from Invitrogen/Life Technologies, Grand Island, NY, USA).
3. Dulbecco's phosphate-buffered saline (DPBS) modified without calcium chloride or magnesium chloride (from ATCC).
4. CellStripper (from Mediatech, Manassas, VA, USA).
5. DMEM without FBS. The serum-free medium will lead to enhanced expression of EGFRs on the cell membrane.
6. All standard cell culture supplies.

14.1.3 LABELING

1. Untreated tissue culture plates, 12, 24, and 96 wells (Corning Inc., Tewksbury MA, USA).
2. Epidermal growth factor, biotin-XX conjugate (biotin EGF) (from Invitrogen/ Life Technologies).
3. Qdot® 655 streptavidin conjugate, 1 μM (see Note 10 in Section 14.3) (Invitrogen/Life Technologies).
4. 50 mM Borate buffer, pH 8.3 (made from 200 mM boric acid and pH adjusted with 200 mM sodium tetraborate, both from Sigma-Aldrich).
5. Microcentrifuge purification column (Ultracel-100YM; Millipore, Billerica, MA, USA).
6. Cacodylate buffer (CB) 0.1 M supplemented with 0.1 M sucrose made from 0.2 M CB stock solution (4.28 g of sodium cacodylate added to 90-mL HPLC grade water, adjusted pH to 7.4 with approximately 5.6 mL of 0.2N hydrochloric acid, volume filled up to 100 mL with water) and 2 M sucrose stock solution (see Note 11 in Section 14.3).
7. 3% Glutaraldehyde in CB (3% GA) (made from 25% GA, EM grade; Ted Pella).
8. Tyrode's salts with sodium bicarbonate, liquid, sterile-filtered, and cell culture tested (Tyrode's buffer (TB)).
9. Tyrode's buffer supplemented with 0.1% albumin from bovine serum ((BSA), biotin-free and molecular biology grade Albumin Fraction V) and 50 mM D-glucose (BSA-TB) (all from Sigma-Aldrich).
10. Phosphate-buffered saline (PBS) (made from 10 × concentrate; Sigma-Aldrich).
11. PBS with 1% BSA (BSA–PBS).
12. 100 mM Glycine (from Sigma-Aldrich) in PBS (GLY–PBS).

14.1.4 MICROSCOPY

1. 35-mm Glass bottom dishes, 35 mm diameter (MatTek Co., Ashland, MA, USA).
2. Microfluidic tubing of 50 or 100 μm inner diameter (PEEK Upchurch Scientific).

3. Instrument glass syringe, 1 mL (Model 1001 TLL-SAL SYR; Hamilton Co., Reno, NV, USA).
4. PBS and 10% PBS (in water).
5. Solution of 50% glycerol and 10% PBS in water.
6. Micropipette for volumes between 0.5 and 10 μL.

14.2 METHODS (SEE NOTE 12 IN SECTION 14.3)

14.2.1 PREPARATION OF THE BASE MICROCHIP FOR THE MICROFLUIDIC CHAMBER

1. For removing the protective coat from the SiN membrane microchips use flat beak tweezers. Use this type of tweezers for all subsequent steps (see Note 13 in Section 14.3).
2. Place the desired number of microchips (usually 4–10), facing upward in a clean glass beaker containing 100–200 mL of acetone. Move the beaker carefully in a slow pivoting motion such that the liquid achieves a rotational movement around the microchips without flipping them. Continue for at least 2 min.
3. Fill another beaker with 100–200 mL ethanol and transfer the microchips to this beaker. The transfer should be done carefully but quickly to avoid drying of the acetone (see Note 14 in Section 14.3). Keep the liquid moving in a similar way for another 2 min.
4. Place the microchips onto a clean room wipe for drying (see Note 15 in Section 14.3).
5. After stripping off the resist, the SiN membrane is hydrophobic. Before they can be used for biological samples, they have to be made hydrophilic. This can be done by plasma cleaning or glow discharging the microchips for approximately 5 min. A good and cost-effective plasma cleaner is the tabletop model of Harrick Plasma, Ithaca, NY, USA. It uses ambient air at a low vacuum and a power of 30 W radio frequency.
6. PLL coating keeps the microchips hydrophilic for about a day and supports cell adherence. Immerse the microchips for 5 min into a well of a 12-well plate filled with 1 mL of 0.01% PLL. Make sure that the SiN window is facing upward and the microchips are kept still such that the coating can settle on the surface. In the volume of 1 mL PLL, a maximum of 10–12 microchips per well are recommended.
7. Fill three wells with 1–2 mL of pure water. Briefly rinse the microchips by a subsequent transfer among the wells.
8. Place each microchip directly into the well of a 96-well plate containing 100 μL of DMEM supplemented with FBS and glutamine.
9. Inspect all microchips in an optical microscope. Discard any microchips with broken or not entirely clean SiN membranes.

14.2.2 PREPARATION OF THE SPACER MICROCHIP OPTION #1 FOR THE MICROFLUIDIC CHAMBER (SEE NOTE 16 IN SECTION 14.3)

1. A microfluidic chamber is formed by assembling a base microchip and a spacer microchip with the flat sides of the SiN windows facing to each other. The gap between the windows is defined by a spacer layer. Two types of spacing can be used. The first type contains a patterned spacer, as shown in Fig. 14.2B. This microchip is prepared in the same way as described in steps 1–5 in Section 14.2.1, but the PLL coating is omitted.

14.2.3 PREPARATION OF THE SPACER MICROCHIP OPTION #2 FOR THE MICROFLUIDIC CHAMBER (SEE NOTE 16 IN SECTION 14.3)

1. A second option is to use a base microchip and place microspheres at the corners. The base microchip is prepared as described in steps 1–4 in Section 14.2.1, but leaving the surface hydrophobic (see Note 17 in Section 14.3).

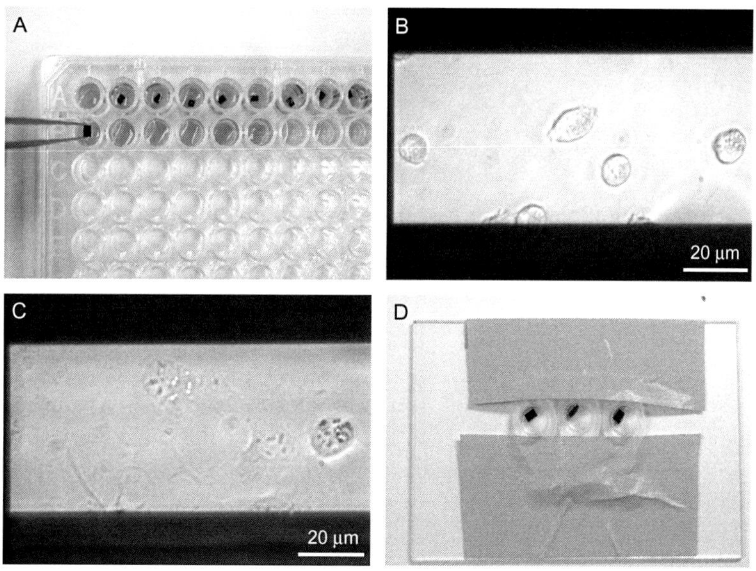

FIGURE 14.2

Pictures recorded during the seeding of cells onto microchips and the labeling of the cells. (A) Microchips in a 96-well plate, where they will be seeded with cells. The cells will subsequently be labeled and fixed. Note the orientation of the tweezers with flat beak gripping a microchip by its sides and not touching its edges or top surface. (B) COS7 cells seeded on the SiN membrane window after ~5 min of incubation. (C) COS7 cells that have adhered to and flattened on an SiN window, after ~1 h of incubation. (D) Microchips in labeling devices are inclined against droplets of labeling solution to reduce the amount of nonspecifically bound label on the samples.

Figure reproduced from Ring, Peckys, Dukes, Baudoin, and de Jonge (2011) with permission from John Wiley & Sons.

2. To prepare the microsphere working solution, the microspheres have to be washed three to five times with HPLC grade water. Add 900 μL of HPLC water to 100 μL of the microsphere stock solution, centrifuge until a pellet is formed and resuspend the pellet with 1 mL of HPLC water. Repeat this step five times.
3. Check the concentration of the microspheres, by placing a 0.2-μL droplet of the washed microsphere solution on a glass microscope slide and examine under a light microscope. The 0.2-μL droplet should contain between 8 and 20 microspheres. Adjust the concentration by dilution with HPLC water.
4. Briefly vortex the working solution of the polystyrene microspheres. Apply a droplet of 0.2 μL (see Note 18 in Section 14.3) containing 8–20 beads to each of the four corners of a microchip. Care has to be taken that the droplets are not placed too close to the SiN window. When the microfluidic device is assembled and the microchips are pressed together with the pressure needed to obtain a vacuum seal, the gap is typically 5–6 μm for 10-μm diameter microspheres at the corners (see Note 19 in Section 14.3). Dry the microchips in a dust-free environment. The number of microspheres deposited on each corner can be checked with reflected light microscopy, also referred to as incident light or epiillumination. Plasma clean or glow discharge the microchips. Store the microchips in a gel box (see Note 20 in Section 14.3).
5. Inspect all microchips under an optical microscope (see Note 21 in Section 14.3).

14.2.4 SEEDING THE CELLS ON THE BASE MICROCHIP

1. Maintain cell culture with standard procedures. Grow cells in cell culture medium at 37 °C in a 5% CO_2 atmosphere.
2. Take the 96-well plate with the prepared base microchips in DMEM (Fig. 14.2A).
3. Harvest confluent cells (see Note 22 in Section 14.3) by rinsing in DPBS (5 mL DPBS for a 25-cm^2 cell culture flask) followed by incubation with cell stripper. For a 25-cm^2 cell culture flask with a 70–100% confluent cell layer, the addition of 0.5-mL cell stripper and a subsequent incubation of 5 min at 37 °C is usually sufficient. Resuspend and dilute the cell suspension with an appropriate volume of DMEM. Usually 10 mL DMEM for a 25-cm^2 flask of 80–100% confluent cells will yield a good density of cells on the microchips.
4. Add one droplet (see Note 23 in Section 14.3) of the cell suspension to each well containing a microchip in DMEM and wait about 5 min.
5. Observe the microchips every minute with an inverted light microscope to see how many cells settle on the windows (Fig. 14.2B). A microchip window should have at least a total of three attached cells ideally an area of 50×50 μm^2 should not bear more than ~1 adhering cell (see Note 24 in Section 14.3).
6. Increase the waiting time as necessary (usually for another 5–10 min) until enough cells have adhered to the SiN window. It needs approximately 5 min from the moment a cell has settled on the microchip until it develops enough adherence to remain on the microchip during the next washing step. When enough cells adhere to the SiN window, transfer the chip into a new well filled with DMEM.

Inspect the chips to verify that the number of adhered cells is still sufficient. If too many cells were flushed off, repeat steps 4–6 (see Note 25 in Section 14.3).

7. The cells should be serum starved through transfer of the microchips into wells filled with DMEM without FBS; this can be done as soon as about 20 min after step 5. Incubate for at least 4 h, best over night, at 37 °C in 5% CO_2, such that the cells fully flatten on the SiN window, as shown in Fig. 14.2C.

14.2.5 PREPARATION OF THE QD LABELS

1. To avoid QD aggregates in the labeling solution, centrifuge the QD stock solution for 4 min at 8000 × *g* and take the supernatant.
2. Couple QDs to EGF via biotin-streptavidin binding using a standard procedure (Lidke et al., 2004): incubate EGF–biotin in a 6:1 molar ratio with streptavidin-QD655 in 50 mM borate buffer, pH 8.3, for 1 h, under rotation or gentle agitation, at 35°C.
3. Remove free/unbound EGF–biotin with a microcentrifuge purification column by following the directions of the supplier.
4. Resuspend EGF–QDs in 30–50 μL of 50 mM borate buffer, pH 8.3, dilute further until the desired final concentration of 5 nM with BSA-TB.

14.2.6 CELL LABELING

1. Prepare 3% GA in CB, 1 mL per microchip.
2. Prepare 96-well plates: for each microchip, prefill wells with 200 μL volumes: 2 × BSA–PBS, 1 × EGF–QD in TB labeling solution (the incubation volume for the labeling step can be reduced to 75 μL if desired), 1 × TB, 5 × PBS, 1 × GLY–PBS, 2 × PBS.
3. In a separate 24-well plate, fill (for each microchip) 2 × CB, and 1 × 3% GA, keep this well plate in a chemical hood. An overview of the solutions needed for the labeling steps is shown as a scheme in Fig. 14.3.
4. Rinse and then incubate the microchips with serum-starved cells for 5 min with BSA–PBS to reduce nonspecific label binding. Rinsing is done by dipping a microchip briefly into a prefilled well without releasing it into the well. Incubation is done by placing (and releasing) the microchip into (another) prefilled well.
5. Incubate each microchip for 5 min in the EGF–QD labeling solution (see Note 26 in Section 14.3).
6. Rinse the microchips 1 × in TB, 2 × in PBS, then transfer the 96-well plate to the chemical hood.
7. Rinse the microchips 1 × in CB already prepared in the 24-well plate (step 3) and incubate for 20 min in 3% GA (at room temperature) for fixation (for optional fixation with formaldehyde, see Note 27 in Section 14.3) (Glauert & Lewis, 1998).

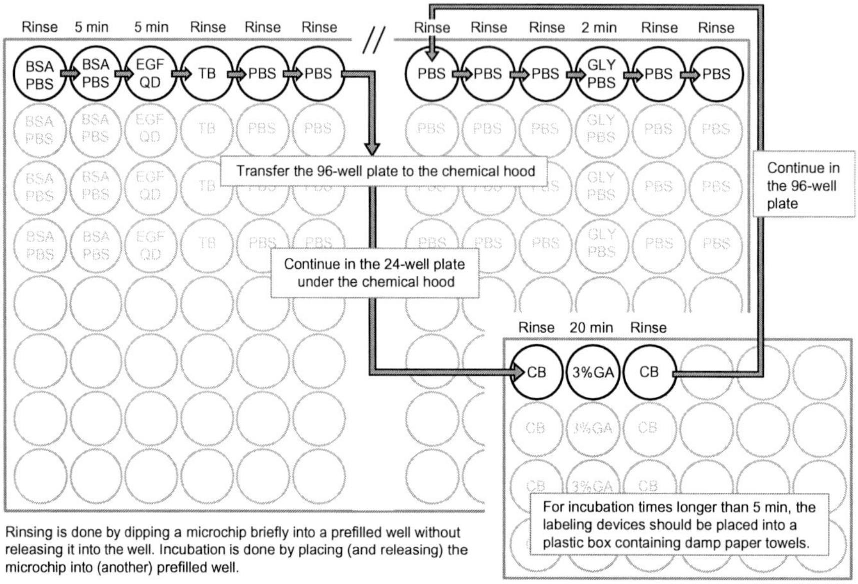

FIGURE 14.3

Schematic overview of the solutions needed for the labeling and fixation steps as described in Section 14.2.6. The sequence of the protocol steps is indicated by the line, first labeling in a 96-well plate proceeds, followed by fixation in a 24-well plate (in the chemical hood), and finalized by rinsing steps in the 96-well plate.

8. To remove excess GA, rinse microchips 1 × in CB and transfer into the prefilled wells of the 96-well plate, rinse 3 × in PBS, by dipping each microchip into a well prefilled with PBS for about 5 s.
9. Quench unreacted aldehyde groups by 2 min incubation in GLY–PBS.
10. Rinse microchips 2 × in PBS.
11. Store the microchips in a new 96-well plate in BSA-PBS at 4 °C until fluorescence imaging (see Note 28 in Section 14.3).

14.2.7 LIGHT MICROSCOPY

1. Fill a 35-mm cell culture dish with thin glass bottom for light microscopy with PBS.
2. Place a microchip with the cells facing downward in the dish (Fig. 14.4A). Handle the microchip carefully. Don't press on the microchip but let it settle onto the glass bottom of the dish. For imaging with an inverted fluorescence microscope equipped with an oil immersion objective, apply the lens oil onto the objective and place the dish on the lens (see Notes 29 and 30 in Section 14.3).
3. Illuminate the microchip with the light source of the fluorescence microscope set to 100% intensity for several minutes to quench the autofluorescence of the glutaraldehyde fixative (see Note 31 in Section 14.3).

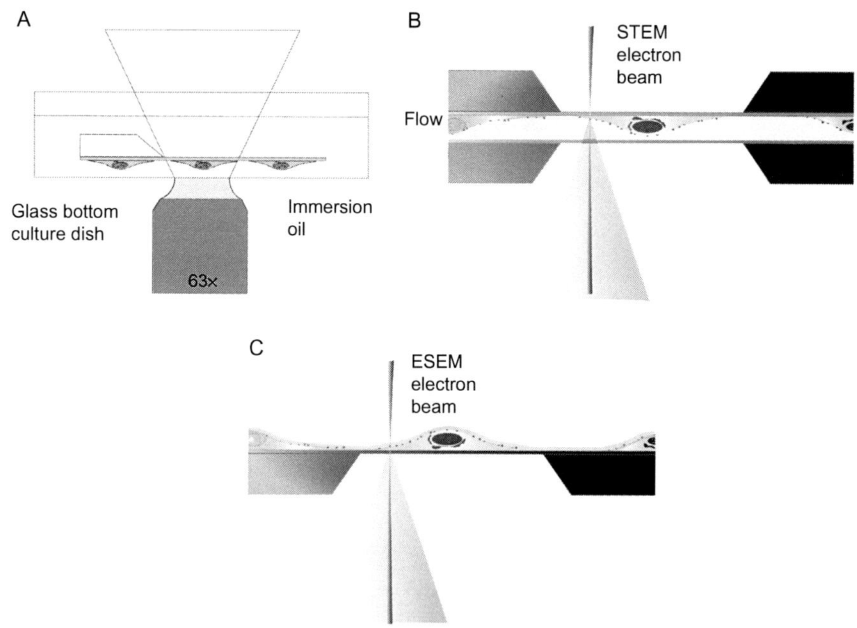

FIGURE 14.4

Schematics of the three microscopy modalities. Images not to scale. (A) The microchip with cells is imaged in liquid in a standard dish with light microscopy using an oil or water immersion lens. (B) Two microchips with electron transparent SiN windows enclosing a cellular sample in liquid can be imaged with liquid scanning transmission electron microscopy (STEM). (C) One microchip with cells under a thin water layer imaged with environmental scanning electron microscopy (ESEM).

Panel (B) is reproduced from Ring et al. (2011) with permission from John Wiley & Sons.

4. Record fluorescence images and phase contrast images (or use direct interference contrast). Use an objective lens with a high (preferably 1.4) numerical aperture and a magnification of at least 40 ×. Make sure that the images capture the outline of the SiN window as this will help to get orientation for correlative microscopy.
5. Grab the microchip gently, avoiding sheer forces on the cell layer. Store the microchip, cell facing up, in a 96-well plate in BSA-PBS.

14.2.8 LIQUID STEM

1. Prepare the liquid flow specimen holder for STEM (for example, the Poseidon system of Protochips Inc., NC, USA, see Fig. 14.5A) (Dukes et al., 2010; Peckys & de Jonge, 2014a). Connect the microfluidic tubing to the syringe pump. First, the liquid flow lines should be flushed with pure water for washing. The supply line should subsequently be flushed with 10% PBS buffer. Make

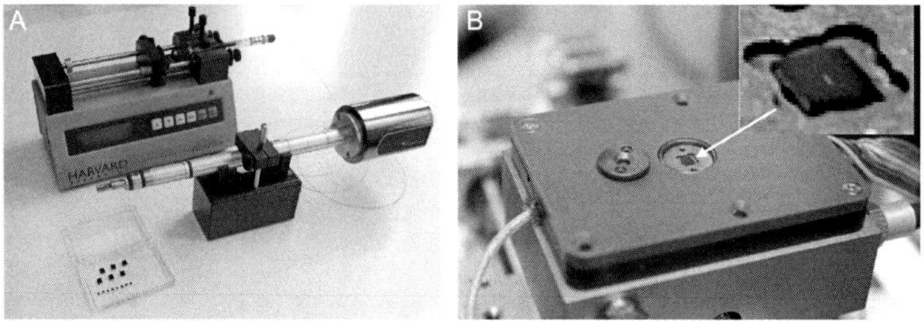

FIGURE 14.5

Systems for electron microscopy in liquid. (A) Liquid flow specimen TEM/STEM holder system ("Poseidon" Protochips Inc., NC, USA). Microchips in a box are shown at the left lower corner. A microfluidic syringe pump is connected to the holder via plastic tubing. (B) Peltier-cooled ESEM stage (Quanta 400 FEG ESEM, FEI, USA). The microchip with an SiN membrane window (a close-up is shown in the inset at the right upper corner) resides in a round copper holder such that it is positioned above an integrated STEM. Prior to the closing of the ESEM specimen chamber, the microchip is fixed with the second copper ring, seen at the left side of the sample.

With permission from Peckys and de Jonge (2014a).

sure all components of the specimen holder are clean and functional before loading the cells.

2. Place a spacer microchip with the SiN surface facing up into the appropriate position within liquid flow holder.
3. Apply a droplet of 0.5 μL of 10% PBS on the microchip (see Note 32 in Section 14.3).
4. Use the tweezers with the fine, curved tips to quickly but carefully place the microchip with the cells facing down on the spacer microchip (Fig. 14.4B).
5. Close the liquid flow specimen holder.
6. It is important to wash out remaining traces of 10% PBS solution outside of the microfluidic chamber, in the etched hole in the silicon at the backside of the microchip. This can be done by pipetting approximately 5 μL of pure water into the etched hole and removing it with the same pipet. This should be repeated a few times with fresh, pure water each time. The washing is needed to avoid obstructing salt traces on the SiN window.
7. Test if the vacuum seals are closing tightly and if the windows are intact by flowing liquid through the system. The flow rate needs to be set depending on the type of system. A typical value is 4 μL per min. A leak or broken window will manifest by the formation of droplets at the position of the SiN window.
8. Load the specimen holder in the STEM. Test for achieving an acceptable vacuum level (see Note 33 in Section 14.3). Move stage to the calibrated position of window position and eucentric height (see Note 34 in Section 14.3).
9. Open the beam valve, begin imaging, and find a corner of the window by moving the stage, adjust eucentric height for standard focus as necessary and record the stage coordinates.

10. Proceed with STEM imaging, move to a new stage position, adjust focus as needed, and acquire images. Image each region only a few times. Record the stage coordinates for each image acquired.
11. To correlate the light microscopy images with those of the STEM, one needs to map the coordinates of the stage on the shape of the SiN window in the light microscopy image. Correlative light microscopy and liquid STEM images are shown in Fig. 14.6.

14.2.9 WET ESEM–STEM

12. Bring the stage (Fig. 14.5B) into a centered position (*x*–*y* axis), set the stage temperature to 3 °C and the working distance between 5.8 and 6.4 mm.
13. For loading a sample into the ESEM stage, use the tweezers with fine, curved tips to quickly but carefully rinse the microchip, with the cells facing up in water. Dip the microchip four times, each time for about a second, into water-filled wells of a 96-well plate, kept on ice. The washing is needed to avoid obstructing salt traces on the cells when the air is evacuated from the microscope chamber. Do not open the tweezers and drop the microchip until step 15.
14. Blot the backside of the microchip briefly on a lint-free clean room wipe.

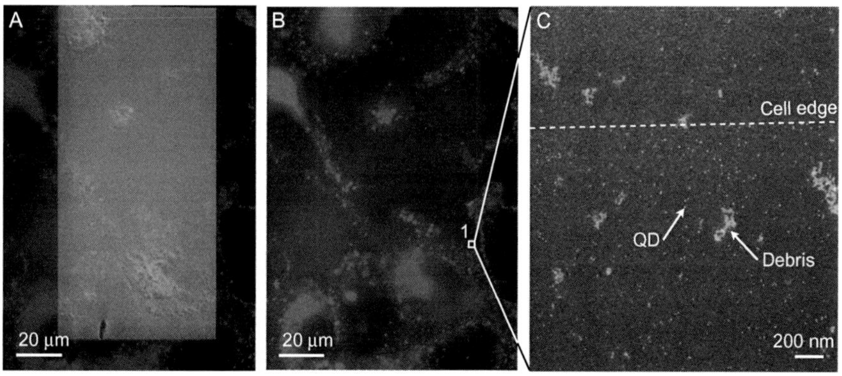

FIGURE 14.6

Correlative light microscopy and liquid STEM of intact fixed eukaryotic cells in saline water. (A) Direct interference contrast (DIC) image (gray) with overlaid fluorescent signal (red) of a microchip with COS7 cells showing the regions with quantum dot (QD)-labeled epidermal growth factor (EGF) receptors. The rectangular shape outlines the silicon nitride window. (B) Fluorescent image (red) showing cellular regions with EGF receptors. Some fluorescence from the fixative is also visible. (C) Liquid STEM image of the region indicated with a square #1 in (B). Individual QDs along the edge of the cell can be discerned as yellow spots on a blue background. Some debris is also visible. The magnification was $M=48{,}000$. The signal intensity was color-coded to increase the visibility of the labels. (See the color plate.)

Figure reproduced from Dukes et al. (2010), with the permission from ACS Publications.

15. Place the microchip into the precooled Peltier stage, while visually controlling that the cells remain wet (a wet sample has a shiny/reflective surface that would loose its shininess upon drying).
16. Pipet 3 μL of cooled water onto the sample surface, make sure not to touch the sample directly with the pipet tip, but only with the water droplet.
17. Close/fix the copper ring that holds the sample in the stage.
18. Place three additional 3 μL water droplets close to the sample on the Peltier stage as water source.
19. Close the specimen chamber and start the ESEM purging sequence, consisting of fivefold pressure cycling between 800 and 1500 Pa.
20. Switch the electron beam (30 kV) on and search the sample with the lowest magnification using the signals from the gaseous secondary electron detector, for the position of the SiN membrane window. If the water layer is too thick, the window may remain obstructed; in this case, start lowering the pressure as described in the next step and try again to find the window after a few minutes of waiting time.
21. Position the SiN membrane window centered. Lower the pressure stepwise within the next 2–5 min, using steps of 10–20, until the desired final pressure of 720–740 Pa is reached. During the lowering of the pressure, the water layer will eventually become thin enough to allow the dark field (DF) segment of the STEM detector to detect signals.
22. Start taking overview DF-STEM images of the cells.
23. Choose a cell and a region for high-resolution imaging and zoom in. When reaching a magnification of 50,000 × use ≥30-μs pixel dwell time and search for QD signals. It is the easiest to start this search at cellular edges (see Note 35 in Section 14.3). An ESEM image with QDs is shown in Fig. 14.7.

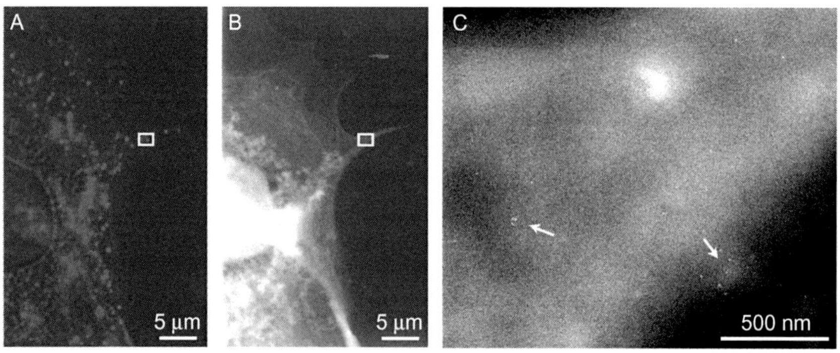

FIGURE 14.7

Correlative fluorescence microscopy and ESEM–STEM. (A) Fluorescence image of COS7 cells showing regions with QD–EGF in red (gray in the print version). (B) ESEM–STEM image of the same cellular region. (C) ESEM–STEM image recorded at the location of the rectangle in (A) and (B) showing the edge of a cell. The white oval shapes represent the individual QDs.

24. To avoid radiation damage-induced artifacts, use other (neighbored) cell areas to adjust the imaging settings than those areas selected for the recording of high-resolution images. Image each cellular region only once.

14.3 NOTES

1. Components needed in multiple sections are listed in the first section only.
2. These dimensions are not critical for performance.
3. These dimensions are critical for performance. The thickness of 50 nm was chosen as balance between the least amount of interaction with the electron beam and the most robust window. The width of the window of 50 μm leads to a bulging of the window of about 1 μm per window when two windows assembled in a microfluidic chamber are placed in the vacuum of the electron microscope. This amount of bulging is acceptable when imaging cells of a thickness of about 5 μm.
4. The protective coating is needed to prevent both contamination of the windows with silicon dust during dicing, and to protect the windows from dust in the environment during transport and storage prior to use.
5. The patterned layer forms a microfluidic channel. A variety of shapes and thicknesses are possible.
6. Different diameters can be used for different gap sizes.
7. All liquids should be of HPLC grade where possible to avoid clogging the microfluidic system with dust particles. Many liquids contain micrometer-sized particles. Solutions of unknown grade should be filtered with a 200-μm filter.
8. All tweezers should be cleaned of dust, grease, and other contamination before use.
9. All glasswares should be clean and be prewashed with the liquid it will contain before use to avoid contamination with dust particles.
10. It is important to first centrifuge the QD stock solution to avoid aggregates.
11. Due to the development of arsenic gas, this buffer should be made and handled in a chemical hood.
12. All procedures are carried out at room temperature unless stated otherwise.
13. For all steps involving the handling of the microchips (except when indicated otherwise), the microchips should be carefully gripped with flat beak tweezers having a Teflon coating, to avoid chipping the edges of the microchips. Different types of tweezers may be used as well, as long as direct contact of the edges with metal is avoided.
14. Transfer between acetone and ethanol should be done carefully but quickly, such that the acetone does not dry on the microchips, which would leave traces that cannot be removed.
15. If possible, use a dust-free hood with a low level of airflow, or use at least a laboratory environment with a low dust level. The microchips on a clean room tissue may also be covered by a glass beaker in such a way that a thin air gap remains.

16. This is not needed when wet ESEM–STEM is used.
17. Note that if the droplets were applied after plasma cleaning, the entire microchip surface would be wetted with the droplets, and the microspheres were not confined to the corners. Microspheres residing on the SiN window will lead to breakage.
18. Adjust the concentration experimentally, such that a droplet of 0.2 μL on a hydrophobic surface leaves well-separated individual droplets after drying.
19. It is assumed that the microchips bend by a small amount when pressurized.
20. If possible, use a dust-free hood with a low level of airflow, or use at least a laboratory environment with a low dust level. The microchips on a clean room tissue may also be covered by a glass beaker in such a way that a thin air gap remains.
21. To avoid the settling of dust, the best is to inspect with an inverted microscope, placing the microchips in an opened gel pack storage box.
22. When the level of cell confluency reaches 80–100% there are typically ~(3–5) × 10^6 COS7 cells in a 25-cm^3 culture flask. Resuspension in 10 mL of media will yield about ~(3–5) × 10^5 COS7 cells/mL, which can also be measured in a hemocytometer or cell Coulter counter.
23. The amount of liquid needs to be experimentally determined for optimal cell coverage.
24. The presence of more cells inhibits them from flattening on the SiN window surface and leads to thick samples.
25. If too many cells are present on the SiN window and cannot be flushed off, the microchip should be removed and stored in cell stripper. After the labeling experiment, it can be tried to retrieve the microchip for another experiment, by carefully applying the cell harvesting procedures, to remove the cells and to clean the microchip for later reuse. However, this involves the risk of breaking the SiN membrane.
26. For incubation times longer than 5 min, the labeling devices should be placed into a humid environment, such as a plastic box containing damp paper towels, to prevent evaporation of the labeling solution.
27. If the GA-induced fluorescence disturbs the fluorescence imaging of the QDs too much, the cells can also be fixed first with 2% formaldehyde in PBS for 10 min, followed by 3 × rinsing in PBS. In this case, steps 8–12 are done directly after the fluorescence imaging.
28. Storage in BSA-PBS is needed to ensure stable attachment of the QDS. Storage in PBS without BSA is not stable. Note also that in case the cells were fixed first with formaldehyde, light microscopy should follow immediately, at least within the next 2 h.
29. A water immersion lens or glycerol immersion lens instead of an oil immersion lens may also be used.
30. One may also use an upright microscope with a water immersion objective. In this case, the cell layer should face up.
31. Quenching is not needed if cells were first fixed with formaldehyde. In this case, the 3% GA fixation is done after the fluorescence imaging.

32. In case this work is done in a dry environment, use of a 50:50 mixture of glycerol and 10% PBS solution, which has a higher viscosity, will prevent drying of the droplet.

33. In case the vacuum level is not good, then there is probably a leak on an O-ring and the specimen holder should be retracted. Leaks on a broken window do not always lead to a lower vacuum because the liquid immediately freezes.

34. The liquid flow holder should first be tested with dry windows in the STEM and the approximate location of the corner of the window and the correct height of the stage should be calibrated.

35. If the QDs are difficult to detect, the right settings for stigmatism correction can first be adjusted with a sample containing 5- or 10-nm diameter gold nanoparticles. One may also apply 5- to 10-nm diameter gold nanoparticles as extra markers on the sample with cells. Gold nanoparticles of similar dimensions as QDs have a stronger contrast with STEM. It is also possible to lower the pressure further to 700 Pa.

ACKNOWLEDGMENTS

We are grateful to Protochips Inc. (NC) for providing the silicon microchips and the specimen holder. We thank M.J. Dukes for his contribution to previously published research, I. Weiss for discussions, and E. Arzt for his continuous support through INM.

REFERENCES

de Jonge, N., Peckys, D. B., Kremers, G. J., & Piston, D. W. (2009). Electron microscopy of whole cells in liquid with nanometer resolution. *Proceedings of the National Academy of Sciences of the United States of America*, *106*, 2159–2164.

Dukes, M. J., Peckys, D. B., & de Jonge, N. (2010). Correlative fluorescence microscopy and scanning transmission electron microscopy of quantum-dot-labeled proteins in whole cells in liquid. *ACS Nano*, *4*(7), 4110–4116. http://dx.doi.org/10.1021/nn1010232.

Gaietta, G., Deerinck, T. J., Adams, S. R., Bouwer, J., Tour, O., Laird, D. W., et al. (2002). Multicolor and electron microscopic imaging of connexin trafficking. *Science*, *296*, 503–507.

Glauert, A. M., & Lewis, P. R. (1998). *Biological specimen preparation for transmission electron microscopy*. London: Portland press.

Hoenger, A., & McIntosh, J. R. (2009). Probing the macromolecular organization of cells by electron tomography. *Current Opinion in Cell Biology*, *21*(1), 89–96. http://dx.doi.org/10.1016/j.ceb.2008.12.003.

Kourkoutis, L. F., Plitzko, J. M., & Baumeister, W. (2012). Electron microscopy of biological materials at the nanometer scale. *Annual Review of Materials Research*, *42*, 33–58.

Lidke, D. S., Nagy, P., Heintzmann, R., Arndt-Jovin, D. J., Post, J. N., Grecco, H. E., et al. (2004). Quantum dot ligands provide new insights into erbB/HER receptor-mediated signal transduction. *Nature Biotechnology*, *22*(2), 198–203. http://dx.doi.org/10.1038/nbt929.

Lippincott-Schwartz, J., & Manley, S. (2009). Putting super-resolution fluorescence microscopy to work. *Nature Methods*, *6*(1), 21–23. http://dx.doi.org/10.1038/nmeth.f.233.

Peckys, D. B., Baudoin, J. P., Eder, M., Werner, U., & de Jonge, N. (2013). Epidermal growth factor receptor subunit locations determined in hydrated cells with environmental scanning electron microscopy. *Scientific Reports*, *3*, 2626. http://dx.doi.org/10.1038/srep02626.

Peckys, D. B., & de Jonge, N. (2011). Visualization of gold nanoparticle uptake in living cells with liquid scanning transmission electron microscopy. *Nano Letters*, *11*, 1733–1738.

Peckys, D. B., & de Jonge, N. (2014a). Liquid scanning transmission electron microscopy: Imaging protein complexes in their native environment in whole eukaryotic cells. *Microscopy and Microanalysis*, *20*(2), 346–365. http://dx.doi.org/10.1017/S1431927614000099.

Peckys, D. B., & de Jonge, N. (2014b). Gold nanoparticle uptake in whole cells in liquid examined by environmental scanning electron microscopy. *Microscopy and Microanalysis*, *20*(1), 189–197. http://dx.doi.org/10.1017/S1431927613013986.

Peckys, D. B., Dukes, M. J., & de Jonge, N. (2014). Correlative fluorescence and electron microscopy of quantum dot labeled proteins on whole cells in liquid. *Methods in Molecular Biology*, *1117*, 527–540. http://dx.doi.org/10.1007/978-1-62703-776-1_23.

Peckys, D. B., Mazur, P., Gould, K. L., & de Jonge, N. (2011). Fully hydrated yeast cells imaged with electron microscopy. *Biophysical Journal*, *100*, 2522–2529.

Ring, E. A., & de Jonge, N. (2010). Microfluidic system for transmission electron microscopy. *Microscopy and Microanalysis*, *16*(5), 622–629.

Ring, E. A., Peckys, D. B., Dukes, M. J., Baudoin, J. P., & de Jonge, N. (2011). Silicon nitride windows for electron microscopy of whole cells. *Journal of Microscopy*, *243*(3), 273–283.

CHAPTER

The Use of a Laser for Correlating Light and Electron Microscopic Images in Thick Tissue Specimens

15

Derron Bishop*, **Ivana Nikic**[†], **Martin Kerschensteiner**[†,‡], **Thomas Misgeld**[‡,§]

**Department of Physiology, Indiana University School of Medicine-Muncie, Muncie, Indiana, USA*

[†]Research Unit Therapy Development, Institute of Clinical Neuroimmunology, Ludwig-Maximilian University, Munich, Nordallee, Germany

[‡]Munich Center for Systems Neurology (SyNergy), Nordallee, Munich, Germany

[§]Institute of Neuronal Cell Biology, Technische Universität München, Nordallee, Munich, Germany

CHAPTER OUTLINE

Methods in Cell Biology, Volume 124, ISSN 0091-679X, http://dx.doi.org/10.1016/B978-0-12-801075-4.00015-X

Abstract

Correlating images between light and electron microscopy is difficult especially in tissue specimens with a substantial *z*-dimension. To facilitate correlated light and electron microscopy (CLEM) in thick tissue specimens, we describe a basic method of using a femto-pulsed near-infrared laser to burn precise three-dimensional fiducial markers that circumscribe cells or regions of interest for easy identification between imaging methods. This rapid and reliable approach permits traditional fixation and avoids the use of electron-dense labeling methods that can obscure ultrastructural details. The versatility of the technique permits CLEM in a variety of tissue specimens to allow interpretation of highly resolved ultrastructural data in the more macroscopic and potentially dynamic context of light microscopy.

INTRODUCTION

A combination of improvements in optical microscopy design and development of fluorescent labeling strategies has greatly enhanced our ability to monitor and visualize cellular events in living animals and tissues. Despite continued incremental improvements in resolution below the diffraction barrier, light microscopy currently lacks sufficient resolution to visualize fine structural details that underpin cellular function. Resolving subcellular details currently requires the far greater resolution of the electron microscope (EM). However, by permitting routine visualization of cellular structures down to the level of macromolecules EM data alone can become difficult to interpret, especially in nonhomogenous cell populations. One way to contextualize high-resolution EM images is to directly correlate them to previous observations made by light microscopy. Such direct correlation provides a macroscopic and potentially dynamic context to interpret highly resolved EM data.

Over the years, a number of methods for correlative light and electron microscopy (CLEM) have emerged, and some commercial products have been developed that make such correlations more routine in cell monolayers. Despite these advances, the ability to precisely and reliably relocate a cell of interest remains technically challenging in monolayers of cells. The problem is exacerbated when working with intact tissues that have a substantial *z*-dimension. In such tissues, it is necessary to section through fixed tissues using serial section reconstruction methods in an attempt to locate the cell of interest. Over the last decade, more automated methods have been developed that make EM reconstruction much more amenable (Briggman & Bock, 2012), but even with these methods it is possible to reconstruct regions containing completely irrelevant cells. It is with this idea in mind that we developed a technique that permits the ability to locate a previously imaged cell or region of interest in thick tissue specimens in the EM using a femtosecond-pulsed near-infrared laser to burn fiducial markers around a cell or region of interest. The technique is reliable and compatible with a variety of tissues as well as serial EM reconstruction methods (serial section transmission electron microscopy, FIBSEM, serial block-face scanning electron microscopy, etc.). In this chapter, we outline the general technique of viewing and branding tissue using a standard multiphoton microscope followed by relocation of a cell of interest using serial section transmission electron microscopy of vibratome-sliced tissue.

15.1 GENERAL FEATURES OF NIRB

Relocating a previously imaged cell from a large tissue volume in the EM has traditionally required selective introduction of the electron-dense precipitate of 3,3-diaminobenzidine tetrahydrochloride (DAB) within the cell of interest. Antibodies conjugated to the enzyme horseradish peroxidase (HRP) have classically been used to bind epitopes and drive the precipitation of DAB selectively at these sites. For instance, HRP-conjugated antibodies to green fluorescent protein (GFP) have been used to relocate neurons expressing cytoplasmic GFP within brain slices (Knott, Holtmaat, Trachtenberg, Svoboda, & Welker, 2009). The main problems with this approach, especially in thick tissue specimens, are the use of mild detergents for antibody penetration and low concentrations of aldehyde fixatives to maintain antigenicity. These two factors result in loss of ultrastructural integrity and degraded image quality. It is also difficult to locate a previously imaged cell if epitope density is high due to the labeling of many similar, but irrelevant cells. An alternative method is to selectively label cells with fluorophores that can be used to photooxidize DAB. Photobleaching some fluorophores generates oxygen radicals that drive the oxidation and deposition of DAB precipitate that replaces the fluorescent label (Lubke, 1993). This technique avoids the problem of permeabilization and weaker fixatives, but also has difficulties. Namely, delivery of some fluorophores in an *in vivo* environment requires surgical access and, in some cases, microinjection of single cells (Gan, Bishop, Turney, & Lichtman, 1999). One way to overcome this problem is to take advantage of genetic labeling that encodes an optical signal, such as fluorescence, that can also generate oxygen radicals to precipitate DAB. A popular approach in the field of neuroscience has been to take advantage of transgenic mice that express variants of GFP in subsets of neurons throughout the nervous system (Feng et al., 2000). Although GFP is not very efficient at generating oxygen radicals while photobleaching, recent reports have demonstrated a level of success even in thick tissue specimens (Grabenbauer, 2012; Horstmann, Vasileva, & Kuner, 2013). Others have successfully used genetic delivery of HRP (Li, Wang, Chiu, & Cline, 2010), miniSOG (Shu et al., 2011), and APEX (Martell et al., 2012) as a means of generating an optical signal that can drive precipitation of DAB, although the latter two have not currently gained widespread use in transgenic animal models. Ultimately, all of these techniques have the problem of depositing the electron-dense signal directly within the cell of interest often obscuring the very subcellular details under examination.

To overcome all of these technical difficulties, we developed a technique called "near-infrared branding" (NIRB) that creates very precise three-dimensional fiducial markers within volumes of tissue to aid in relocation of previously imaged cells in the EM (Bishop et al., 2011). The technique relies upon a high numerical aperture objective lens that focuses a femtosecond-pulsed near-infrared laser, such as those used for multiphoton microscopy, to a fine point to burn away small amounts of tissue. Using standard scan mirrors available on most commercially available multiphoton microscopes, it is possible to burn away tissue leaving indelible points or lines at arbitrary positions or angles. To demonstrate the precision of the NIRB markings, we sequentially selected line scans of a femtosecond-pulsed titanium-sapphire laser in different orientations to carefully burn the letters N, I, R, and B into tissue fixed with 2.5% glutaraldehyde + 2.0% paraformaldehyde (see Fig. 15.1). A single *XZ* slice through the middle of the tissue volume demonstrates the letters N and R were burned at a depth of 15 μm, while the letters I and B were burned at a depth of 30 μm (Fig. 15.1B). The width and depth of the burn marks can be controlled by changing a number of parameters that control the dose of pulsed light. In particular, the number of "swipes" of the laser in line scan mode dramatically influences the degree of burning where the width and depth of the burned tissue increases with successive swipes of the laser (Fig. 15.1C).

To relocate a cell of interest, the choice of fiducial shapes is based upon the application and volume of tissue under consideration. From our experiences locating axons and dendrites in the nervous system, we choose to burn both a large and a small orientation rectangle. The larger orientation rectangle is burned 3–5 μm superficial to the cell of interest and has sides that are ~25 μm. The size of the rectangle is for easy identification in subsequent processing steps. In addition, this rectangle serves

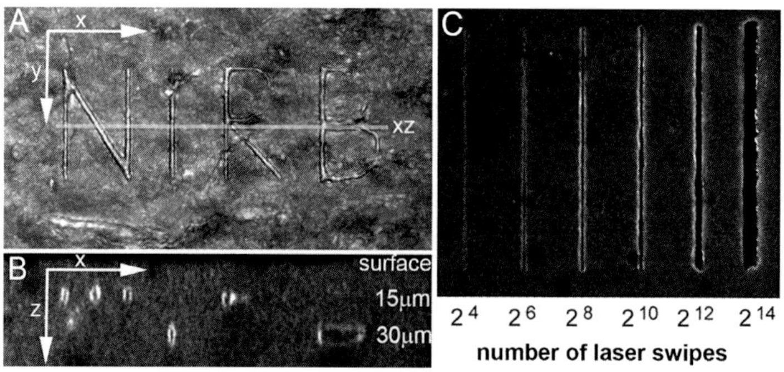

FIGURE 15.1

Near-infrared branding introduces precise three-dimensional fiducial markers in thick tissue specimens. (A) A maximum pixel intensity two-photon projection overlayed with a transillumination image of a vibratomed slice of mouse cerebral cortex where the letters NIRB have been burned into the tissue using a femto-pulsed near-infrared laser. A gray line has been introduced to the *x–z* plane illustrated in panel B. (B) A single *x–z* two-photon optical slice from gray line in panel (A) shows that letters N and R are burned at a depth of 15 μm, while letters I and B are burned at a depth of 30 μm. (C) Maximum pixel intensity two-photon projection shows that increasing the number of laser swipes increases the width (and depth) of the NIRB marks. Each vertical line is 106 μm.

Figure modified from Bishop et al. (2011).

as a marker during thick sectioning of the plastic-block for EM that the target volume is within a few microns in the *z*-dimension. The smaller orientation rectangle is burned at the same z-height as the cell of interest and as close to the cell of interest as possible. In Fig. 15.2, we targeted a single dendritic spine in the cerebral cortex of mice and burned the smaller orientation box with dimensions of 5 μm. In some cases, we add a final spot (using a point scan) directly adjacent to the structure of interest but away from the area chosen for electron microscopic reconstruction to provide a confirmation for successful reidentification. Both boxes will be visible as a tissue defect using transmitted light and relatively low NA objective lenses (~0.5 NA). Following the NIRB markings, we reimage the area of interest and use the image stack to serve as a map for tissue sectioning by providing a rough distance from the tissue surface to the object of interest.

The multiphoton fiducial marks are autofluorescent at the burned edges and can photooxidize DAB. After illuminating the tissue in DAB solution with 530–560 nm band-pass illumination, the autofluorescence will be completely replaced by an electron-dense precipitate (Fig. 15.2C). This step is not necessary, but can serve several purposes. First, the larger photooxidized orientation rectangle is easily visible with the aid of a dissecting microscope (Fig. 15.3). Accordingly, the tissue block can be substantially trimmed prior to resin embedding without inadvertently removing the region of interest. Second, after resin embedding, the larger orientation box can be easily identified in 1 μm unstained sections during semi-thin sectioning (Fig. 15.3C). This step is important for reducing the number of serial sections to

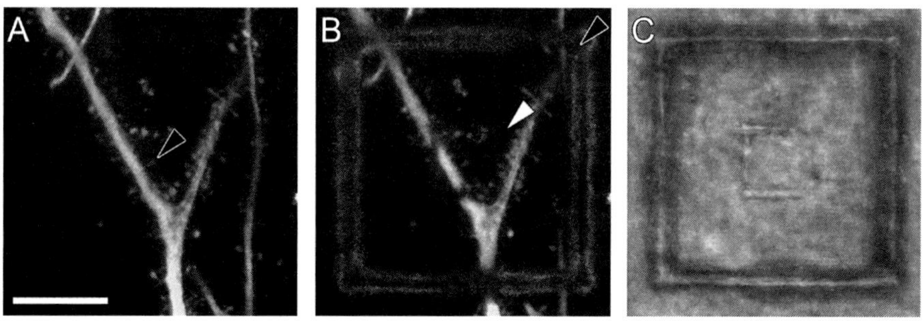

FIGURE 15.2

NIRB markings and photooxidation. (A) A confocal maximum pixel intensity projection of an apical dendrite (green) from a mouse cortex. A single dendritic spine (arrowhead) is targeted for EM correlation. (B) The confocal projection from panel (A) has been marked with two orientation rectangles. The larger orientation rectangle (dark arrowhead) is 25 μm on each side and is burned 5 μm superficial to the dendritic spine of interest. The smaller orientation rectangle (light arrowhead) is 6.5 μm on each side and is burned at the same *z*-position as the dendritic spine of interest. (C) Photooxidation of the autofluorescent edges of the NIRB markings results in deposition of electron-dense 3,3-diaminobenzidine precipitate. Scale bar for all three panels is 10 μm. (See the color plate.)

Figure modified from Bishop et al. (2011).

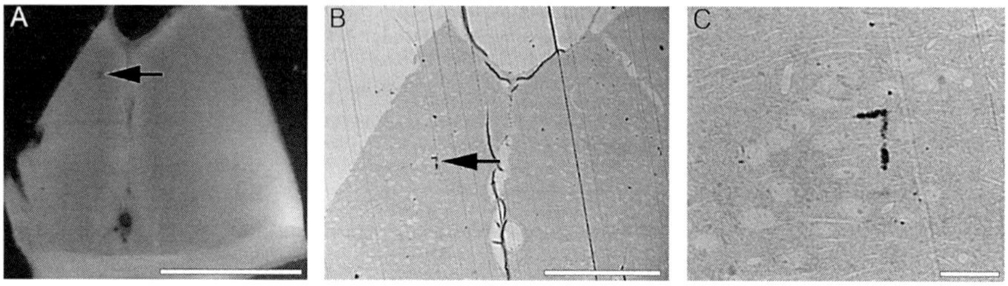

FIGURE 15.3

Reidentification of NIRB markings after tissue embedding. (A) An image from a dissecting microscope reveals the relative position of the larger photooxidized orientation rectangle (dark arrow) within the trimmed tissue prior to osmication and embedding in resin. Scale bar, 1 mm. (B) A 1.0-μm unstained thick section reveals the edge of the photooxidized larger orientation rectangle. Scale bar, 0.5 mm. (C) Higher magnification view of 1.0 μm section from panel (B) shows two edges of the larger orientation rectangle. Scale bar, 25 μm.

Figure modified from Bishop et al. (2011).

be cut to section through the volume of interest. The tissue can be osmicated, dehydrated, and infiltrated with resin using standard techniques.

The detailed method we describe below is a reliable method for locating individual cells or small regions of interest in a variety of fixed tissues from different animal models. The examples provided are in the context of relocating a single dendritic spine within a fixed 100 μm vibratome slice from mouse cerebral cortex using a multiphoton and transmission EM.

15.2 METHOD

15.2.1 FIXATION

1. Prior to obtaining tissue samples from animals, be sure to adhere to all institutional and ethical guidelines for animal experiments. After deep anesthesia, transcardially perfuse mouse with ~25 mL heparinized Tyrode's solution to exsanguinate followed immediately by ~200 mL of 2.5% glutaraldehyde + 2.0% paraformaldehyde in 0.1 M sodium phosphate buffer (pH 7.4) containing 4 mM Ca^{2+} and 2 mM Mg^{2+}.
2. Carefully dissect the organ and tissue from the perfused animal. Tissues can be postfixed in the same fixative up to 24 h without altering the ability to generate NIRB fiducial markers.
3. After postfixation, rinse very well with buffer (at least 5 ×) and store at 4 °C.

15.2.2 VIBRATOME AND MOUNTING OF TISSUE SLICE

1. Prepare phosphate buffer ice cubes 1 day prior to sectioning to place in the vibratome boat.

2. Prepare some agar blocks to prevent the tissue from rolling over when cutting: Add 5 g of agar into heated distilled water and stir to dissolve. Pour the agar into a petri dish, at the height of the tissue, and cut out blocks as needed.
3. Prior to sectioning: add ice cold buffer and few buffer ice cubes to the vibratome boat to cool specimen blocks, blades, etc.
4. Mount an agar block that is roughly the same height as the tissue onto the slicing stage using tissue glue or cyanoacrylate glue.
5. Mount the tissue just in front of the agar block using tissue glue, cyanoacrylate glue, etc. If the tissue has been previously imaged *in vivo*, mount the tissue in the same orientation as the *in vivo* imaging session.
6. Set the vibratome thickness to ~200 μm. The thickness setting is greater than what is desired, but the intent is to capture the first slice so it completely contains the region of interest. Orient the blade angle so that it is reasonably close to parallel to the surface of the region of interest. The blade angle is less an issue for a successful cut, so focus on trying to make the blade parallel, thereby increasing the chances of getting the region of interest on the first slice. If it appears the blade will "skim the surface," retract the blade and reapproach the block. This technique usually results in a ~100–150 μm thick slice of tissue containing the region of interest. Try to prevent any bending or twisting of the slice when transferring in and out of buffer as they break through the surface tension of the bath. Typically, fine paint brushes work well to prevent any twisting. To keep track of orientation, carefully cut a single notch in one corner of the slice away from the region of interest (upper left or upper right) using either fine Vannas scissors or a microscalpel.
7. Maintaining orientation, carefully mount the slice cut side down on a glass slide with a few drops of buffer. Gently place the tissue on the slide using fine paint brushes to avoid and torsion or twisting of the slice.
8. Mount a #1.5 glass coverslip onto the slide. With thick vibratome slices, it is often necessary to add a spacer by cutting a parafilm rectangle that is the same dimensions of the coverslip. Cut a square out from the inside of the parafilm spacer so it resembles a picture frame that is the same size as the coverslip. Place this parafilm "frame" on the glass slide, add a drop or two of buffer, add the slice to the buffer, and place the coverslip on top of the spacer.
9. Seal the edges of the coverslip with nail polish. At this point, take images of the slice with a confocal or multiphoton microscope for documentation purposes. There will be a great deal of background autofluorescence due to the glutaraldehyde fixation. Do not use any chemical autofluorescence reduction (i.e., sodium borohydride) since this step tends to complicate photooxidation if this step is desired. Further fluorescent signals tend to decrease after glutaraldehyde fixation.

15.2.3 NIRB SETTINGS AND MARKINGS

1. Basic settings for successful branding are dependent upon fixation, tissue, depth, pixel dwell time, and laser power. The following settings work well in a variety of

tissues fixed with 2.0% paraformaldehyde and 2.5% glutaraldehyde at a depth ≤75 μm. At greater depths, the tissue begins to attenuate laser power requiring lengthening of pixel dwell time or "swipes" of the laser (see below). Even with known settings in familiar tissues, it is usually advisable to run a few test brandings in parts of the sample far away from the region of interest.

   a. Set the multiphoton microscope to the following settings: $\lambda = 910$ nm, 200 mW laser power at the back focal plane, line scan, 5 μs/pixel dwell time
   b. Use a 1.4 NA lens (~60×, oil immersion).

2. Generating NIRB fiducial markers
   a. The number of "swipes" of the laser in line scan mode determines both the width and depth of the NIRB markings. Fewer swipes make thinner burn lines in all three dimensions reducing the potential area of interest (see Fig. 15.1C). The smaller lines are especially important for the smaller box that will be used to identify the cell of interest in the EM. We found that ~15 swipes produced the smallest discernable line that was nearly 4 μm in the *z*-dimension. For a rectangle with sides of 4 μm, this restricts the total volume of interest to ~65 μm^3.
   b. The larger orientation rectangle (~25 μm) is burned by *z*-focusing to ~3–5 μm superficial (closer to the coverslip) to the region of interest. With the above settings, 250 swipes of the laser for each 25 μm side of the box will make an easily discernable rectangle (see larger box in Fig. 15.2B).
   c. The smaller orientation box (~5 μm) is burned at the same *z*-level as the cell of interest. The goal for this box is to introduce it as close as possible without interfering with the cell of interest. Fifteen to thirty swipes of the laser at the above settings produce a thin line (see smaller box in Fig. 15.2B).
3. Collect a new set of "post-NIRB" images through the volume of interest plus the new autofluorescent NIRB mark. This set of images is useful for reidentifying the cell of interest at the EM level.

15.2.4 PHOTOOXIDATION OF NIRB MARKINGS (OPTIONAL)

1. After burning NIRB marks, carefully remove the coverslip with a razor blade and transfer the brain slice to a Sylgard-lined 35-mm petri dish containing phosphate buffer. Carefully pin the slice with fine minutien pins avoiding the area near NIRB markings.
2. Bathe the preparation in 1.5 mg/mL 3,3-diaminobenzidine (DAB) in 0.1 M Tris buffer (pH 7.4). Load the DAB solution into a 10-mL syringe and filter it through a 0.22-μm filter prior to adding to the dish.
3. Use either a 63× 0.95 NA or 40× 0.8 NA dipping cone immersion lenses to photobleach the edges of the autofluorescent NIRB marks using standard DsRed or rhodamine filter sets. Since these objective lenses will be placed in DAB solution (a known carcinogen), it is advisable to set aside a single lens that is dedicated for this purpose. We have successfully used both lenses to photooxidize DAB with the higher NA lens driving the reaction quicker.

a. The photoconversion process will take ~20 min. The DAB polymerization will not simply replace the fluorescence linearly. The autofluorescence will completely disappear prior to appearance of the DAB reaction product. A number of photooxidation protocols bubble oxygen during the reaction, but we typically omit this process since in this preparation it does not dramatically improve the signal/noise ratio and can cause DAB solution to contaminate the microscope stage.
b. After photoconversion, rinse multiple times (at least 5×) with phosphate buffer.

15.2.5 EMBEDDING AND RELOCATING THE NIRB MARKINGS FOR EM

1. Trimming
   a. After photooxidation, the larger orientation rectangle will be readily visible with the aid of a dissecting scope. Figure 15.3A shows an example of a vibratomed brain slice where the larger orientation rectangle has been photooxidized. If no DAB precipitate has been introduced, the NIRB markings should still be visible using intense transmitted light illumination on a compound microscope. Fine Vannas scissors or a microscalpel can be used to trim the tissue as small as desired.
   b. Take a picture of the final trimmed slice using a dissecting microscope to show the relative position of the photooxidized rectangle within the entire slice (Fig. 15.3A). There is no need for taking coordinates, blood vessel maps, etc., but these can be recorded as an additional level of assurance.
2. Tissue processing
   a. Place tissue in ferrocyanide-reduced osmium (1.25% ferrocyanide + 1.0% osmium) for 15 min followed by 1.0% osmium for 45 min. The tissue will turn black after this step and the NIRB fiducials will no longer be visible in either the dissecting microscope or compound microscope using transmitted light unless the vibratome slice thickness is ≤65 μm (see Section 15.4).
   b. Place the tissue in a glass vial and dehydrate with an ascending ethanol series finishing with 100% ethanol maintained on a molecular sieve. After transitioning through two 10 min exchanges of propylene oxide, infiltrate the block on a tissue rotator with ascending resin:propylene oxide solutions (30:70, 50:50, 70:30, 100:0 for a minimum of 1 h). The NIRB technique is not dependent upon using any particular resin and has been successful using standard available resins (Spurr's, Araldite, EMBed, Durcopan, etc.).
   c. After infiltration, flat embed the tissue by first cutting the cap off a BEEM capsule as well as the pointed end to get a BEEM cap and cylinder. Pour resin into BEEM cap and carefully transfer tissue into BEEM cap using a wooden toothpick. Using a dissecting microscope, orient the tissue so the outside surface of the tissue (i.e., the imaged surface) is placed against the inside surface of the cap. Attach the BEEM cylinder to the cap and very slowly backfill the cylinder being careful to not disturb the tissue and cause it to tilt. Add a printed label to the block and polymerize at 60 °C for 2 days.

3. Sectioning
   a. Remove the BEEM mold from the polymerized tissue block and trim the block to a mesa containing the area of interest in preparation for cutting semi-thin sections.
   b. Cut semi-thin (1.0 μm) sections from the block using glass knives. Once close to the *z*-depth and area of interest, mount sections onto slides and examine with transmitted light. If the NIRB markings have been photooxidized, simply examine unstained sections for the DAB precipitate from the larger orientation rectangle. The larger photooxidized orientation rectangle will be unmistakable in the semi-thin sections if no staining is used (see Fig. 15.3B and C). If no photooxidation has been used, stain mounted sections with toluidine blue and examine them for the larger orientation rectangle.
   c. Once the larger orientation rectangle has been located, retrim the block for serial ultrathin sectioning. The position of the orientation block can be determined by careful examination of the last stained/unstained semi-thin section with transmitted light.
   d. Cut serial ultrathin sections from the retrimmed block and mount on formvar or pioloform-coated slot grids for subsequent staining in 2.0% aqueous uranyl acetate and Reynold's lead citrate. Assuming a section thickness of ~60 nm and the orientation block 3 μm superficial to the volume of interest, ~50 serial sections would need to be cut before entering the volume of interest.
4. Locating the area of interest in the EM
   a. The larger orientation rectangle will most likely still be apparent through the majority of the sections at the ultrastructural level (see Fig. 15.4A). The *z*-thickness of the larger orientation rectangle will be ~7 μm in the *z*-dimension with 250 swipes of the laser. If the larger orientation rectangle is introduced ~3–5 μm superficial to the cell of interest, the rectangle will still be present after cutting serial sections through the cell.
   b. After identifying the larger orientation rectangle, examine subsequent serial sections for the appearance of the smaller rectangle that is inside the larger orientation box. The cell of interest will be oriented the same way in the small box as documented by light microscopy (see Fig. 15.4B).

15.3 MATERIALS

15.3.1 FIXATION

a. Reagents
   i. 0.1 M phosphate buffered saline (pH 7.4)
   ii. EM grade glutaraldehyde (25% vol/vol) (Electron Microscopy Sciences cat. #16220)
   iii. EM grade paraformaldehyde (16% vol/vol) (cat. #15710)
   iv. Double-distilled water

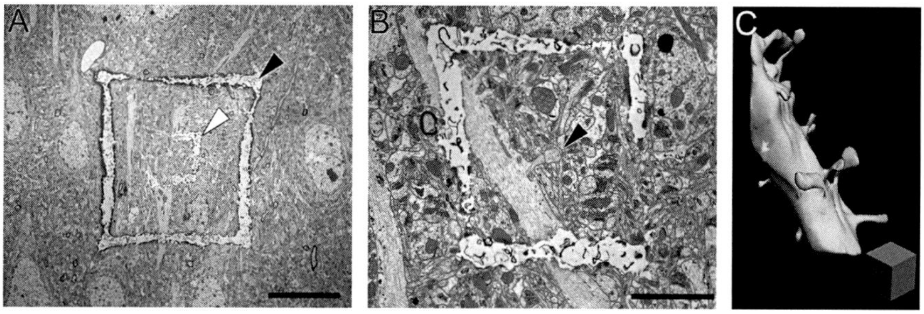

FIGURE 15.4

Identification of NIRB markings at the EM level. (A) A transmission electron micrograph from mouse cerebral cortex in Figs. 15.2 and 15.3 showing the larger orientation rectangle (dark arrowhead). The edges of the smaller orientation rectangle are just visible (light arrowhead). Scale bar, 10 μm. (B) A transmission electron micrograph shows the smaller orientation rectangle as well as the pseudocolored (green) dendrite of interest. A dark arrowhead points to the targeted dendritic spine of interest. Scale bar, 2.0 μm. (C) A surface reconstruction of the apical dendrite from serial section transmission electron micrographs shows the targeted dendrite and spine of interest. Scale cube, 1.0 μm. (See the color plate.)

Figure modified from Bishop et al. (2011).

b. Equipment
- **i.** 50-mL syringe with 25 gauge needle or perfusion peristaltic pump
- **ii.** Dissection tools (forceps, surgical scissors, dissecting tray, etc.)

15.3.2 VIBRATOME AND MOUNTING OF TISSUE SLICE

a. Reagents
- **i.** Agarose (Sigma-Aldrich, cat. #A9414)
- **ii.** Cyanoacrylate or tissue glue

b. Equipment
- **i.** Vibratome and blades
- **ii.** Glass microscope slides
- **iii.** Petri dish
- **iv.** Fingernail polish
- **v.** #1.5 glass coverslips
- **vi.** Fine paint brushes (Ted Pella, Inc., cat. #11860)
- **vii.** Vannas scissors (Fine Science Tools, cat. #15000-08)
- **viii.** Parafilm M (Sigma-Aldrich, cat. #P7793)

15.3.3 NIRB SETTINGS AND MARKINGS

a. Equipment
- **i.** Multiphoton microscope with femtosecond-pulsed titanium-sapphire laser

15.3.4 PHOTOOXIDATION OF NIRB MARKINGS

a. Reagents
   i. Sylgard silicone elastomer (World Precision Instruments, cat. #SYLG184)
   ii. 3,3-diaminobenzidine (Sigma-Aldrich, cat. #D4168)
b. Equipment
   i. 35-mm petri dish
   ii. Minutien pins (Fine Science Tools, cat. #26002-10)
   iii. 63 × 0.95 NA or 40 × 0.8 NA dipping cone objective lens
   iv. Epifluorescence microscope with DsRed or rhodamine filter cube

15.3.5 EMBEDDING AND RELOCATING THE NIRB MARKINGS FOR EM

a. Reagents
   i. Osmium tetroxide (Electron Microscopy Sciences, cat. #19110)
   ii. Potassium ferrocyanide (Sigma-Aldrich, cat. #455989)
   iii. Ethanol (Spectrum Chemical, cat. #ET107)
   iv. Propylene oxide (Electron Microscopy Sciences, cat. #20401)
   v. Molecular Sieve (Delta Absorbents, cat. #MS3A4801)
   vi. Araldite-Embed 812 (Electron Microscopy Sciences, cat. #13940) or similar resin kit
   vii. Toluidine blue
   viii. Uranyl acetate
   ix. Reynold's lead citrate
b. Equipment
   i. 80–120 kV transmission EM
   ii. Glass vials (Electron Microscopy Sciences, cat. #72631)
   iii. Dissecting microscope
   iv. BEEM capsules (Electron Microscopy Sciences, cat. #70000-B)
   v. Single edge razor blades
   vi. Tissue rotator (Ted Pella, Inc., cat. #1050)
   vii. Ultramicrotome
   viii. Glass knives and glass knife maker (Leica EM KRM3)
   ix. Diamond knife (Diatome)
   x. Formvar-coated slot grids

15.4 DISCUSSION AND TROUBLESHOOTING

NIRB is a versatile and reliable tool for relocating a cell of interest in a variety of thick tissue specimens. The procedure is relatively rapid (an hour at most) and does not obscure the cell or region of interest with electron-dense precipitates. There are, however, some potential difficulties that may be encountered when using the technique. Two of the main difficulties relate to relocating the cell or region of interest due to (1) autofluorescence generated by fixation in high glutaraldehyde concentrations and (2) the loss of fluorescence signal from the fluorophore. The former problem is usually encountered when using epifluorescence to examine fixed tissue to relocate the cell of interest to begin NIRB markings (Section 15.2.3, steps 2a–c). This problem can be solved by using confocal or multiphoton microscopes to gain optical sections that are less susceptible to the background autofluorescence to relocate the cell of interest rather than standard epifluorescence. With regard to the latter problem, we determined with 2.5% glutaraldehyde fixation that the cytoplasmic GFP signal decreased by 30% over background fluorescence 1 h after fixation (Bishop et al., 2011). Despite this decreased signal-to-noise, cytoplasmic GFP expression remains easily identifiable. However, this remains a problem in samples with punctate staining with less bright fluorophores. To overcome this problem, we initially lower the glutaraldehyde concentration during transcardial perfusion (e.g., 0.2% glutaraldehyde instead of 2.5%). We continue with the method as described until the NIRB fiducial markings are introduced and documented. After these markings are in place, we postfix the vibratome slice with 2.5% glutaraldehyde for 12 h and proceed with the remainder of the method. Using this approach, we have been able to relocate less-brightly labeled samples while maintaining ultrastructural integrity.

A second potential problem is relocating the NIRB markings during semi-thin sectioning (Section 15.2.5, step 3b). We recommend burning the larger orientation box, not only for identification in the EM, but also for easy identification at the light microscopic level during semi-thin sectioning. While the photooxidation step is not required, the DAB precipitate in the larger orientation box greatly prevents missing the region of interest in unstained semi-thin sections. A second approach that can help relocate the NIRB markings, especially if photooxidation is not desired, is to collect thinner vibratome slices (Section 15.2.2, step 6). With proficiency in vibratome sectioning, it is possible to collect thinner tissue slices that contain the region of interest. The larger orientation rectangle can be visualized using transmitted light and a dissecting microscope in slices that are ≤ 65 μm in the z-axis, even after osmication and embedding in resin. Since the NIRB markings can be visualized in the resin block, it is possible to either reexamine the block during semi-thin sectioning to verify the region of interest has not been sliced through, or trim the block to contain only the region of interest. The potential drawback to this strategy is the possibility of damaging the region of interest during collection of the vibratome section if it is near the sliced surface of the section.

Perhaps the greatest challenge with this approach as described is the ability to manually obtain serial sections for transmission electron microscopy. Even in the hands of an experienced ultramicrotomist, manual serial sectioning is difficult at best. This problem can be overcome by taking advantage of some of the block-face scanning electron microscopy techniques that take much of the drudgery out of the process to provide fast and more reliable serial section datasets (Denk & Horstmann, 2004; Knott, Marchman, Wall, & Lich, 2008). While these approaches have been mostly been used to reconstruct large tissue volumes (Helmstaedter et al., 2013), serial block scanning can be used in conjunction with NIRB to focus reconstructions to small regions of interest using nearly any fluorophore for CLEM in tissue volumes. With only minor modifications to the approach outlined in this chapter (Maco et al., 2013), NIRB has been successfully implemented in conjunction with FIBSEM to identify synaptic sites between light and EM within mouse cortex (Grillo et al., 2013; Mostany et al., 2013).

CONCLUSIONS

Using NIRB to introduce precise three-dimensional fiducial markers overcomes many of the technical obstacles that hinder CLEM in thick tissue volumes. While many approaches to CLEM require both skilled expertise and specialized equipment, we have outlined a basic procedure using scientific instruments that inhabit many modern day imaging facilities. As continued improvements in both specimen fixation and reconstruction methods become more routine at the EM level, the NIRB approach to contextualizing EM data in both macroscopic and dynamic contexts from *in vivo* light microscopy will become more important. Without such context, EM is likely to once again be maligned as a tool of only the most descriptively minded.

REFERENCES

Bishop, D., Nikic, I., Brinkoetter, M., Knecht, S., Potz, S., Kerschensteiner, M., et al. (2011). Near-infrared branding efficiently correlates light and electron microscopy. *Nature Methods*, *8*(7), 568–570.

Briggman, K. L., & Bock, D. D. (2012). Volume electron microscopy for neuronal circuit reconstruction. *Current Opinion in Neurobiology*, *22*(1), 154–161.

Denk, W., & Horstmann, H. (2004). Serial block-face scanning electron microscopy to reconstruct three-dimensional tissue nanostructure. *PLoS Biology*, *2*(11), e329.

Feng, G., Mellor, R. H., Bernstein, M., Keller-Peck, C., Nguyen, Q. T., Wallace, M., et al. (2000). Imaging neuronal subsets in transgenic mice expressing multiple spectral variants of GFP. *Neuron*, *28*(1), 41–51.

Gan, W. B., Bishop, D. L., Turney, S. G., & Lichtman, J. W. (1999). Vital imaging and ultrastructural analysis of individual axon terminals labeled by iontophoretic application of lipophilic dye. *Journal of Neuroscience Methods*, *93*(1), 13–20.

Grabenbauer, M. (2012). Correlative light and electron microscopy of GFP. *Methods in Cell Biology*, *111*, 117–138.

Grillo, F. W., Song, S., Teles-Grilo Ruivo, L. M., Huang, L., Gao, G., Knott, G. W., et al. (2013). Increased axonal bouton dynamics in the aging mouse cortex. *Proceedings of the National Academy of Sciences of the United States of America*, *110*(16), E1514–E1523.

Helmstaedter, M., Briggman, K. L., Turaga, S. C., Jain, V., Seung, H. S., & Denk, W. (2013). Connectomic reconstruction of the inner plexiform layer in the mouse retina. *Nature*, *500*(7461), 168–174.

Horstmann, H., Vasileva, M., & Kuner, T. (2013). Photooxidation-guided ultrastructural identification and analysis of cells in neuronal tissue labeled with green fluorescent protein. *PLoS One*, *8*(5), e64764.

Knott, G. W., Holtmaat, A., Trachtenberg, J. T., Svoboda, K., & Welker, E. (2009). A protocol for preparing GFP-labeled neurons previously imaged *in vivo* and in slice preparations for light and electron microscopic analysis. *Nature Protocols*, *4*(8), 1145–1156.

Knott, G., Marchman, H., Wall, D., & Lich, B. (2008). Serial section scanning electron microscopy of adult brain tissue using focused ion beam milling. *The Journal of Neuroscience*, *28*(12), 2959–2964.

Li, J., Wang, Y., Chiu, S. L., & Cline, H. T. (2010). Membrane targeted horseradish peroxidase as a marker for correlative fluorescence and electron microscopy studies. *Frontiers in Neural Circuits*, *4*, 6.

Lubke, J. (1993). Photoconversion of diaminobenzidine with different fluorescent neuronal markers into a light and electron microscopic dense reaction product. *Microscopy Research and Technique*, *24*(1), 2–14.

Maco, B., Holtmaat, A., Cantoni, M., Kreshuk, A., Straehle, C. N., Hamprecht, F. A., et al. (2013). Correlative *in vivo* 2 photon and focused ion beam scanning electron microscopy of cortical neurons. *PLoS One*, *8*(2), e57405.

Martell, J. D., Deerinck, T. J., Sancak, Y., Poulos, T. L., Mootha, V. K., Sosinsky, G. E., et al. (2012). Engineered ascorbate peroxidase as a genetically encoded reporter for electron microscopy. *Nature Biotechnology*, *30*(11), 1143–1148.

Mostany, R., Anstey, J. E., Crump, K. L., Maco, B., Knott, G., & Portera-Cailliau, C. (2013). Altered synaptic dynamics during normal brain aging. *The Journal of Neuroscience*, *33*(9), 4094–4104.

Shu, X., Lev-Ram, V., Deerinck, T. J., Qi, Y., Ramko, E. B., Davidson, M. W., et al. (2011). A genetically encoded tag for correlated light and electron microscopy of intact cells, tissues, and organisms. *PLoS Biology*, *9*(4), e1001041.

CHAPTER

Correlative *In Vivo* 2-Photon Imaging and Focused Ion Beam Scanning Electron Microscopy: 3D Analysis of Neuronal Ultrastructure

16

Bohumil Maco*,†, Anthony Holtmaat‡, Anne Jorstad†, Pascal Fua†
Graham W. Knott*

**BioEM Facility, Centre of Electron Microscopy, EPFL, Lausanne, Switzerland*
†Computer Vision Laboratory, EPFL, Lausanne, Switzerland
‡Department of Neuroscience, University Medical Center (CMU), University of Geneva, Geneva, Switzerland

CHAPTER OUTLINE

Methods in Cell Biology, Volume 124, ISSN 0091-679X, http://dx.doi.org/10.1016/B978-0-12-801075-4.00016-1

Abstract

This protocol describes how dendrites and axons, imaged *in vivo*, can subsequently be analyzed in 3D using focused ion beam scanning electron microscopy (FIBSEM). The fluorescent structures are identified after chemical fixation and their position highlighted using the 2-photon laser to burn fiducial marks around the region. Once the section has been stained and resin embedded, a small block is trimmed close to these marks. Serially aligned EM images are acquired through this region, using FIBSEM, and the neurites of interest then reconstructed semi-automatically using the Ilastik software (ilastik.org). This fast and reliable imaging and reconstruction technique avoids the use of specific labels to identify the features of interest in the electron microscope and optimizes their preservation for high-quality imaging and 3D analysis.

INTRODUCTION

Visualizing the morphology and connectivity of neurons is crucial to understanding how they function. The increasing use of live-imaging methods has enabled neuroscientists to study neurons in their native environment. However, this has also demanded greater use of electron microscopy to give further interpretation to the changes seen with the light microscope. With its higher resolution and ability to give more structural details, electron microscopy remains the only technique that is able to see fine cellular architecture and synaptic connectivity of neurons. In recent years, the development of new EM methods has enabled large volumes of tissue to be imaged with only limited human interaction. These have opened up many different opportunities for exploring the brain.

This chapter gives a detailed explanation of one technique that has recently been used in a number of high-profile studies combining the imaging of neurons in the live brain, using 2-photon microscopy, with electron microscopy using the focused ion beam scanning electron microscope (Allegra Mascaro et al., 2013; Cane, Maco, Knott, & Holtmaat, 2014; Canty et al., 2013; Grillo et al., 2013; Mostany et al., 2013). These studies carried out extensive series of live-imaging experiments showing how the morphology of axons and dendrites of fluorescently labeled neurons changed during aging and in response to injury.

Previous correlative light and electron microscopy (CLEM) experiments on *in vivo*-imaged neurons had used serial-section transmission electron microscopy (Holtmaat, Wilbrecht, Knott, Welker, & Svoboda, 2006; Knott, Holtmaat, Trachtenberg, Svoboda, & Welker, 2009; Knott, Holtmaat, Wilbrecht, Welker, & Svoboda, 2006; Trachtenberg et al., 2002; Zito, Parnas, Fetter, Isacoff, & Goodman, 1999), however, these studies were rare, requiring considerable amounts of time and skill to manually section large volumes of tissue that is then imaged one section at a time to locate the region of interest. Maintaining a complete series of sections, in perfect condition, is critical for this method to work. The first study of this kind appeared soon after green fluorescent protein (GFP) started to be used in live-cell imaging, and was undertaken in *Drosophila* (Zito et al., 1999) to understand how the neuromuscular junction developed and synapses inserted into the growing boutons. This pioneering work required the axons of interest to be found in the large series of sections without any label being used to highlight its whereabouts. Later studies of neuronal plasticity in the mammalian adult brain used antibodies against the GFP in a preembedding protocol that revealed the location of the structure of interest in the resin-embedded tissue. This also allows specific neurites to be found in the bewilderingly complex networks of the central nervous system.

Labeling fluorescent cells, so that they can be found later with electron microscopy, however, risks being counterproductive as the label itself may obscure the ultrastructural details. For this reason, the method outlined here has been developed using the near-infrared branding (NIRB) method (Bishop et al., 2011; Maco et al., 2013) to make fiducial marks inside the tissue close to the structures of interest, but not associated with them. Using a 2-photon laser, the fixed tissue, with the fluorescent structures still visible, is branded with easily identifiable marks. These are seen both in the resin-embedded blocks and in the electron microscope and are used to locate the position of the features of interest. This region is then imaged with FIBSEM, making sure that the entire volume indicated by the NIRB marks is captured.

This approach has several advantages to previous correlative methods. Without using any immunolabeling converting the fluorescence into an electron-dense signal ensures that optimal fixation protocols can be used without compromising the quality of the ultrastructure. Using FIBSEM provides an automated imaging method requiring only limited human intervention to obtain large series of uninterrupted images. This block-face scanning approach not only gives aligned images, but as the ion beam is able to remove a few nanometers from the block surface, image stacks with isotropic voxels of only a few nanometers in each dimension are possible (see, e.g., Fig. 16.5). As well as giving a more accurate view of 3D structures, it also provides better opportunities to use computer vision algorithms for automatically segmenting and reconstruction of certain features (Kreshuk et al., 2011; Straehle, Köthe, Knott, & Hamprecht, 2011).

With the continued development of block-face scanning electron microscopy, detailed analysis using manual methods on ever-larger series of images is becoming unfeasible. The advances in computer vision algorithms have made significant strides toward automated methods for recognizing and reconstructing different features, particularly in serial FIBSEM images (Becker, Ali, Knott, & Fua, 2013; Lucchi, Smith, Achanta, Knott, & Fua, 2012; Lucchi, Smith, Achanta, Lepetit, & Fua, 2010). This includes structures such as mitochondria, synapses, and neurites. This chapter describes how, once the correct region has been imaged, axons and dendrites can be reconstructed automatically from the serial images using the Ilastik software (www.ilastik.org). This task would take a few days of work if attempted by hand. Instead, it is now possible in a few hours to generate a computer model of the structures of interest. This is then ready for further analysis. The final part of this chapter describes how measurements can be made of these models using a set of software tools that run on the open-source 3D imaging platform Blender (www.blender.org). Measuring the sizes of structures like synaptic contacts, dendritic spines, and axonal boutons is an important means of comparing different neurites, to understand how they function. These tools offer an accurate and rapid means to make morphometric measurements in a 3D workspace, with rapid access to the serial images from which the models were derived.

This complete protocol represents a significant advance in the field of correlative light and electron microscopy. As well as using optimal fixation and staining conditions for this type of material, the imaging is automated, as well as the reconstruction process. This is a method that is therefore not only more reliable but faster and more accurate.

16.1 METHOD

16.1.1 FIXATION AND VIBRATOME SECTIONING

The fixation of the living brain is the most crucial part of the entire procedure to ensure high-quality EM images. The fixative is perfused through the circulatory system as soon after opening the chest cavity as possible, with the animal deeply anesthetized throughout. The brain is therefore exposed to fixatives without it experiencing hypoxic conditions for more than 30 s. Any delay in this part will result in alterations to the ultrastructure (Tao-Cheng, Gallant, Brightman, Dosemeci, & Reese, 2007). The fixation conditions are considered very standard and have changed little from the original EM fixation studies in the 1960s (Schultz & Karlsson, 1965). The process is complete after 2 h, and therefore, the brain can be removed and vibratome sectioned shortly after fixation.

Vibratome sections need to be cut parallel to the imaging plane of the 2-photon microscope to ensure that the orientation of the neurites in the fixed material is similar to those seen in the live brain. If the orientation is too dissimilar, recognizing the same neurites in the final EM image stack is too difficult.

The protocol begins once the live 2-photon imaging has been completed with the animal deeply anesthetized, nonreflexive, and ready to be perfusion fixed.

Animal perfusion fixation

1. At the end of the 2-photon *in vivo* imaging session, take a low-magnification image of the blood vessel vasculature through the cranial window (Fig. 16.1A) to ensure that the cell of interest can be located relative the blood vessel pattern.
2. Pin the mouse to the dissection board, exposing its underside, and then, without delay, cut into the chest cavity to reveal the heart and lungs using surgical scissors.
3. Push the blunt perfusion needle (1″ Straight Cannula Blunt End Tip, 15 gauge; Fisnar) into the left ventricle, taking care not to damage the walls of the heart. Fix the needle in place with a heart clamp.
4. Perfuse the animal with 10 ml of isotonic PBS, with a perfusion pump, at rate of 12 ml/min followed immediately with fixative solution. To avoid stopping and restarting the pump to change solutions, backfill the tubing with PBS, after the entire tubing is initially filled with fixative.
5. Perfuse the animal with 300 ml of 2.5% glutaraldehyde and 2% paraformaldehyde in 0.1 M phosphate buffer (pH 7.4). The perfusion takes

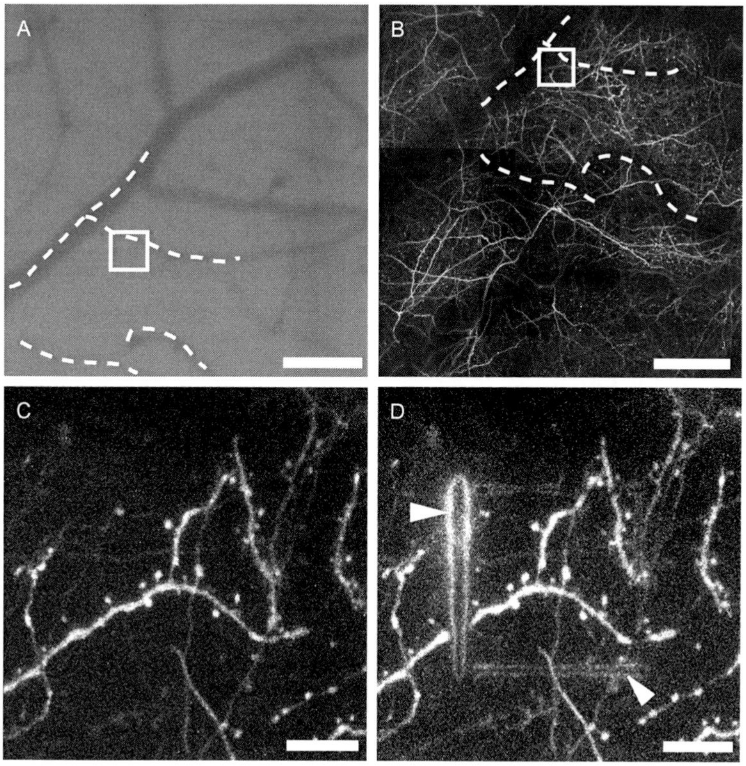

FIGURE 16.1

In vivo imaging and near-infrared branding (NIRB) of the fiducial laser marks. (A) Blood vasculature on the surface of the brain, seen through the cranial window. The pattern of the blood vessels (white dotted lines) is seen in 2-photon microscope (B) and is used to locate the region of interest (white square). (C) High-magnification image of the region of interest with GFP-labeled neurites indicated by white square in (A) and (B). (D) 2-Photon laser was used to burn the fiducial laser marks (white arrowheads) around the structure of interest on the fixed vibratome section containing the region of interest previously imaged *in vivo*. Scale bar in (A) and (B) is 100 μm, and in (C) and (D) is 10 μm.

Modified from Maco et al. (2013); reproduced with permission.

approximately 25 min, and once completed, the animal is then left for a further 90 min at room temperature, and sealed inside a plastic bag.

6. Dissect the brain from the skull using thin, blunt forceps, taking care not to damage the brain tissue. Once the brain is removed immerse in PBS.

Vibratome sectioning

7. Sink the brain in molten agarose (5%, in PBS) that has been cooled to 30 °C, and manipulate its position until the region of the cranial window is approximately

positioned uppermost. Then place in the fridge for 10 min or until the agarose sets.

8. Trim the agarose block, containing the brain, so that the area of cortex imaged through the cranial window is parallel to the upper surface.
9. Glue the agarose block to the vibratome support using cyanoacrylate glue.
10. Vibratome section 60-μm thick slices (Leica VT1200S, Leica Microsystems) from the region of the cranial window making sure that the sections are cut in the same plane as the focal plane as the 2-photon microscope. Cut sections picked up with a paint brush and place sequentially in a 24-well culture plate.

16.1.2 LASER FIDUCIAL MARKS AROUND REGION OF INTEREST (NIRB TECHNIQUE)

As the protocol relies on making fiducial marks in the fixed tissue, it is crucial that there is enough visible fluorescent signal after using this type of fixative with a high concentration of glutaraldehyde (2.5%). Without any label to see the presence of the fluorescent signal in the electron microscope, the neurites of interest need to be pinpointed together with fiducial marks. This gives their relative position allowing them to be found in the serial images.

The number of NIRB marks made depends on how many features are already present in the slice. If there is a clear pattern of blood vessels, for example, it would only be necessary to make a single square around the neurites. However, if natural landmarks are less obvious, it may be necessary to include larger burn marks around the first. These help only in finding the small set of marks in the resin-embedding section. It may also be useful to make NIRBs that form an asymmetric shape which will indicate from which side the section should be imaged and attached to the block resin slab.

16.1.2.1 Method

Localization of neurites of interest on the fixed vibratome section

1. Using a dissecting microscope (Leica M205C), capture images at approximately 12 × zoom to visualize the blood vessels. These are most prominent in the first sections cut from the surface of the brain.
2. Overlay each image, into a stack, and align them (e.g., Photoshop, ImageJ, or GIMP). This stack will reveal the complete pattern of blood vessels on the brain's surface.
3. Match the image of the blood vessels from the fixed sections with the image of the blood vessel pattern taken of the live brain through the cranial window. This will reveal in which section to find the fluorescent structures of interest.
4. Place the identified section on a glass microscope slide, with adequate PBS to ensure that it does not dry. Place a coverslip on top and seal with nail varnish.

Making the NIRB marks

5. Identify the neurites of interest (Fig. 16.1B and C) using the 2-photon microscope. Then set the laser to a wavelength of 850 nm, with a power of

300 mW at the back focal plane of the objective. Here, a 2-photon laser scanning microscope equipped with a pulsed titanium sapphire laser (Chameleon Ultra II, Coherent) was used.

6. In the line scan mode, with a speed of approximately 2 ms/line, burn stripes in the fixed tissue around the region of interest. Between 500 and 2000 sweeps were used until the marks became visible from the autofluorescence generated from the line edges (Fig. 16.1D). The more sweeps used, the wider the burnt stripe, and the further apart the two fluorescent edges. Make sure that the photomultipliers are switched off for this marking phase.
7. NIRB marks are made around the neurites of interest in the form of a square or rectangle approximately 15–24 μm along one side (Fig. 16.1D). These are also placed approximately 3 μm above the neurites of interest in the *Z*-axis.
8. Image this region, with the NIRB marks, using the 2-photon laser, and then again using transmitted light, to capture its position in relation to other features, such as blood vessels and cell bodies.

16.1.3 STAINING AND RESIN EMBEDDING

The staining and resin embedding part of this protocol uses a standard method with conventional reagents. This approach has been used for years by many laboratories. The most important consideration is to stain the membranes as fully as possible to give good contrast with FIBSEM, while being careful not to overstain the tissue that would make it difficult to view the laser marks in the embedded section using transmitted light microscopy. Therefore, the timings are important. Although thinner sections could be considered as a solution, these often do not remain flat during the dehydration process and end up with an undulating appearance in the final hardened resin. This gives regions of interest that are difficult to mount horizontally on the blank resin slab.

16.1.3.1 Method

Staining, dehydration, and resin embedding of the fixed sections

1. Wash the section 3 × for 5 min each change in 0.1 M sodium cacodylate buffer (pH 7.4), in a 20 ml glass scintillation vial mixing continuously on a rotating mixer.
2. Stain in 1% osmium tetroxide and 1.5% ferrocyanide in 0.1 M sodium cacodylate buffer (pH 7.4) for 40 min.
3. Stain with 1% osmium tetroxide in 0.1 M sodium cacodylate buffer (pH 7.4) for 40 min.
4. Rinse the sections in ddH_2O, 2 × for 5 min each rinse.
5. Stain section with 1% uranyl acetate in ddH_2O for 40 min.
6. Rinse section in ddH_2O, 5 min.
7. Dehydrate in increasing concentrations of ethanol: 2 × in 50%, 1 × in 70%, 1 × in 90%, 1 × in 95%, and 2 × in 100% ethanol, 3 min each change.

8. Infiltrate the sections with 1:1 mixture of Durcupan resin and 100% ethanol for 30 min.
9. Infiltrate the section for 2 × 30 min with fresh resin each change.
10. Exchange for fresh resin and agitate for 4 h.
11. Arrange the section onto treated glass slide (coated with mold separating agent), add more resin and overlay a second glass slide to sandwich the section between.
12. Polymerize the resin and section in 65 °C oven for 24 h and then separate the glass slides from the resin using a razor blade.

16.1.4 PREPARATION OF THE BLOCK FACE AND MOUNTING THE BLOCK FOR FIBSEM

The goal of the block preparation process is to have a block face in the microscope on which the observer can pinpoint the exact position of the NIRB marks and without having to undertake long periods of milling before the final image series is collected. This requires trimming with glass knives mounted in the ultramicrotome so that the NIRB marks lie close to the edge of the resin block and within ~4 μm of its surface (Fig. 16.3F).

To help guide the trimming process, a laser-dissecting microscope can also be used to etch the block surface and mark the position of the NIRB marks within the block (Fig. 16.2C). These surface marks are visible with the stereomicroscope and also in the ultramicrotome. This means that the block does not need to be removed from its mounting and imaged under a transmitted light microscope to check each time the position of the required region.

To locate the exact region to be imaged with FIBSEM, a small indentation is also made each side of the final imaging face (Fig. 16.3C), as well as score marks running across the top of the block (Fig. 16.3D). The position of these marks are also carefully measured, in relation to the NIRB marks, as they will be the only indicators with which to place the final imaging window for the collection of the FIBSEM image stack.

16.1.4.1 Method

1. Image the resin-embedded section in the light microscope to pinpoint the position of the NIRB marks (Fig. 16.2B).
2. Etch the shape of the NIRB onto the surface of the resin section, using the pulsed laser of the laser dissection microscope (Fig. 16.2C). These marks serve to indicate the position of the NIRB marks inside the resin block.
3. Cut a 4 × 20 mm strip of blank resin, with thickness of 1 mm, using a jeweler's saw, and rinse with water and then dry, to remove all the dust particles.
4. Glue the section of interest, containing the NIRB mark, to the resin slab with a small drop of Durcupan resin and polymerize in the oven. The drop of resin should be small so that it does not raise the section too much from the surface of the slab (Figs. 16.2D and 16.3A).

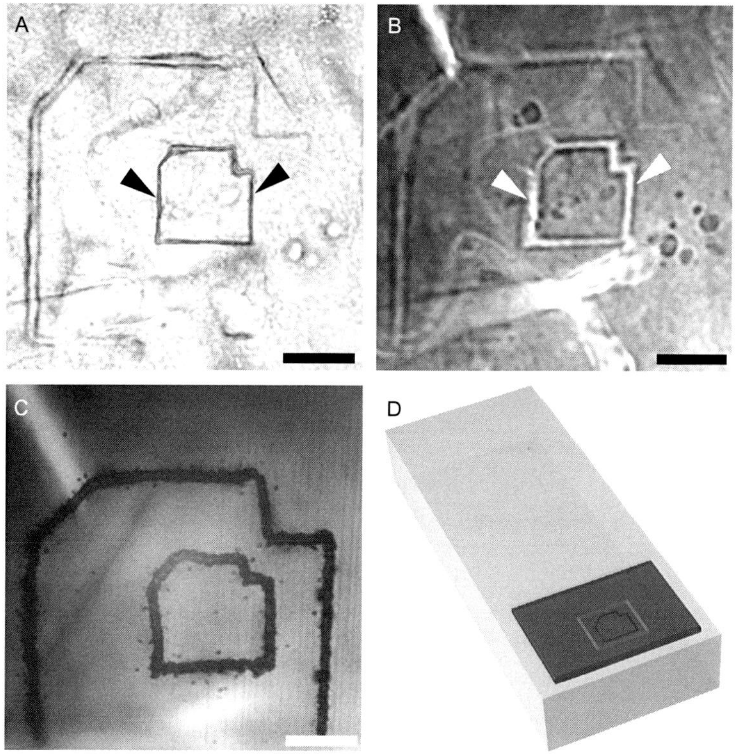

FIGURE 16.2

Tissue preparation for the FIBSEM. (A) Fiducial laser marks can be seen inside the fixed tissue section (black arrowheads) as well as after staining and embedded in resin (B, white arrowheads). (C) Using a laser-dissecting microscope, the original NIRB marks were replicated on to the top surface of resin-embedded sample to help indicate the region of interest during the trimming process. (D) The piece of resin, with the NIRB marked region of interest, was cut from the section and mounted onto a flat slab of empty resin providing a thicker support for easy manipulation during the trimming process. Note the asymmetric shape of the 2-photon laser marks used to distinguish the correct orientation, as well as the larger laser marks, used to facilitate the search and identification of the region of interest on the resin-embedded section. Scale bar: 20 μm.

5. Clamp the block in the ultramicrotome so that the embedded section with NIRB mark is held vertically.
6. Trim away the resin, using a glass knife, so that the region of interest marked by the NIRB marks lies less than 10 μm from this front face (Fig. 16.3B). Use the etched surface of the slab to guide the trimming process, and avoid too much resin being removed.
7. Rotate the block so that the NIRB region is lying horizontally (Fig. 16.3C) and trim an indentation of <50 μm to the right and left of the region of interest,

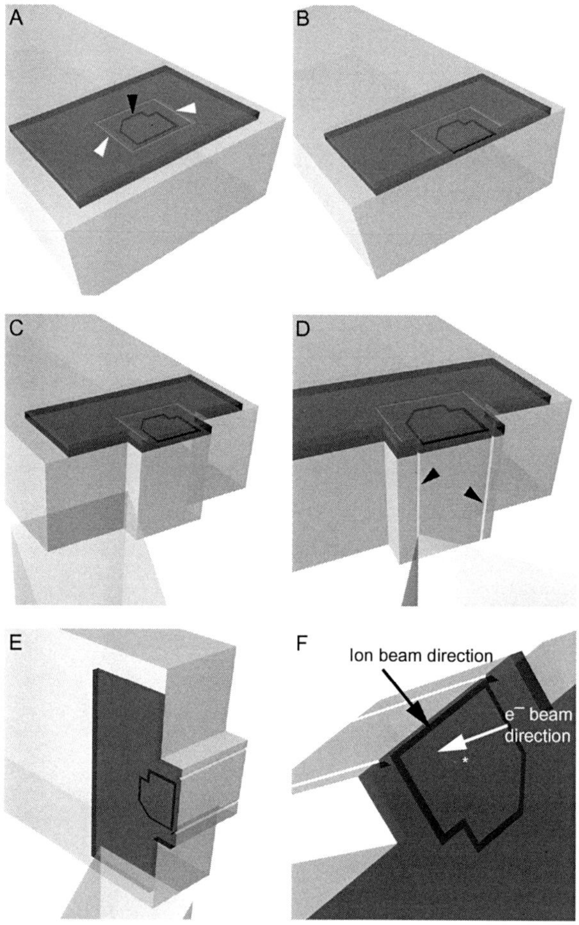

FIGURE 16.3

Trimming process and preparation of the block face for the FIBSEM. (A) Piece of resin-embedded section with 2-photon laser marks (black arrowhead) replicated on the top surface (white arrowheads) by the laser dissection microscope and mounted on the slab of empty resin. (B) The slab is then mounted on the ultramicrotome, and using the glass knife, one side is trimmed away until the NIRB marks are exposed at the edge. (C) To locate the region of interest inside the FIBSEM, indentations are made either side of the laser marks and (D) two score marks are made down the front edge (black arrowheads) serving as additional fiducial marks. These are easily seen in the electron microscope. (E) The block is then orientated to allow any excess resin above the NIRB marks to be removed. (F) The block is then mounted on a 45° inclined aluminum SEM stub, and coated with a protective layer of gold (approximately 30 nm thick). Inside the FIBSEM, the focused ion beam is oriented parallel to the block face (black arrow) and about 54° relative to the electron beam (white arrow).

about 50 μm from the NIRB marks (Fig. 16.3C). Additionally, score marks can be made down the face of the block marking the position of the NIRB mark in the block face. This is achieved using the edge of the glass knife held at an angle of 45° (Fig. 16.3D).

8. Hold the resin block in the jaws of the ultramicrotome holder so that the surface of the block, parallel to the NIRB mark, is also parallel to the edge of the glass knife. Trim away the resin from above the NIRB marks (Fig. 16.3E). Ensure that only approximately 5 μm of resin remains above the NIRB marks.

Mounting the block to SEM stub and gold coating

9. Photograph this final block, using the light microscope, to show the position of the NIRB mark inside the resin block relative to the block's edges, the indentation made on the right-hand side of the block, and the score marks on the front edge.
10. Match this final block image with the 2-photon image of the NIRB marks that also includes the structures of interest. These two images are used to precisely position the imaging face in the FIBSEM.
11. Glue the final trimmed block to the scanning electron microscopy (SEM) stub, on the included surface, using carbon conductive cement.
12. Place the stub in the sputter coater and deposit a 30 nm layer of gold over the surface.

16.1.5 FIBSEM IMAGING

FIBSEM images are taken from the resin-embedded tissue, at the same angle as the 2-photon images were collected in the live brain. This means that the block face is milled with the ion beam at the same plane as the focus plane of the 2-photon microscope (Fig. 16.3F). This results in imaging of the neurites of interest at the same orientation as they were imaged live.

The initial work in the FIB microscope is to roughly mill the block face according to the position indicated by the score marks, described above. Once the underlying tissue is exposed milling continues until other fiducial marks are seen, such as the NIRB marks. Once this has been achieved the block face can then be finally polished and a protective layer is deposited on the top surface of the block. This layer of platinum or carbon ensures a consistent milling without streaking and the curtaining effect.

Unfortunately, at this point it is not advised to immediately start capturing the final image series. During these milling steps, to prepare the final imaging window, the block has been heated and will drift while a thermal equilibrium is reached. For this reason, it is advisable to switch off the electron beam, close the valve of the ion beam, and leave the microscope for at least 2 h. During this time, a thermal and mechanical stability will have been reached and less drift of the block will occur.

During the imaging phase, accurate focus and consistent milling thickness is paramount. Although the thickness is given by the initial settings of the ion beam, in reality this can be different and needs to be calculated *post hoc*, using the cylindrical diameters method (Fiala & Harris, 2001).

16.1.5.1 Method

1. Load the sample into the FIBSEM microscope.
2. Position the ion beam so that it is parallel to the block face. As the face is already inclined at 45°, the SEM stub needs only to be tilted by around 10°, using the microscope stage to ensure the ion beam is parallel to this face. This imaging face is scanned with the electron beam during image acquisition.
3. Locate the position of the NIRB marks along the imaging face using the images acquired during the block preparation.
4. Mill this region of the face, about 60 μm wide, with an ion beam current of approximately 6 nA. The region of interest should be positioned in the center. Continue milling into the block until some underlaying features in the tissue appear when imaging the face in the back-scattered imaging mode. These features, like blood vessels, or cells bodies, or even the NIRB marks themselves are used to calculate the exact position of serial milling and image acquisition.
5. Deposit a protective layer of platinum (or carbon, 1-μm thick) onto the top surface of the block, above the region of interest. This patch of deposited material is approximately 30 μm wide and 10 μm deep (Fig. 16.4C).
6. Finely mill the imaging face with a weaker ion beam of current 0.7–1.5 nA. This creates a more vertical and smooth face.
7. Choose the exact region for sequential milling and imaging (approximately 15 × 10 μm) to capture the neurites of interest (Fig. 16.4).
8. Set the ion beam current to 700 pA with a voltage of 30 kV and set the width of the milling window to three times wider than the imaged face. This ensures that no redeposition occurs on the face to be imaged.
9. With an electron beam current of 800 pA, voltage of 1.5 kV, and grid tension on the ESB detector (Zeiss microscopes only) of 1.3 kV start imaging the block face.
10. Acquire images with pixel size of 4–5 nm per pixel and dwell times of around 10 μs. Mill the face with a thickness of around 5–10 nm each slice.
11. Use microscope dynamic focus, image tracking, and tilt correct functions to keep the images focused while the block face retreats during sequential milling, and in focus down the entire face of the tilted sample.

16.1.6 3D RECONSTRUCTION AND ANALYSIS

The analysis part begins with finding the neurites in the serial images. This can be achieved simply by overlaying the first EM image onto the fluorescent NIRBs image that also includes the neurites of interest. This will show the location, in x and y, at which the different neurites will be seen in the EM image stack.

The 3D models of the neurites of interest are generated from the image stack with the Ilastik software (Ilastik.org). This uses advanced watershed algorithms and allows the user to carry out segmentations of selected features, and importantly, to rapidly make corrections to the generated 3D models when mistakes are made (Ilastik.org; Fig. 16.5A and B). The small spacing in the Z direction optimizes the accuracy of the reconstruction and reduces the number of errors. Here, the protocol explains

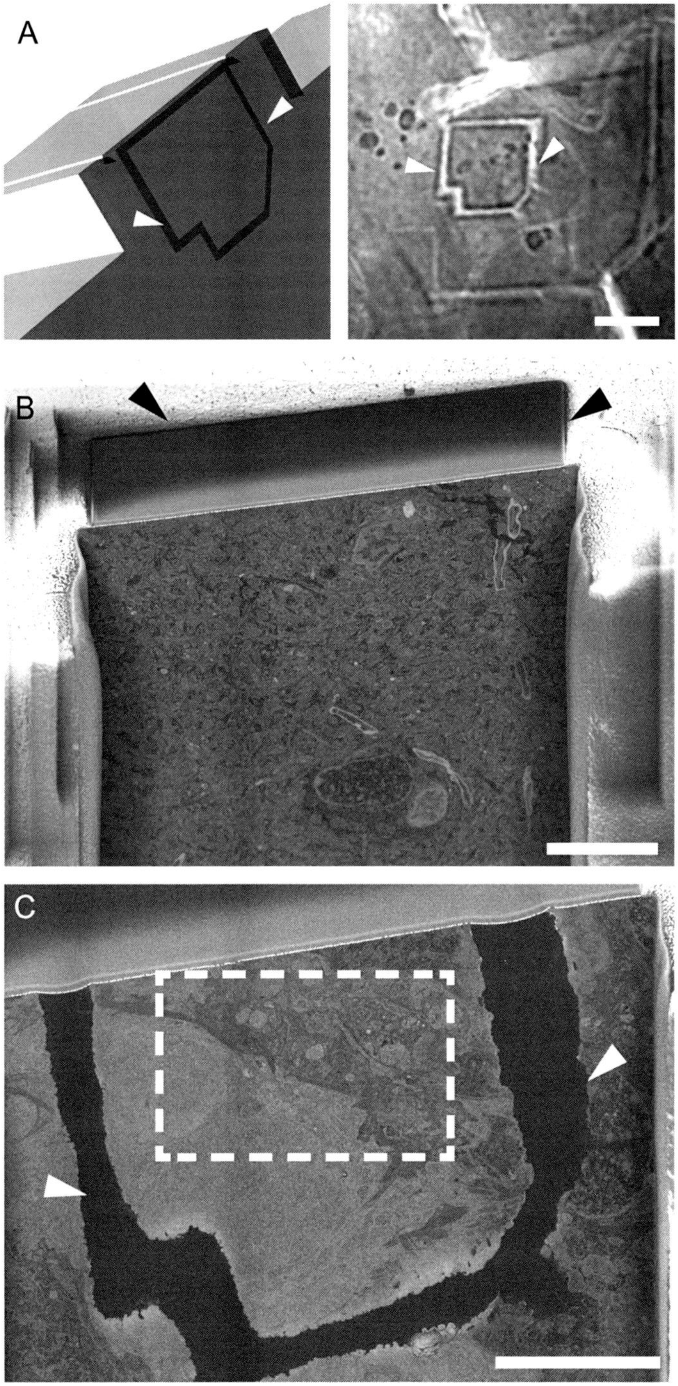

FIGURE 16.4

FIBSEM imaging. (A) A model representing the orientation of the block inside the FIBSEM (left) and resin-embedded tissue (right) with visible NIRB laser marks (white arrowheads). (B) The block face showing the edge of the block and the trench milled out by the ion beam imaged by back-scattered electrons showing the embedded tissue in the block. The top face of the block is coated with a layer of carbon (arrowheads). (C) The milled face covers the entire region of the NIRB marks (arrowheads) and the images were collected from a smaller area contained within (indicated with the dotted square). Scale bars: in (A) is 20 μm, and in (B) and (C) is 10 μm.

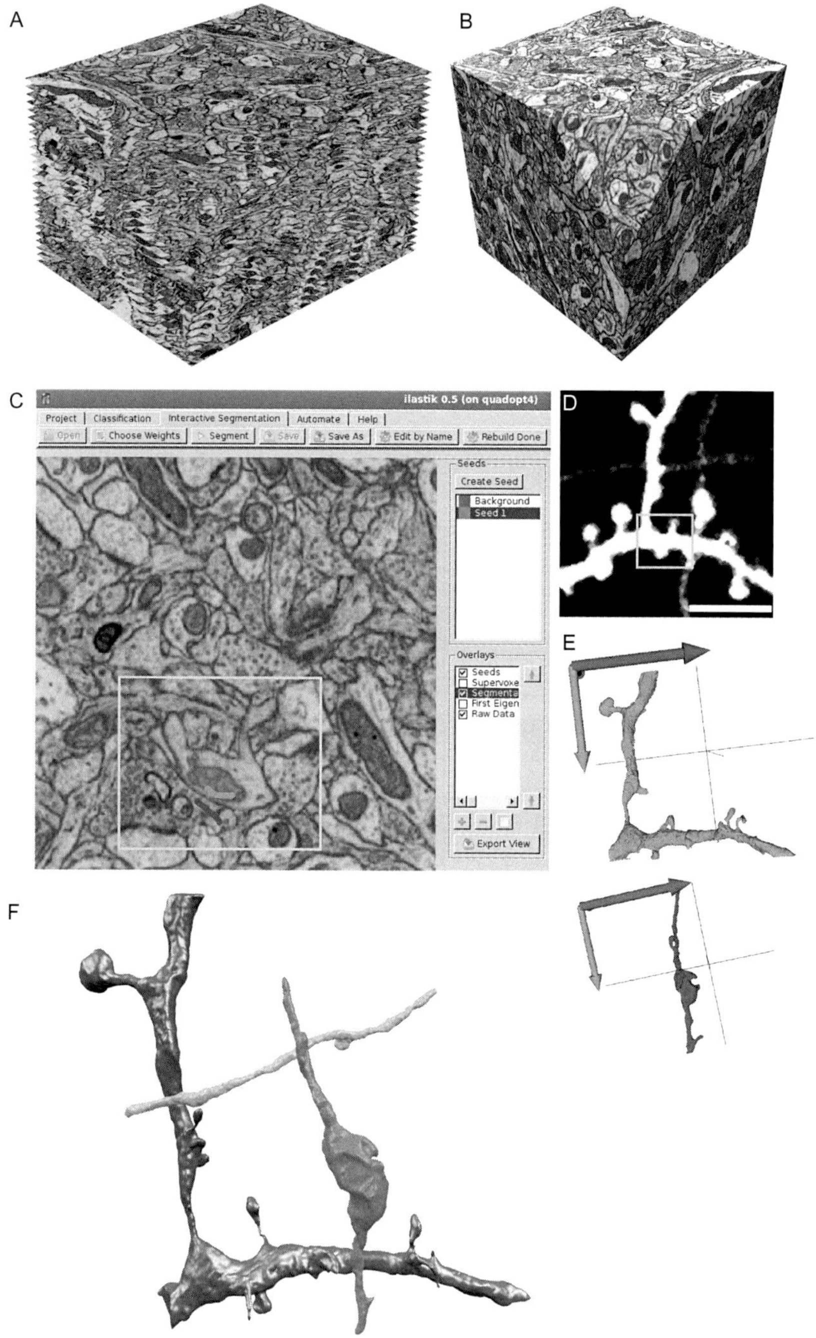

FIGURE 16.5

Segmentation and 3D reconstruction. FIBSEM imaging results in aligned serial images (A) producing the image volume with near-isotropic voxel properties (B) through which the how seeds are scribbled into the neurites of interest on a single image from any orthogonal plane, with background seeds added to indicate the regions for exclusion. The program grows these seeds until they fill the entire volume of the selected neurite and a reconstruction is then produced within a few seconds. This can be refined and corrected by adding additional seeds.

The final 3D model is exported in a wavefront .obj file format that can be imported into the Blender modeling software (www.blender.org). A set of tools have recently been constructed for this software, named NeuroMorph, that enable the measurement and analysis on the models in the 3D space (http://cvlab.epfl.ch/NeuroMorph; http://wiki.blender.org/index.php/Extensions:2.6/Py/Scripts/Neuro_tool). The toolset is composed of three tools or add-ons: Import, Measurement tools, and Image Stack Interactions. The Import tool opens the models in the Blender software with the correct size, calibrated to known units (microns). The Measurement tool allows the models to be analyzed and such parameters as surface area, volume, and length to be made of any part. The measurements are stored together with each object, and when exported appear in a single-character separated values (CSV) text file (.txt), hierarchically organized. Each value appears as a child of the original object that can be easy ordered in a spreadsheet program.

Finally, the Image Stack Interactions tool allows the models to be viewed in 3D, but also with the original EM images. These are placed at the correct position relative to the model. The user can scroll through the images and all images are shown at their current position in the *z* depth of the volume. Additionally, the image plane can also be used to select other objects that may have been reconstructed, but are not visible. This allows the user to search for other objects, using the EM images are maps.

16.1.6.1 Method

Alignment of the serial images

1. Using open-source free FIJI software package (http://fiji.sc/wiki/index.php/Fiji), combine individual sequential images in tiff image format into a single stack by selecting "File"—"Import"—"Image Sequence" command.
2. Select image slice in the middle of the stack as reference for stack alignment by selecting "Plugins"—"Registration"—"StackReg" command using the transformation option "translation only."

FIGURE 16.5—Cont'd features of interest can be visualized using arbitrary imaging planes. (C) The Ilastik program was used to interactively segment the structures of interest using seeded watershed segmentation. An image from the stack shows green marks made by the user to indicate the structures of interest (object) and red marks for the background. The yellow rectangle indicates the part of the segmented structure of GFP-labeled dendrite shown on 2-photon image (D). (E) After the segmentation process, the 3D models of the reconstructed dendrite (green) and axon (blue) are displayed inside the Ilastik program. (F) The final model of the GFP-labeled neurites is rendered using the Blender 3D modeling software (in gray is the dendrite; axons are in green and orange). Scale bar in (D) is 5 μm. (See the color plate.)

Modified from Maco et al. (2013); reproduced with permission

3. If necessary, crop the aligned stack by using Rectangular selection tool and "Image"—"Crop" command.
4. Invert image contrast by selecting "Edit"—"Invert" command and adjust image contrast by selecting "Image"—"Adjust"—"Brightness/Contrast" command.
5. Save processed stack as sequence of individual 8 bit, gray-scale images by selecting "File"—"Save As"—"Image Sequence" command.
6. Determine the exact section thickness to measure the cylindrical object (e.g., mitochondria) diameter by dividing the number of slices it spans using the cylindrical diameters method (Fiala & Harris, 2001).
7. Identify the exact position of the neurites of interest within the image volume by overlaying and aligning their 2-photon image including the NIRB marks with the known position of the NIRB marks on the block face.

Semi-automated segmentation with Ilastik

8. Segment the structure of interest and create a 3D model using the seeded watershed segmentation algorithm within the Ilastik software (version 0.5.12; http://ilastik.org). Create the new project by selecting "Project" tab—"New." In "Project" window, select "Load Stack" and choose the folder containing the sequential images of the aligned stack. The entire stack is displayed in orthogonal view with the possibility to scroll through the stack in any of the x–y, x–z, or y–z planes.
9. To specify a set of features required to identify the membrane boundaries select the "Classification" tab—"Select Features" button. In the opened "Spatial Features" window, select "Texture" (includes the Eigenvalues of the Hessian matrix as a membrane boundary indicator). Set the scale of the features to "Medium" (representing the sigma value of 1.0). The selected set of features for the entire image stack is calculated.
10. Switch to "Seeded Watershed" tab and import previously calculated Eigenvalues of Hessian matrix of Gaussian Sigma Channel 2 as chosen "Input weights." In the "Select Border Indicator" window, specify whether the boundaries in "Calculated Weights" are dark or bright. The supervoxels graph used for watershed segmentation for the entire image stack is calculated.
11. Select "Seed 1" (as default in green color) in "Seeds" preview window, adjust brush size ("M" key increase and "N" key decrease) and draw a line inside the structure of interest. Select the "Background" (as default in red color) and mark the outside of the selected structure or background seed (Fig. 16.5C). Note that the lines representing the seeds can be placed from any displayed planes of the orthogonal view. Additional seeds (in different colors) can be added using the "Create Seed" button or deleted by right-clicking on the selected seed in the list on the "Seeds" preview window.
12. Select the "Segment" button to start the segmentation. The selected structure is colorized as "Segmentation" overlay in the "Overlays" preview window. The segmentation can be iteratively refined by placing more foreground and background seeds, or adding seeds for other objects. The "Bias" dialogue field

allows to tune the growth of the background seeds more (values around 0.8) or less (values around 0.97) aggressively than the default.

13. Display the rendered 3D model of the segmented structure by right-clicking on the "Segmentation" overlay and select "Display in 3D" (Fig. 16.5E).
14. Export the final model as a wavefront OBJ mesh file by selecting the "Export" button to render high-quality images, or movies, using the Blender 3D modeling software (version 2.69; Blender Foundation; http://www.blender.org) (Fig. 16.5F).

3D analysis using NeuroMorph tools in Blender

15. Install the NeuroMorph tools into the Blender by selecting the "File"—"User Preferences." Under the "Addons" tab, select "Install from File" button and locate where the addon files are stored. Select newly installed addon by checking the right-hand box. Click "Save User Preferences" button and close the "User Preferences" window. Repeat this procedure for all addon files.

Import tool

16. Under the "Scene" context panel, calibrate the scene by checking left-hand box "Microns Per Pixel" and define the pixel size in microns (Fig. 16.6A).
17. Check the left-hand box "Use Remesh" to apply the octree remesh modifier to imported object. This modifier reduces the number of vertices and memory allocation, but still preserves the object shape. This can be tuned in "Remesh resolution" field (higher number generates denser meshes and increases the memory allocation).
18. To start import, click the "Import" button and locate the folder with stored OBJ files. User can select multiple files which are then loaded and displayed in Blender.
19. Imported objects are listed in the "Outliner" context panel. Select the object and in the "Properties" context panel select "Modifier" tab. Further "Remesh" modifier optimization can be tuned. By pressing "Apply" button the modifier is applied.

Surface area/volume measurement

20. Define the type of structure to be measured in the "Measurement Tools" under "Object Name" select the "Object type" pop-up menu with predefined categories: spine, bouton, synapse, dendrite, axon, and other to define any new category (Fig. 16.6B and C).
21. To measure the surface area and/or volume of the chosen structure, with the right mouse button (RMB) select the object and switch to "Edit" mode using the "Tab" key.
22. Activate "Vertex select" button and deactivate "Limit selection to visible" button (both located at the bottom panel of the 3D view window).

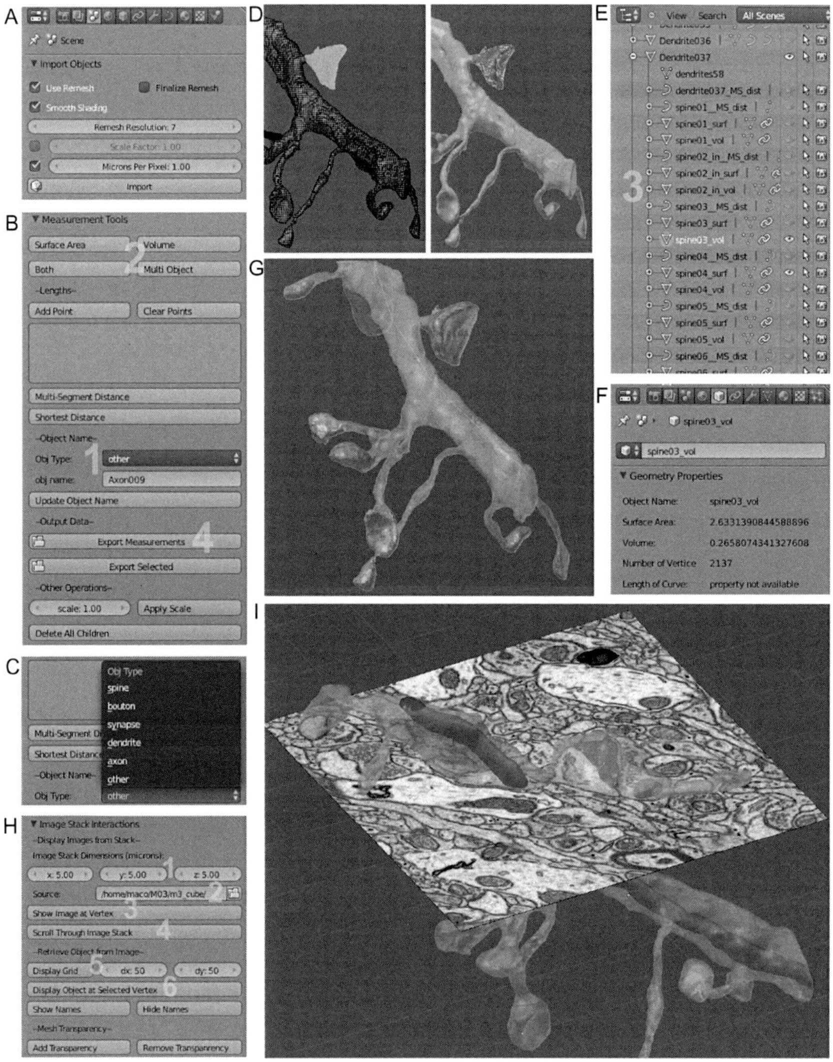

FIGURE 16.6

NeuroMorph import, Measurement and Image Stack Interaction toolset for Blender. (A) The mesh files created by Ilastik are imported into Blender using the Import tool that also calibrates the size of the models into a micron scale. During the import process, an algorithm (remesh modifier) is applied to reduce the number of vertices that comprises the mesh object while also preserving the objects shape. (B) The Measurement tool is able to name the objects being analyzed (here the names of neuronal structures are used: spine, bouton, synapse, dendrite, axon, or other; indicated with yellow 1 (light gray in the print version), and in C). After selecting the vertices of the chosen structure, the type of measurement is selected (yellow 2 (light gray in the print version)). The results can be exported into a text file for further analysis (yellow 4 (light gray in the print version)). (D) A model of dendrite with selected

23. Using lasso selection, by simultaneously pressing "Ctrl" key and the left mouse button (LMB), select vertices of the chosen structure. Alternatively, "Border" selection can be used by pressing "B" key and with LMB select the region. Selected vertices are highlighted in orange (Fig. 16.6D).
24. In the "Measurement Tools" the user can select the type of measurement by activating "Surface Area" or "Volume" buttons separately or the "Both" button to perform both measurements simultaneously (Fig. 16.6B). Note that for each measurement category a new child object is created and listed in the "Outliner" context panel (Fig. 16.6E). For the surface area, an open mesh named "surf" and for the volume a close mesh named "vol" are created. The measurement values are displayed in the "Properties" context panel under the "Object" tab and in the "Geometry Properties" section (Fig. 16.6F).

Length measurement

25. To measure the length, select the object with RMB and switch to "Edit" mode "Tab" key.
26. Activate both "Vertex select" and "Limit selection to visible" buttons (both located at the bottom panel of the 3D view window).
27. Select points (vertices) along the object to be measured, using RMB, adding each point into the list by selecting "Add Point" button (or "Ctrl S" shortcut).
28. Activate "Multi-Segment Distance" button to measure the length from multiple points. The new child object named "MS_dist" is created and listed in the "Outliner" context panel. The length values are displayed in the "Properties"

FIGURE 16.6—Cont'd vertices that define a dendritic spine (highlighted in orange (light gray in the print version), left panel) is measured, creating a new mesh object (3) listed in the Outliner window (E). Other structures (colorized in magenta (gray in the print version)) on this dendrite model, and shown in the Object mode, can be measured (G), to form many mesh objects (shown in the Outliner in (E)). (F) The result of these measurements is displayed in the "Properties" context panel under the "Object" tab and in the "Geometry Properties" section. (H) The Image Stack Interaction tool is able to display the EM image (size is defined in microns; yellow 1 (light gray in the print version), and from the correct directory; yellow 2 (light gray in the print version)), from which the model is derived, at the correct position in the model (displayed in Edit mode after clicking on the button indicated with yellow 1 (light gray in the print version)). User can also scroll through images of the entire stack (yellow 4 (light gray in the print version)). The displayed image can also be used to retrieve the 3D model of any structure (object) by selecting the points density (in pixels) and displaying the grid over the image (5). Then by selecting any grid point(s), the corresponding object(s) is displayed at selected vertex (6). (I) Retrieved objects based on the selection the grid points are displayed: presynaptic bouton (in transparent green (light gray in the print version)), axonal mitochondrion (in green (dark gray in the print version)), synapse (in red (dark gray in the print version)), and dendritic mitochondrion (in magenta (dark gray in the print version)). The initial dendrite is shown in transparent magenta (light gray in the print version).

context panel under the "Object" tab and in the "Geometry Properties" section under Length of Curve. Alternatively, for shorter or relatively straight segments, adding two points (start–end) to the list is sufficient for length measurement by selecting "Shortest distance" button. The new child object named "2p_dist" is created and listed in the "Outliner" panel (Fig. 16.6E).

29. When finished with length measurement, remove the points from the list with "Clear points" button.

Export measurement results

30. Measurement results can be exported into a single CSV text file (.txt) for further analysis with a spreadsheet program (e.g., Excel, Open Office, Numbers). By activating the "Export Measurements" button in the "Measurement Tools" under "Output Data" section, user specifies the destination folder and the file name. All measurement results for all analyzed objects present in the blender file are exported and written to the text file.
31. Alternatively, the measurement results for only selected object(s) can be exported by activating the Export Selected" button.

Image Stack Interaction tool

32. This tool allows the interaction of the reconstructed model with the original EM images (Fig. 16.6H and I). In the "Display Images from Stack" section under "Image Stack Dimensions," specify the reconstructed volume (in microns) size in *x*, *y*, and *z* dimensions in appropriate fields.
33. In the "Source" section, locate and activate folder button (right-hand side) to select the storage folder with the sequential images of the volume stack used for the reconstruction.
34. To overlay the model with the image, select the object with RMB and switch to edit mode pressing "Tab" key.
35. Select any vertex from the mesh and activate "Show image at Vertex" button. The image from the volume stack at the position of selected vertex is placed to the scene (called "Empty" with EM image as a texture). Note: user can selected multiple vertices (RMB with "Shift" key) in which case for each selected point, the image plane is added.
36. To see the EM image through the object, transparency can be changed by selecting the object with RMB and activating the "Add Transparency" button in the "Mesh Transparency" section. In the field "Alpha," the opacity intensity can be adjusted as well as the color can be changed from the "Color Palette" button (next to "Alpha" button). To remove the transparency of the object, select "Remove Transparency" button (Fig. 16.6H).
37. By selecting the image plane (or Empty), the user is able to scroll the image plane up and down throughout the whole volume by activating the "Scroll Through Image Stack" and by using mouse middle button. This shows each

image from the original stack at its correct position relative to the model. Pressing the "Esc" key deactivates this scrolling function.

38. The image plane can be used to display any additional object in the scene based on the current image. Select the image plane (RMB) and define the grid density (distance between grid points in pixels) in x- and y dimensions in the "dx" and "dy" fields located under "Retrieve Object" from Image section.
39. By activating the "Display Grid" button, the mesh grid is overlayed over the image plane. Select any point (RMB) or multiple points (RMB with "Shift" key).
40. To display the object based on grid point selection, activate the "Display Object at Selected Vertex" button.

CONCLUSION

Recent years have seen significant advances in EM imaging methods for capturing volumetric data on a large scale. In particular, the rise of block-face scanning approaches now offers many opportunities for automated serial imaging. These are likely to improve, with more efficient detectors achieving better signal to noise ratios giving faster imaging rates. More automated procedures that ensure consistent image quality for longer periods will improve. All this will allow users to image more samples and allow these microscopes to be available to more users. Correlating light microscopy data with structural information from the electron microscope will, therefore, become increasingly more prevalent.

The protocol presented here circumvents a number of the steps that have made CLEM methods using transmission electron microscopy an exception rather than the norm. In many places, it reduces considerably the need of highly skilled hands to carry out such tasks as cutting, collecting, and imaging many hundreds of serial sections. It also describes how to use a freely available set of software tools, useable on any computer platform, to rebuild automatically, in the computer, the various targeted structures, as well as carry out detailed morphometric analysis.

The computer vision research providing the segmentation and reconstruction tools, described here is, however, dependent on large and varied images, as well as ground truth segmentations to develop and test their algorithms. As more laboratories use these imaging methods, more datasets becomes available and better software will lead to more opportunities to analyze larger image volumes, and reducing the time needed to proofread and correct the results. This will inevitably lead to greater insights into the structure and function of biological systems.

ACKNOWLEDGMENTS

This work was supported by the Swiss National Foundation Synergia project grant CRF II313470/1 (G.W.K.), EU ERC MicroNano Project (P.F.) and project grant 31003A_135631 (A.H.).

REFERENCES

Allegra Mascaro, A. L., Cesare, P., Sacconi, L., Grasselli, G., Mandolesi, G., Maco, B., et al. (2013). In vivo single branch axotomy induces GAP-43-dependent sprouting and synaptic remodeling in cerebellar cortex. *Proceedings of the National Academy of Sciences of the United States of America*, *110*, 10824–10829.

Becker, C., Ali, K., Knott, G., & Fua, P. (2013). Learning context cues for synapse segmentation. *IEEE Transactions on Medical Imaging*, *32*, 1864–1877.

Bishop, D., Nikic, I., Brinkoetter, M., Knecht, S., Potz, S., Kerschensteiner, M., et al. (2011). Near-infrared branding efficiently correlates light and electron microscopy. *Nature Methods*, *8*, 568–570.

Cane, M., Maco, B., Knott, G., & Holtmaat, A. (2014). The relationship between PSD-95 clustering and spine stability in vivo. *The Journal of Neuroscience*, *34*, 2075–2086.

Canty, A. J., Huang, L., Jackson, J. S., Little, G. E., Knott, G., Maco, B., et al. (2013). *In-vivo* single neuron axotomy triggers axon regeneration to restore synaptic density in specific cortical circuits. *Nature Communications*, *4*, 2038.

Fiala, J. C., & Harris, K. M. (2001). Cylindrical diameters method for calibrating section thickness in serial electron microscopy. *Journal of Microscopy*, *202*, 468–472.

Grillo, F. W., Song, S., Teles-Grilo Ruivo, L. M., Huang, L., Gao, G., Knott, G. W., et al. (2013). Increased axonal bouton dynamics in the aging mouse cortex. *Proceedings of the National Academy of Sciences of the United States of America*, *110*, E1514–E1523.

Holtmaat, A., Wilbrecht, L., Knott, G. W., Welker, E., & Svoboda, K. (2006). Experience-dependent and cell-type-specific spine growth in the neocortex. *Nature*, *441*, 979–983.

Knott, G. W., Holtmaat, A., Trachtenberg, J. T., Svoboda, K., & Welker, E. (2009). A protocol for preparing GFP-labeled neurons previously imaged *in vivo* and in slice preparations for light and electron microscopic analysis. *Nature Protocols*, *4*, 1145–1156.

Knott, G. W., Holtmaat, A., Wilbrecht, L., Welker, E., & Svoboda, K. (2006). Spine growth precedes synapse formation in the adult neocortex *in vivo*. *Nature Neuroscience*, *9*, 1117–1124.

Kreshuk, A., Straehle, C. N., Sommer, C., Koethe, U., Cantoni, M., Knott, G., et al. (2011). Automated detection and segmentation of synaptic contacts in nearly isotropic serial electron microscopy images. *PLoS One*, *6*, e24899.

Lucchi, A., Smith, K., Achanta, R., Knott, G., & Fua, P. (2012). Supervoxel-based segmentation of mitochondria in em image stacks with learned shape features. *IEEE Transactions on Medical Imaging*, *31*, 474–486.

Lucchi, A., Smith, K., Achanta, R., Lepetit, V., & Fua, P. (2010). A fully automated approach to segmentation of irregularly shaped cellular structures in EM images. *Medical Image Computing and Computer-Assisted Intervention*, *13*, 463–471, International Conference on Medical Image Computing and Computer-Assisted Intervention.

Maco, B., Holtmaat, A., Cantoni, M., Kreshuk, A., Straehle, C. N., Hamprecht, F. A., et al. (2013). Correlative *in vivo* 2 photon and focused ion beam scanning electron microscopy of cortical neurons. *PLoS One*, *8*, e57405.

Mostany, R., Anstey, J. E., Crump, K. L., Maco, B., Knott, G., & Portera-Cailliau, C. (2013). Altered synaptic dynamics during normal brain aging. *Journal of Neuroscience*, *33*, 4094–4104.

Schultz, R. L., & Karlsson, U. (1965). Fixation ot the central nervous system for electron microscopy by aldehyde perfusion. II. Effect of osmolarity, pH of perfusate, and fixative concentration. *Journal of Ultrastructure Research*, *12*, 187–206.

Straehle, C. N., Köthe, U., Knott, G., & Hamprecht, F. A. (2011). Carving: Scalable interactive segmentation of neural volume electron microscopy images. *Medical Image Computing and Computer-Assisted Intervention*, *14*, 653–660.

Tao-Cheng, J. H., Gallant, P. E., Brightman, M. W., Dosemeci, A., & Reese, T. S. (2007). Structural changes at synapses after delayed perfusion fixation in different regions of the mouse brain. *The Journal of Comparative Neurology*, *501*, 731–740.

Trachtenberg, J. T., Chen, B. E., Knott, G. W., Feng, G., Sanes, J. R., Welker, E., et al. (2002). Long-term *in vivo* imaging of experience-dependent synaptic plasticity in adult cortex. *Nature*, *420*, 788–794.

Zito, K., Parnas, D., Fetter, R. D., Isacoff, E. Y., & Goodman, C. S. (1999). Watching a synapse grow: Noninvasive confocal imaging of synaptic growth in Drosophila. *Neuron*, *22*, 719–729.

CHAPTER

17

Integrated Light and Scanning Electron Microscopy of GFP-Expressing Cells

Christopher J. Peddie*, Nalan Liv[†], Jacob P. Hoogenboom[†], Lucy M. Collinson*

**Electron Microscopy Unit, London Research Institute, Cancer Research UK, London, United Kingdom*

[†]Department of Imaging Physics, Faculty of Applied Sciences, Delft University of Technology, Delft, The Netherlands

CHAPTER OUTLINE

Methods in Cell Biology, Volume 124, ISSN 0091-679X, http://dx.doi.org/10.1016/B978-0-12-801075-4.00017-3

Abstract

Integration of light and electron microscopes provides imaging tools in which fluorescent proteins can be localized to cellular structures with a high level of precision. However, until recently, there were few methods that could deliver specimens with sufficient fluorescent signal and electron contrast for dual imaging without intermediate staining steps. Here, we report protocols that preserve green fluorescent protein (GFP) in whole cells and in ultrathin sections of resin-embedded cells, with membrane contrast for integrated imaging. Critically, GFP is maintained in a stable and active state within the vacuum of an integrated light and scanning electron microscope. For light microscopists, additional structural information gives context to fluorescent protein expression in whole cells, illustrated here by analysis of filopodia and focal adhesions in Madin Darby canine kidney cells expressing GFP-Paxillin. For electron microscopists, GFP highlights the proteins of interest within the architectural space of the cell, illustrated here by localization of the conical lipid diacylglycerol to cellular membranes.

INTRODUCTION

Correlative light and electron microscopy (CLEM) makes use of two separate imaging modalities to link functional information derived from fluorescent marker proteins to their structural localization within cells and tissues. As CLEM has become more widely adopted, there has been a drive to improve the technique by integrating the two imaging modalities, thereby simplifying and expediting the correlative process. Early attempts to integrate a light microscope into an electron microscope date from the 1980s (Wouters & Koerten, 1982; Wouters, Koerten, Bonnet, Daems, & Ploem, 1986) and improved systems have been presented in recent years that are now commercially available (Agronskaia et al., 2008; Nishiyama et al., 2010; Zonnevylle et al., 2013). In this chapter, we focus on the integration of a fluorescence microscope *in situ* in a scanning electron microscope (SEM), which allows observation of the same region of the specimen with both microscopes. Here, the fluorescence microscope operates in an epiconfiguration, using a single objective lens to illuminate the specimen and collect emitted fluorescent signal (Liv et al., 2013; Zonnevylle et al., 2013). The integrated light and scanning electron microscope (ILSEM) can be perceived in two ways.

First, taking the viewpoint of a light microscopist, the ILSEM can be operated as a fluorescence microscope with the added benefit that the distribution patterns of fluorescent markers can be enhanced with details of cell surface structure. The focus here is on inspection of individual cells without or with only moderate additional contrast enhancement for electron imaging. Typical applications in the domain of light microscopy include the study of membrane topology in relation to intracellular processes, cellular adhesion and motility, and cell–cell interactions.

Alternatively, the ILSEM can be viewed as an electron microscope with added fluorescence imaging capability, which complements electron contrast with information from biomolecular labels. In this electron microscopist's perspective, the most pursued in CLEM applications, the focus is on inspection of subcellular ultrastructure in ultrathin sections of heavy metal-stained cells and tissues to reveal details at the nanometer scale. Applications commonly explored in the electron microscopy (EM) domain include the study of intracellular trafficking, organelle biogenesis, and intracellular structure–function assays.

However, in order to fully exploit an integrated imaging system, specimens must contain both fluorescence and electron signals and must be compatible with observation under vacuum. Until recently, preparing specimens for this purpose was a significant challenge. Here, we report preservation of green fluorescent protein (GFP) for *in vacuo* imaging in two different cellular systems. The first uses chemical fixation and air drying of whole Madin Darby canine kidney (MDCK) cells stably expressing GFP-Paxillin to examine cellular structural variations at focal adhesion complexes. The second uses an in-resin fluorescence (IRF) protocol to prepare HeLa cells expressing a GFP-C1a-C1b construct to examine the subcellular distribution of the fusogenic lipid diacylglycerol (DAG; Domart et al., 2012; Peddie et al., 2014).

17.1 RATIONALE

In routine CLEM, fluorescence microscopy is performed on live or fixed specimens using widefield, confocal, or super-resolution light microscopes. The specimen is then prepared for imaging in the electron microscope using a protocol adapted to the biological question under investigation. To image biomarkers on the cell surface, whole cells can be chemically fixed, dehydrated, and observed directly in a SEM. To image biomarkers within cells and tissues, the specimen is chemically fixed, stained with heavy metals, dehydrated, resin-embedded, and sectioned for transmission electron microscopy (TEM). However, it is important to note that processing specimens in this way will quench organic fluorophores. In addition, a number of artifacts can be introduced that complicate registration between light and electron images. Cells shrink in response to dehydration; sectioning reduces the physical slice thickness of the electron image compared to the virtual slice thickness of the fluorescence (confocal) image by at least an order of magnitude; and the collected sections suffer from compression. Together, these factors significantly reduce the accuracy with which proteins can be localized to structure in a nonlinear fashion.

Integrated microscopes facilitate CLEM by allowing both microscopy techniques to be carried out directly on a single prepared specimen, with no further manipulation necessary when switching between imaging modalities. As a result, the risk of introducing artifacts during transfer between microscopes is substantially reduced. Furthermore, artifacts associated with specimen preparation are present in both light and electron images and so have little effect on overlay accuracy. Several protocols suitable for integrated microscopy, where fluorescent signals are preserved while simultaneously retaining electron contrast for EM, have been reported in recent years. These can be divided into several categories.

The first category covers fluorescent dyes that are stable and active within the vacuum of an electron microscope. Fluorescent markers such as DAPI and Alexa

Fluor have been shown to be compatible with integrated light and EM methods (Faas et al., 2013; Karreman et al., 2009, 2012; Liv et al., 2013; Zonnevylle et al., 2013). Usually targeted via conjugation to appropriate ligands or antibodies, these markers are suitable for use in live cells as endocytic tracers, on fixed cells for surface labeling, or on ultrathin resin or Tokuyasu sections to enable subcellular localization. However, as they have no native electron contrast, correlation is achieved solely by overlay of light and electron images.

The second category covers inorganic dual-contrast probes that can be visualized in both light and electron microscopes. Because these probes can be imaged directly, they increase localization accuracy in the electron microscope rather than relying solely on positional correlation with the fluorescent signal. Colloidal quantum dots (Chan & Nie, 1998; Deerinck, Giepmans, Smarr, Martone, & Ellisman, 2007; Giepmans, Deerinck, Smarr, Jones, & Ellisman, 2005) and fluoronanogold particles (Powell et al., 1997; Robinson & Takizawa, 2009; Takizawa & Robinson, 2000; Takizawa, Suzuki, & Robinson, 1998) are fluorescent and electron dense, but must be conjugated to ligands or antibodies for targeting. Contrast enhancement may also be required in order to visualize the probe above the background cellular structure in the electron image.

The third category covers genetically encoded dual-contrast probes, which combine increased localization accuracy in the electron microscope with the major advantage of probe expression within living cells and tissues. Ligands or antibodies are therefore not required for targeting, but the genes of interest must be cloned into appropriate vectors and comprehensively characterized. Chemical conversion is required for horseradish peroxidase (Connolly, Futter, Gibson, Hopkins, & Cutler, 1994) and APEX (Martell et al., 2012), while photoconversion is required for miniSOG (Shu et al., 2011). The need for an additional conversion step means that these probes are incompatible with cryo-preparation techniques like high-pressure freezing (HPF) and plunge freezing, which deliver improved structural preservation of cells and tissues.

The fourth category covers genetically encoded fluorescent markers, frequently derived from the GFP family (Shaner, Steinbach, & Tsien, 2005). Again, expression of the probes in living cells and tissues is a major advantage, and ligands or antibodies are unnecessary for targeting. The main disadvantage of this approach is the masking or deactivation of the organic fluorophores during specimen preparation, as discussed earlier. However, recent descriptions of methods to preserve organic fluorophore activity throughout all EM specimen preparation steps neatly avoid this problem. These protocols have been demonstrated on a range of biological systems including viruses and yeast (Faas et al., 2013; Kukulski, Schorb, Kaksonen, & Briggs, 2012; Kukulski et al., 2011), *Caenorhabditis elegans* (Sims & Hardin, 2007; Watanabe et al., 2011), zebrafish (Luby-Phelps, Ning, Fogerty, & Besharse, 2003; Nixon et al., 2009), *Drosophila* embryos (Fabrowski et al., 2013), and plants (Bell, Mitchell, Paultre, Posch, & Oparka, 2013). In common with the markers described earlier in category one, these genetically encoded probes have no significant native electron contrast, and correlation is achieved solely by overlay of light and electron images. Thus, the more accurate the overlay, the more precise the correlation of fluorophore to structure.

The choice of probe for integrated light and electron microscopy (ILEM) ultimately depends on the biological model system, the nature of the scientific question under investigation, and the availability of suitable ligands, antibodies, and genetic constructs. In this chapter, we describe our specimen preparation protocol for ILSEM of cells expressing organic fluorophores (category four), which allows scientists to take advantage of the global catalogue of cells and organisms already engineered to express GFP. In particular, our work adds an IRF protocol for cultured cells that preserves GFP in such a way that the fluorophore remains stable and active in the vacuum of the ILSEM (Peddie et al., 2014).

17.2 METHODS

17.2.1 PREPARATION OF WHOLE CELLS EXPRESSING GFP FOR ILSEM

17.2.1.1 Principle

Culturing cells on glass is a common procedure for light microscopy applications. However, because glass is a strong insulating material, unwanted charging artifacts can be a significant problem for SEM imaging. To reduce charging, the specimen itself can be coated with a metal such as platinum to create a conductive path, but this may quench fluorophores at the cell surface and/or obstruct the light path. The use of glass cover slips coated with a thin layer of indium-tin oxide (ITO), which is both conductive and optically transparent, removes the need for metal coating of thin specimens. Layers of ITO with a surface resistivity of 70–100 Ω/sq provide enough conductivity for SEM imaging, and a variety of cell types have been demonstrated to adhere to and grow on these cover slips (Pluk, Stokes, Lich, Wieringa, & Fransen, 2009). Though cells can be cultured directly on ITO-coated cover slips, it is suggested that some form of extracellular matrix is applied to promote cell adherence. Here, we used poly-L-lysine, which does not adversely affect light or electron imaging. Processing procedures commonly used for cell culture and fixation in routine light microscopy can be followed; care must only be taken to avoid fixatives, stains, and complete dehydration that will quench the GFP fluorescence. As proof of principle, we use whole MDCK cells stably expressing GFP-Paxillin. Paxillin is a focal adhesion-associated protein that has a coordinative role in signaling pathways regulating cell shape, motility, and spreading (Schaller, 2001). Using ILSEM, variations in cellular structure at Paxillin-labeled focal adhesion complexes can be directly visualized.

17.2.1.2 Protocol

Cell culture and fixation

1. ITO-coated cover slips were cleaned prior to cell seeding by sonication for 2 min in absolute ethanol, and rinsing with ddH_2O. Where problems arose with adhesion of cells to the ITO substrate, an additional O_2 plasma cleaning step was used (250 W for 15 min) to make the cover slip surface more hydrophilic. After plasma cleaning, it was important to carry out steps 2 and 3 immediately.

2. The cleaned cover slips were placed in culture dishes with the conductive side facing up. Where an adherent matrix was used, the cover slips were incubated in 0.01% poly-L-lysine for 30 min at 37 °C. Care was taken to ensure the conductive ITO coating remained facing up; where necessary, orientation was checked by measuring surface conductivity with a multimeter.
3. All ITO cover slips (with or without adherent matrix) were washed with culture medium prior to cell seeding. The GFP-expressing MDCK cell line was grown in Dulbecco's modified Eagle's medium (DMEM)+penicillin/streptomycin and L-glutamine, with 10% fetal calf serum (FCS). The cells were washed twice with phosphate-buffered saline (PBS), pH 7.4, trypsinized, and seeded directly onto the ITO-coated cover slips. The cells were cultured for 1–2 days in normal medium, before switching to medium containing 1% FCS for the last 6 h.
4. The cells were fixed by adding an equal volume of 4% paraformaldehyde in PBS to the culture medium, before dehydration by passing the cover slips through an ascending series of ethanol (20%, 50%, 70%, 90%, and 100%), incubating for at least 2 min in each concentration, and air drying.

17.2.2 PREPARATION OF RESIN-EMBEDDED CELLS FOR ILSEM

17.2.2.1 Principle

To access the interior of the cell, specimens are routinely embedded in resin and cut into ultrathin sections, but, as noted above, many of the steps used in traditional processing schedules quench fluorophores. The aim of the following protocol was to preserve GFP while simultaneously introducing enough heavy metal staining for membrane visualization in the EM.

The use of HPF gives excellent ultrastructural preservation to a depth of ~200 μm and bypasses the need for chemical fixatives such as glutaraldehyde, which quenches fluorophores and is in itself slightly autofluorescent. The cryo-fluorescence screening step is optional but can provide valuable information about cell density, transfection efficiency, and fluorophore preservation when working with cultured cells; likewise, fluorophore preservation and location can also be assessed within tissues. Screening a proportion of frozen specimens in this way is a useful checkpoint before proceeding to full-scale specimen processing. The dogma of freeze substitution methods feature many extended protocols, but these were found to quench GFP and mCherry fluorescence in cultured HeLa cells. For this reason, the recent quick freeze substitution (QFS) protocol of McDonald and Webb (2011) was adopted. We suggest that the significantly shorter solvent incubations, as well as the addition of a small percentage of water to the substitution medium, may help to maintain a hydration shell around the fluorophores. The incorporation of a low concentration of uranyl acetate into the substitution medium generates sufficient electron contrast for subsequent electron imaging using an SEM fitted with a high sensitivity backscatter electron detector (BSED). The protocol detailed here focuses on the use of HM20 as an embedding medium, but successful processing using K4M and LR White is also possible (Peddie et al., 2014). IRF sections can be imaged using separate light and electron microscopes. However, to fully exploit this technique, integrated imaging systems should be used. For ILSEM, an array of serial ultrathin sections is collected on an ITO-coated glass cover slip, and sequentially imaged for light and electron signals *in situ* following the principle of array tomography (Wacker & Schroeder, 2013).

As proof of principle, we performed postembedding CLEM and ILSEM on resin-embedded HeLa cells expressing a probe for GFP-C1a-C1b from PKCε to examine the subcellular distribution of the lipid DAG. A probe for mCherry-H2B was included for nuclear localization. DAG, shown to be a modulator of membrane dynamics and a second messenger within the cell, plays a critical role in cellular membranes (Domart et al., 2012; Dumas et al., 2010; Larijani, Barona, & Poccia, 2001; Peddie et al., 2014). Lipids such as DAG are, however, extremely difficult to localize at the EM level using conventional labeling methods. For this reason, we developed an integrated imaging workflow, thereby significantly improving localization precision over standard CLEM methods, and further extending our understanding of the role of DAG in membrane formation and morphology.

17.2.2.2 Protocol

High-pressure freezing

1. Cells were grown in an appropriate culture medium in 10 cm dishes. In preparation for HPF, the cells were dissociated from the dishes using trypsin, and spun gently in a falcon tube to form a pellet. The supernatant was removed, and the cells were resuspended approximately 1:1 in a volume of the same medium containing 20% bovine serum albumin. Prior to freezing, the resuspended cells were maintained at 37 °C.
2. A modified version of the method described by McDonald et al. (2010) was used to transfer the cells into membrane carriers supported in the Rapid Transfer System (RTS) of the EMPACT2 high-pressure freezer (HPF; Leica Microsystems, Vienna; Fig. 17.1). The end of a 200 μl pipette tip was blocked using a small quantity of resin (arrow; Fig. 17.1A), and the tip was shortened to fit into a 1.5 ml Eppendorf tube (Fig. 17.1B). The tip was pierced to form a breather hole in order to prevent a buildup of air pressure within the tip during attachment to the pipette, which would otherwise rapidly eject the contents (arrow; Fig. 17.1C). Thirty microliters of the cell suspension was added to the prepared tip (Fig. 17.1D), which was then placed within an Eppendorf tube and spun briefly (~15 s at ~3000 × *g*) to concentrate the cells within the tip (arrow; Fig. 17.1E). Spinning the cells for longer time periods resulted in a densely packed cell pellet that was difficult to work with.
3. To load cells, the resin blockage was carefully removed using a single-edged razor blade, and the tip was attached to a 20 μl pipette with the breather hole uncovered (Fig. 17.1F). The breather hole was then closed with a finger to seal the tip, and a sufficient quantity of cells to produce a slightly convex meniscus was transferred into the carrier while viewing under a dissecting microscope. Where necessary, the cells were quickly redistributed manually using the end of

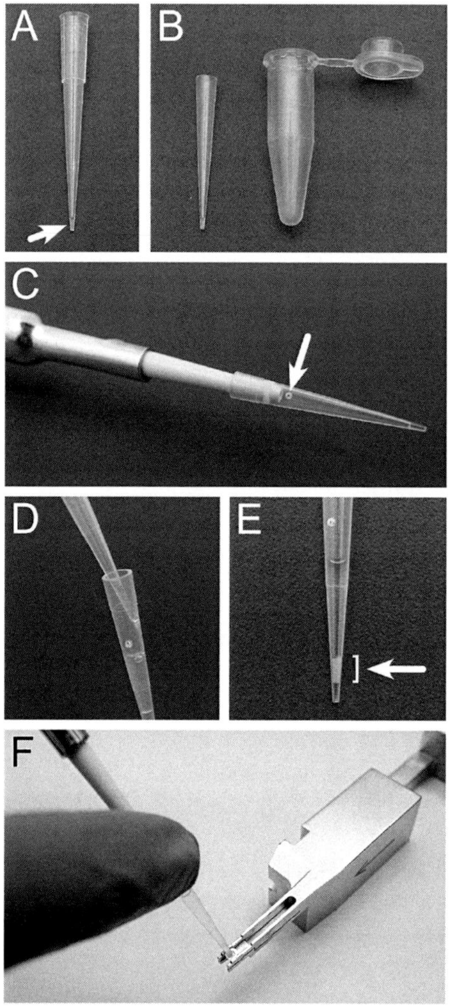

FIGURE 17.1

Loading cells for high-pressure freezing. (A) The end of a 200 μl pipette tip is first blocked with resin (arrow), or similar after McDonald et al. (2010). (B) The tip is shortened to fit inside a 1.5 ml Eppendorf tube. (C) A small breather hole is placed in the tip just below the end of the pipette to prevent pressure and vacuum build up (arrow). (D) A small quantity (~30 μl) of resuspended cells is added to the tip below the breather hole. (E) The tip is placed into the Eppendorf tube and spun briefly to concentrate the cells within the tip (arrow). (F) To load the cells into membrane carriers, the tip blockage is removed with a razor blade, and the cells are pipetted into the carrier before immediately being high-pressure frozen.

another 200 μl pipette tip to ensure full coverage within the carrier. To prevent the vacuum formed when releasing the pipette plunger from resuspending the cells within the tip, the finger covering the breather hole was always released before the thumb holding the pipette plunger.

4. The rapid loader holding the specimen carrier was then immediately transferred to the EMPACT2+RTS, and the cells were high-pressure frozen. Frozen membrane carriers were stored in labeled cryo-vials under liquid nitrogen until required for further processing.

Cryo-fluorescence microscopy

5. Cryo-vials were removed from storage, and the membrane carriers were transferred under liquid nitrogen from the vial to the precooled CMS-196 cryo-correlative stage (Linkam Scientific Instruments, Chilworth; Fig. 17.2A). Membrane carriers were either placed directly onto the cooled viewing platform or placed in a specifically designed cassette via a loading tool (Linkam Scientific Instruments, Chilworth; Fig. 17.2B). Up to three membrane carriers could be screened per transfer (Fig. 17.2C).
6. The cryo-stage was mounted on a widefield epifluorescence light microscope (Axio Scope.A1; Zeiss, Cambridge). Membrane carriers were imaged both at low magnification (N-Achroplan 10×/0.25) to give an overview of fluorescence preservation and cell number (Fig. 17.2D) and at higher magnifications to check the quality of cell preservation (EC Plan-Neofluar 40×/0.75; EC Epiplan-Neofluar 100×/0.75; Fig. 17.2E–H). After screening, membrane carriers were either returned to the cryo-vials in liquid nitrogen storage or processed for freeze substitution.

Quick freeze substitution

7. QFS was performed using a modified version of the method described by McDonald and Webb (2011) and Peddie et al. (2014). A metal block suitable for holding cryo-tubes was placed inside a polystyrene box. Liquid nitrogen was added to cool the block to − 196 °C, and a temperature probe mounted within an acetone filled tube was placed in the block. Refer to McDonald and Webb (2011) for further details of the QFS apparatus.
8. Membrane carriers containing frozen cells were transferred to new cryo-tubes containing substitution medium (5% H_2O in acetone, containing 0.1% uranyl acetate diluted from a 20% stock in methanol) under liquid nitrogen and placed with the cells facing the medium. The liquid nitrogen in the tube was carefully poured away, and the tubes were capped and placed in the cooled block. Tubes with a screw cap and rubber sealing ring were used to ensure a tight seal during the substitution process.
9. To initiate substitution, the liquid nitrogen was carefully poured out of the polystyrene box and replaced with dry ice. The temperature of the cooled block was closely monitored, and the dry ice was removed once the block reached

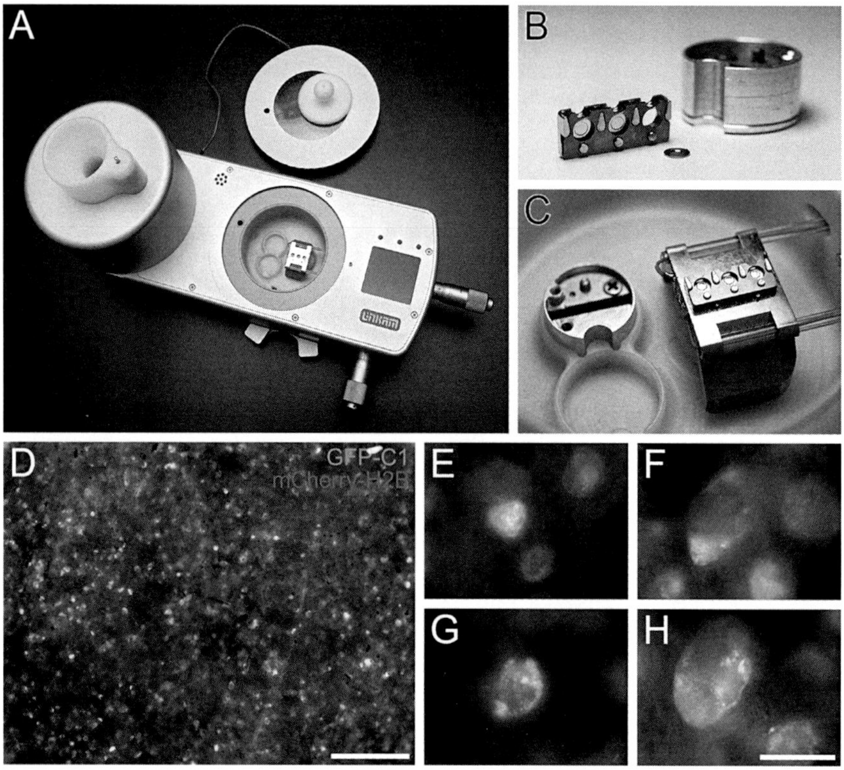

FIGURE 17.2

Cryo-fluorescence microscopy of frozen specimens. (A) Cryo-fluorescence screening of specimens was performed using a cryo-correlative stage. (B) Frozen specimens were loaded into a specially designed cassette (Linkam Scientific) under liquid nitrogen. (C) The cassette was transferred to the cryo-stage for viewing with a widefield epifluorescence microscope. (D) Image showing a low-magnification overview of GFP and mCherry fluorescence within HeLa cells in a high-pressure frozen membrane carrier. (E–H) Images showing high magnification views of individual cells from the carrier shown in (D). Scale bars: (D) 200 μm and (E–H) 25 μm. (See the color plate.)

−85 °C, which took 80–100 min depending on the room temperature (Fig. 17.3). The speed at which the block warmed did not appear to significantly influence specimen quality. At this point, the block was turned on its side within the polystyrene box and placed on a rotary shaker set at approximately 100 rpm.

10. For embedding in HM20 resin, the tubes were transferred to an automated freeze substitution unit (AFS2; Leica Microsystems, Vienna) once the block reached −50 °C and were held at this temperature until the total substitution run-time reached 3 h.

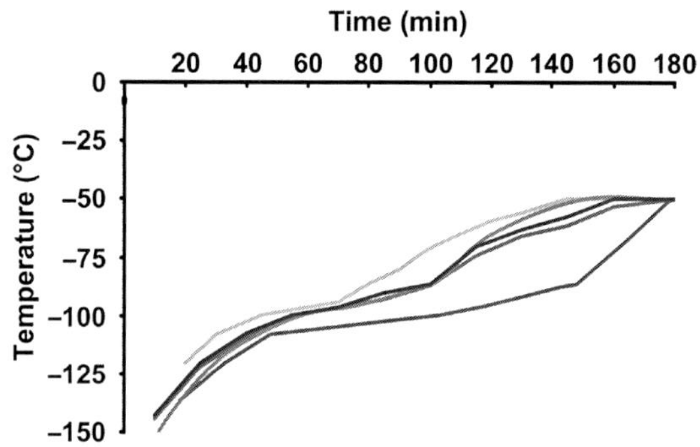

FIGURE 17.3

Four representative temperature profiles recorded during quick freeze substitution. Each experiment reached the target temperature of −50 °C by the end of the 3-h substitution time period, but clear variations in the warm up temperature profile were observed. This variability was attributed to differences in room temperature and humidity on the day of each experiment and did not appear to significantly influence the quality of fluorescence preservation or level of electron contrast. (See the color plate.)

QFS in an automated freeze substitution unit

11. Substitution was also performed using the AFS2, programmed to mimic an average temperature profile calculated across several substitution experiments using a polystyrene box (Fig. 17.3 and Table 17.1). Screw-cap cryo-tubes were loaded as detailed above and transferred directly to the chamber of the AFS2, which had been precooled using liquid nitrogen. The program was then started and allowed to proceed to the appropriate holding temperature.

Resin embedding

12. To prepare for resin infiltration, membrane carriers were transferred from the substitution media to molds filled with 100% acetone and incubated for 15 min. A pair of fine-tipped tweezers was used to transfer the carriers, as tipping the carriers directly out of the cryo-tubes was found to dislodge the cells. Further 3 × 15 min washes with 100% acetone were carried out prior to resin infiltration over 3 h through 20%/40%/60%/80% dilutions of resin. After incubation overnight in 100% resin, four changes of fresh resin were carried out the following day, prior to polymerization under 360 nm UV light over 48 h, and subsequent warming to room temperature.

Resin sectioning and mounting

13. Polymerized blocks were trimmed from the molds and stored at room temperature in the dark. Fluorescent signal was clearly evident within the blocks prior to sectioning (Fig. 17.4A–C). The membrane carriers were next carefully

Table 17.1 Program for automation of quick-freeze substitution

Step	T_{start} (°C)	T_{end} (°C)	Slope (°C/h)	Time	Reagent
1	−140	−107	66	0:30	FS media/acetone
2	−107	−93	21	0:40	FS media/acetone
3	−93	−72	42	0:30	FS media/acetone
4	−72	−60	35.9	0:20	FS media/acetone
5	−60	−50	24.1	0:25	FS media/acetone
6	−50	−50	0	0:35	FS media/acetone[a]
7	−50	−50	0	1:00	100% acetone[b]
8	−50	−50	0	0:45	20% HM20
9	−50	−50	0	0:45	40% HM20
10	−50	−50	0	0:45	60% HM20
11	−50	−50	0	0:45	80% HM20
12	−50	−50	0	16:00	100% HM20
13	−50	−50	0	8:00	100% HM20[c]
14	−50	−50	0	48:00	100% HM20[d]
15	−50	20	4.4	16:00	

[a]*Holding step to reach 3 h substitution time.*
[b]*Four changes over 60 min.*
[c]*Four changes over 8 h.*
[d]*UV illumination turned on.*

trimmed away by hand. The blocks were then either cut and trimmed perpendicular to the cell layer to allow examination of the full depth of the layer or trimmed so that sections could be taken parallel to the cell layer (Fig. 17.5A).

14. Sections of 70 and 200 nm thickness were cut using a large diamond knife (Ultra Jumbo 45°; Diatome Ltd., Biel/Bienne; Fig. 17.5B and C) on a UCT ultramicrotome (Leica Microsystems, Vienna), collected on glass slides, and screened to confirm the presence of fluorescence using a light microscope.
15. For specimens exhibiting sufficient signal, additional ribbons of serial sections were collected on ITO-coated glass cover slips (200 nm thick for ILSEM-based analysis; Fig. 17.5D) or 1.5% formvar-coated finder grids or slot grids (70 and 200 nm thick for TEM analysis; Fig. 17.5E).

CLEM of IRF sections using separate light and electron microscopes

16. Ribbons of sections collected on finder grids were screened for fluorescence using a widefield epifluorescence light microscope (Axio Scope.A1; Zeiss, Cambridge; Fig. 17.6A and B), and images of cells of interest were acquired using air objective lenses (EC plan-Neofluar 40×/0.75; EC Epiplan-Neofluar 100×/0.75) and an MRm CCD camera (Zeiss, Cambridge). Where embedding was less than optimal, it proved necessary to image sections in a hydrated state to minimize reflections caused by holes and tears in the sections.

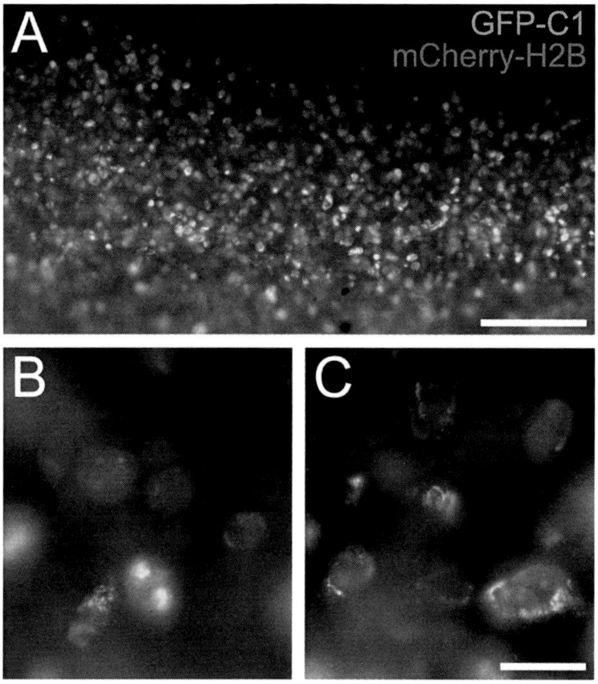

FIGURE 17.4

Preservation of fluorescent signal in resin-embedded HeLa cells. (A) Image showing an overview of GFP-C1 and mCherry-H2B fluorescence recorded directly from the surface of an HM20 resin block trimmed parallel to the cell layer. (B and C) Images at higher magnification showing individual cells from the block shown in (A). Signal for mCherry was restricted to the nucleus, whereas the signal for GFP was cytoplasmic. Scale bars: (A) 300 μm and (B and C) 25 μm. (See the color plate.)

For this, grids were mounted between a glass slide and a cover slip using phosphate-buffered glycerol, pH 7.4. However, under optimal conditions, this step was unnecessary, and the fluorescent signal could be collected directly from grids placed atop a glass slide (Fig. 17.6B).

17. For electron imaging, the grids were imaged using a Tecnai G2 Spirit BioTWIN 120 keV TEM (FEI Company, Eindhoven) with Orius CCD camera (Gatan Inc., Pleasanton). The cells and structures of interest were relocated using the corresponding grid references from light microscopic images (Fig. 17.6C), and electron micrographs at a variety of magnifications were acquired.
18. Composite overlays of separate light and electron images were generated using Adobe Photoshop (Fig. 17.6D–H). Images of the fluorescent signal were cropped to the area of interest and interpolated linearly to a magnification sufficient to match that of the electron image before overlaying with appropriate final linear adjustments to scale, and adjustments in rotation. Each of the images

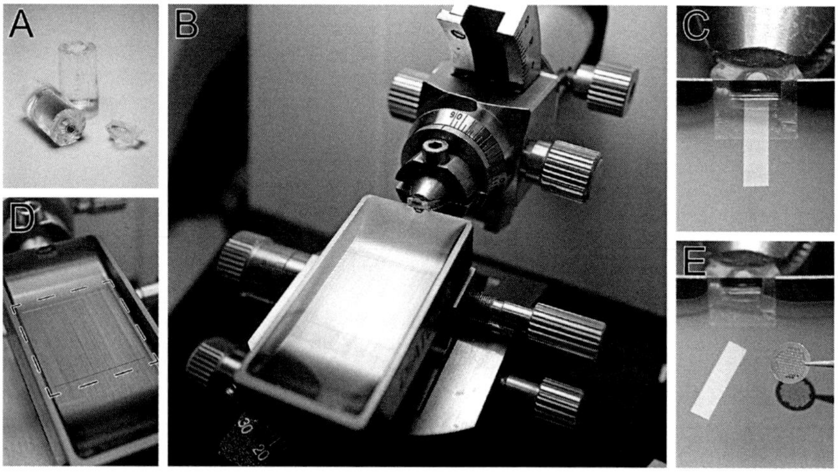

FIGURE 17.5

Resin trimming and sectioning with a Jumbo diamond knife. (A) Resin blocks were trimmed in one of two orientations for thin sectioning. Left block: trimmed parallel to the cell layer. Right block: trimmed perpendicular to the cell layer. (B) Resin-embedded cell layers were sectioned using an Ultra Jumbo 45° diamond knife (Diatome). (C) Serial ultrathin sections of 70 nm and 200 nm thicknesses were cut. (D) For ILSEM, sections were collected on ITO-coated glass. The cover slips were immersed in the water bath of the diamond knife (dashed line), and after sectioning, the water level was gradually reduced until the sections settled onto the glass. (E) For correlative imaging using separate light and electron microscopes, sections were collected on 1.5% formvar-coated finder grids.

was adjusted for contrast, brightness, and RGB levels to closely match the original signal as viewed on the epifluorescence microscope. Electron micrographs were adjusted linearly for brightness and contrast and were sharpened using unsharp mask. In comparison to routine CLEM methods (Peddie et al., 2014), postembedding light and EM of ultrathin sections significantly improved the localization precision of GFP-C1 due to physical section thickness and allowed the tight correlation of fluorescent structures at the light microscopic level with subcellular features at the EM level (Fig. 17.6E–H). Fluorescent signal for DAG was particularly associated with membranous structures including endoplasmic reticulum, Golgi, patches of vesicles, and the nuclear envelope, further supporting a role for DAG in localized membrane shaping (Domart et al., 2012; Peddie et al., 2014).

17.2.3 IMAGING IN THE ILSEM

17.2.3.1 Principle

Integrated light and electron microscopes offer fast retrieval of ultrastructural information from regions of interest that have been identified using fluorescence

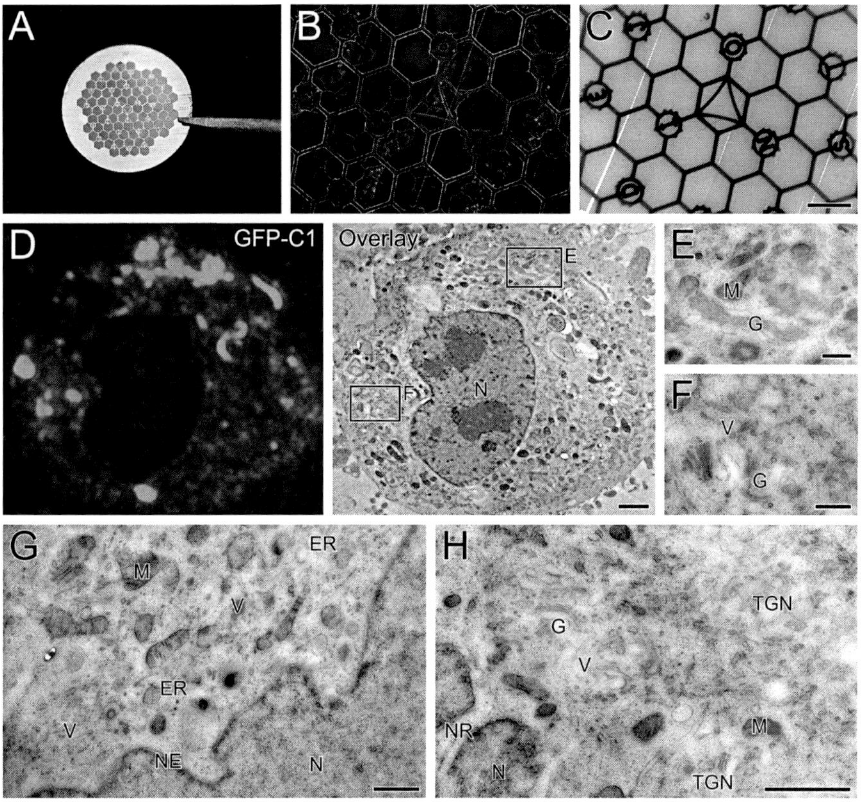

FIGURE 17.6

Screening of GFP-C1 and mCherry-H2B fluorescence in sections collected on finder grids, and CLEM of GFP-C1 in HeLa cells using postembedding light microscopy and transmission electron microscopy. (A) Serial 70 nm sections collected on a 1.5% formvar-coated finder grid. Images showing the fluorescent signal in ultrathin sections (B), and finder coordinates to relocate specific areas of interest in the electron microscope (C). (D) Images showing the GFP signal alone, and overlaid onto the transmission electron micrograph of a cell expressing a moderately high level of GFP-C1. Fluorescent signal is associated with the nuclear envelope and with structures in the surrounding cytoplasm. (E and F) Boxed detail from (D) showing intense cytoplasmic fluorescent signal corresponding to Golgi stacks and vesicles. (G and H) Images from cells expressing a lower level of GFP-C1 highlighting localization of fluorescence to the nuclear envelope and membranous structures within the cytoplasm, including endoplasmic reticulum and vesicles (G) and Golgi stacks and trans-Golgi networks (H). ER, endoplasmic reticulum; G, Golgi; M, mitochondrion; N, nucleus; NE, nuclear envelope; NR, nucleoplasmic reticulum; TGN, trans-Golgi network; V, vesicles. Scale bars: (C) 200 μm, (D) 2 μm, (E and F) 500 nm, and (G and H) 1 μm. (See the color plate.)

microscopy. With no intermediate specimen preparation or manipulation steps, the risk of contamination or damage during transfer between modalities is also significantly reduced. In the SECOM system (Delmic B.V., Delft), an inverted epifluorescence microscope is positioned underneath the stage of a SEM (Fig. 17.7). A specimen mounted in the SECOM microscope can thus be observed using epifluorescence from below, through the ITO-coated glass cover slip, and with the SEM from above. High-resolution fluorescence microscopy can be performed using vacuum-compatible immersion oil (Delmic B.V., Delft). As both microscopes share the same field of view, higher magnification SEM images can be recorded from any position, and a new field of view can be selected by translating the specimen remotely, which allows for inspection of a large number of regions of interest. The overlaid data can be directly visualized in the SECOM software interface, but composite overlay images can also be created *a posteriori* as described later. A fully automated procedure to register and generate the overlaid images directly without additional user input is under development, but was not fully implemented at the time of writing.

Imaging whole cells in the ILSEM enables structural detail to be acquired from regions highlighted by the fluorescent probe. Fluorescence information is gathered from the whole cell, whereas structural information is limited to the surface and thinner edges of the cell. Thus, fluorophores inside the cell are less affected by bleaching during initial screening with the electron beam. In comparison, resin sections are much more susceptible to fluorophore bleaching and contamination from the electron beam. It is therefore not usually possible to use the electron beam for the screening of sections without severely compromising the specimen for subsequent

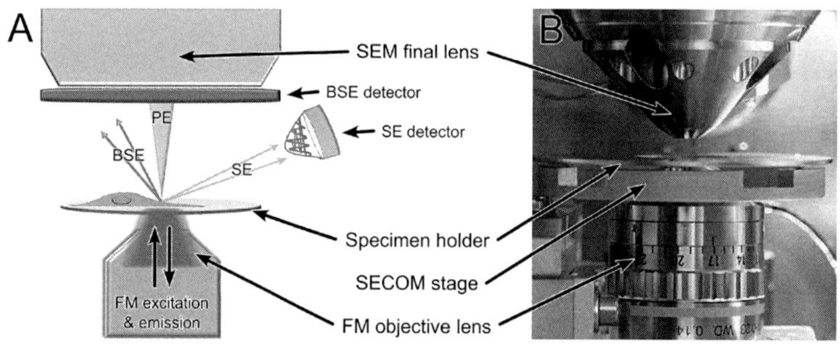

FIGURE 17.7

SECOM-integrated light and scanning electron microscope design. (A) Schematic representation of the SECOM-integrated microscope showing an inverted epifluorescence microscope positioned inside an SEM. The SEM can detect backscattered electrons (BSEs) and secondary electrons (SEs), which are generated by the beam of primary electrons (PEs). The light path for fluorescence imaging is also indicated. (B) Image of the SECOM platform installed within the scanning electron microscope vacuum chamber. In this configuration, a BSE detector is not fitted and the SEM final lens obscures the SE detector. BSE, backscattered electron; FM. fluorescence microscope; PE, primary electron; SE, secondary electron; SEM, scanning electron microscope. (See the color plate.)

fluorescence and electron imaging. For this reason, it was important to image sections sequentially, collecting fluorescence data before collecting the electron image.

17.2.3.2 Protocol

ILSEM analysis of whole cells

1. The glass cover slips were attached to the underside of the SECOM specimen holder using conductive carbon tape with the ITO coating contacting the holder (Fig. 17.8A–C). Alternative mountants such as silver adhesive can also be used.
2. A drop of vacuum-compatible immersion oil (Delmic B.V., Delft) was applied to the objective lens, and the specimen holder was placed on the integrated microscope (underside facing the objective lens, upperside with cell monolayer facing the SEM column). The objective lens was then raised to bring the oil droplet into contact with the cover slip before closing the SEM door and pumping down the SEM chamber (Fig. 17.8E and F).
3. Making use of the large field of view of the SEM, a low-magnification image covering a field of view of several square millimeters was acquired at 10 keV using the Everhart–Thornley detector (ETD) with 30 μs dwell time (Fig. 17.9A). Using this reference image, a specific region was selected for focusing the fluorescence and electron microscopes. For this, the specimen was translated to bring the chosen region into the center of the field of view, and focusing procedures were carried out.
4. Next, the low-magnification reference SEM image was used to select specific regions of interest, e.g., areas containing only a few individual or touching cells. The specimen was then translated to bring the first region of interest into the center of the SEM field of view, and images of the fluorescent signal were acquired (Fig. 17.9B and G). An LED light source with a wavelength of 485 nm and power rating of 172 mW (Spectra, Lumencor) was used for excitation of GFP, and the emission was detected through a 520/10-nm bandpass filter (Thorlabs) with a CCD camera (Clara, Andor). For image collection, an exposure time of 1 s with LED power of 10 mW was sufficient in most cases.
5. After FM image acquisition, electron micrographs were acquired at higher magnifications to capture the finer details of filopodia and other cellular extensions (Fig. 17.9C–F). Care was taken to record all necessary fluorescence information before imaging the area by SEM as at higher magnifications, exposure to the electron beam quenched fluorescence. The SEM was operated both in secondary electron (SE) and backscattered electron (BSE) modes using the ETD and the solid-state BSED, respectively.
6. SEs are low-energy electrons (<50 eV) generated by inelastic collisions when the incident electron beam hits a specimen, and escape from a shallow depth. This makes the SE signal particularly sensitive to specimen topography, whereas higher energy BSEs reveal information about the internal features of a specimen, related to atomic number, in addition to the topographical information. The electron acceleration voltage also influences the penetration depth of incident electrons and thus higher voltage electrons can probe a larger

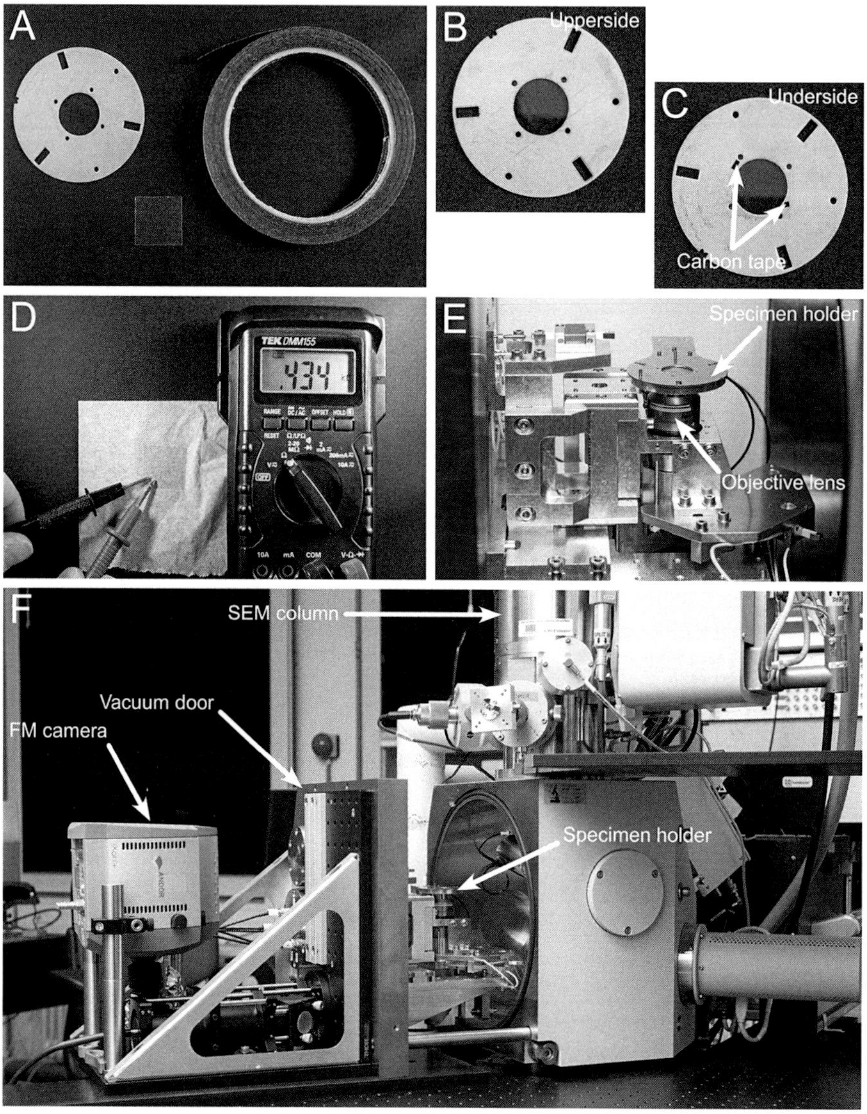

FIGURE 17.8

Mounting specimens for integrated light and scanning electron microscopies. (A) Image showing the SECOM specimen holder, an ITO-coated glass cover slip, and roll of carbon tape. (B and C) ITO-coated cover slips are attached to the specimen holder using small segments of conductive carbon tape, and the coating must contact the holder. (D) The conductive side of the ITO-coated cover slip can be easily determined with a quick resistivity measurement, placing both tips of a multimeter at distant positions on the cover slip. The ITO coating has a finite resistance (measured in Ω or kΩ). (E) The specimen holder is placed onto the stage of the SECOM-integrated microscope with the underside facing the objective lens and specimen facing the SEM column. (F) Image to show the integrated SECOM microscope mounted on the SEM. During operation, the specimen and objective lens assembly are inserted into the SEM vacuum chamber, while all other light microscope components remain outside. FM, fluorescence microscope; SEM, scanning electron microscope.

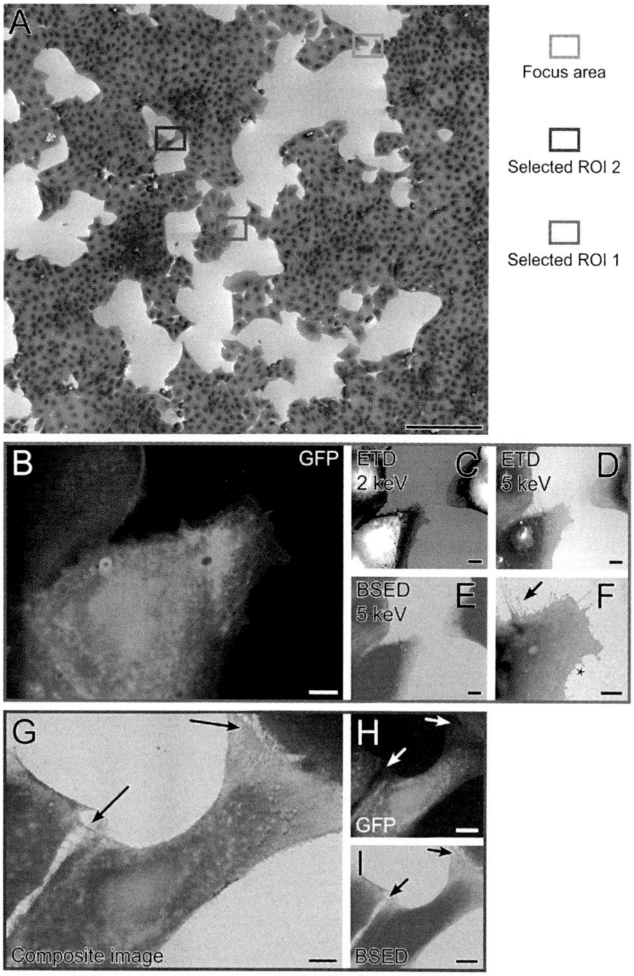

FIGURE 17.9

Integrated light and scanning electron microscopy of whole MDCK cells expressing GFP-Paxillin. (A) Low-magnification SEM reference image. Regions chosen for fine focusing of the microscopes and subsequent correlative imaging investigation are marked with colored boxes. (B) Fluorescence image showing the GFP signal from ROI 1. (C and D) SEM images acquired using the ETD at electron voltages of 2 keV (C) and 5 keV (D). (E) SEM image of the BSED signal recorded at 5 keV. (F) Higher magnification BSED image highlighting finer details of the filopodial network at the cell periphery, with predominantly short, rounded tip extensions at sites distant from adjacent cells (asterisk), and longer extensions close to neighboring cells (arrow). (G–I) Composite overlay image from ROI 2 (G), with individual GFP (H), and BSED (I) signals. Arrows mark the two areas where complex connections between neighboring cells can be seen. BSED, backscatter electron detector; ETD, Everhart–Thornley detector; MDCK, Madin Darby canine kidney; ROI, region of interest; SEM, scanning electron microscopy. Scale bars: (A) 200 μm, (B) 5 μm, (C–E) 10 μm, and (F–I) 5 μm. (See the color plate.)

depth. This is illustrated in Fig. 17.9C and D, which shows SE images acquired using the ETD at acceleration voltages of 2 and 5 keV, respectively. For comparison, Fig. 17.9E shows the image recorded with the BSED at an accelerating voltage of 5 keV, which contains less surface information. As the mean atomic number of the ITO-coated substrate is higher than that of the cellular features, the latter are recorded with negative contrast in the BSED images. The thicker and/or denser parts of the specimen, such as the nucleus and surrounding area, appear dark in the BSED image (Fig. 17.9E). It is also important to note that the ITO-coating thickness, in addition to accelerating voltage, significantly influences BSED image contrast (Pluk et al., 2009). Structural detail at the cell periphery shows that filopodia in cells distant from neighboring cells (asterisk; Fig. 17.9F) are shorter and more rounded compared to those extending toward adjacent cells (arrow; Fig. 17.9F). Also notable was the increased frequency of focal adhesions at the periphery of adjacent cells.

7. Optimized SEM imaging parameters for examining MDCK cell–cell connections on ITO-coated glass cover slips (Fig. 17.9G–I) were as follows: the acceleration voltage was set at 5 keV (beam current 0.4 nA), and images were recorded with a 10 μs dwell time (4096 × 3536 pixels) at a working distance of 8 mm, under high vacuum. In the overlay (Fig. 17.9G), fluorescence (Fig. 17.9H) and BSED (Fig. 17.9I) images, two regions with an intricate network of cell–cell connections were observed (marked with arrows), showing how ILSEM can enhance understanding of the fluorescence image by providing additional detailed structural information.

ILSEM analysis of IRF sections

8. Before mounting the specimen or commencing imaging, the alignment of light and electron paths was checked to ensure accurate and reproducible stage positioning. Sections of 200 nm thickness collected on ITO-coated glass cover slips were then screened for suitable fluorescent signal (Fig. 17.10A–G) and attached to the underside of a SECOM specimen holder (as described earlier) in readiness for insertion into the SECOM platform.
9. Images of the fluorescent signal were acquired from cells of interest at both 200 and 40 Pa chamber pressure; an exposure time of 2 s with a laser power of 10 mW was sufficient in most cases. Fluorophore excitation was achieved using a 488 nm laser light source, and emission was recorded over the range of 509.5–538.5 nm. Imaging of nearby fluorescent regions was completed before switching to SEM image acquisition, as prolonged exposure to the electron beam quenched the fluorescent signal.
10. During electron imaging, 200 nm resin sections quickly became contaminated and damaged by the electron beam. To minimize unnecessary exposure, the first section of each series was used to map the cells of interest identified during fluorescence imaging. This facilitated rapid relocation on subsequent serial sections. Electron dose during search and relocation phases was minimized by reducing overall frame size and keeping the pixel dwell time to an absolute minimum.

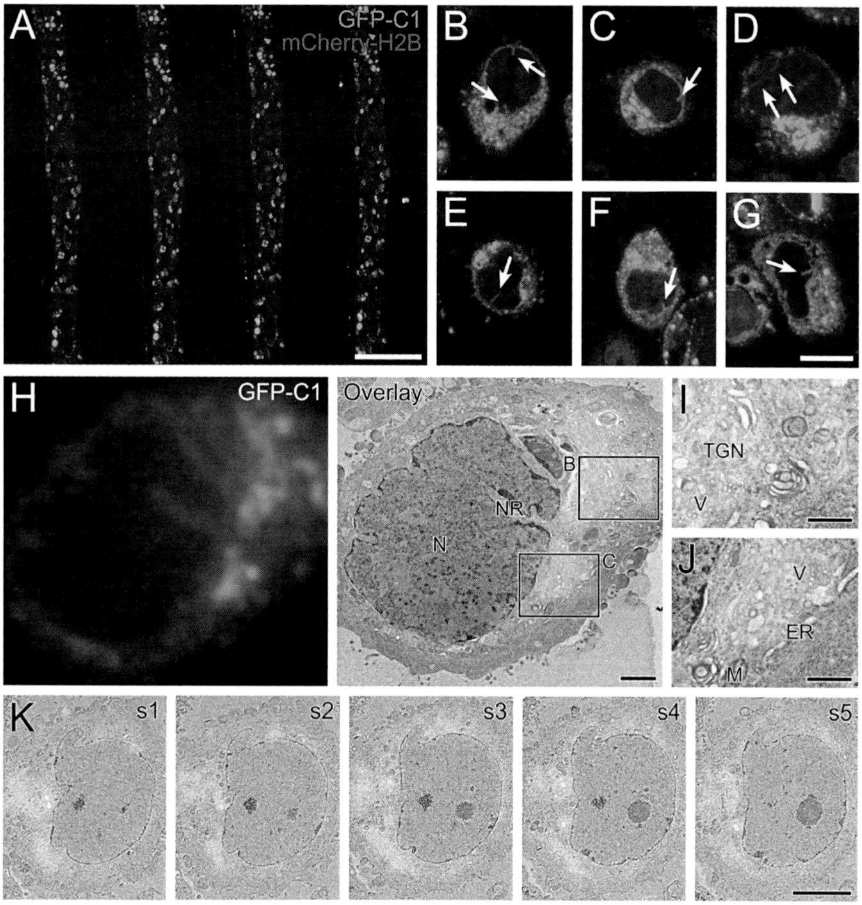

FIGURE 17.10

Screening of GFP-C1 and mCherry-H2B fluorescence in sections collected on an ITO-coated cover slip, and integrated light and scanning electron microscopies of resin-embedded HeLa cells expressing GFP-C1. (A) Image showing fluorescent signal for GFP and mCherry within serial sections collected on an ITO-coated glass cover slip. (B–G) Images of individual cells recorded at higher magnification highlighting GFP and mCherry fluorescence within resin-embedded HeLa cells. Signal for mCherry was generally restricted to the nucleus, whereas the signal from GFP was associated with structures within the cytoplasm, the nuclear envelope, and nucleoplasmic reticulum (arrows). (H) Images showing the GFP signal alone and overlaid onto the scanning electron micrograph of a cell expressing a comparatively high level of GFP-C1. Fluorescent signal is evident associated with the nuclear envelope and reticulum, and within the surrounding cytoplasm. (I and J) Boxed detail from (H) showing intense cytoplasmic fluorescent signal corresponding to complex vesicular structures, trans-Golgi networks, and endoplasmic reticulum. (K) Images from five serial sections through a cell expressing a lower level of GFP-C1. The fluorescent signal corresponds to the nuclear envelope, and other membranous structures within the cytoplasm. ER, endoplasmic reticulum; M, mitochondrion; N, nucleus; NR, nucleoplasmic reticulum; TGN, trans-Golgi network; V, vesicles. Scale bars: (A) 200 μm, (B–G) 10 μm, (H) 2 μm, (I and J) 250 nm, and (K) 5 μm. (See the color plate.)

11. Optimal electron image quality was achieved by selecting an accelerating voltage appropriate to the thickness of the specimen, subject to backscatter detector sensitivity. For 200-nm sections collected onto ITO-coated glass cover slips, the following conditions were used: 2.5 keV, spot size 3.5, 30 μm aperture, 60 μs dwell time (1536 × 1024 pixels per image), working distance 5.8 mm, chamber pressure set at 40 Pa, or high vacuum.
12. Composite overlays of light and electron images were generated as detailed earlier (Point 18 in Section 17.2.2.2), with the notable exception of any requirement for rotational adjustments (Fig. 17.10H–K). When compared to postembedding light microscopy and TEM, the integration of imaging modalities in a single instrument significantly streamlined acquisition. GFP-C1 localized to the nuclear envelope, nucleoplasmic reticulum, trans-Golgi network, and vesicles (Fig. 17.10H–J). Collection of serial sections allowed an array tomography-like workflow to be followed (Fig. 17.10K).

17.3 MATERIALS

17.3.1 EQUIPMENT

Cell culture equipment: Cell culture hood, incubator (37 °C, 5% CO_2), water bath, refrigerator, freezer (−20 °C), inverted microscope, and autoclave;
Centrifuges for 15-ml falcon tubes and 1.5-ml Eppendorf tubes;
Liquid nitrogen storage vessels;
High pressure freezer (e.g., EMPACT2 with RTS; Leica Microsystems, Vienna);
(Optional) Cryo-fluorescence stage (e.g., CMS-196; Linkam Scientific Instruments, Chilworth);
Widefield epifluorescence light microscope (e.g., Axio Scope.A1; Zeiss, Cambridge);
QFS apparatus: rotary shaker, polystyrene box, and metal cryo-tube rack;
USB data logger (EL-USB-TC-LCD; Lascar Electronics Ltd., Salisbury), t-type thermocouple probe (621–2209; RS Components Ltd., Corby), and appropriate t-type connector;
Freeze substitution unit with UV lamp (e.g., AFS2; Leica Microsystems, Vienna);
Ultramicrotome (e.g., UC7; Leica Microsystems, Vienna);
TEM (e.g., 120 keV Tecnai G2 Spirit BioTWIN; FEI Company, Eindhoven);
(Optional) Multimeter for testing ITO-coated slides;
SECOM (Delmic B.V., Delft) ILSEM, with high sensitivity BSED (e.g., Quanta 250 FEG SEM; FEI Company, Eindhoven).

17.3.2 MATERIALS

Liquid nitrogen;
Dry ice;

Fine-tipped forceps and cryo-forceps;
Two milliliter cryo-vials (430489; Corning, Fisher Scientific, Loughborough);
1.5 mm membrane carriers (Leica Microsystems, Vienna);
Gilson pipettes (20 and 200 μl) and tips (200 μl);
1.5 ml Eppendorf tubes;
Uranyl acetate (20% w/v stock solution in methanol);
Acetone (100% and anhydrous);
Ethanol (absolute);
Deionized water;
4% Paraformaldehyde PBS, pH 7.4;
(Optional) Vacuum-compatible immersion oil (Delmic B.V., Delft);
Carbon tape;
Cell culture consumables: aspiration vacuum pump, pH meter, cell culture flasks, glass pipettes, pipettor, waste containers, medium (DMEM), antibiotics (penicillin/streptomycin), media supplement (L-glutamine), FCS (10%), and six-well plates;
Two-milliliter skirted cryo-tubes and caps with rubber sealing ring (e.g., E1420-2340 and E1480-0100; Starlab (UK) Ltd., Milton Keynes);
Freeze substitution consumables suitable for processing membrane carriers (Leica Microsystems, Vienna);
Lowicryl HM20 (Polysciences, Inc.; Park Scientific Ltd., Northampton);
Ultra Jumbo 45° diamond knife (Diatome Ltd., Biel/Bienne);
1.5% Formvar-coated 3.05-mm 100-mesh TEM finder grids (AGG247; Agar Scientific, Stansted) and/or 2-mm × 1-mm slot grids (AGG2500C; Agar Scientific, Stansted);
Glass microscope slides and cover slips (22 mm × 22 mm × 0.17 mm);
ITO-coated glass cover slips (22 mm × 22 mm × 0.17 mm; Delmic B.V., Delft);
SECOM specimen holders (Delmic B.V., Delft).

17.4 DISCUSSION

Integration of light and electron microscopes and parallel advances in specimen preparation are heralding a new era of faster and more accurate correlative microscopy. In this chapter, we have described a workflow for imaging of GFP in whole and resin-embedded cells *in vacuo* in an ILSEM, which is an essential proof of principle for wider adoption of integrated systems.

Preservation of GFP fluorescence after processing for EM has been reported in several recent studies, but these use separate light and electron microscopes for imaging and often require intermediate preparation steps to switch between imaging modalities. The protocol described here requires no additional specimen treatment and delivers *in situ* sequential fluorescence and electron imaging within a single instrument. Furthermore, both low and high expression levels of GFP can be detected *in vacuo*, even in ultrathin sections that contain only a small fraction of the fluorophore population when compared with a whole cell. This is in stark contrast to previously reported methods, in which fluorescence from GFP constructs could only be detected after sectioning when expressed at high levels, which can be detrimental to cell morphology and function. We postulate that the stability and activity of GFP fluorescence in-resin is a result of the limited exposure of the fluorophore to dehydration by solvents. As such, the protocol is expected to be applicable to other GFP-expressing cell lines for a wide range of structure–function studies. Further work is now underway to test the protocol on tissues and model organisms expressing GFP and related fluorophores.

Concomitant with advances in specimen preparation, ILSEM technology is rapidly evolving. Advances include refinements to the light microscope in the ILSEM in order to achieve lateral super resolution, a natural extension of recent work on super-resolution CLEM (Watanabe et al., 2011). For super-resolution CLEM, errors resulting from specimen distortion and image registration may become dominant over the actual super-resolution localization error. This effect may be circumvented using integrated imaging. It should also be noted that in the integrated protocol, there is no need to incorporate fiducial markers into the specimen or onto the substrate to assist in navigation and image overlay. While we currently rely on manual procedures to generate the overlaid images, we anticipate that automated methods for image registration and overlay will soon be available.

Integrated imaging of large volumes could also be achieved more quickly by further automation of the sectioning process using a system like the automated tape ultramicrotome (Schalek et al., 2011; Tapia et al., 2012). Furthermore, as the fluorophores of interest are visible at the surface of the resin block (Fig. 17.4), it may ultimately be possible to combine integrated imaging methods with *in situ* sectioning approaches like serial block face SEM (Denk & Horstmann, 2004) or focused ion beam SEM (Heymann et al., 2006; Knott, Marchman, Wall, & Lich, 2008) to deliver 3D-registered light and electron images over large tissue volumes.

ACKNOWLEDGMENTS

We thank Dr. Marie-Charlotte Domart and Prof. Banafshe Larijani for the samples and biological questions resulting in Figs. 17.2, 17.4, 17.6, and 17.10 and Martijn de Gruijter, Gert van Cappelle, and Prof. Adriaan Houtsmuller for the samples and biological questions resulting in Fig. 17.9. We also thank Helmut Gnaegi and Diatome Ltd. for the loan of a Jumbo Ultra diamond knife, Kent McDonald for valuable and insightful discussions, and Robert Moerland for photographs in Figs. 17.7 and 17.8. C. J. P. and L. M. C. are grateful for support from Cancer Research UK. The authors would also like to acknowledge funding from the MRC, BBSRC, and EPSRC under Grant award MR/K01580X/1 to L. M. C.

Statement of financial interest: The SECOM-integrated light microscope and sample holder were obtained from Delmic B.V. J. P. H. is a shareholder in Delmic B.V.

REFERENCES

Agronskaia, A. V., Valentijn, J. A., van Driel, L. F., Schneijdenberg, C. T., Humbel, B. M., van Bergen en Henegouwen, P. M., et al. (2008). Integrated fluorescence and transmission electron microscopy. *Journal of Structural Biology*, *164*(2), 183–189. http://dx.doi.org/10.1016/j.jsb.2008.07.003.

Bell, K., Mitchell, S., Paultre, D., Posch, M., & Oparka, K. (2013). Correlative imaging of fluorescent proteins in resin-embedded plant material. *Plant Physiology*, *161*(4), 1595–1603. http://dx.doi.org/10.1104/pp. 112.212365.

Chan, W. C., & Nie, S. (1998). Quantum dot bioconjugates for ultrasensitive nonisotopic detection. *Science*, *281*(5385), 2016–2018.

Connolly, C. N., Futter, C. E., Gibson, A., Hopkins, C. R., & Cutler, D. F. (1994). Transport into and out of the Golgi complex studied by transfecting cells with cDNAs encoding horseradish peroxidase. *The Journal of Cell Biology*, *127*(3), 641–652.

Deerinck, T. J., Giepmans, B. N., Smarr, B. L., Martone, M. E., & Ellisman, M. H. (2007). Light and electron microscopic localization of multiple proteins using quantum dots. *Methods in Molecular Biology*, *374*, 43–53pii: 1-59745-369-2:43.

Denk, W., & Horstmann, H. (2004). Serial block-face scanning electron microscopy to reconstruct three-dimensional tissue nanostructure. *PLoS Biology*, *2*(11), e329. http://dx.doi.org/10.1371/journal.pbio.0020329.

Domart, M. C., Hobday, T. M., Peddie, C. J., Chung, G. H., Wang, A., Yeh, K., et al. (2012). Acute manipulation of diacylglycerol reveals roles in nuclear envelope assembly & endoplasmic reticulum morphology. *PLoS One*, *7*(12), e51150. http://dx.doi.org/10.1371/journal.pone.0051150.

Dumas, F., Byrne, R. D., Vincent, B., Hobday, T. M., Poccia, D. L., & Larijani, B. (2010). Spatial regulation of membrane fusion controlled by modification of phosphoinositides. *PLoS One*, *5*(8), e12208. http://dx.doi.org/10.1371/journal.pone.0012208.

Faas, F. G., Barcena, M., Agronskaia, A. V., Gerritsen, H. C., Moscicka, K. B., Diebolder, C. A., et al. (2013). Localization of fluorescently labeled structures in frozen-hydrated samples using integrated light electron microscopy. *Journal of Structural Biology*, *181*(3), 283–290. http://dx.doi.org/10.1016/j.jsb.2012.12.004.

Fabrowski, P., Necakov, A. S., Mumbauer, S., Loeser, E., Reversi, A., Streichan, S., et al. (2013). Tubular endocytosis drives remodelling of the apical surface during epithelial morphogenesis in Drosophila. *Nature Communications*, *4*, 2244. http://dx.doi.org/10.1038/ncomms3244.

Giepmans, B. N., Deerinck, T. J., Smarr, B. L., Jones, Y. Z., & Ellisman, M. H. (2005). Correlated light and electron microscopic imaging of multiple endogenous proteins using quantum dots. *Nature Methods*, *2*(10), 743–749. http://dx.doi.org/10.1038/nmeth791, pii: nmeth791.

Heymann, J. A., Hayles, M., Gestmann, I., Giannuzzi, L. A., Lich, B., & Subramaniam, S. (2006). Site-specific 3D imaging of cells and tissues with a dual beam microscope. *Journal of Structural Biology*, *155*(1), 63–73. http://dx.doi.org/10.1016/j.jsb.2006.03.006, pii: S1047-8477(06)00078-5.

Karreman, M. A., Agronskaia, A. V., van Donselaar, E. G., Vocking, K., Fereidouni, F., Humbel, B. M., et al. (2012). Optimizing immuno-labeling for correlative fluorescence and electron microscopy on a single specimen. *Journal of Structural Biology*, *180*(2), 382–386. http://dx.doi.org/10.1016/j.jsb.2012.09.002.

Karreman, M. A., Agronskaia, A. V., Verkleij, A. J., Cremers, F. F., Gerritsen, H. C., & Humbel, B. M. (2009). Discovery of a new RNA-containing nuclear structure in UVC-induced apoptotic cells by integrated laser electron microscopy. *Biology of the Cell, 101*(5), 287–299. http://dx.doi.org/10.1042/BC20080076.

Knott, G., Marchman, H., Wall, D., & Lich, B. (2008). Serial section scanning electron microscopy of adult brain tissue using focused ion beam milling. *The Journal of Neuroscience, 28*(12), 2959–2964. http://dx.doi.org/10.1523/JNEUROSCI.3189-07.2008, pii: 28/12/2959.

Kukulski, W., Schorb, M., Kaksonen, M., & Briggs, J. A. (2012). Plasma membrane reshaping during endocytosis is revealed by time-resolved electron tomography. *Cell, 150*(3), 508–520. http://dx.doi.org/10.1016/j.cell.2012.05.046.

Kukulski, W., Schorb, M., Welsch, S., Picco, A., Kaksonen, M., & Briggs, J. A. (2011). Correlated fluorescence and 3D electron microscopy with high sensitivity and spatial precision. *The Journal of Cell Biology, 192*(1), 111–119. http://dx.doi.org/10.1083/jcb.201009037.

Larijani, B., Barona, T. M., & Poccia, D. L. (2001). Role for phosphatidylinositol in nuclear envelope formation. *The Biochemical Journal, 356*(Pt. 2), 495–501.

Liv, N., Zonnevylle, A. C., Narvaez, A. C., Effting, A. P., Voorneveld, P. W., Lucas, M. S., et al. (2013). Simultaneous correlative scanning electron and high-NA fluorescence microscopy. *PLoS One, 8*(2), e55707. http://dx.doi.org/10.1371/journal.pone.0055707.

Luby-Phelps, K., Ning, G., Fogerty, J., & Besharse, J. C. (2003). Visualization of identified GFP-expressing cells by light and electron microscopy. *The Journal of Histochemistry and Cytochemistry, 51*(3), 271–274.

Martell, J. D., Deerinck, T. J., Sancak, Y., Poulos, T. L., Mootha, V. K., Sosinsky, G. E., et al. (2012). Engineered ascorbate peroxidase as a genetically encoded reporter for electron microscopy. *Nature Biotechnology, 30*(11), 1143–1148. http://dx.doi.org/10.1038/nbt.2375.

McDonald, K., Schwarz, H., Muller-Reichert, T., Webb, R., Buser, C., & Morphew, M. (2010). "Tips and tricks" for high-pressure freezing of model systems. *Methods in Cell Biology, 96*, 671–693. http://dx.doi.org/10.1016/S0091-679X(10)96028-7.

McDonald, K. L., & Webb, R. I. (2011). Freeze substitution in 3 hours or less. *Journal of Microscopy, 243*(3), 227–233. http://dx.doi.org/10.1111/j.1365-2818.2011.03526.x.

Nishiyama, H., Suga, M., Ogura, T., Maruyama, Y., Koizumi, M., Mio, K., et al. (2010). Atmospheric scanning electron microscope observes cells and tissues in open medium through silicon nitride film. *Journal of Structural Biology, 169*(3), 438–449. http://dx.doi.org/10.1016/j.jsb.2010.01.005.

Nixon, S. J., Webb, R. I., Floetenmeyer, M., Schieber, N., Lo, H. P., & Parton, R. G. (2009). A single method for cryofixation and correlative light, electron microscopy and tomography of zebrafish embryos. *Traffic, 10*(2), 131–136. http://dx.doi.org/10.1111/j.1600-0854.2008.00859.x.

Peddie, C. J., Blight, K., Wilson, E., Melia, C., Marrison, J., Carzaniga, R., et al. (2014). Correlative and integrated light and electron microscopy of in-resin GFP fluorescence, used to localise diacylglycerol in mammalian cells. *Ultramicroscopy, 143*, 3–14. http://dx.doi.org/10.1016/j.ultramic.2014.02.001.

Pluk, H., Stokes, D. J., Lich, B., Wieringa, B., & Fransen, J. (2009). Advantages of indium-tin oxide-coated glass slides in correlative scanning electron microscopy applications of uncoated cultured cells. *Journal of Microscopy, 233*(3), 353–363. http://dx.doi.org/10.1111/j.1365-2818.2009.03140.x.

Powell, R. D., Halsey, C. M., Spector, D. L., Kaurin, S. L., McCann, J., & Hainfeld, J. F. (1997). A covalent fluorescent-gold immunoprobe: Simultaneous detection of a pre-mRNA splicing factor by light and electron microscopy. *The Journal of Histochemistry and Cytochemistry*, *45*(7), 947–956.

Robinson, J. M., & Takizawa, T. (2009). Correlative fluorescence and electron microscopy in tissues: Immunocytochemistry. *Journal of Microscopy*, *235*(3), 259–272. http://dx.doi.org/10.1111/j.1365-2818.2009.03221.x.

Schalek, R., Kasthuri, N., Hayworth, K., Berger, D., Tapia, J., Morgan, J., et al. (2011). Development of high-throughput, high-resolution 3D reconstruction of large-volume biological tissue using automated tape collection ultramicrotomy and scanning electron microscopy. *Microscopy and Microanalysis*, *17*(Suppl. S2), 966–967.

Schaller, M. D. (2001). Paxillin: A focal adhesion-associated adaptor protein. *Oncogene*, *20*(44), 6459–6472. http://dx.doi.org/10.1038/sj.onc.1204786.

Shaner, N. C., Steinbach, P. A., & Tsien, R. Y. (2005). A guide to choosing fluorescent proteins. *Nature Methods*, *2*(12), 905–909. http://dx.doi.org/10.1038/nmeth819.

Shu, X., Lev-Ram, V., Deerinck, T. J., Qi, Y., Ramko, E. B., Davidson, M. W., et al. (2011). A genetically encoded tag for correlated light and electron microscopy of intact cells, tissues, and organisms. *PLoS Biology*, *9*(4), e1001041. http://dx.doi.org/10.1371/journal.pbio.1001041.

Sims, P. A., & Hardin, J. D. (2007). Fluorescence-integrated transmission electron microscopy images: Integrating fluorescence microscopy with transmission electron microscopy. *Methods in Molecular Biology*, *369*, 291–308. http://dx.doi.org/10.1007/978-1-59745-294-6_14.

Takizawa, T., & Robinson, J. M. (2000). FluoroNanogold is a bifunctional immunoprobe for correlative fluorescence and electron microscopy. *The Journal of Histochemistry and Cytochemistry*, *48*(4), 481–486.

Takizawa, T., Suzuki, K., & Robinson, J. M. (1998). Correlative microscopy using FluoroNanogold on ultrathin cryosections. Proof of principle. *The Journal of Histochemistry and Cytochemistry*, *46*(10), 1097–1102.

Tapia, J. C., Wylie, J. D., Kasthuri, N., Hayworth, K. J., Schalek, R., Berger, D. R., et al. (2012). Pervasive synaptic branch removal in the mammalian neuromuscular system at birth. *Neuron*, *74*(5), 816–829. http://dx.doi.org/10.1016/j.neuron.2012.04.017.

Wacker, I., & Schroeder, R. R. (2013). Array tomography. *Journal of Microscopy*, *252*(2), 93–99. http://dx.doi.org/10.1111/jmi.12087.

Watanabe, S., Punge, A., Hollopeter, G., Willig, K. I., Hobson, R. J., Davis, M. W., et al. (2011). Protein localization in electron micrographs using fluorescence nanoscopy. *Nature Methods*, *8*(1), 80–84. http://dx.doi.org/10.1038/nmeth.1537.

Wouters, C. H., & Koerten, H. K. (1982). Combined light microscope and scanning electron microscope, a new instrument for cell biology. *Cell Biology International Reports*, *6*(10), 955–959.

Wouters, C. H., Koerten, H. K., Bonnet, J., Daems, W. T., & Ploem, J. S. (1986). Quantitative DNA measurements in an instrument combining scanning electron microscopy and light microscopy. *Journal of Microscopy*, *141*(Pt. 1), 41–53.

Zonnevylle, A. C., Van Tol, R. F., Liv, N., Narvaez, A. C., Effting, A. P., Kruit, P., et al. (2013). Integration of a high-NA light microscope in a scanning electron microscope. *Journal of Microscopy*, *252*(1), 58–70. http://dx.doi.org/10.1111/jmi.12071.

CHAPTER

Correlation of the Same Fields Imaged in the TEM, Confocal, LM, and MicroCT by Image Registration: From Specimen Preparation to Displaying a Final Composite Image

18

Douglas R. Keene*, Sara F. Tufa*, Melissa H. Wong[†], Nicholas R. Smith[†], Lynn Y. Sakai*, William A. Horton*

**Research Center, Shriners Hospital for Children, Portland, Oregon, USA*

[†]Department of Cell and Developmental Biology, Oregon Health Sciences University, Portland, Oregon, USA

CHAPTER OUTLINE

Methods in Cell Biology, Volume 124, ISSN 0091-679X, http://dx.doi.org/10.1016/B978-0-12-801075-4.00018-5

Abstract

Correlated imaging is the process of imaging a specimen with two complementary modalities and then registering and overlaying the fields obtained in each modality to create a composite view. One of the images is made somewhat transparent, allowing detail in the underlying image to be visible and assisting in the registration of the two images. As an example, an image localizing a specific tissue component by fluorescence may be overlaid atop a TEM image of the same field. The resulting composite image would demonstrate specific ultrastructural features in the high-resolution TEM field, which are colorized in the overlay. Other examples include composites from MicroCT or soft X-ray images overlaid atop light microscopy or TEM images. Automated image registration may be facilitated by a variety of sophisticated computer programs utilized by high-throughput laboratories. This chapter is meant for the more occasional user wishing to align images manually. ImageJ is a public domain, image processing program developed at the National Institutes of Health and is available to anyone as a free download. ImageJ performs marvelously well for the purpose of image registration; therefore, step-by-step instructions are included here. Specimen handling, including fixation and choice of embedding media, is not straightforward for correlative imaging. A step-by-step description of the protocols which work in our laboratory is included for simultaneous localization in LM, EM and micro-CT, as well as maintaining GFP emission in tissue embedded for TEM.

INTRODUCTION

No single imaging technology can reveal all the structural details within a biological specimen, although the resolution gap between two of the more common methods, light microscopy (LM) and electron microscopy, is certainly narrowing with the introduction of "super resolution" microscopes. Built upon a platform of conventional fluorescent microscopy, these technologies allow the positional mapping of light emitted from a sample, but with nanometer resolution. Still, the images produced by these methods lack structural detail; it is only by correlation with additional imaging methods that a specific region emitting fluorescence may be identified. This correlation is dependent on recognizing common features in images collected with each instrument. The aim of this chapter is to suggest specimen preparation and imaging tools aiding in the registration of two or more image fields to produce a single aligned composite image.

The need to correlate LM images with higher magnification transmission electron microscopy (TEM) images has long been recognized by microscopists. Deemed "CLEM" (correlative light and electron microscopy), a brief PubMed search yields many examples of the methodology. From the standpoint of aligning images and producing an image overlay from two different modalities, correlating images from the same section observed by fluorescence microscopy (FM) and TEM is straight forward. Sims and Hardin (2005) presented a method to overlay laser scanning confocal microscopy (LSCM) images of a GFP construct localizing AJM-1 to the apical junctions of *Caenorhabditis elegans* embryos with TEM images collected from the same ultrathin sections. Using Adobe Photoshop, the resulting images were aligned and

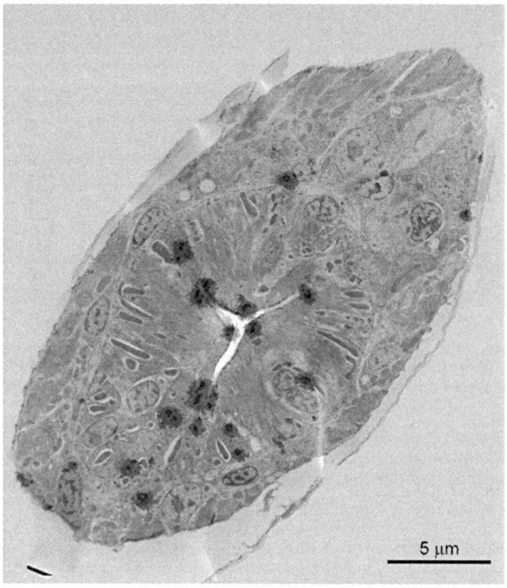

FIGURE 18.1

Caenorhabditis elegans embryo expressing GFP fusion protein. An LSCM image is overlaid with a TEM image collected from the same grid to produce an integrated image. (See the color plate.)

Reprinted with permission from Sims and Hardin (2007).

overlaid to produce a "fluorescence-integrated TEM" image (Fig. 18.1). Describing further refinements in methodology, Sims and Hardin (2007) published a step-by-step sequence of their protocol. Building upon this work, our laboratory (Keene, Tufa, Lunstrum, Holden, & Horton, 2008) developed an alternative protocol for stabilizing GFP- and YFP-expressing constructs in cell culture systems. We stabilized the cells in ice-cold 4% paraformaldehyde/1% glutaraldehyde for 30 min followed by progressive lowering temperature (PLT) dehydration (Newman & Hobot, 1987) to 90% EtOH, followed by embedding in London Resin White (LR White) media. Adequately bright images were obtained in our Leica SPS SP2 laser scanning confocal system from 0.5-μm sections; however, GFP and YFP emission was not of sufficient intensity to image in ultrathin sections. We therefore collected fluorescent images from 0.5-μm thick sections and TEM images from adjacent ultrathin sections. The registration of images was complicated by the fact that the sectioning artifacts in 0.5-μm sections mounted on glass are not consistent with sectioning artifacts in 80-nm sections mounted on formvar-coated grids; therefore, the two image sets could not be registered without the need to "skew" or distort one image relative to the other. Despite these challenges, the resulting composite images clearly demonstrated localization of GFP and YFP constructs at the resolution of the electron microscope (Fig. 18.2). As an aid in initial field alignment, we routinely adjusted the brightness and contrast of the FM image to visualize cell silhouettes as these were easily identified in low-magnification TEM images. This important step allowed

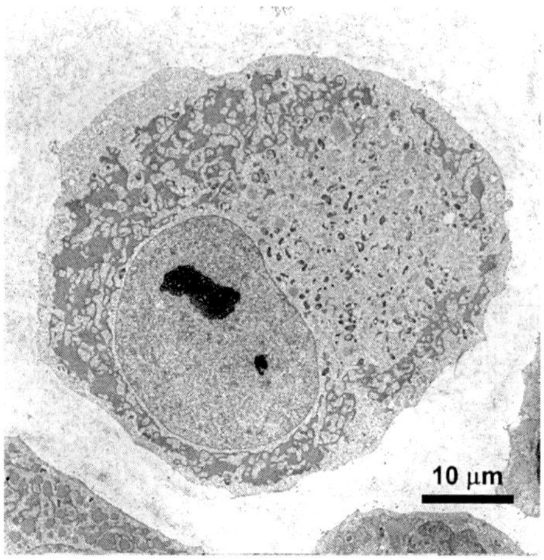

FIGURE 18.2

An LSCM image from a 0.5-μm section was overlaid with a TEM image from a serial ultrathin section to produce this composite overlay of a human chondrocyte expressing YFP::mCOMP. (See the color plate.)

From Keene et al. (2008).

rotational alignment so that the two images were in the same axial plane. Additional features common to both imaging modalities could be recognized at progressively higher magnifications.

As an initial evaluation of fluorescence emission intensity, imaging the block face of an embedded sample is often a valued starting point. As an example, only a fraction of the cultured cells shown in Fig. 18.2 expressed the YFP::mCOMP construct. Areas within intact samples could be selected by placing the block surface facedown on a large cover slip and imaged by FM or LSCM using the 20 × objective lens of an inverted microscope (Fig. 18.3). Similarly, using a 63 × water dipping objective lens, Bell, Mitchell, Paultre, Posch, and Oparka (2013) were able to detect fluorescent proteins in optical sections up to 40 μm below the surface of the block face in LR White-embedded plant tissue. With the knowledge that a feature of interest could be mapped via optical coordinates within a "*z*" stack relative to the block face, the position within the block from which 0.5-μm sections would be cut and later overlaid with TEM images could be determined with precision and correlated back to the intact sample.

Correlation between other imaging modalities, such as LM or TEM with micro-computed X-ray tomography (micro-CT), may be similarly accomplished. Micro-CT is conventionally used to generate high-resolution 3D images for the nondestructive evaluation of mineralized tissues such as insect exoskeletons, shells, and skeletal tissues (Neues & Epple, 2008; Oest, Jones, Hatfield, & Prater, 2008). It was not until the utilization of stains such as inorganic iodine, phosphotungstic acid, and osmium (compared by Metscher, 2009a) that micro-CT use became widespread for the

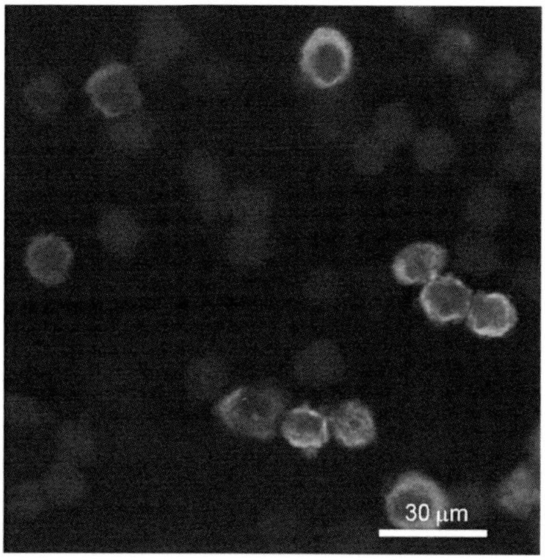

FIGURE 18.3

The "faced" block of an intact sample may be imaged to select a field for further sampling.

evaluation of soft tissues. The critical elements in these stains have high atomic numbers (Z) with correspondingly high X-ray attenuations (Henke, Gullikson, & Davis, 1993) favorable to micro-CT imaging (iodine $Z=53$; osmium $Z=76$; lead $Z=82$, and uranium $Z=92$). With the exception of iodine, these elements have been long favored as contrasting stains in TEM. Their use as contrast agents for micro-CT allows high-volume images of intact organisms. Brown fat and red blood cells are particularly well contrasted with iodine allowing analysis of blood-containing vascular anatomy. Muscle fiber architecture may also be evaluated nondestructively with iodine staining (Jeffery, Stephenson, Gallagher, Jarvis, & Cox, 2010). Osmium is recognized to differentially stain lipids within white and brown adipose tissue, also contrasting the lipoproteins of most bilayered cell and organelle membranes (Hayat, 1970). This broad-staining characteristic allows detailed imaging of both external and internal regions of the embryos and organ systems (Johnson, Hansen, Wu, Healy, & Johnson, 2006). Using this technology, we (Sengle, Tufa, Sakai, Zulliger, & Keene, 2013) proposed to evaluate the fat content of mice harboring a mutation in the connective tissue component *fibrillin-2*. We prepared intact mouse limbs as if for TEM and then mapped small fat deposits in *x*, *y*, and *z* relative to the block surface by micro-CT, without physically cutting into the sample. Following identification of fat deposits in the micro-CT, we used an ultramicrotome to accurately remove tissue sections so that the deposits might be precisely approached. Similar to the method of overlaying fluorescent and TEM images, micro-CT images were registered and overlaid atop LM images from half-micron sections (Fig. 18.4). To determine if the deposits were white or brown fat, micro-CT images were overlaid onto TEM images collected from ultrathin sections (Fig. 18.5). Further exemplifying micro-CT as a complementary method to LM and TEM, Handschuh, Baeumler, Schwaha, & Ruthensteiner (2013) imaged the fixed and embedded mollusk *Mytilus*

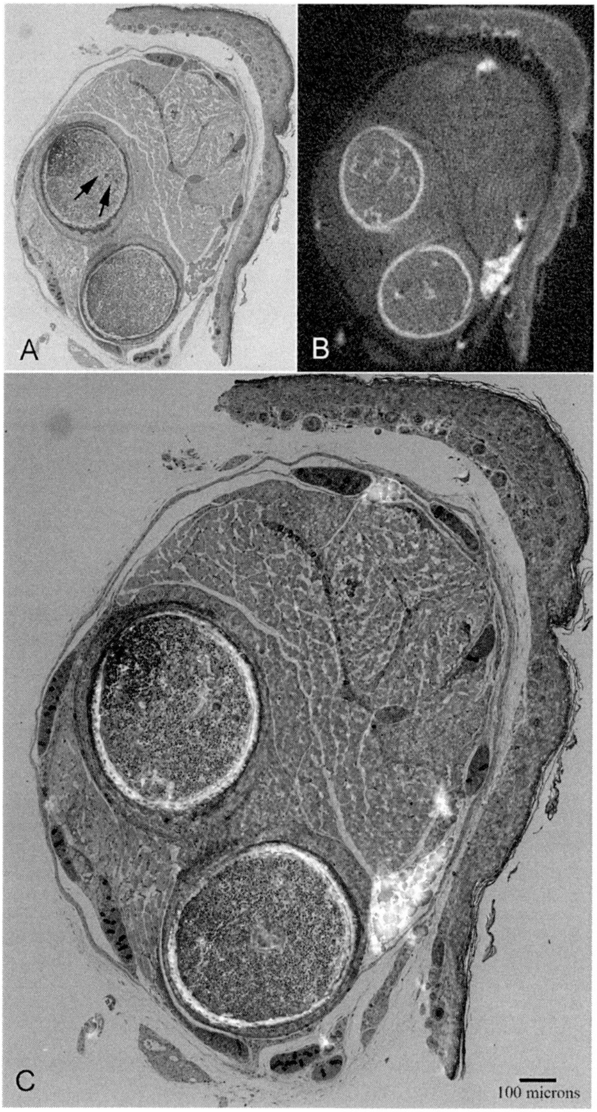

FIGURE 18.4

A virtual slice of a region of high X-ray attenuation (B) is selected from the micro-CT data set, then registered with an LM image (A) mapped to the same region. The regions of high X-ray attenuation correlate to adipose tissue (C).

From Sengle et al. (2013).

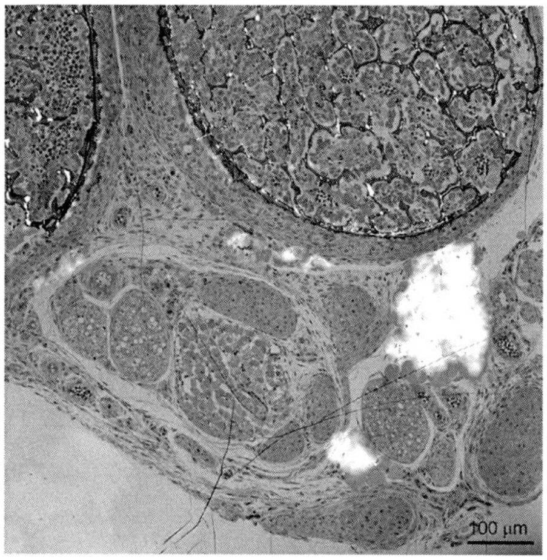

FIGURE 18.5

A virtual slice within a micro-CT data set is overlaid with a TEM image of the same region.

From Sengle et al. (2013).

galloprovincialis by micro-CT. Serial 0.5-μm LM sections ($n = 585$) were then cut through the entire organism. The micro-CT dataset was visualized by volume rendering using AMIRA. The renopericardial system, the nervous system, the digestive system, muscles, and the gill vessels were segmented. Coregistration of the micro-CT and LM images was accomplished by "affine registration" in AMIRA. The capabilities of this method are impressive, as any plane within the volume of the organism may be segmented out, with corresponding LM images selected within AMIRA. Any one of the sections collected for LM can be remounted and cut for TEM, allowing visualization of virtually any chosen region within the organism by high-resolution microscopy. The method is particularly valuable, as it may be applied to the evaluation of archived TEM sample blocks. Similarly, Smith et al. (2014) correlated structural data from soft X-ray tomography of intact, high-pressure frozen *Schizosaccharomyces pombe* cells with internal localization of a GFP construct collected using a cryorotation stage fitted within an LSCM system.

The recognition of the same structures in the LM and EM is often not trivial, and one must rely on additional markers to coregister image sets. As demonstrated by Micheva, Busse, Weiler, O'Rourke, and Smith (2010), DAPI images collected in the epifluorescent microscope may be aligned with condensed heterochromatin imaged in the TEM. Yuan, Li, Hong, and Hong (2014) aligned IF and TEM images to localize a GFP::mitochondrial fusion protein on the surface of semi-thin sections prepared from high-pressure frozen, Lowicryl HM20 embedded tissue samples. A secondary antibody conjugate with both a fluorescent and an ultrasmall (0.6-nm) gold component (nanogold) was utilized. The registered overlays of IF and TEM images demonstrated that localization was within fields of mitochondria;

however, it was the immunogold that resolved the localization to the outer membrane of the mitochondria. The method was complicated only by the need to "enhance" the ultrasmall gold particles to be large enough to visualize in the TEM. Molecular Probes now markets a variety of secondary antibody conjugates with fluorescent and large (6 or 10 nm) gold components, eliminating the need for enhancement. We have used these conjugates to successfully localize lysozyme to RER and Paneth cell granules on the surface of sections from mouse small intestine embedded in LR White (Fig. 18.6). A detailed protocol for specimen preparation, imaging, and alignment leading to this composite image is presented later in Section 18.1.

In our laboratory, an often employed strategy to aid in the initial recognition of features in both the IF or LSCM and TEM relies on mounting ultrathin sections on 1 × 2-mm single-hole, formvar-coated grids so that the entire section fits within the grid window. We shape the block face so that it is nonsymmetrical, allowing mapping of an ROI relative to a distinctive edge recognizable in both imaging modes. Sections may also be mounted on "finder" grids. This initial mapping saves time in locating a single cell among many others. Should the fate of an individual live cultured cell be followed, for example after microinjection, the cells may be grown on cover slips that are photoetched so that markings visible and recorded in the LM remain in the ultrathin section (Reddick & Alto, 2012), or cells may be grown on aclar films having recognizable patterns etched into them using the pulsed laser of a microdissection microscope (Spiegelhalter et al., 2010).

A method which results in detail immediately apparent in both imaging modalities involves the use of a "miniSOG," a fluorescent protein which may be genetically tagged to virtually any protein. When illuminated by blue light in fixed tissue, photogenerated singlet oxygen locally polymerizes diaminobenzidine resulting in a

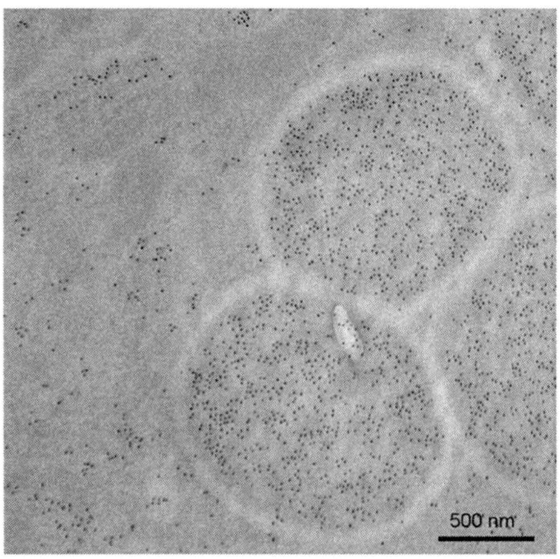

FIGURE 18.6

Molecular Probes Alexa 488/10-nm gold secondary conjugate allows simultaneous IF and TEM localization of the same tissue component. (See the color plate.)

precipitate that can be readily visualized in the electron microscope (Ellisman, Deerinck, Shu, & Sosinsky, 2012; Shu et al., 2011). Image alignment between the two modalities is complicated only since initial illumination is done prior to osmication and embedding, and EM is done on ultrathin sections collected from the bulkier sample.

At higher magnifications, the use of dedicated correlation markers (fiducial markers) visible in the different imaging modalities is particularly useful for image registration. Schellenberger et al. (2014) registered cryo-FM and cryo-EM fields according to finder-grid marks and gross sample features, then utilized TetraSpeck (Invitrogen) 200-nm diameter multichannel fluorescent microspheres to monitor shifts between multicolor fluorescent image channels. These particles were also visible in phase-contrast cryo-EM projection images and could be used to assign coordinates between LM and EM to automatously record tomograms of structures initially identified by cryo-FM. Kukulski et al. (2012) describe a similar method in tissues prepared by HPF, freeze substitution, and embedding in Lowicryl. Kopek, Shtengel, Grimm, Clayton, and Hess (2013) utilized photoactivated localization microscopy (PALM) and later interferometric PALM (iPALM) (Kopek, Shtengel, Xu, Clayton, & Hess, 2012), to fantastically localize proteins within the submitochondrial structure known as the nucleoid. iPalm localizes photoactivatable fluorescent proteins in the image plain to approximately 20-nm resolution, and further extracts the *z* position to approximately 10-nm resolution. Tokuyasu cryosections were prepared from fixed cell cultures supported on a cover glass coated with 80-nm gold particles; then, the sections were then stained, covered with cyanoacrylate, and imaged by FIB-SEM. The 80-nm gold particles were visible in both image modalities, and together with a precise, automated registration program (POLYWARP 1 image transformation from the IDL image processing language, http://www.exclisvis.com/docs/POLYWARP.html), the average registration error approximated only 5-nm (Fig. 18.7). The use of these 80-nm dedicated fiducial markers was absolutely critical to the registration of images and to precise localization within this mitochondrial domain.

18.1 METHODS

18.1.1 GENERAL DISCUSSION

For the preservation of native structure, the method of choice for the immobilization of tissue components begins with cryostabilization by high-pressure freezing (reviewed by McDonald, 2009). A significant advantage to freezing is that a biological event can be followed using the live stage of an LM, then immobilized by fast freezing within a few seconds of surveillance. Imaging cryopreparations at low temperature may be accomplished using low-cost home-fabricated (Carlson & Evans, 2011) or more sophisticated devices (McDonald, 2009; Schwartz, Sarbash, Ataullakhanov, McIntosh, & Nicastro, 2007; van Driel, Valentijn, Valentijn, Koning, & Koster, 2009; Verkade, 2008). As the problems associated with

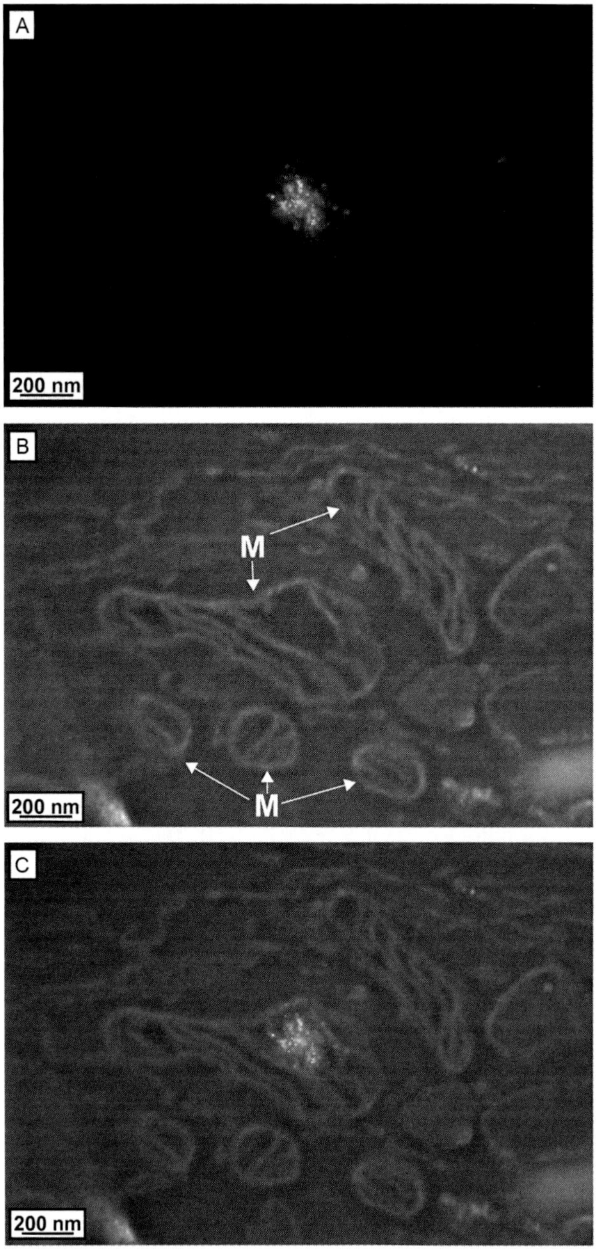

FIGURE 18.7

Panel (A) shows the PALM image of cells expressing TFAM-mEos2 and panel (B) shows mitochondria (M) in a backscatter image collected by SEM of the same area. Panel (C) is the integrated PALM/SEM image showing localization to the mitochondria nucleoloid. (See the color plate.)

Reprinted with permission from Kopek et al. (2013).

postembedding methods are avoided using cryofixation following the Tokuyasu cryo immunogold labeling technique (Tokuyasu, 1973), the method appears to have the best success for maintaining emission of GFP fusion proteins; however, subcellular details are often obscured in cryosections due to poor tissue contrast.

Cryostabilization followed by freeze substitution and embedding may result in improved morphology, but fluorescent proteins are quenched by the acidic, dehydrated, and oxidizing conditions required for fixation and polymer embedding of the specimen. Hence, the emission intensities of many flourophores may not be adequate following embedding. Reviewed by Campbell and Choy (2001), the pH of wild-type GFP (wtGFP) is stable from pH 6 to 10, decreases at pH less than 6, and increases from pH 10 to 12 (Haupts, Maiti, Schwille, & Webb, 1998). Different GFP variants have different pH sensitivities, with current GFP mutants having pK_as ranging from 4.8 to 8.0. In the context of this paragraph, it is interesting to note that the pH sensitivity of GFP variants may serve as molecular reporters of shifting pH environments (Bizzarri, Serresi, Luin, & Beltram, 2009). The need to monitor and adjust pH during processing of tissues carrying GFP constructs is essential, and knowledge of the peculiarities of the GFP variant may be invaluable. In our hands, YFP seems to endure fixation, dehydration, and embedding media somewhat better than wtGFP. Brown, Fetter, Tkachuk, and Clayton (2010) used PALM microscopy to localize mitochondrial proteins in HPF-cultured cells, which were subsequently freeze substituted in 95% EtOH containing 1–2% glutaraldehyde, then infiltrated in LR White with 95% EtOH at −20 °C, then cold-polymerized at −20 °C using benzoyl peroxide-catalyzed resin. In a detailed protocol, the authors point out the importance of storing the catalyzed resin for no more than 1 month, and further discuss a method for adjusting the pH of LR White to near-neutral using ethanolamine. Watanabe et al. (2011) also used PALM and simulated emission depletion microscopy together with electron microscopy to localize fluorescently tagged proteins at high resolution. This comprehensive study in methodology points out the need to avoid absolute dehydration, the need to adjust the pH of LR White from the typical 5.5 to just under 7.0, and also the variability of LR White in polymerization. In our experiments between 2005 and 2007, we repeatedly localized YFP and GFP constructs within several tissues using LR White without adjustment of pH. But within the past few years, we have been much less successful in localizing stable GFP constructs in many of these same cells and tissues newly embedded in LR White. A qualitative assessment of the emission intensity of GFP, observed throughout the protocol, revealed that emission remained stable through fixation and Tris–Glycine, then decreased slightly but progressively through graded dehydration to 90% EtOH. There was a marked decrease in emission during infiltration in 1:1, 1:2, and 1:3 of 90% EtOH:LR White, then emission strongly decreased during infiltration in 100% LR White. This is true despite having adjusted the pH of the media and varying fixation and dehydration conditions. Many report that the GFP-expressing system of their choice is sensitive to higher percentages of glutaraldehyde and limit exposure to 0.1%; however, we have experimented with up to 1% glutaraldehyde with 4% paraformaldehyde without significant decrease in GFP emission.

Despite failures with LR White, we have had success in parallel experiments using LR Gold polymerized at −20 °C. In our hands, LR Gold does not section as well as LR White and the ultrastructure suffers considerably. McDonald, Sharp, and Rickoll (2012) report good results using Lowicryl HM20 polymerized at low temperature. From the varying reports in the literature and from our own experience, it is clear that for each experimental system a compromise must be found between preservation of GFP fluorescence and tissue morphology.

Methods of localization using primary and secondary antibody conjugates circumvent the issues of preserving GFP fluorescence *en bloc*. The size of primary antibodies (about 19 nm in length) limits the ultimate resolution of localization particularly when secondary antibody conjugates are also used, but the method may be preferred when GFP-expressing systems are not available or when GFP is not visible after embedding. As an example of the latter, sections cut from GFP-expressing tissue may be labeled with primary antibody to GFP followed by a secondary antibody conjugate. Several commercial antibodies do well in labeling exposed GFP epitopes on the surface of lightly fixed, LR White-embedded tissue including Abcam (#ab290) and Life Technologies (#A11122). Figure 18.8 demonstrates localization of mCOMP::YFP to dilated RER in a cultured cell embedded in LR White followed by a −6-nm gold secondary antibody conjugate. Although we favor LR White for labeling antigens exposed at the surface of sections, others including Fabig et al. (2012) report success using the methacrylate Lowicryl K4M for immunolabeling nonfluorescing GFP domains. In theory, virtually any primary

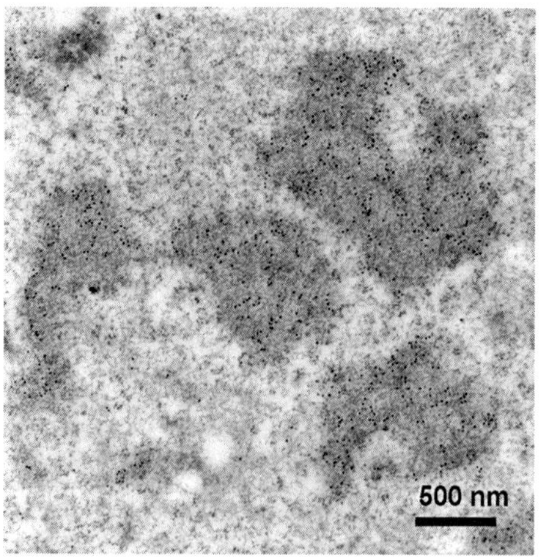

FIGURE 18.8

Ultrathin sections cut from cells expressing mCOMP::YFP fusion protein were immunolabeled with a primary antibody to GFP followed by a secondary −6 nm gold conjugate, resulting in high-density labeling to dilated RER.

Reprinted from Keene et al. (2008).

antibody can be used successfully to localize an antigen using a postembedding protocol. The screenshots shown in Workflow for Overlaying Images Using FIJI presented in Section 18.1.8 are from mouse small intestine fixed in 4% paraformaldehyde/0.5% glutaraldehyde and embedded in LR White. Ultrathin sections were immunolabeled using rabbit antibody to lysozyme. A 10-nm gold/Alexa 488 goat anti-rabbit secondary conjugate facilitated imaging by FM or LSCM; then the same grid was examined for gold particle distribution in the TEM. Secretory granules within Paneth cells were intensely florescent, and corresponding structures could be found easily in the TEM.

18.1.2 SAMPLE PREPARATION

Notes

- Cultured cells are grown on Lux "Thermanox" 13-mm diameter cover slips which are "cell culture treated" on one side, sterile, and impervious to acetone, propylene oxide and embedding media.
- Costar 24-well culture clusters may be used as both the growth and processing chamber for the cover slips up to and including 90% ethanol. They are manufactured from polystyrene and will not tolerate acetone, propylene oxide or embedding media.
- Sarstedt 50-ml *polypropylene* centrifuge tubes may serve as processing chambers for individual cover slips. Polypropylene will tolerate acetone, propylene oxide and embedding media, and the round sides and conical bottom ensure that the culture surface of the cover slips is protected from mechanical damage.
- Embedding is accomplished using polypropylene Wheaton "snap caps" which have been baked a minimum of 24 h in a 60–70 °C oven. Do not use paper labels. Embed the discs with the culture surface facing up.
- LR White will not polymerize in the presence of atmospheric oxygen. For thermal polymerization of LR White, include one or two small thermoses of liquid nitrogen introduced into the oven just prior to the samples. The oxygen-rich atmosphere in the oven will be displaced with an oxygen-depleted atmosphere (Fig. 18.9).
- LR Gold also will not polymerize in the presence of atmospheric oxygen. For polymerization at −20 to −35 °C, we utilize the chamber of a Leica AFS (automatic freeze substitution) device, which ensures a nitrogen-rich atmosphere during polymerization using an UV light source. Alternatives to the AFS are gelatin embedding capsules which have been warmed in a 70 °C oven prior to use, driving out excess moisture from the gelatin. The gelatin capsules are filled with media, then capped with as little air as possible. Do not use Beem capsules since the UV light will not penetrate these capsules. Polymerization can be accomplished using a "black light" bulb over a foil-lined box placed in the freezer.
- For best results, use cryofixation (high-pressure freezing) followed by low-temperature embedding in Lowicryl HM20. If this is not possible, the next

FIGURE 18.9

A thermos filled with liquid nitrogen is placed in the polymerization oven to insure an oxygen-depleted atmosphere.

best approach is to use PLT (progressive lowering of temperature) dehydration, followed by infiltration and embedding with Lowicryl HM20, LR Gold, or LR White.

18.1.3 *EN BLOC* GFP LOCALIZATION IN CULTURED CELLS AND TISSUES

- Tissues should be processed as quickly after sacrifice as possible. Cultured cells are grown on 13-mm diameter Lux Permanox cover slips.

GFP emission may be monitored during the protocol by sacrificing a disc which is placed cell side down on a cover glass, then imaged by FM or LSCM with an inverted microscope.

- Within a few minutes of removing the culture from the incubator, rinse the cells in warm, serum-free Dulbecco's modified eagle media (SFM) twice, each for 5 min. Using a stereoscope, scratch a mark onto the disc so that the cell side of the disc can be recognized during embedding. Then transfer the processing chamber to an ice bucket. Fresh tissue should be rinsed briefly in ice-cold SFM.

All further processing of cell cultures should be done using ice-cold solutions and the samples kept on ice for the duration of the procedure (see PLT protocol for tissues).

- Fix cultured cells for 30 min in ice-cold 4% paraformaldehyde/1% glutaraldehyde buffered with SFM. *Tissue* should be cut into pieces no larger than 1×2 mm and fixed 30–60 min, dependent on size.
- Rinse cultured cells two times over 5 min; rinse *tissues* three times over 15 min in SFM.

- Incubate cultures in 0.15 M Tris–HCl, pH 7.4, containing 0.05 M Glycine for 60 min on ice. *Tissues* should incubate 4 h overnight in ice-cold solution.
- Dehydrate cultures and tissues in ice-cold ethanol: 15 min each in 30%, 50%, 70%, and 90% ethanol in water.
- Infiltrate cultures and tissues for 30 min each in ice-cold 1:1, 1:2, and 1:3 of 90% EtOH:LR Gold (including 0.5% benzoin methyl ether).

If possible for tissues, dehydrate at progressively lower temperature beginning on ice with 30% ethanol then cooling to −10 °C over 15 min, then at −10 °C with 50% ethanol; than in 70% ethanol at −10 °C cooling to −20 °C over 15 min, then continuing all additional steps to 100% embedding media at −20 °C.

Achieving the low temperatures necessary for PLT dehydration and low-temperature polymerization of embedding media may be accomplished using the Leica AFS device. If this tool is not available, temperatures from −17 to −77 ° may be achieved using a cooling bath consisting of dry ice with varying ratios of ethylene glycol and ethanol. For example, a temperature of −28 ° may be reached with ethylene glycol at 80% and 100% ethanol at 20% (Lee & Jensen, 2000). As suggested by Dr. Kent McDonald (U.C. Berkeley), temperatures in this range may also be empirically acquired by positioning a rack above dry ice (personal communication).

- Infiltrate cultures for 30 min each in three changes of ice-cold 100% LR Gold (including 0.5% benzoin methyl ether) or HM20 monostep (requiring no catalyst). Incubate tissues for 60 min each change. For tissues, the last change may continue overnight. For cultured cells, avoid an overnight incubation in media.

Note: LR Gold requires the addition of 0.50% benzoin methyl ether to polymerize at low temperature. Mix the benzoin methyl ether into solution using nitrogen bubbles. Use only glass or polypropylene pipettes when handling LR Gold or LR White.

- For polymerization of cultured cells in the Leica AFS device, cool the sample chamber to −10 °. Scratch an identifying mark onto the outer surface of the Wheaton snap cap (marker ink solubilizes upon contact with media). Place these empty snap caps into the chamber and fill half way with 100% media. Quickly transfer tissues or cultured cell discs, cell side up and submerge into the media. Hopefully, a recognizable mark (such as an "a," "b," "L," but not an "*x*") was scratched onto the cell side of the disc; otherwise, the cells are difficult to see and the cell side will need to be determined using a stereoscope during which time condensation will form on the warming surfaces, potentially causing problems with polymerization. After the samples are introduced, fill the caps to the rim with media.
- Decrease the temperature of the chamber to the polymerization temperature (−20 to −35 °). This will cool the chamber and also fill it with a nitrogen-rich atmosphere. Cover the sample chamber with glass then polymerize using the UV light included with the system. Polymerize for 24 h and if necessary, continue polymerization at 4 ° for another 4 h.

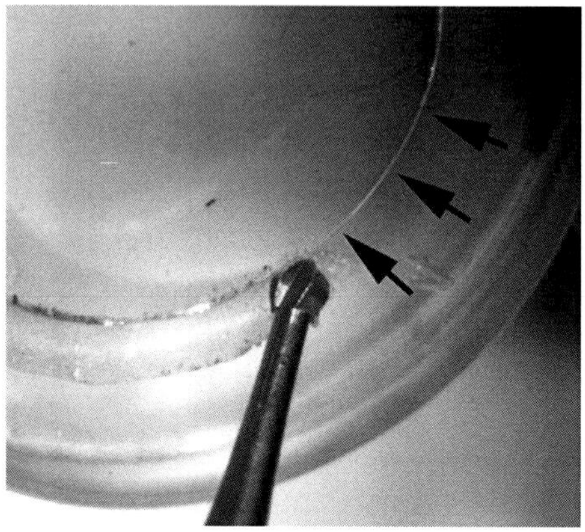

FIGURE 18.10

A partial edge of a round thermanox cover slip embedded in LR White is delineated with arrows; the remaining edge of the cover slip has been exposed by a trench cut using a Dremel 3000 variable speed rotary tool fitted with a Dremel #106 engraving cutter. The cover slip will separate easily after the entire edge of the disc is exposed by the trench.

- After polymerization, there may be a thin film of liquid media at the surface of the polymerized discs; wipe it away with 100% ethanol. As the samples are released from the polypropylene snap caps, score an identifying mark into the polymerized media; marking pens will not be permanent. To ensure easy release of the culture discs from polymerized media, use a high-speed rotary tool (Dremel) fitted with a Dremel 106 1/16″ Engraving Cutter, carving a trench around the edge of the disc (Fig. 18.10), then the disc will separate easily. Use a fine coping saw to cut out individual blocks from the culture.

18.1.4 EMBEDDING CULTURES AND TISSUES FOR POSTEMBEDDING, SECTION-SURFACE LABELING

Prepare cultured cells and tissues with the same protocol listed above for *en bloc* GFP localization; however, the aldehyde concentration of the primary fixative may need to be reduced to as little as 1.5% paraformaldehyde/0.1% glutaraldehyde. In the initial pilot experiment, try different aldehyde concentrations to determine the best compromise between labeling density and ultrastructure. For reference, the tissue shown in Fig. 18.6 was fixed in 4% paraformaldehyde/0.5% glutaraldehyde. Our best results are with the embedding media LR White without pH adjustment. Others (McDonald et al., 2012) report success with Lowicryl HM20. Swirl the contents within the bottle containing media prior to use, and add a nitrogen cap when the bottle is stored. Polymerization is accomplished in polypropylene snap caps in a 60 ° oven with an opening (thermometer, vacuum vent valve), door held closed, containing one or two small thermoses of liquid nitrogen (Fig. 18.9). Two Styrofoam coffee cups may be stacked together to contain liquid nitrogen. A large beaker of dry ice is an alternative. Do not open the oven for several hours to ensure an oxygen-depleted atmosphere.

18.1.5 IMMUNOLABELING THE SURFACE OF ULTRATHIN SECTIONS

Protocols for postembedding labeling vary; McDonald et al. (2012) describe a method which works well in their hands. We use the following protocol for virtually all epitopes exposed at the surface of an ultrathin section.

- Sections are cut using a diamond knife with a low water level in the boat, just enough to keep the diamond knife edge wet. An antistatic device is used to help keep the block face dry. Sections are cut 60–90 nm thick and mounted onto formvar-coated 1 × 2-mm nickel slot grids, recently made hydrophilic by glow-discharge. We use a homemade device for discharge (after Aebi & Pollard, 1987). This step helps ensure that the section lies flat onto the formvar. Use nonmagnetic forceps when handling nickel grids. Dry the grids thoroughly before proceeding.
- Use a PTFE Immunostaining Pad (Pelco) with 8 × 5 depressions. You may also use parafilm. Float the grids, section side down, onto the surface of each solution. Place the immunostaining pad in a glass Petri dish with wet filter paper at the bottom to ensure a humid environment. Move the grids from one solution to another using nonmagnetic forceps or a wire loop. All incubations are at ambient temperature. Incubate in 30-μl of each of the following solutions:
- 0.15 M Tris–HCl, pH 7.5, for 15 min
- 0.15 M Tris–HCl, pH 7.5, with 0.05 M glycine for 60 min
- 2% nonfat dry milk, 0.5% ovalbumin, and 0.5% fish gelatin in Tris–HCl for 30 min
- Quick rinse in 0.15 M Tris–HCl
- Primary antibody with appropriate dilution: typically, we dilute concentrated antibody (1 mg/ml) 1:50 in Tris–HCl. This is the dilution we use for the Abcam polyclonal antibody 290 recognizing GFP. Incubate for 2.5 h.
- Three drops Tris–HCl over 15 min
- Secondary antibody with appropriate dilution: typically, we dilute Aurion secondary gold conjugate 1:10; we also dilute Molecular Probes Alexa 488/10nm gold dual conjugate 1:10 in Tris–HCl. Incubate for 60 min.
- Three drops of Tris–HCl over 15 min
- Three drops of dH_2O over 15 min
- Wick away excess water and dry the grids.

18.1.6 ULTRAMICROTOMY AND OBSERVATION OF SEMI-THIN AND ULTRATHIN SECTIONS USING CONFOCAL OR EPIFLUORESCENT IMAGING

- For CLEM, blocks are trimmed with a nonsymmetrical block shape. The block is faced off with a glass or diamond knife; then it may be placed facedown on a cover slip and evaluated for positive fluorescent emission by either FM or LSCM using an inverted microscope.

- Semi-thin (0.5–1.0 μm) sections are picked from the water surface of a glass or diamond knife boat using an eyelash stick, then transferred to a small drop of water on a 24 × 50-mm No.1 glass cover slip. The sections are air-dried and examined unstained and without mounting media. Ultrathin sections are mounted on single-hole 1 × 2-mm formvar-coated slot grids, then placed section side down on the surface of a cover glass supported by the stage of an inverted microscope. Images may be collected using dry objectives; when imaged using a 63 ×, water immersion 1.2 NA lens, a drop of water is added to the lens side of the cover slip but *not* the grid side. Any aberrations in the imaging path resulting from the tiny air interface under the grid are negligible.
- Initial assessment by IF or LSCM is done as quickly as possible to minimize photo bleaching. Low-magnification, low-resolution images are collected to determine a region of interest. These low-magnification images are invaluable as "section maps," as a first step in correlating LM imaging fields to EM fields. It is also helpful to record on paper (draw) the relative position of a region of interest within the space of the nonsymmetrical section. Image collection is a compromise between photo bleaching and high-resolution recording.
- For the final image used in CLEM, using LSCM we typically collect a 2048 × 2048 image with a scan speed of 400 and line averaging = 4 with no field averaging. We collect an image of the ROI first with a Plan Apo 20 × dry objective (NA = 0.8), then with a Plan Apo 63 × water immersion objective (NA = 1.2). We are sure to wet the 63 × lens prior to an imaging session so that the section need not be moved to expose the lens to water. Aperture and other beam settings are typically set to maximize resolution dependent on the objective lens used. If the goal is to image the same ultrathin section by both FM and TEM, it is important to image the section first by FM, since damage caused by the electron beam will extinguish the fluorophore. If the goal is to correlate a semi-thin section with a serial ultrathin section, we cut the ultrathin section first and then cut the semi-thin section. This seems a reliable sequence for collecting an acceptable ultrathin section.

18.1.7 REGISTRATION OF LM AND EM IMAGES

- Correlation of LM and EM imaging fields is typically accomplished in several steps, "macro to micro." LM images are collected prior to the EM imaging session.
- A 20 × (or lower magnification) image collected in the LM is displayed on a monitor while in the EM suite; with the brightness and contrast adjusted so that cell profiles are imaged. Open the LM image using a program that allows not only the control of brightness and contrast but also rotational alignment and flipping of the image (i.e., Adobe Photoshop, Microsoft Office 2010 Picture Manager, IrfanView, QuickTime).

- During initial stages of observation in the EM, locate the approximate area from which the LM images were collected by observing the entire section at low magnification. Find the region relative to the shape of the section noted during LM observation. Control the computer displaying the LM image with a cordless mouse from the TEM work surface so that the image may be easily manipulated.
- Once a corresponding region of interest is recognized on both monitors, rotate the LM image field so that it matches the live-field imaged in the EM. It may be that the EM grid is upside down relative to the section imaged in the LM. If so, "flip" the LM image using the imaging software to avoid handing the grid any more than needed. Once the region imaged in the LM is recognized in the EM, collect progressively higher magnification images of the region of interest. Progressively higher magnifications will ensure no beam damage in the lower magnification images. These TEM images will be used as the background on which LM images are overlaid when composing the registered composite images.
- You may wish to make initial EM observations on unstained sections. If the region photographed in the LM is found to be defective in the EM (dirt, wrinkles), you may then return to the LM to take additional images. Should your EM imaging system have the ability to collect image montages, these can be enormously useful in observing LM data overlaid atop large image fields tolerating substantial digital enlargement.
- LR White, LR Gold, and Lowicryl HM20 may be stained for EM using uranyl acetate in 50% EtOH for 5 min, 5 min in water, then stain in quarter-strength Reynold's lead citrate for 15 s before a final 5-minute water wash. The use of "grid sticks" allows the staining of multiple grids simultaneously.

18.1.8 WORKFLOW FOR OVERLAYING IMAGES USING FIJI (IMAGEJ VERSION 1.480)

Download Fiji (http://fiji.sc/wiki/index.php/Downloads). Fiji is a distribution of ImageJ together with Java, Java 3D, and several plugins organized to assist research in life sciences, targeting image registration, stitching, segmentation, feature extraction, and 3D visualization, among others. After installing Fiji, check to be sure it is the latest version (Fiji/Help/Update Fiji).

- Open Fiji
- Open the images to be overlaid. The images should be in .tiff format. Open the background image (TEM) first and keep it open, then open the image to be overlaid (LM) (File/Open/double click on image). Images may also be dragged into the work bar of Fiji to open them.
- Manipulate one of the images so that fields are approximately oriented with respect to one another. This may require that one image be flipped (Image/Transform/Flip horizontally or vertically) and/or rotated (Image/Transform/Rotate).

- Crop the images so that almost the same region remains in each image (click on one image, select the Rectangular selection on the far left of the tool bar, crop the image (Image/Crop); then do the same for the other image.
- Adjust images so that they are both 8 bit (Image/Type/8bit). It is not important if color remains in the image to be overlaid.
- Open the TurboReg Plugin (Plugins/Registration/TurboReg).

TurboReg (Thévenaz, Ruttimann, & Unser, 1998) is used to register one image (Source) to a second image (Target); Fiji is then used to overlay the registered image atop the Target image.

- Within the TurboReg window, confirm that the Source is the image to be overlaid (LM) and that the Target is the image on which the overlaid image is placed (TEM). If each image was opened in the order suggested above, this will be correct. Change them if necessary. Check the boxes for "Bilinear," "Save on Exit," and "Accurate."
- Each image will now have a green cross, a magenta cross, a blue cross, and a yellow cross. Place the green cross on a feature at the upper left quadrant of one image. Looking to the second image, move the green cross to the corresponding feature. If you wish to magnify either image, you may select the magnifying tool in the Fiji tool bar but remember to go back to the select tool when placing a cross. You may also adjust the contrast and brightness of either image (Image/Adjust/Brightness/Contrast). Move the magenta, blue, and yellow crosses to common features on each image in the other three quadrants (these crosses are difficult to see in the screenshot below) (Fig. 18.11; colored arrows point to the smaller crosses).
- Select the TurboReg window again; press the "Manual" button. A new 32-bit grayscale image will appear which is automatically named "Output."
- If you wish, save the "landmarks" selected in the previous window, giving it a file name.
- The Output image produced in TurboReg may now be manipulated in Fiji.
- Select the "Output" image. There are actually two images here; each image is accessible via the slider. Move the slider to the black image (Mask) and delete it (Image/Stacks/Delete Slice); now only the Data image remains. Match the bit size of the Output file to the Target image (Image/Type/8bit).
- Next, merge the Output with the Target image (Image/Color/Merge Channels). Choose the color you prefer for the data contained in the overlaid image. If you would like your overlaid image to be blue on top of a grayscale (EM) image, then set red to "none," green to "none," blue to "output," gray to the file name of the background EM image, and cyan, magenta, and yellow to "none." Check the boxes "create composite" and "keep source images," then press OK (Fig. 18.12).

When choosing a color for your overlay, consider that approximately 8% of males and 0.4% of females are red–green color blind (Albrecht, 2010, also see http://en.wikipedia.org/wiki/Color_blindness). Individuals with red–green color

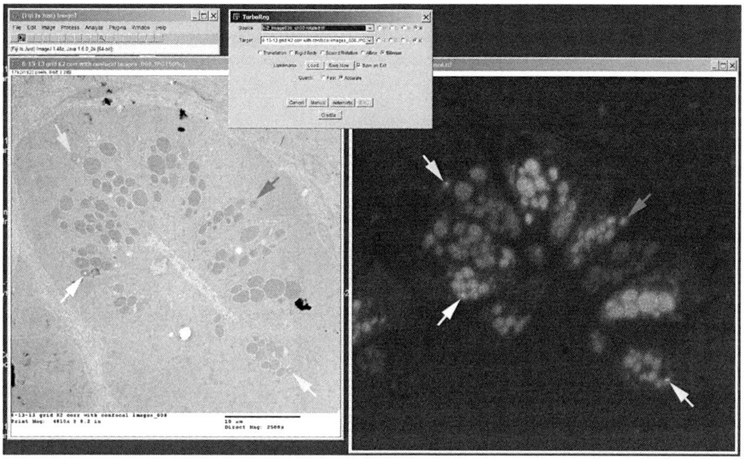

FIGURE 18.11

In this screenshot, both a confocal image (on the right) and a TEM image of the same region (on the left) were opened in FIJI. The TurboReg plugin was initialized and "Bilinear," "Save on Exit," and "Accurate" were selected. The green, pink, blue, and yellow crosses were moved to the same features on each image. Next, "Manual" will be selected to form an aligned "Output" image (Fig. 18.12). (See the color plate.)

blindness cannot discern between the two colors; both appear yellow–brown. To simulate how a color image is perceived by individuals with specific color insensitivities, open the image in FIJI (FIJI/Image/Color/Simulate Color Blindness). FIJI may be used to simulate the perception of the full spectrum of colors. Importantly, magenta is perceived as a hue of blue by color blind individuals; therefore Magenta/Green images are perceived as Blue/Yellow. FIJI allows one-button conversion of a Red/Green image to Magenta/Green (FIJI/Image/Color/Convert Red to Magenta) may be changed using Adobe Photoshop/Image/Adjustment/Hue/Saturation.

- This creates a "Composite" File, composed of two images.
- Importantly, both the underlying and overlying images can be adjusted for contrast and brightness (use the slider to select the layer/Image/Adjust/). This adjustment affects the opacity of the overlaid image and is a critical adjustment. Demonstrated by Fig. 18.14, when the overlaid CF image is made increasingly opaque, only lysozyme labeling within the Paneth cell secretory granules is realized. Making the overlaid CF image increasingly less opaque reveals labeling not only to Paneth cell secretory granules but also to the RER within the Paneth cells (Fig. 18.13).
- The images composing the Composite file may not be perfectly aligned. This could be due to any number of reasons, including sectioning artifacts such as wrinkles or compression. Images may be manually aligned relative to each other by selecting one image of the Composite (usually the overlaid image), then distorting this image relative to the background image (Fiji/Plugins/Transform/

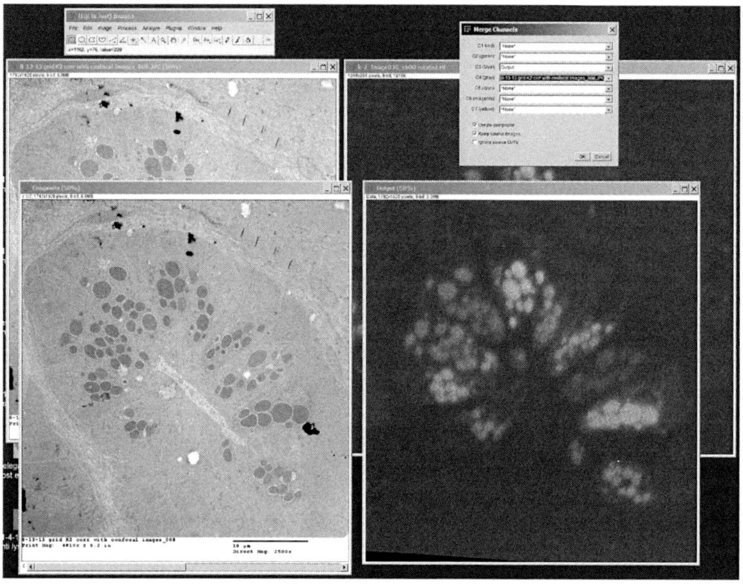

FIGURE 18.12

The initial "Output" image is composed of an aligned image and a black mask, selectable using the slider. Select the black mask and delete it (Image/Stacks/Delete slice), leaving only the aligned image as in this screenshot. Next, select the TEM image and match it to the bit depth of the "Output" image (Image/Type/-bit). The images may now be merged (Image/Color/Merge Channels). In this example, the "Output" image was assigned the blue channel and the TEM image was assigned the gray channel, producing an aligned "Composite" image. (See the color plate.)

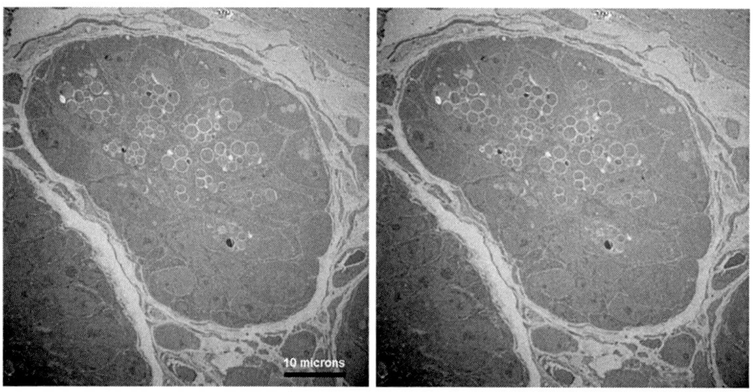

FIGURE 18.13

The "Composite" image has two components, selectable using the slider, each of which may be further adjusted for contrast/brightness. Careful adjustment of contrast and brightness is critical. Improper adjustment of this image (left) would miss cytoplasmic localization (right) verified by immunogold labeling (shown in Fig. 18.6). (See the color plate.)

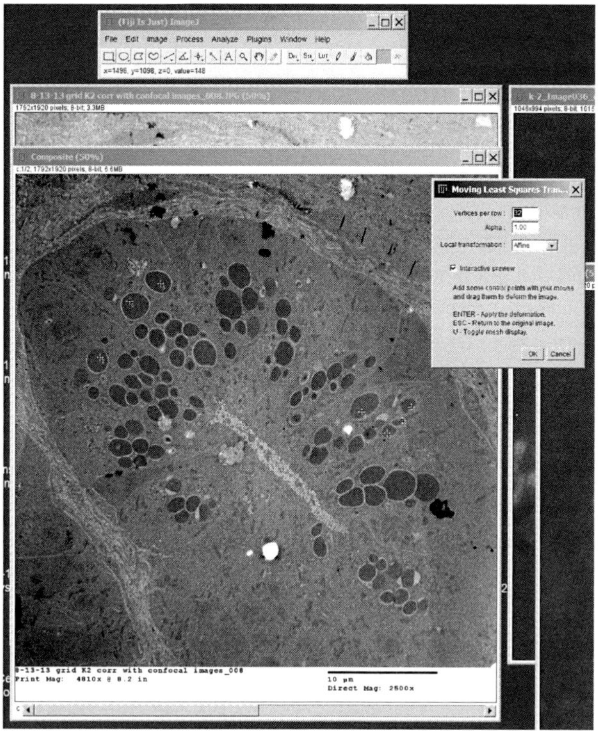

FIGURE 18.14

Should the alignment of two images composing the "Composite" image be unsatisfactory, one of the images may be transformed, or skewed, relative to the other. In this screenshot, the confocal portion of the image was manually dragged to match the TEM portion of the image at each of the seven points selected. (See the color plate.)

Interactive Moving Least Squares). Choose Affine, choose interactive preview. Left-click on *more than three* misaligned regions, then select (hold down left mouse button) and drag to the correct alignment. You may select as many regions as necessary to deform the image to a perfect alignment. On the keyboard, pressing "u" will toggle in grid that can be used in determining the extent of deformation throughout the entire image (Fig. 18.14).

- Save the Composite as a .tiff (File/Save As/Tiff. . .name the image); it can be opened latter in Fiji to make further adjustments. To open the fully adjusted image in other programs, convert the Composite to a single flattened RGB image (Image/Type/RGB Color) and save is as a .tiff.

18.1.8.1 Materials

Fiducial markers: AU nanospheres: Microspheres-Nanospheres, Cold Spring, NY (http://www.microspheres-nanospheres.com/Microspheres/Inorganic/Metals/Gold.htm)

Fluorescent polymer-coated Au gold particles: http://www.nanopartz.com/fluorescence_gold_nanoparticles.asp
Embedding media: LR White resin kit, Ted Pella, Inc. (#18181); LR Gold resin kit, Electron Microscopy Sciences (#14370); Lowicryl HM20 embedding media monostep, Electron Microscopy Sciences (#14345)
Primary antibodies: Abcam GFP antibody, Abcam (#ab290); Life Technologies GFP antibody, Life Technologies (#A11122)
Secondary antibodies: Aurion gold conjugates, Electron Microscopy Sciences; goat anti-rabbit 6 nm, EM grade (#25103); goat anti-rabbit 10 nm, EM grade (#25108); Dual fluorescent-gold conjugate (Molecular Probes Alexafluor 488 with goat anti-rabbit IgG 10-nm gold conjugate), Life Technologies (#31566).
Tools and materials: Lux Thermanox cover slips, 13 mm diameter, Electron Microscopy Sciences (#72281); Corning Costar cell culture plates, Fisher Scientific (#07-200-84); Sarstedt 50-mL conical tubes, Phenix Research (#SS-7004); Wheaton snap caps, Fisher Scientific (#06-450-201); Gelatin embedding capsules, Electron Microscopy Sciences (#71000); 2 mm × 1 mm slot grids, Electron Microscopy Sciences: copper-palladium (#G2010-CP special order); nickel (#G2010-Ni); NRD Static master ionizer, Rice Lake Weighing Systems (#76676); PELCO PTFE immunostaining pad, Ted Pella, Inc. (#10526-1); 24 mm × 50 mm cover glass, Fisher Scientific (#12-548-5M); Anti-magnetic tweezers, Electron Microscopy Sciences (#72700-D); Grid stick, Electron Microscopy Sciences (#71175); Sylvania Blacklight Blue F15T8/BLB 15W (#21625); Spiral saw blades, Electron Microscopy Sciences (#72015-02)
Fixatives: 10% aqueous formaldehyde, Tousimis (#1008C); 50% EM grade aqueous glutaraldehyde, Electron Microscopy Sciences (#16316)
Chemicals: Dulbecco's Modification of Eagle's Medium (DMEM), Cellgro (#10-013-CV); TRIS, Electron Microscopy Sciences (#11720); Glycine, Sigma-Aldrich (#410225-50g); Carnation nonfat dry milk (grocery store); Chicken egg albumin, Sigma-Aldrich (A5503); Cold water fish gelatin, Electron Microscopy Sciences (#25560); Benzoin methyl ether, Electron Microscopy Sciences (#11290).

REFERENCES

Aebi, U., & Pollard, T. D. (1987). A glow discharge unit to render electron microscope grids and other surfaces hydrophilic. *Journal of Electron Microscopy Technique*, *7*, 29–33.

Albrecht, M. (2010). Color blindness. *Nature Methods*, *7*, 775.

Bell, K., Mitchell, S., Paultre, D., Posch, M., & Oparka, K. (2013). Correlative imaging of fluorescent proteins in resin-Embedded plant material. *Plant Physiology*, *161*, 1595–1603.

Bizzarri, R., Serresi, M., Luin, S., & Beltram, F. (2009). Green fluorescent protein based pH indicators for in vivo use: A review. *Analytical and Bioanalytical Chemistry*, *393*, 1107–1122.

Brown, T. A., Fetter, R. D., Tkachuk, A. N., & Clayton, D. A. (2010). Approaches toward super-resolution fluorescence imaging of mitochondrial proteins using PALM. *Methods*, *51*, 458–463.

Campbell, T. N., & Choy, F. Y. M. (2001). The effect of pH on green fluorescent protein: A brief review. *Molecular Biology Today*, *2*, 1–4.

Carlson, D. B., & Evans, J. E. (2011). Low-cost cryo-light microscopy stage fabrication for correlated light/electron microscopy. *Journal of Visualized Experiments*, *52*, 2909.

Ellisman, M. H., Deerinck, T. J., Shu, X., & Sosinsky, G. E. (2012). Picking faces out of a crowd: Genetic labels for identification of proteins in correlated light and electron microscopy imaging. In T. Müller-Reichert & P. Verkade (Eds.), *Methods in cell biology: Vol. 111. Correlative light and electron microscopy* (pp. 75–93): Academic Press.

Fabig, G., Kretschmar, S., Weiche, S., Eberle, D., Ader, M., & Kurth, T. (2012). Labeling of ultrathin resin sections for correlative light and electron microscopy. In T. Müller-Reichert & P. Verkade (Eds.), *Methods in cell biology: Vol. 111. Correlative light and electron microscopy* (pp. 75–93): Academic Press.

Handschuh, S., Baeumler, N., Schwaha, T., & Ruthensteiner, B. (2013). A correlative approach for combining microCT, light and transmission electron microscopy in a single 3D scenario. *Frontiers in Zoology*, 10–44.

Haupts, U., Maiti, S., Schwille, P., & Webb, W. W. (1998). Dynamics of fluorescence fluctuations in green fluorescent protein observed by fluorescence correlation spectroscopy. *Proceedings of the National Academy of Sciences of The United States of America*, *95*, 13573–13578.

Hayat, M. A. (1970). Principles and techniques of electron microscopy. In *Biological applications:: Vol. 1*. New York, NY: Van Nostrand Reinhold Company.

Henke, B. L., Gullikson, E. M., & Davis, J. C. (1993). X-ray interactions: Photoabsorption, scattering, transmission, and reflection at E = 50-30000 eV, Z = 1-92. *Atomic Data Nuclear Data Tables*, *54*, 181–342.

Jeffery, N. S., Stephenson, R. S., Gallagher, J. A., Jarvis, J. C., & Cox, P. G. (2010). Micro-computed tomography with iodine staining resolves the arrangement of muscle fibers. *Journal of Biomechanics*, *44*, 189–192.

Johnson, J. T., Hansen, M. S., Wu, I., Healy, L. J., & Johnson, C. R. (2006). Virtual histology of transgenic mouse embryos for high-throughput phenotyping. *PLoS Genetics*, *2*, 471–477.

Keene, D. R., Tufa, S. F., Lunstrum, G. P., Holden, P., & Horton, W. A. (2008). Confocal/TEM overlay microscopy: A simple method for correlating confocal and electron microscopy of cells expressing GFP/YFP fusion proteins. *Microscopy and Microanalysis*, *14*, 342–348.

Kopek, B. G., Shtengel, G., Grimm, J. B., Clayton, D. A., & Hess, H. F. (2013). Correlative photoactivated localization and scanning electron microscopy. *PLoS One*, *8*, e77209.

Kopek, B. G., Shtengel, G., Xu, C. S., Clayton, D. A., & Hess, H. F. (2012). Correlative 3D superresolution fluorescence and electron microscopy reveal the relationship of mitochondrial nucleoids to membranes. *Proceedings of the National Academy of Sciences*, *109*, 6136–6141.

Kukulski, W., Schorb, M., Welsch, S., Picco, A., Kaksonen, M., & Briggs, J. A. (2012). Precise, correlated fluorescence microscopy and electron tomography of lowicryl sections using fluorescent fiducial markers. In T. Müller-Reichert & P. Verkade (Eds.), *Methods in cell biology: Vol. 111. Correlative light and electron microscopy* (pp. 75–93): Academic Press.

Lee, D. W., & Jensen, C. M. (2000). Dry-ice bath based on ethylene glycol mixtures. *Journal of Chemical Education*, *77*, 629.

McDonald, K. L. (2009). A review of high-pressure freezing preparation techniques for correlative light and electron microscopy of the same cells and tissues. *Journal of Microscopy*, *235*, 273–281.

McDonald, K. L., Sharp D. J., Rickoll W. (2012). Postembedding immunolabeling of thin sections of Drosophila tissues for transmission electron microscopy. In Sullivan et al. (Eds.), *Drosophila protocols*, *Cold Spring Harbor Protocols*, *7*, 810–812. Cold Spring Harbor, NY: CSHL Press.

Metscher, B. D. (2009a). Micro-CT for comparative morphology: Simple staining methods allow high-contrast 3D imaging of diverse non-mineralized animal tissues. *BMC Physiology*, *9*, 11.

Micheva, K. D., Busse, B., Weiler, N. C., O'Rourke, N., & Smith, S. J. (2010). Single-synapse analysis of a diverse synapse population: Proteomic imaging methods and markers. *Neuron*, *68*, 639–653.

Neues, F., & Epple, M. (2008). X-ray microcomputer tomography for the study of biomineralized endo- and exoskeletons of animals. *Chemical Reviews*, *108*, 4734–4741.

Newman, G. R., & Hobot, J. A. (1987). Modern acrylics for post-embedding immunostaining techniques. *The Journal of Histochemistry and Cytochemistry*, *35*, 971–981.

Oest, M. E., Jones, J. C., Hatfield, C., & Prater, M. R. (2008). Micro-CT evaluation of murine fetal skeletal development yields greater morphometric precision over traditional clear-staining methods. *Birth Defects Research. Part B, Developmental and Reproductive Toxicology*, *83*, 582–589.

Reddick, L. E., & Alto, N. M. (2012). Correlative light and electron microscopy (CLEM) as a tool to visualize microinjected molecules and their eukaryotic sub-cellular targets. *Journal of Visualized Experiments*, *63*, e3650.

Schellenberger, P., Kaufmann, R., Siebert, C. A., Hagen, C., Wodrich, H., & Grünewald, K. (2014). High-precision correlative fluorescence and electron cryo microscopy using two independent alignment markers. *Ultramicroscopy*, *143*, 41–51.

Schwartz, C. L., Sarbash, V. I., Ataullakhanov, F. I., McIntosh, J. R., & Nicastro, D. (2007). Cryo-fluorescence microscopy facilitates correlations between light and cryo-electron microscopy and reduces the rate of photobleaching. *Journal of Microscopy*, *227*, 98–109.

Sengle, G., Tufa, S. F., Sakai, L. Y., Zulliger, M. A., & Keene, D. R. (2013). A correlative method for imaging identical regions of samples by micro-CT, light microscopy, and electron microscopy: Imaging adipose tissue in a model system. *The Journal of Histochemistry and Cytochemistry*, *61*, 263–271.

Shu, X., Lev-Ram, V., Deerinck, T. J., Qi, Y., Ramko, E. B., Davidson, M. W., et al. (2011). A genetically encoded tag for correlated light and electron microscopy of intact cells, tissues, and organisms. *PLoS Biology*, *4*, e1001041.

Sims, P. A., & Hardin, J. D. (2005). Visualizing green (GFP) and red (DsRed) fluorescent proteins in thin sections with laser scanning confocal and transmission electron microscopy. *Microscopy and Microanalysis*, *11*(Suppl. 2), 6–7.

Sims, P. A., & Hardin, J. D. (2007). Fluorescence-integrated transmission electron microscopy images: Integrating fluorescence microscopy with transmission electron microscopy. *Methods in Molecular Biology*, *369*, 291–308.

Smith, E. A., Cinquin, B. P., Do, M., McDermott, G., Le Gros, M. A., & Larabell, C. A. (2014). Correlative cryogenic tomography of cells using light and soft x-rays. *Ultramicroscopy*, *143*, 33–40.

Spiegelhalter, C., Tosch, V., Hentsch, D., Koch, M., Kessler, P., Schwab, Y., et al. (2010). From dynamic live cell imaging to 3D ultrastructure: Novel integrated methods for high pressure freezing and correlative light-Electron microscopy. *PLoS One*, *5*, e9014.

Thévenaz, P., Ruttimann, U. E., & Unser, M. (1998). A pyramid approach to subpixel registration based on intensity. *IEEE Transactions on Image Processing*, *7*, 27–41.

Tokuyasu, K. T. (1973). A technique for ultracryotomy of cell suspensions and tissues. *The Journal of Cell Biology*, *57*, 551–565.

van Driel, L. F., Valentijn, J. A., Valentijn, K. M., Koning, R. I., & Koster, A. J. (2009). Tools for correlative cryo-fluorescence microscopy and cryo-electron tomography applied to whole mitochondria in human endothelial cells. *European Journal of Cell Biology*, *11*, 669–684.

Verkade, P. (2008). Moving EM: The rapid transfer system as a new tool for correlative light and electron microscopy and high throughput for high-pressure freezing. *Journal of Microscopy*, *230*, 317–328.

Watanabe, S., Punge, A., Hollopeter, G., Willig, K. I., Hobson, R. J., Davis, M. W., et al. (2011). Protein localization in electron micrographs using fluorescence nanoscopy. *Nature Methods*, *8*, 80–84.

Yuan, Y., Li, M., Hong, N., & Hong, Y. (2014). Correlative light and electron microscopic analyses of mitochondrial distribution in blastomeres of early fish embryos. *The FASEB Journal*, *28*(2), 577–585.

FURTHER READING

Metscher, B. D. (2009b). Micro-CT for developmental biology: A versatile tool for high-contrast 3D imaging at histological resolutions. *Developmental Dynamics*, *238*, 632–640.

Index

Note: Page numbers followed by *f* indicate figures, *t* indicate tables and *b* indicate boxes.

D

E

N

O

P

Y

Z

Volumes in Series

Founding Series Editor
DAVID M. PRESCOTT

Volume 1 (1964)
Methods in Cell Physiology
Edited by David M. Prescott

Volume 2 (1966)
Methods in Cell Physiology
Edited by David M. Prescott

Volume 3 (1968)
Methods in Cell Physiology
Edited by David M. Prescott

Volume 4 (1970)
Methods in Cell Physiology
Edited by David M. Prescott

Volume 5 (1972)
Methods in Cell Physiology
Edited by David M. Prescott

Volume 6 (1973)
Methods in Cell Physiology
Edited by David M. Prescott

Volume 7 (1973)
Methods in Cell Biology
Edited by David M. Prescott

Volume 8 (1974)
Methods in Cell Biology
Edited by David M. Prescott

Volume 9 (1975)
Methods in Cell Biology
Edited by David M. Prescott

Volume 10 (1975)
Methods in Cell Biology
Edited by David M. Prescott

Volume 11 (1975)
Yeast Cells
Edited by David M. Prescott

Volume 12 (1975)
Yeast Cells
Edited by David M. Prescott

Volume 13 (1976)
Methods in Cell Biology
Edited by David M. Prescott

Volume 14 (1976)
Methods in Cell Biology
Edited by David M. Prescott

Volume 15 (1977)
Methods in Cell Biology
Edited by David M. Prescott

Volume 16 (1977)
Chromatin and Chromosomal Protein Research I
Edited by Gary Stein, Janet Stein, and Lewis J. Kleinsmith

Volume 17 (1978)
Chromatin and Chromosomal Protein Research II
Edited by Gary Stein, Janet Stein, and Lewis J. Kleinsmith

Volume 18 (1978)
Chromatin and Chromosomal Protein Research III
Edited by Gary Stein, Janet Stein, and Lewis J. Kleinsmith

Volume 19 (1978)
Chromatin and Chromosomal Protein Research IV
Edited by Gary Stein, Janet Stein, and Lewis J. Kleinsmith

Volume 20 (1978)
Methods in Cell Biology
Edited by David M. Prescott

Advisory Board Chairman
KEITH R. PORTER

Volume 21A (1980)
Normal Human Tissue and Cell Culture, Part A: Respiratory, Cardiovascular, and Integumentary Systems
Edited by Curtis C. Harris, Benjamin F. Trump, and Gary D. Stoner

Volume 21B (1980)
Normal Human Tissue and Cell Culture, Part B: Endocrine, Urogenital, and Gastrointestinal Systems
Edited by Curtis C. Harris, Benjamin F. Trump, and Gray D. Stoner

Volume 22 (1981)
Three-Dimensional Ultrastructure in Biology
Edited by James N. Turner

Volume 23 (1981)
Basic Mechanisms of Cellular Secretion
Edited by Arthur R. Hand and Constance Oliver

Volume 24 (1982)
The Cytoskeleton, Part A: Cytoskeletal Proteins, Isolation and Characterization
Edited by Leslie Wilson

Volume 25 (1982)
The Cytoskeleton, Part B: Biological Systems and *In Vitro* Models
Edited by Leslie Wilson

Volume 26 (1982)
Prenatal Diagnosis: Cell Biological Approaches
Edited by Samuel A. Latt and Gretchen J. Darlington

Series Editor
LESLIE WILSON

Volume 27 (1986)
Echinoderm Gametes and Embryos
Edited by Thomas E. Schroeder

Volume 28 (1987)
***Dictyostelium discoideum:* Molecular Approaches to Cell Biology**
Edited by James A. Spudich

Volume 29 (1989)
Fluorescence Microscopy of Living Cells in Culture, Part A: Fluorescent Analogs, Labeling Cells, and Basic Microscopy
Edited by Yu-Li Wang and D. Lansing Taylor

Volume 30 (1989)
Fluorescence Microscopy of Living Cells in Culture, Part B: Quantitative Fluorescence Microscopy—Imaging and Spectroscopy
Edited by D. Lansing Taylor and Yu-Li Wang

Volume 31 (1989)
Vesicular Transport, Part A
Edited by Alan M. Tartakoff

Volume 32 (1989)
Vesicular Transport, Part B
Edited by Alan M. Tartakoff

Volume 33 (1990)
Flow Cytometry
Edited by Zbigniew Darzynkiewicz and Harry A. Crissman

Volume 34 (1991)
Vectorial Transport of Proteins into and across Membranes
Edited by Alan M. Tartakoff

Selected from Volumes 31, 32, and 34 (1991)
Laboratory Methods for Vesicular and Vectorial Transport
Edited by Alan M. Tartakoff

Volume 35 (1991)
Functional Organization of the Nucleus: A Laboratory Guide
Edited by Barbara A. Hamkalo and Sarah C. R. Elgin

Volume 36 (1991)
***Xenopus laevis:* Practical Uses in Cell and Molecular Biology**
Edited by Brian K. Kay and H. Benjamin Peng

Series Editors
LESLIE WILSON AND PAUL MATSUDAIRA

Volume 37 (1993)
Antibodies in Cell Biology
Edited by David J. Asai

Volume 38 (1993)
Cell Biological Applications of Confocal Microscopy
Edited by Brian Matsumoto

Volume 74 (2004)
Development of Sea Urchins, Ascidians, and Other Invertebrate Deuterostomes: Experimental Approaches
Edited by Charles A. Ettensohn, Gary M. Wessel, and Gregory A. Wray

Volume 75 (2004)
Cytometry, 4th Edition: New Developments
Edited by Zbigniew Darzynkiewicz, Mario Roederer, and Hans Tanke

Volume 76 (2004)
The Zebrafish: Cellular and Developmental Biology
Edited by H. William Detrich, III, Monte Westerfield, and Leonard I. Zon

Volume 77 (2004)
The Zebrafish: Genetics, Genomics, and Informatics
Edited by William H. Detrich, III, Monte Westerfield, and Leonard I. Zon

Volume 78 (2004)
Intermediate Filament Cytoskeleton
Edited by M. Bishr Omary and Pierre A. Coulombe

Volume 79 (2007)
Cellular Electron Microscopy
Edited by J. Richard McIntosh

Volume 80 (2007)
Mitochondria, 2nd Edition
Edited by Liza A. Pon and Eric A. Schon

Volume 81 (2007)
Digital Microscopy, 3rd Edition
Edited by Greenfield Sluder and David E. Wolf

Volume 82 (2007)
Laser Manipulation of Cells and Tissues
Edited by Michael W. Berns and Karl Otto Greulich

Volume 83 (2007)
Cell Mechanics
Edited by Yu-Li Wang and Dennis E. Discher

Volume 84 (2007)
Biophysical Tools for Biologists, Volume One: In Vitro Techniques
Edited by John J. Correia and H. William Detrich, III

Volume 85 (2008)
Fluorescent Proteins
Edited by Kevin F. Sullivan

Volume 86 (2008)
Stem Cell Culture
Edited by Dr. Jennie P. Mather

Volume 87 (2008)
Avian Embryology, 2nd Edition
Edited by Dr. Marianne Bronner-Fraser

Volume 88 (2008)
Introduction to Electron Microscopy for Biologists
Edited by Prof. Terence D. Allen

Volume 89 (2008)
Biophysical Tools for Biologists, Volume Two: In Vivo Techniques
Edited by Dr. John J. Correia and Dr. H. William Detrich, III

Volume 90 (2008)
Methods in Nano Cell Biology
Edited by Bhanu P. Jena

Volume 91 (2009)
Cilia: Structure and Motility
Edited by Stephen M. King and Gregory J. Pazour

Volume 92 (2009)
Cilia: Motors and Regulation
Edited by Stephen M. King and Gregory J. Pazour

Volume 93 (2009)
Cilia: Model Organisms and Intraflagellar Transport
Edited by Stephen M. King and Gregory J. Pazour

Volume 94 (2009)
Primary Cilia
Edited by Roger D. Sloboda

Volume 95 (2010)
Microtubules, in vitro
Edited by Leslie Wilson and John J. Correia

Volume 96 (2010)
Electron Microscopy of Model Systems
Edited by Thomas Müeller-Reichert

Volume 97 (2010)
Microtubules: In Vivo
Edited by Lynne Cassimeris and Phong Tran

Volume 98 (2010)
Nuclear Mechanics & Genome Regulation
Edited by G.V. Shivashankar

Volume 99 (2010)
Calcium in Living Cells
Edited by Michael Whitaker

Volume 100 (2010)
The Zebrafish: Cellular and Developmental Biology, Part A
Edited by: H. William Detrich III, Monte Westerfield and Leonard I. Zon

Volume 101 (2011)
The Zebrafish: Cellular and Developmental Biology, Part B
Edited by: H. William Detrich III, Monte Westerfield and Leonard I. Zon

Volume 102 (2011)
Recent Advances in Cytometry, Part A: Instrumentation, Methods
Edited by Zbigniew Darzynkiewicz, Elena Holden, Alberto Orfao, William Telford and Donald Wlodkowic

Volume 103 (2011)
Recent Advances in Cytometry, Part B: Advances in Applications
Edited by Zbigniew Darzynkiewicz, Elena Holden, Alberto Orfao, Alberto Orfao and Donald Wlodkowic

Volume 104 (2011)
The Zebrafish: Genetics, Genomics and Informatics 3rd Edition
Edited by H. William Detrich III, Monte Westerfield, and Leonard I. Zon

Volume 105 (2011)
The Zebrafish: Disease Models and Chemical Screens 3rd Edition
Edited by H. William Detrich III, Monte Westerfield, and Leonard I. Zon

Volume 106 (2011)
Caenorhabditis elegans: Molecular Genetics and Development 2nd Edition
Edited by Joel H. Rothman and Andrew Singson

Volume 107 (2011)
Caenorhabditis elegans: Cell Biology and Physiology 2nd Edition
Edited by Joel H. Rothman and Andrew Singson

Volume 108 (2012)
Lipids
Edited by Gilbert Di Paolo and Markus R Wenk

Volume 109 (2012)
Tetrahymena thermophila
Edited by Kathleen Collins

Volume 110 (2012)
Methods in Cell Biology
Edited by Anand R. Asthagiri and Adam P. Arkin

Volume 111 (2012)
Methods in Cell Biology
Edited by Thomas Müler Reichart and Paul Verkade

Volume 112 (2012)
Laboratory Methods in Cell Biology
Edited by P. Michael Conn

Volume 113 (2013)
Laboratory Methods in Cell Biology
Edited by P. Michael Conn

Volume 114 (2013)
Digital Microscopy, 4th Edition
Edited by Greenfield Sluder and David E. Wolf

Volume 115 (2013)
Microtubules, *in Vitro*, 2nd Edition
Edited by John J. Correia and Leslie Wilson

Volume 116 (2013)
Lipid Droplets
Edited by H. Robert Yang and Peng Li

Volume 117 (2013)
Receptor-Receptor Interactions
Edited by P. Michael Conn

Volume 118 (2013)
Methods for Analysis of Golgi Complex Function
Edited by Franck Perez and David J. Stephens

Volume 119 (2014)
Micropatterning in Cell Biology Part A
Edited by Matthieu Piel and Manuel Théry

Volume 120 (2014)
Micropatterning in Cell Biology Part B
Edited by Matthieu Piel and Manuel Théry

Volume 121 (2014)
Micropatterning in Cell Biology Part C
Edited by Matthieu Piel and Manuel Théry

Volume 122 (2014)
Nuclear Pore Complexes and Nucleocytoplasmic Transport - Methods
Edited by Valérie Doye

Volume 123 (2014)
Quantitative Imaging in Cell Biology
Edited by Jennifer C. Waters and Torsten Wittmann

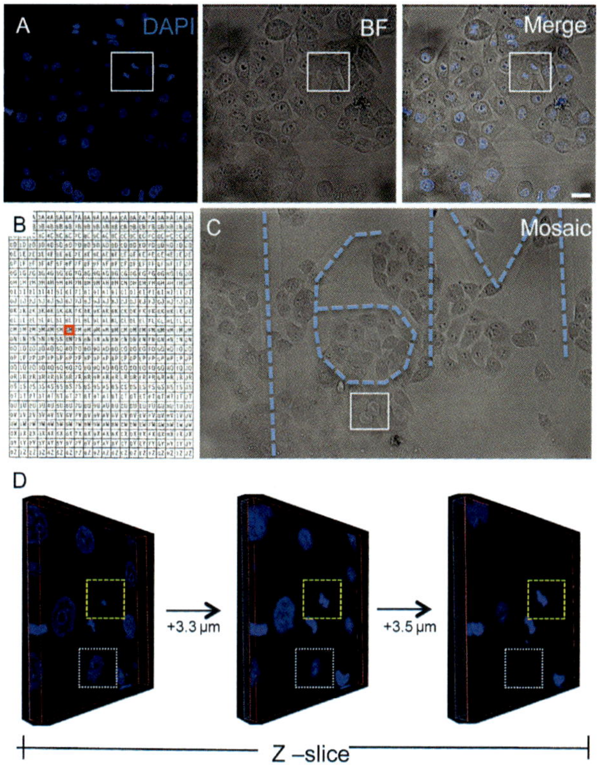

PLATE 1 (Fig. 1.5 on page 13 of this volume).

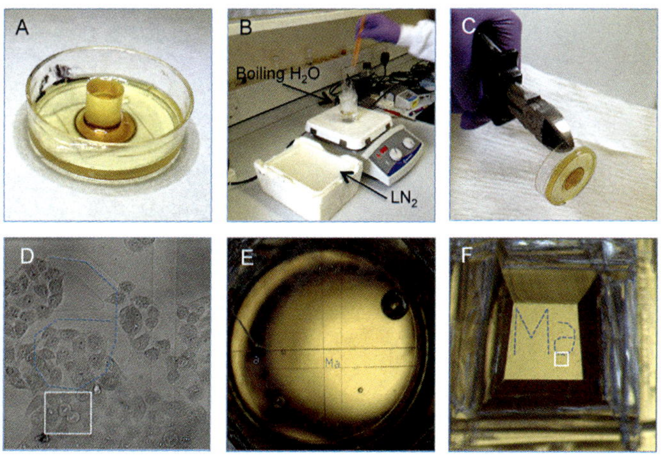

PLATE 2 (Fig. 1.6 on page 14 of this volume).

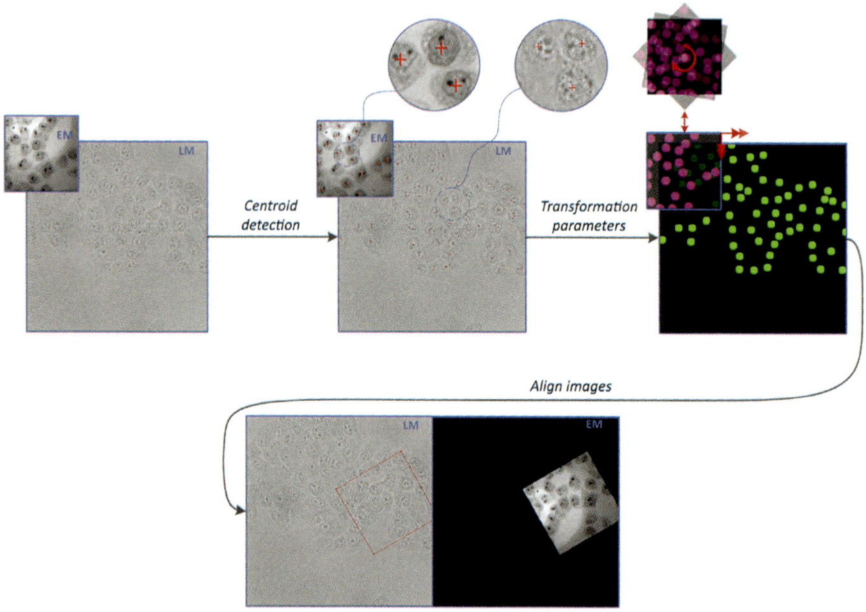

PLATE 3 (Fig. 1.7 on page 15 of this volume).

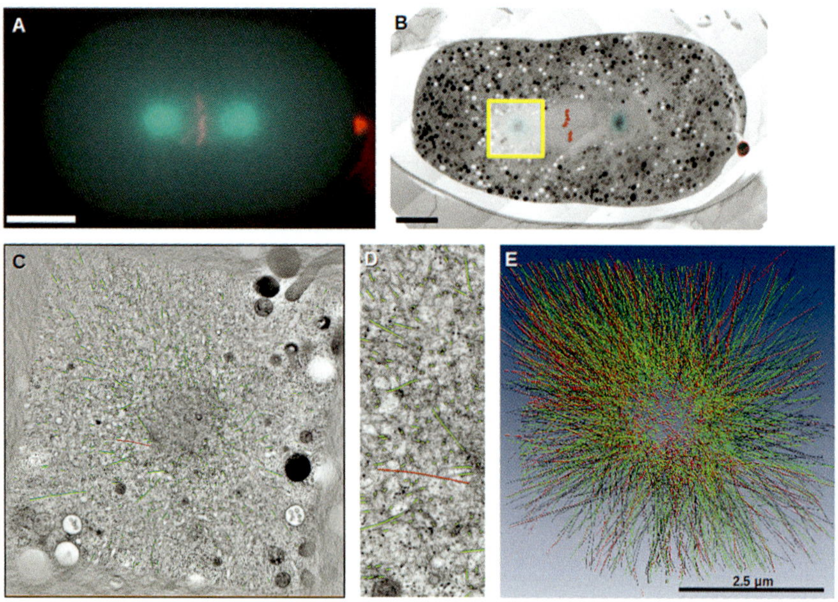

PLATE 4 (Fig. 2.2 on page 33 of this volume).

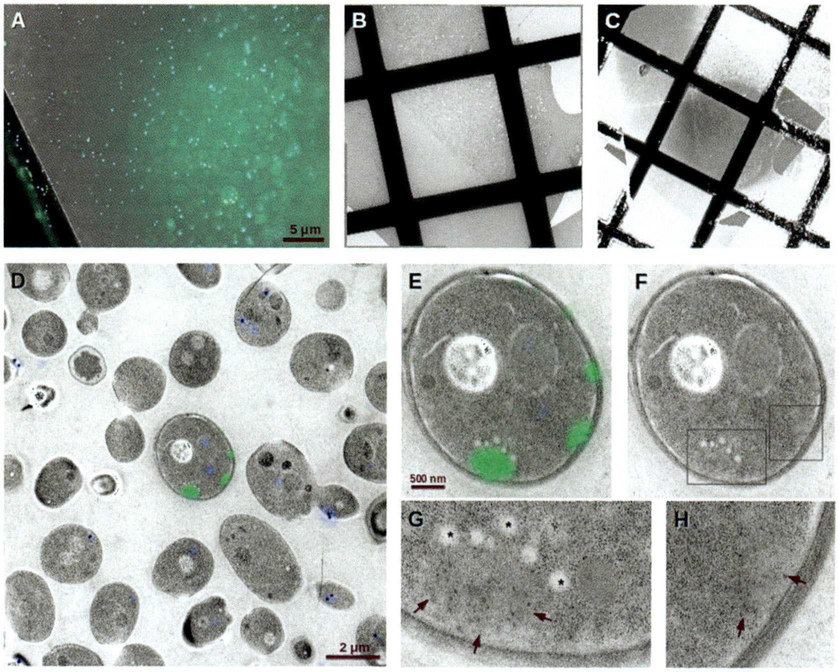

PLATE 5 (Fig. 2.4 on page 39 of this volume).

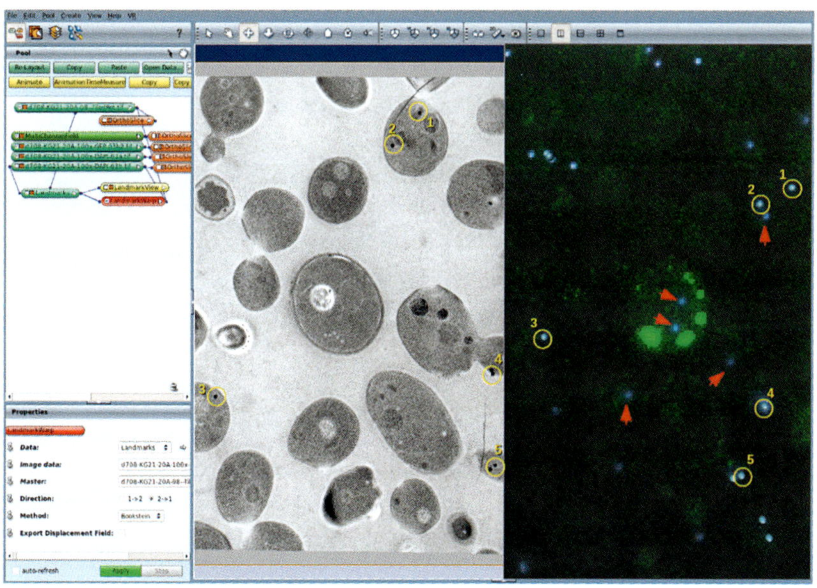

PLATE 6 (Fig. 2.5 on page 40 of this volume).

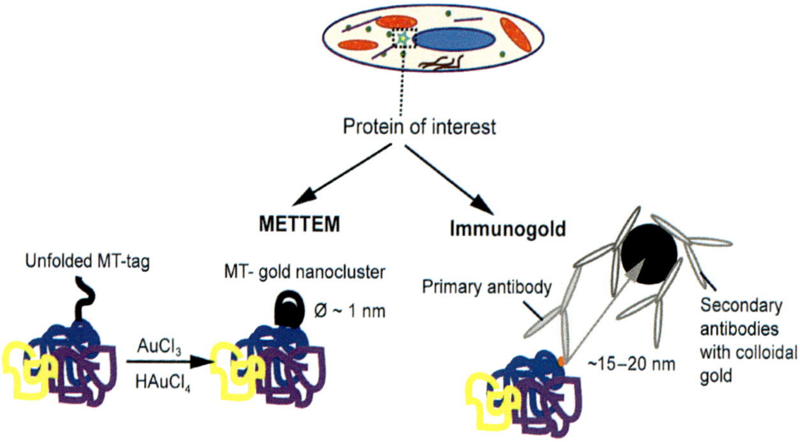

PLATE 7 (Fig. 3.1 on page 57 of this volume).

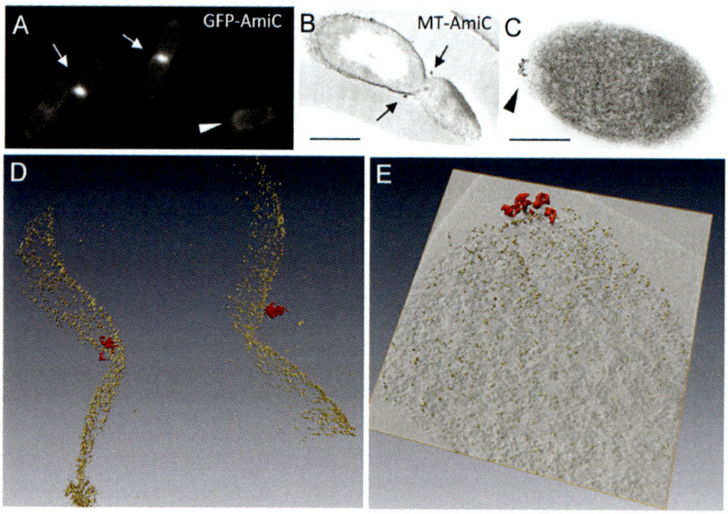

PLATE 8 (Fig. 3.3 on page 60 of this volume).

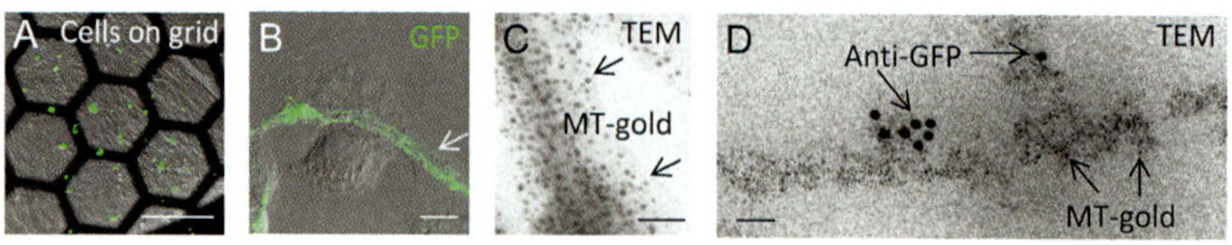

PLATE 9 (Fig. 3.5 on page 62 of this volume).

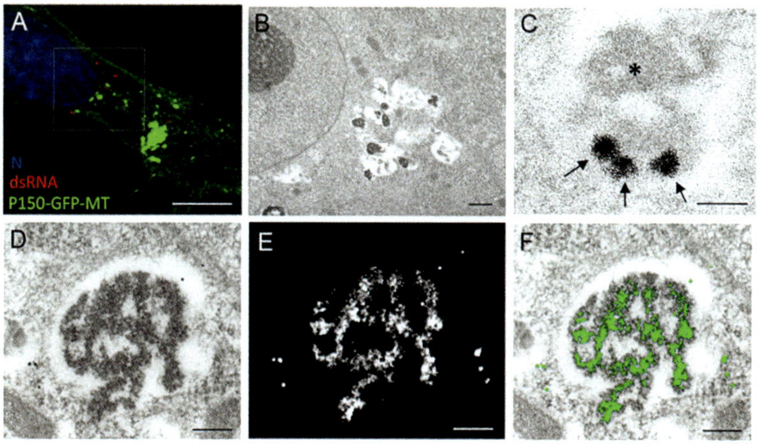

PLATE 10 (Fig. 3.6 on page 63 of this volume).

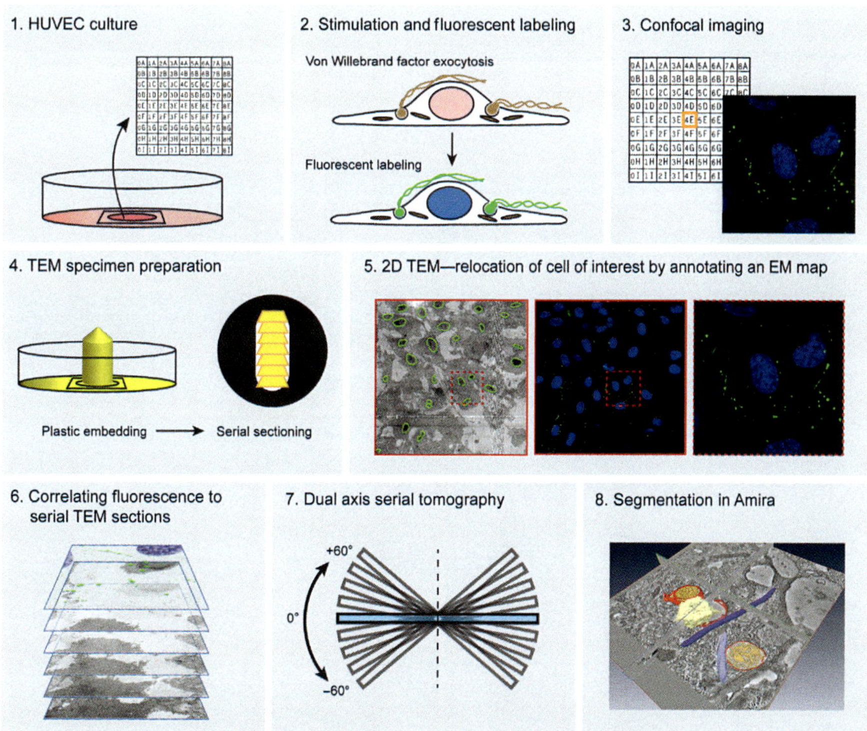

PLATE 11 (Fig. 4.1 on page 75 of this volume).

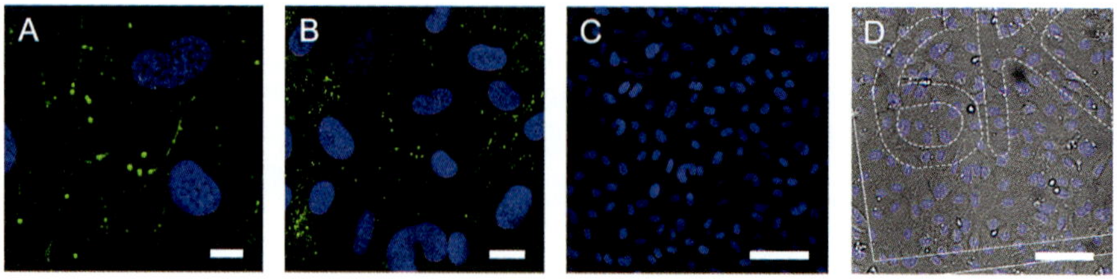

PLATE 12 (Fig. 4.2 on page 77 of this volume).

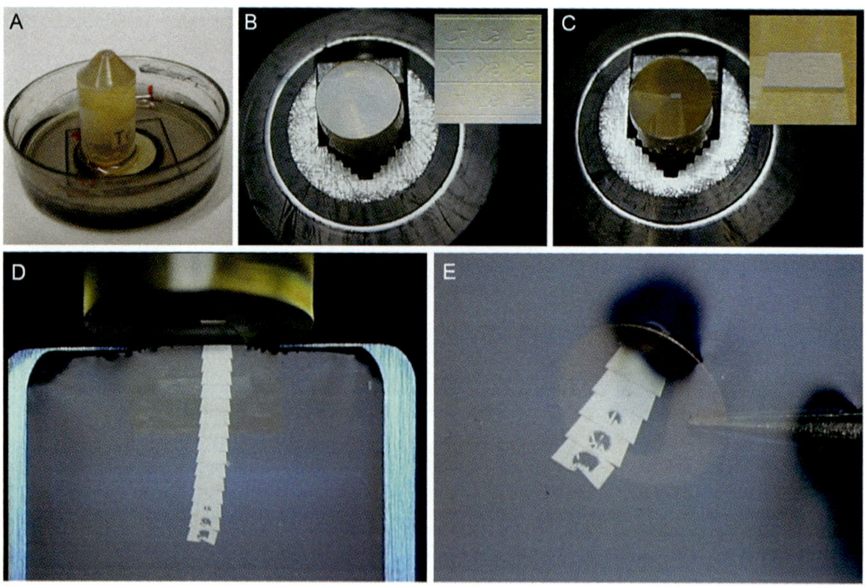

PLATE 13 (Fig. 4.3 on page 79 of this volume).

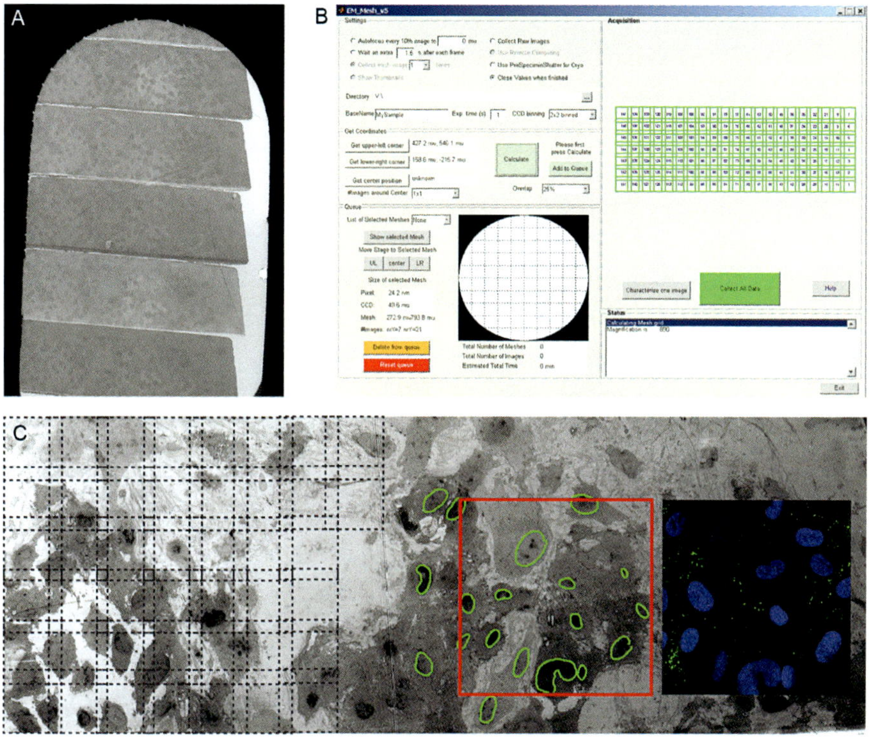

PLATE 14 (Fig. 4.4 on page 81 of this volume).

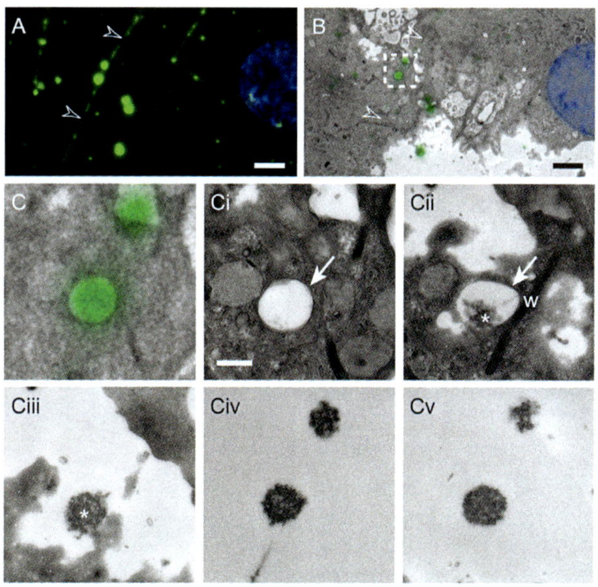

PLATE 15 (Fig. 4.5 on page 83 of this volume).

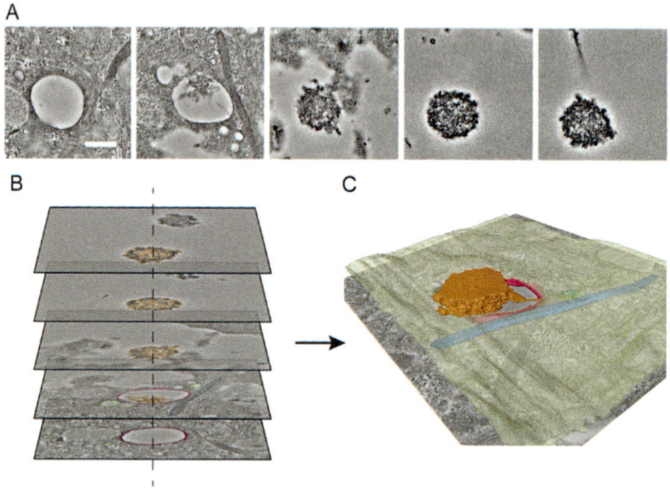

PLATE 16 (Fig. 4.6 on page 85 of this volume).

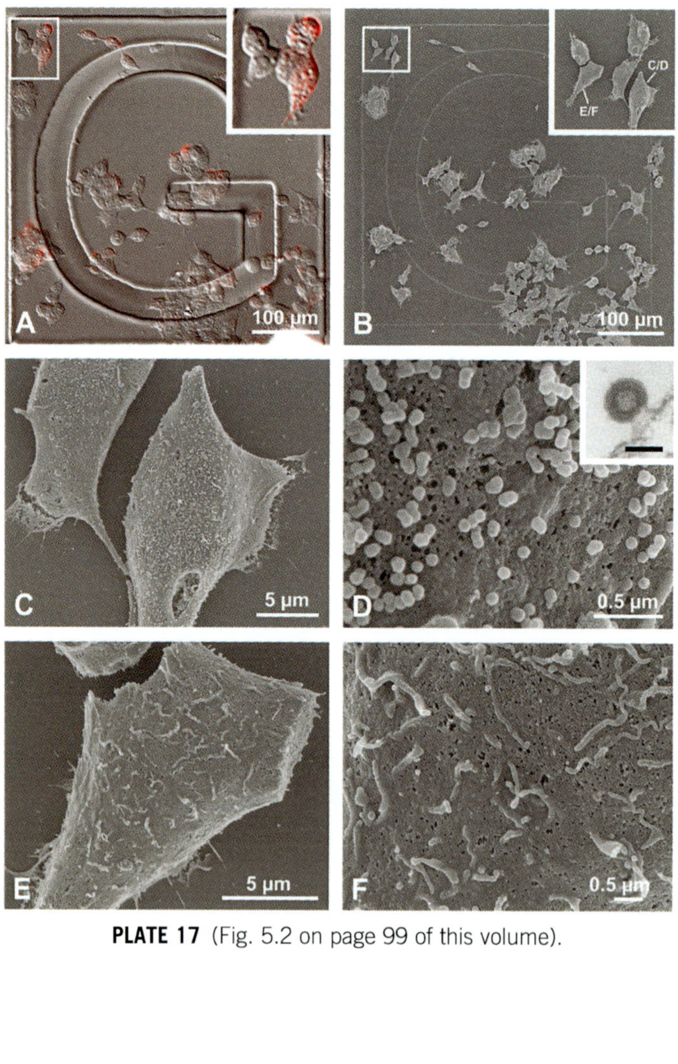

PLATE 17 (Fig. 5.2 on page 99 of this volume).

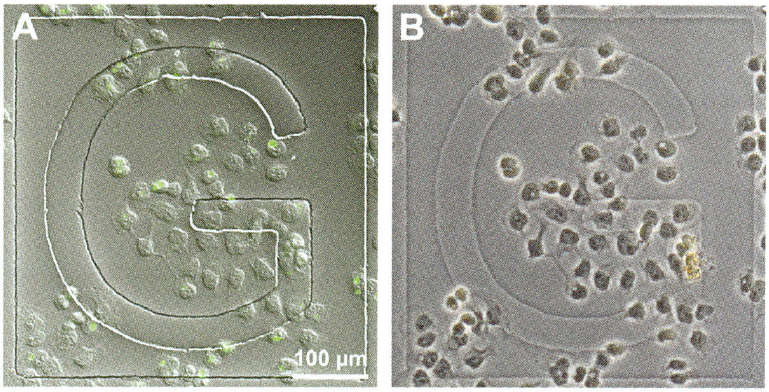

PLATE 18 (Fig. 5.3 on page 101 of this volume).

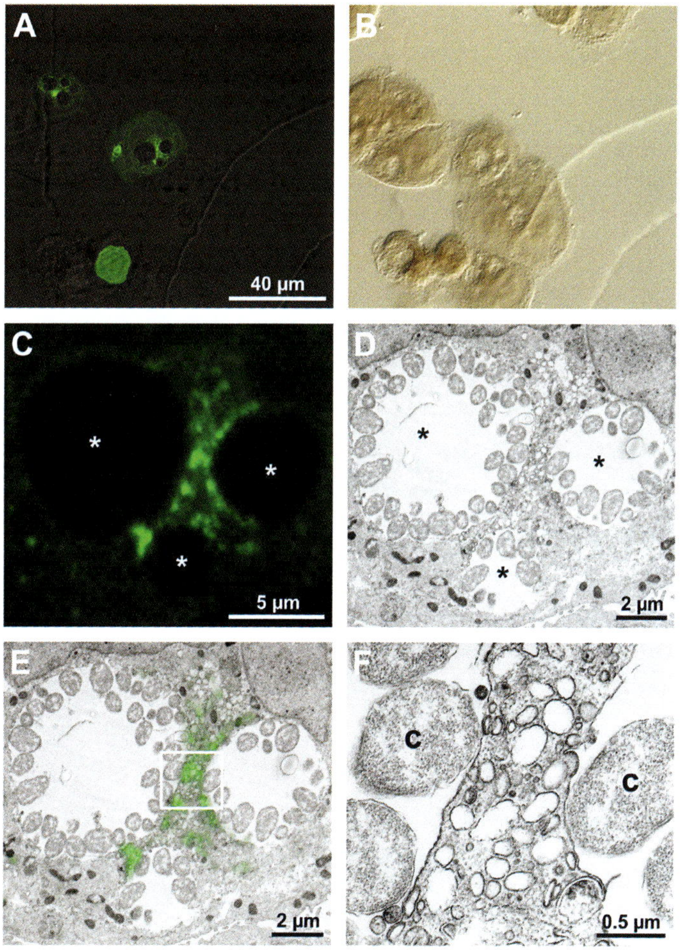

PLATE 19 (Fig. 5.4 on page 102 of this volume).

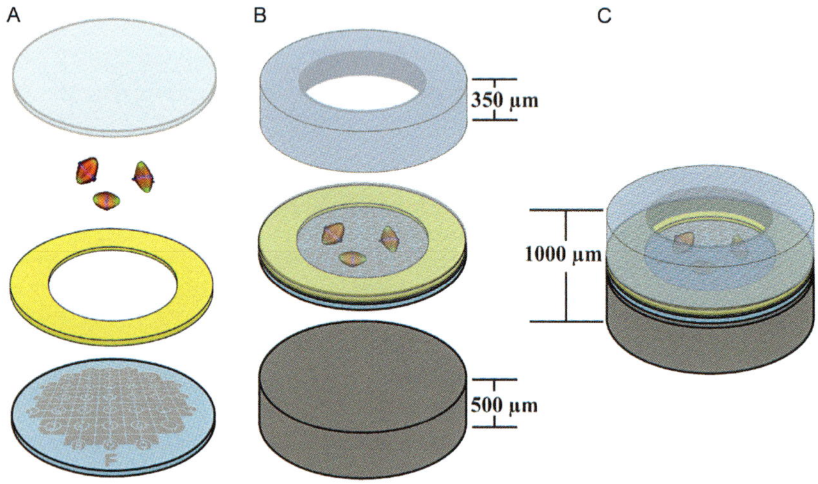

PLATE 20 (Fig. 6.1 on page 114 of this volume).

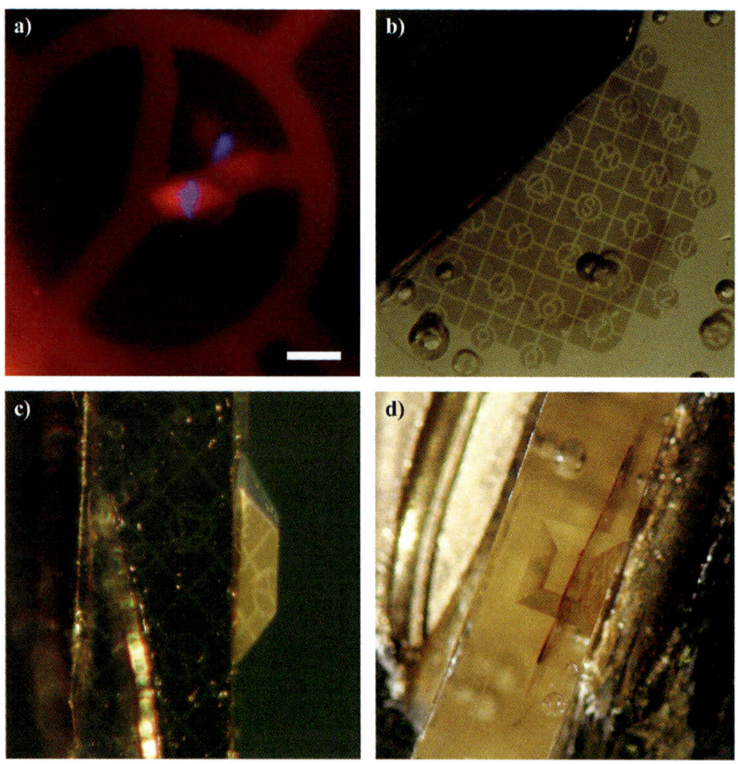

PLATE 21 (Fig. 6.7 on page 124 of this volume).

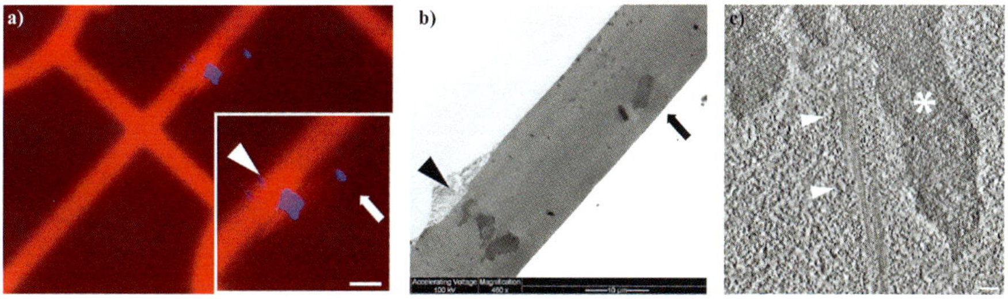

PLATE 22 (Fig. 6.9 on page 126 of this volume).

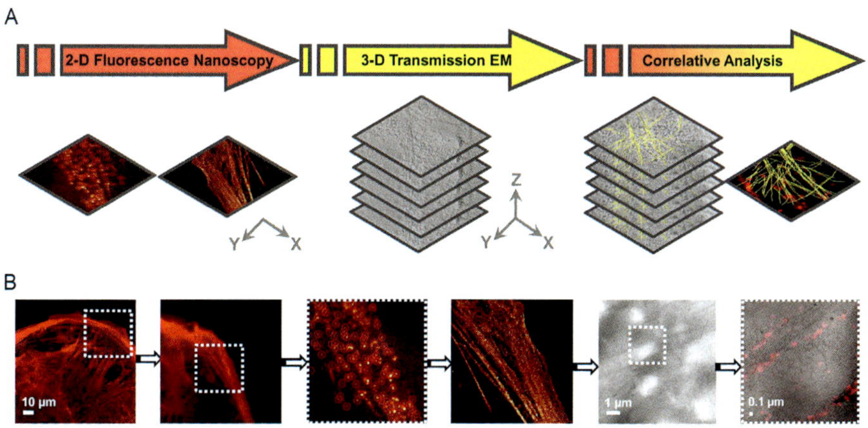

PLATE 23 (Fig. 7.1 on page 132 of this volume).

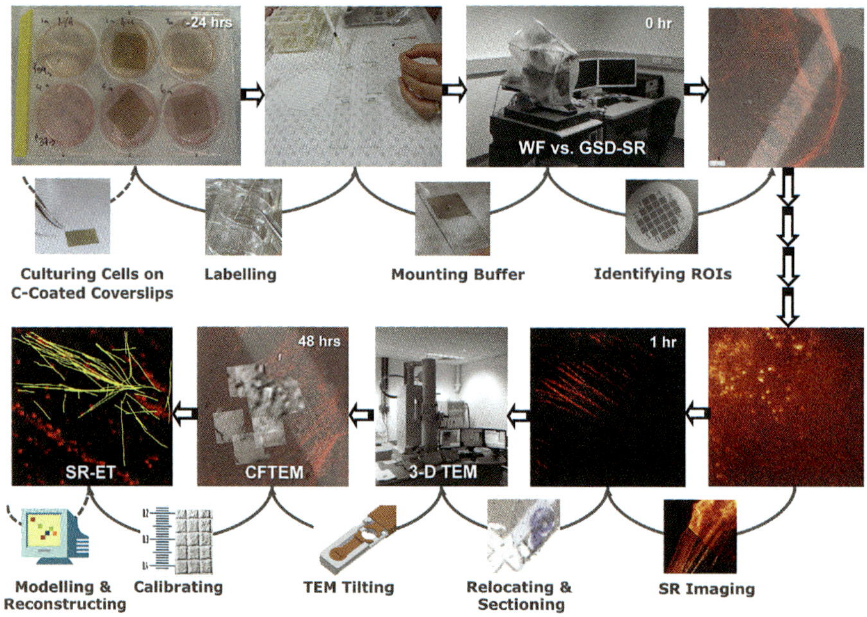

PLATE 24 (Fig. 7.2 on page 133 of this volume).

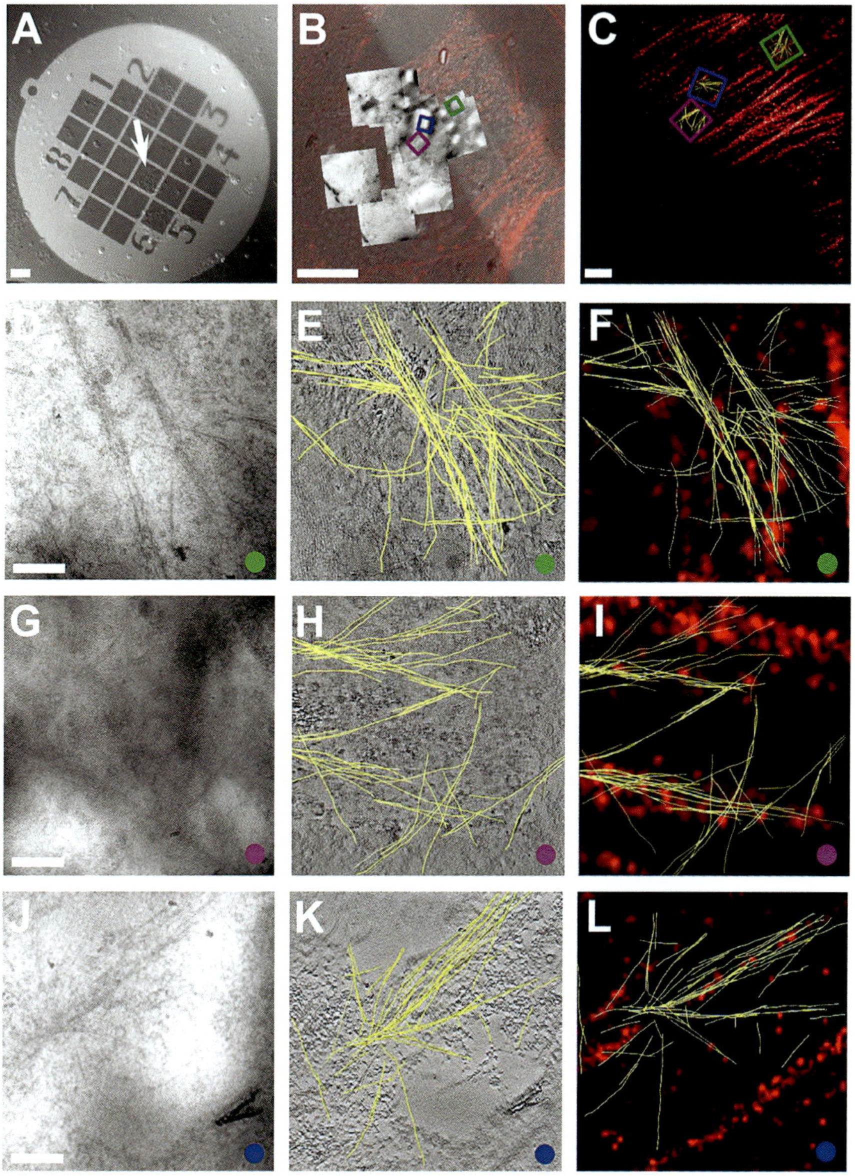

PLATE 25 (Fig. 7.3 on page 136 of this volume).

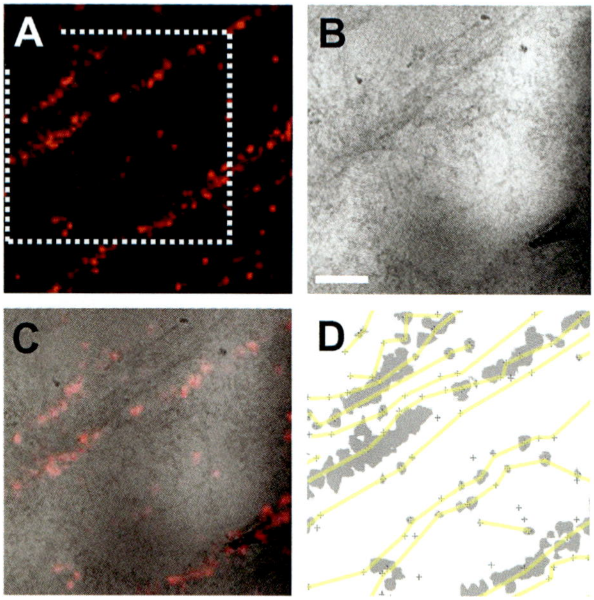

PLATE 26 (Fig. 7.4 on page 138 of this volume).

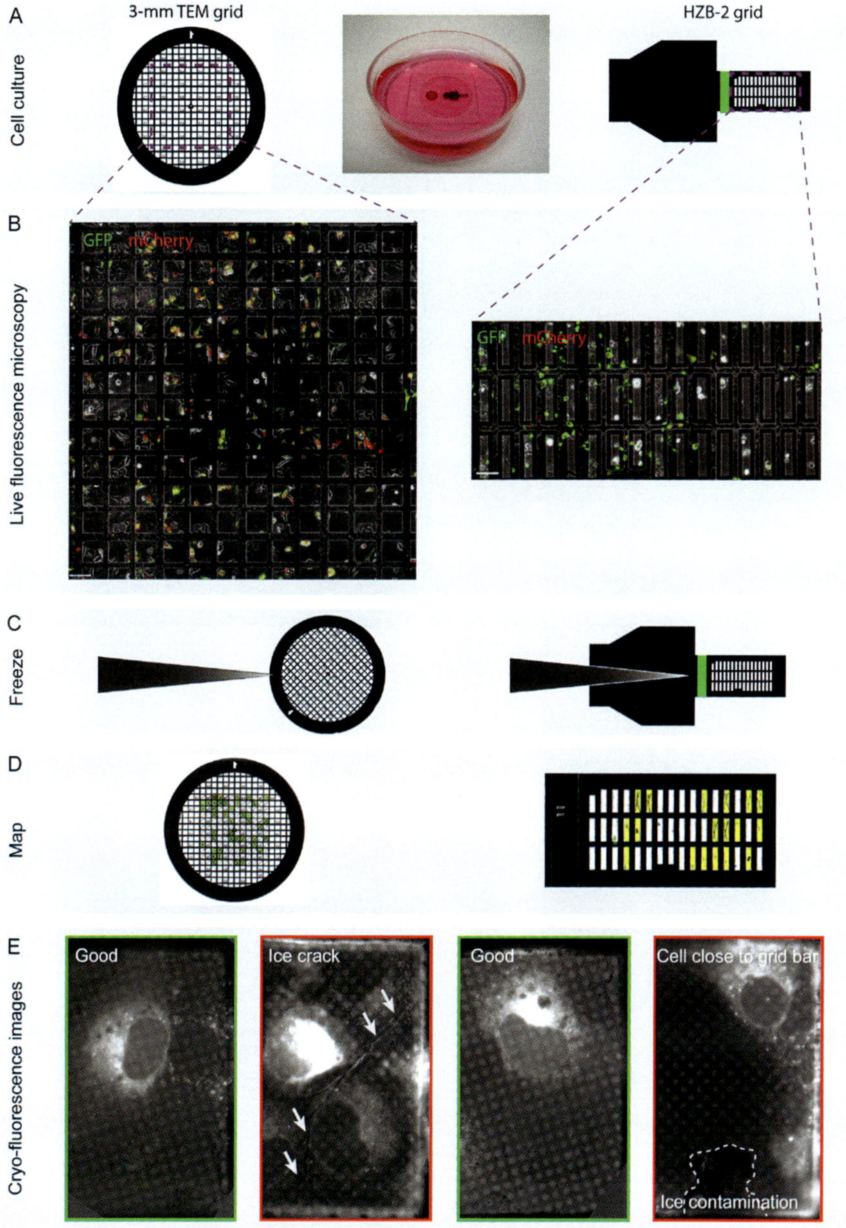

PLATE 27 (Fig. 8.2 on page 159 of this volume).

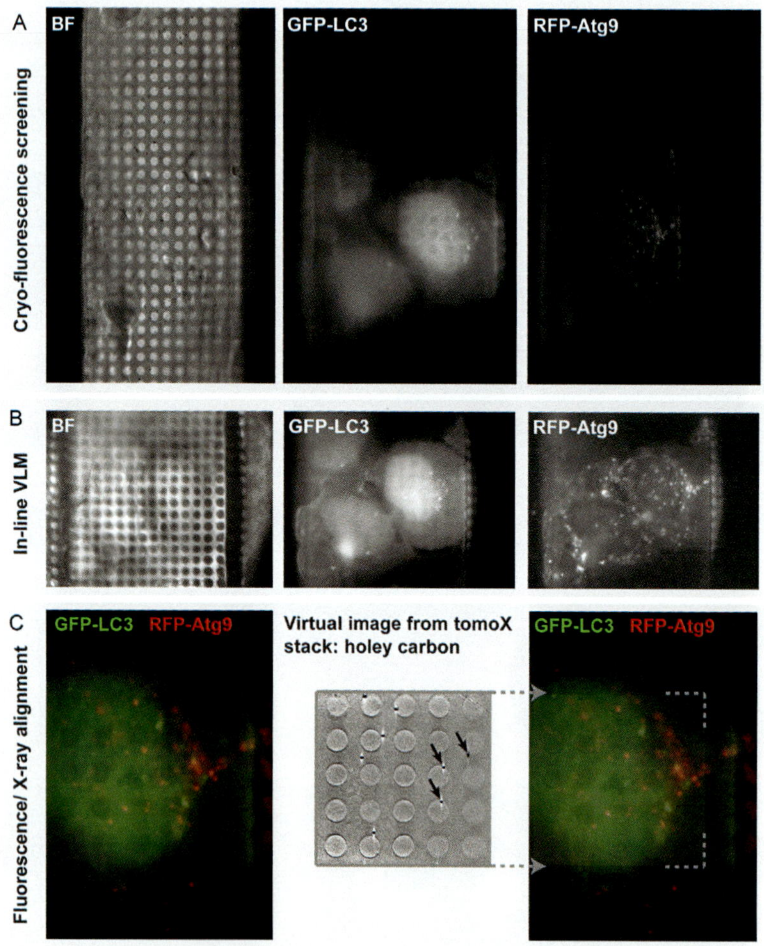

PLATE 28 (Fig. 8.6 on page 172 of this volume).

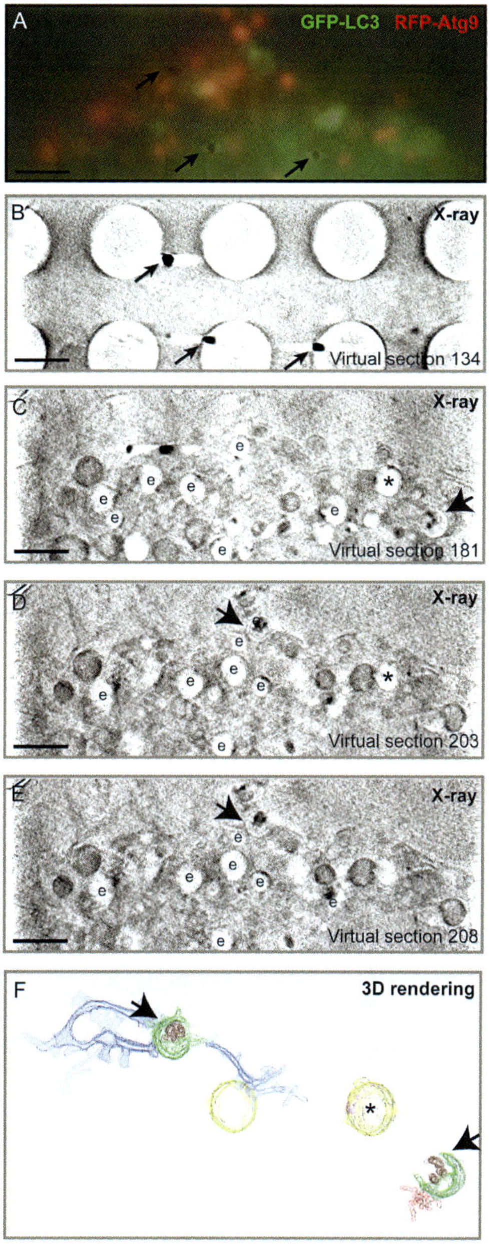

PLATE 29 (Fig. 8.7 on page 173 of this volume).

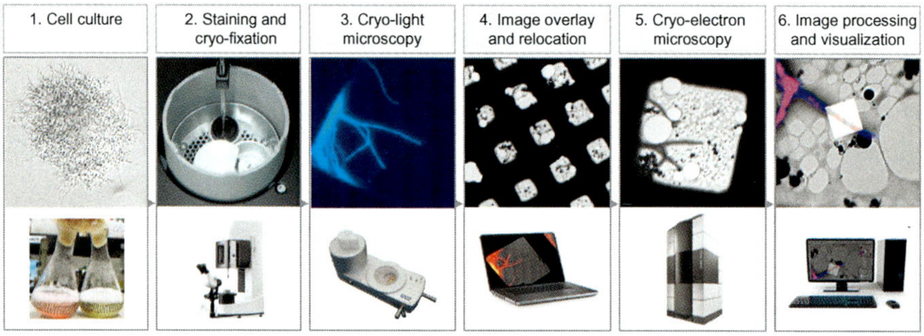

PLATE 30 (Fig. 10.1 on page 222 of this volume).

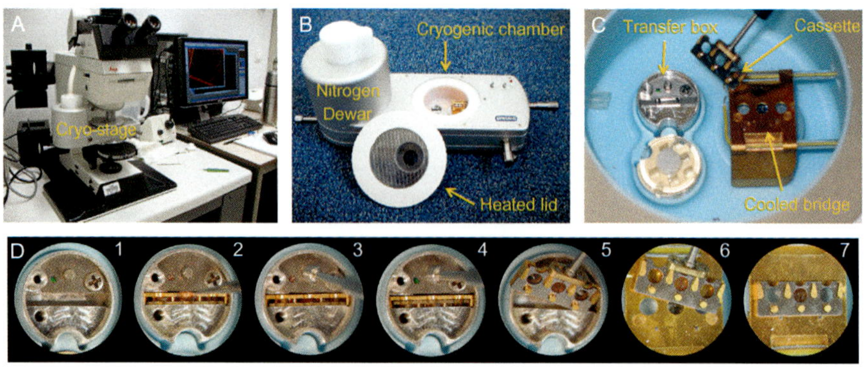

PLATE 31 (Fig. 10.2 on page 224 of this volume).

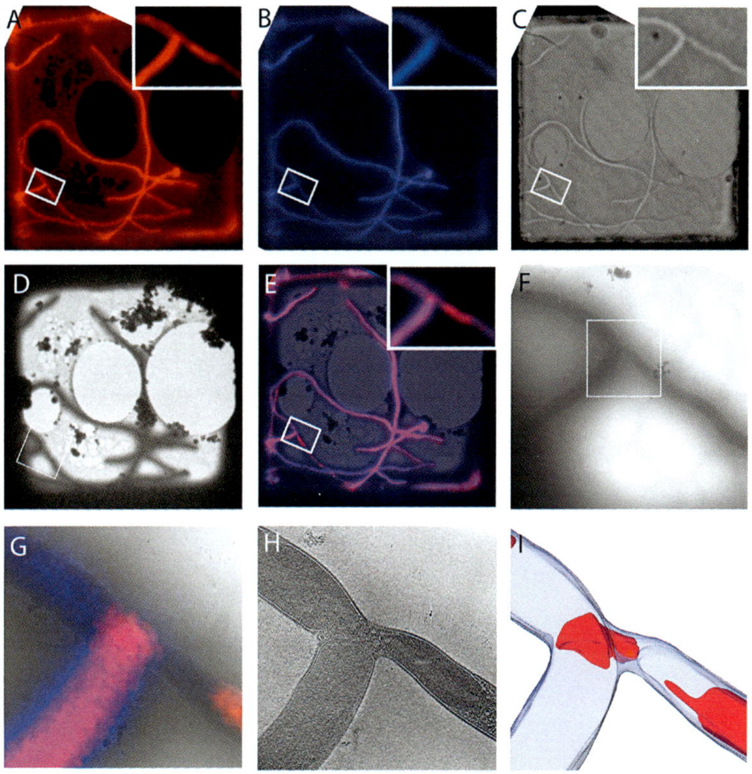

PLATE 32 (Fig. 10.3 on page 234 of this volume).

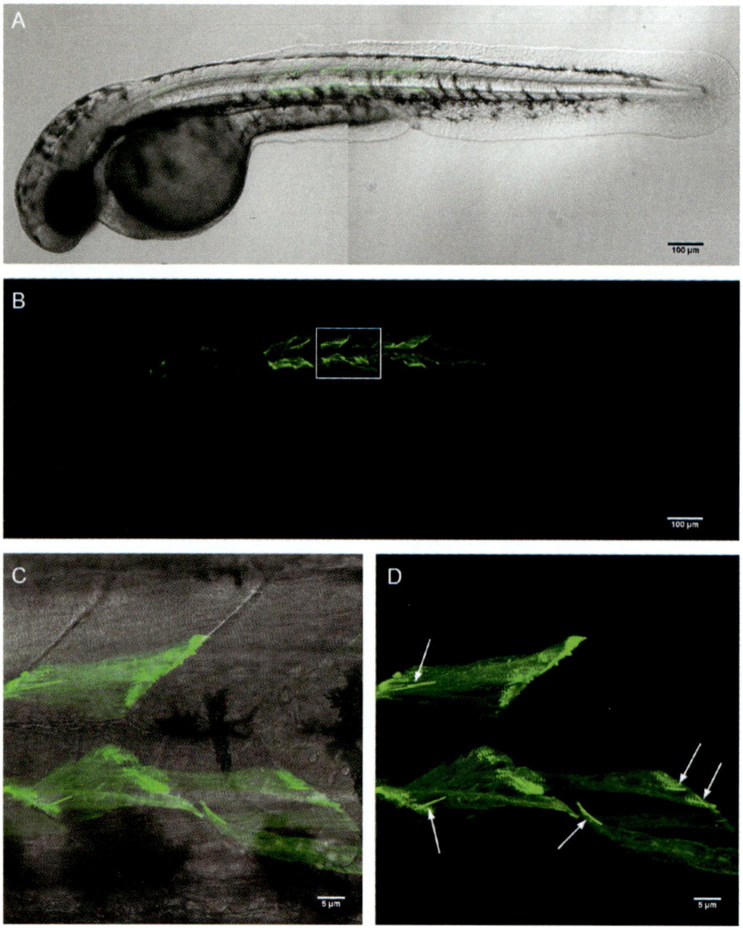

PLATE 33 (Fig. 11.1 on page 245 of this volume).

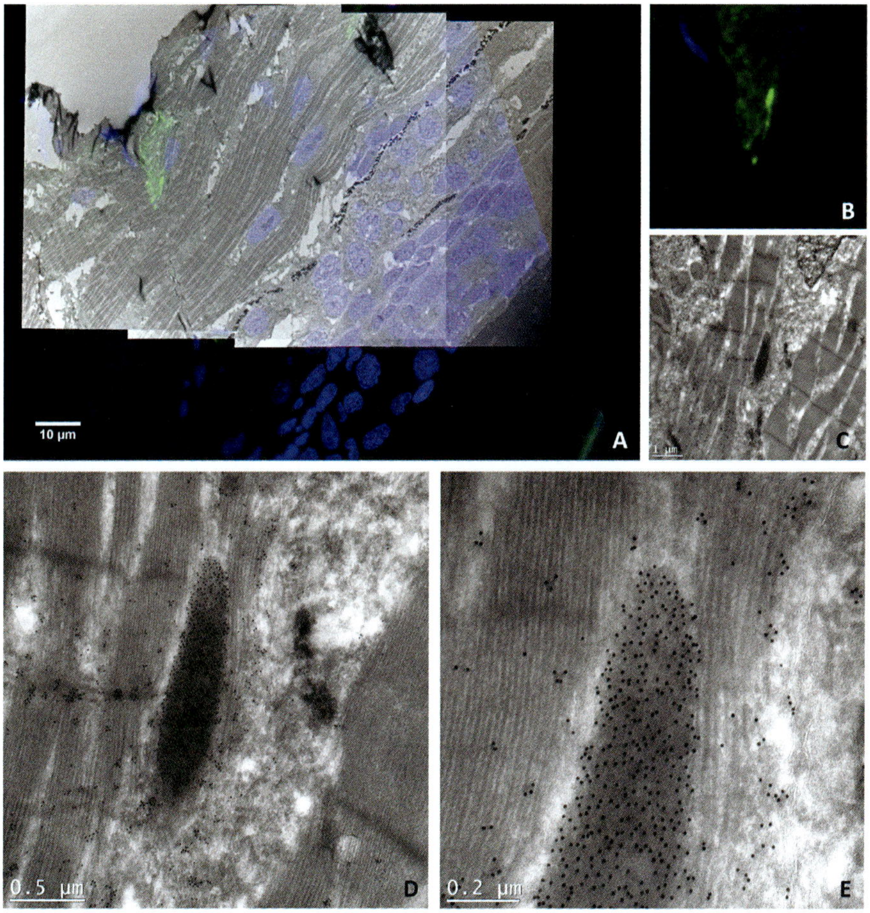

PLATE 34 (Fig. 11.4 on page 252 of this volume).

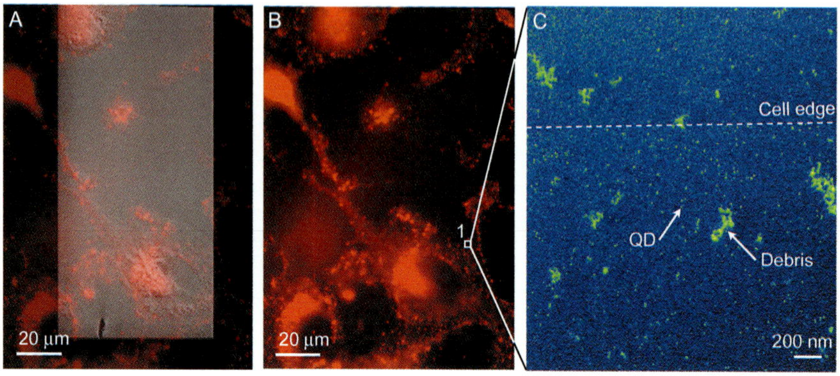

PLATE 35 (Fig. 14.6 on page 317 of this volume).

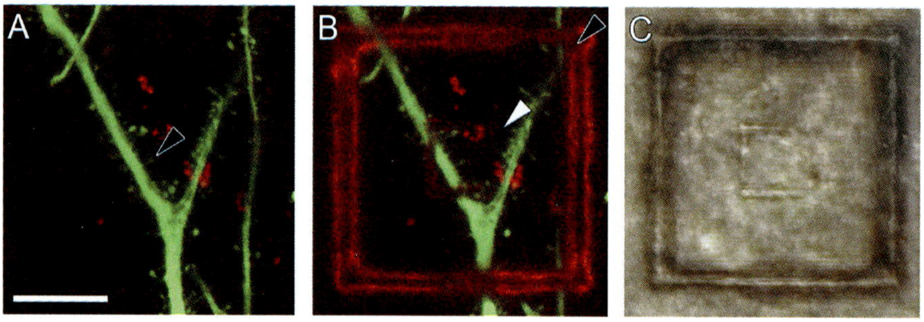

PLATE 36 (Fig. 15.2 on page 327 of this volume).

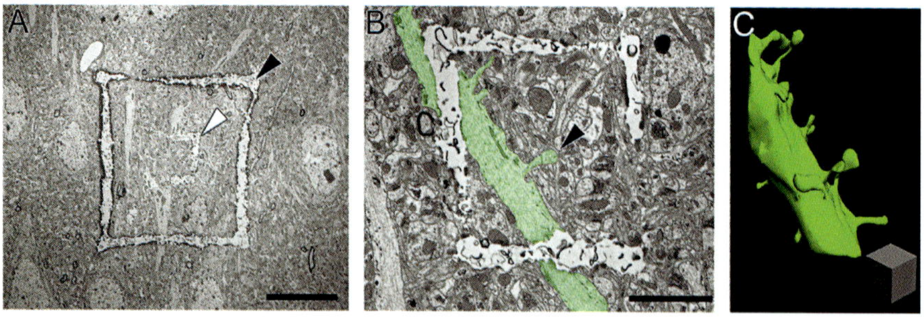

PLATE 37 (Fig. 15.4 on page 333 of this volume).

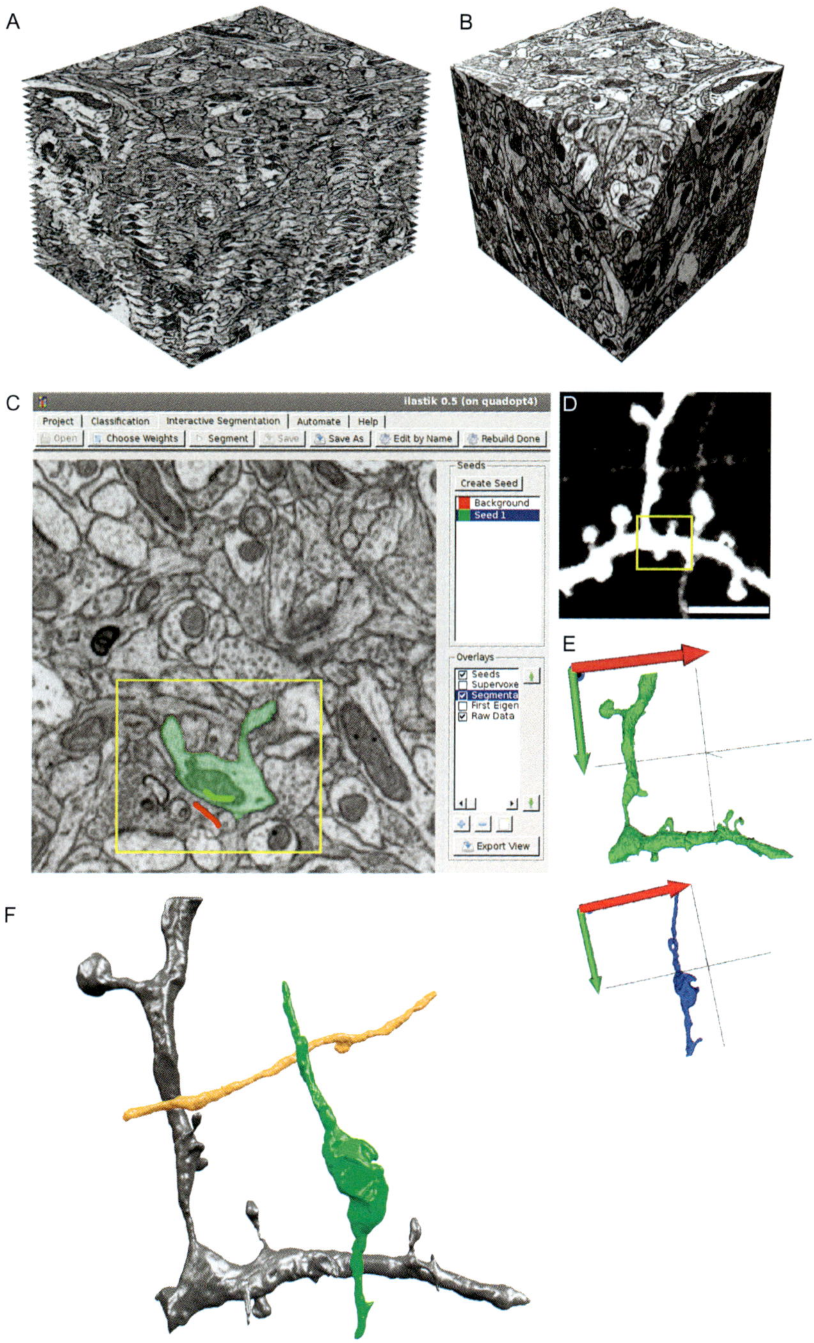

PLATE 38 (Fig. 16.5 on page 352 of this volume).

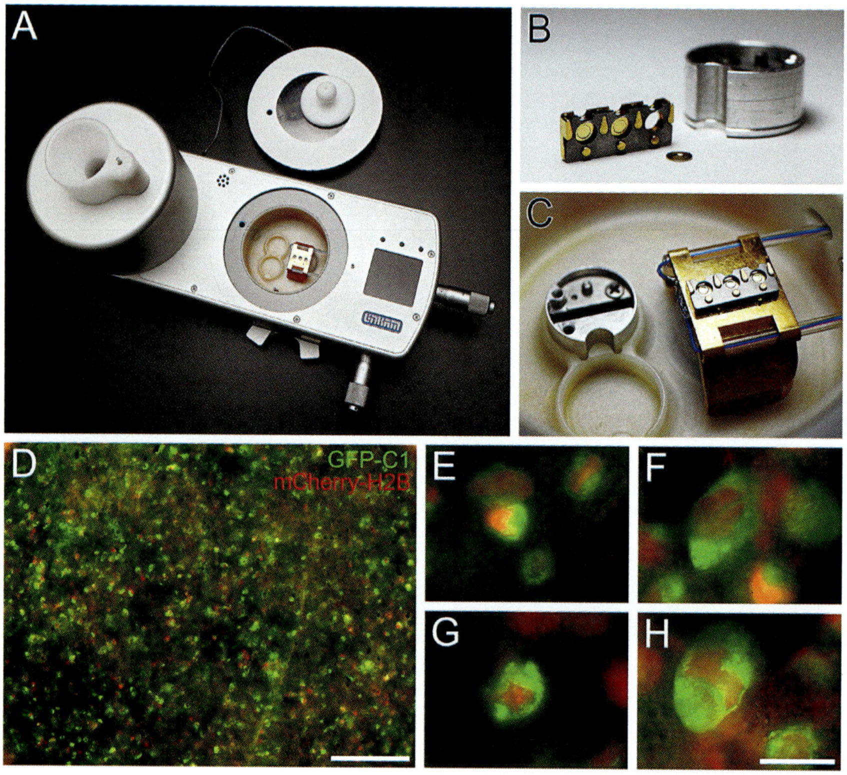

PLATE 39 (Fig. 17.2 on page 372 of this volume).

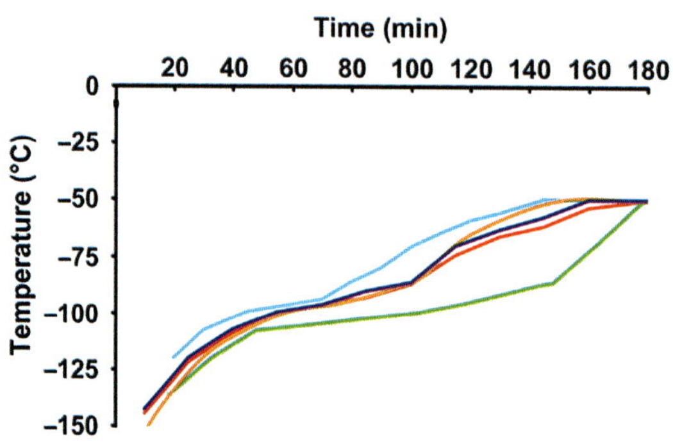

PLATE 40 (Fig. 17.3 on page 373 of this volume).

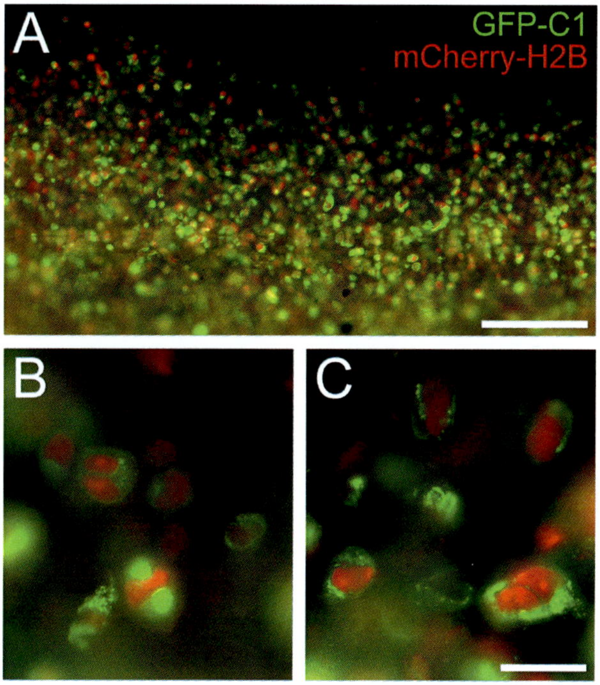

PLATE 41 (Fig. 17.4 on page 375 of this volume).

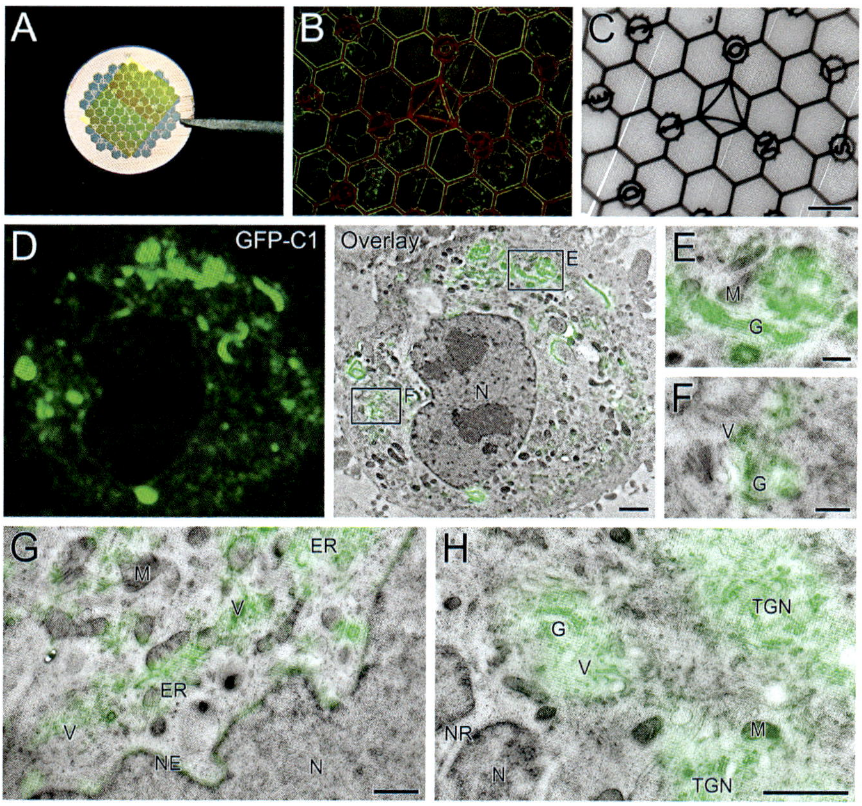

PLATE 42 (Fig. 17.6 on page 377 of this volume).

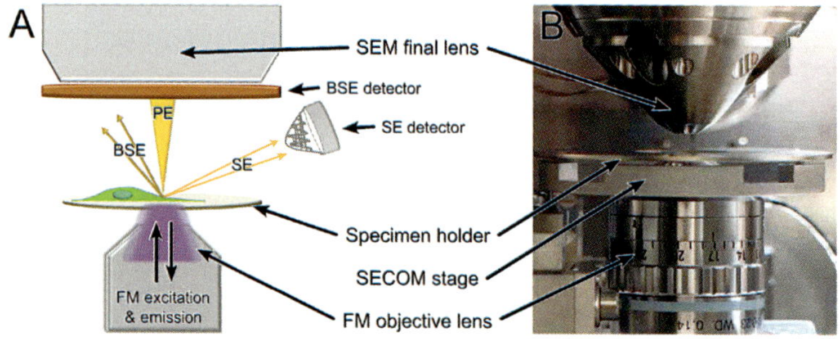

PLATE 43 (Fig. 17.7 on page 378 of this volume).

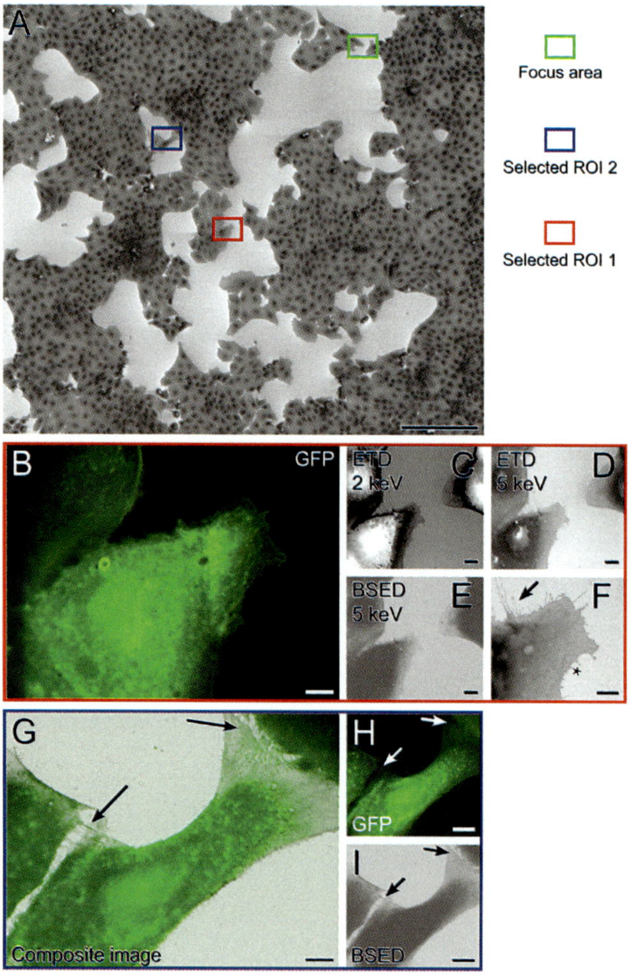

PLATE 44 (Fig. 17.9 on page 381 of this volume).

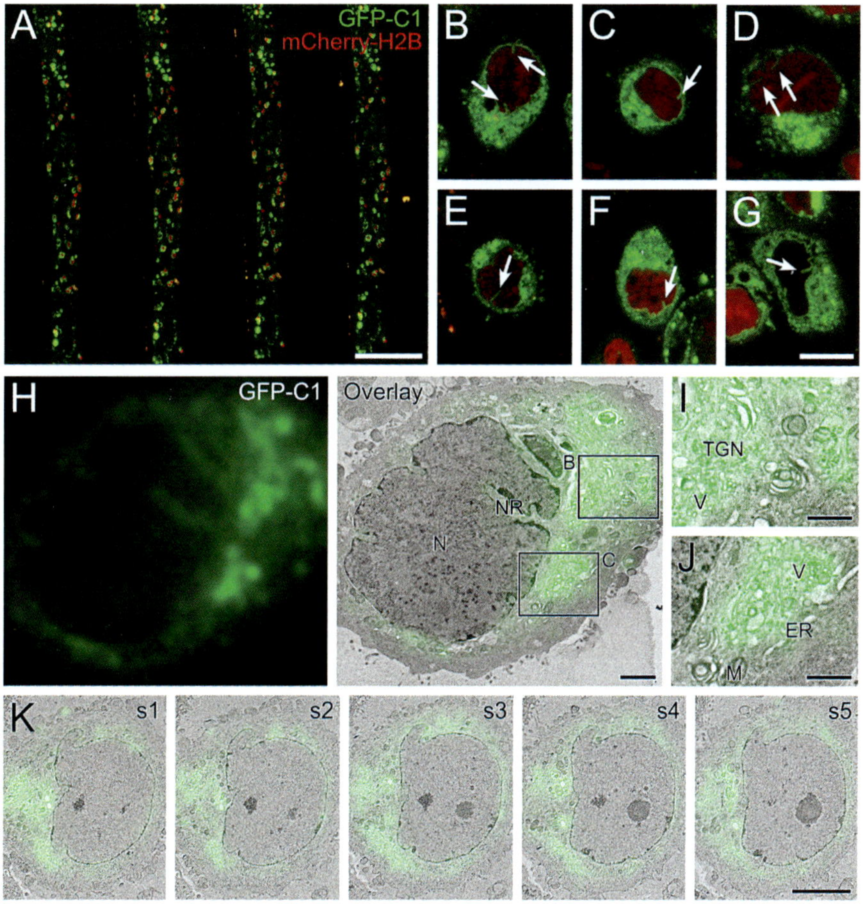

PLATE 45 (Fig. 17.10 on page 383 of this volume).

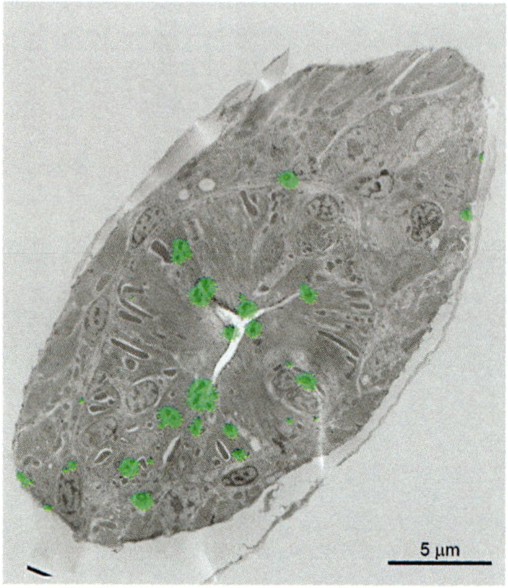

PLATE 46 (Fig. 18.1 on page 393 of this volume).

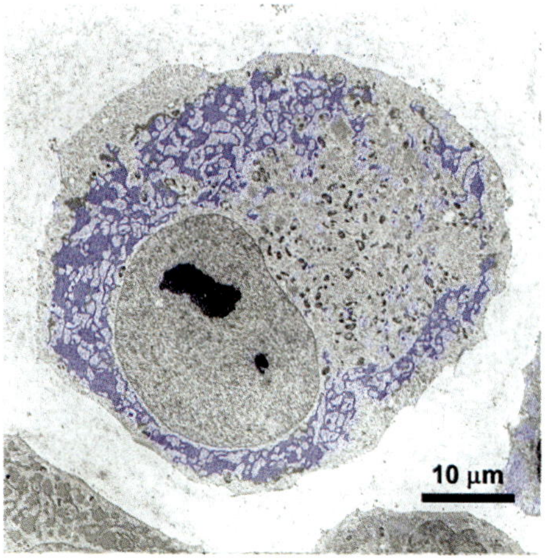

PLATE 47 (Fig. 18.2 on page 394 of this volume).

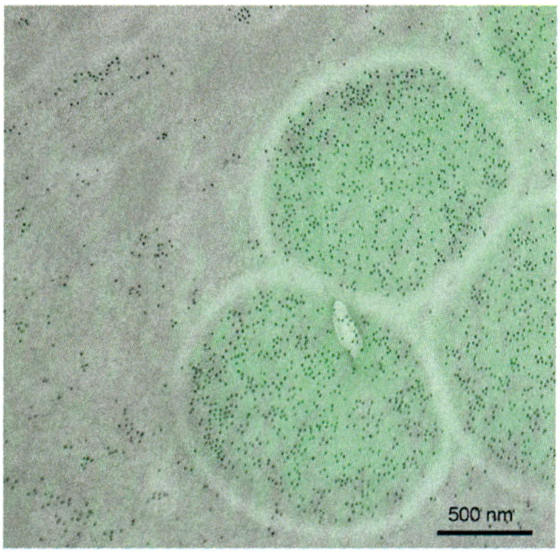

PLATE 48 (Fig. 18.6 on page 398 of this volume).

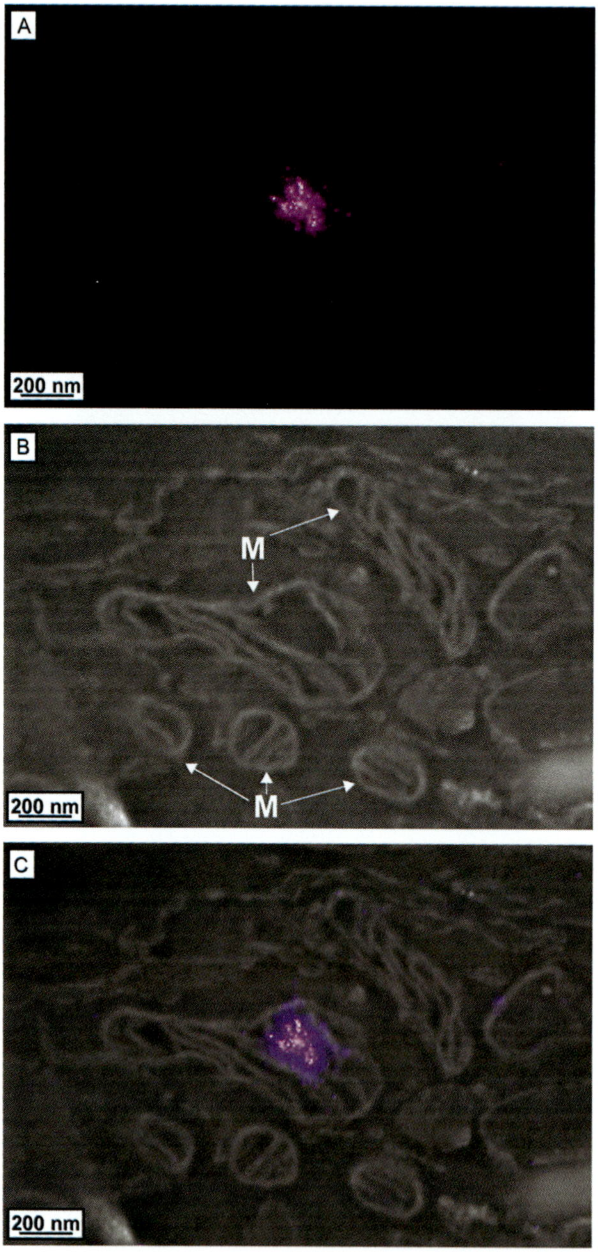

PLATE 49 (Fig. 18.7 on page 400 of this volume).

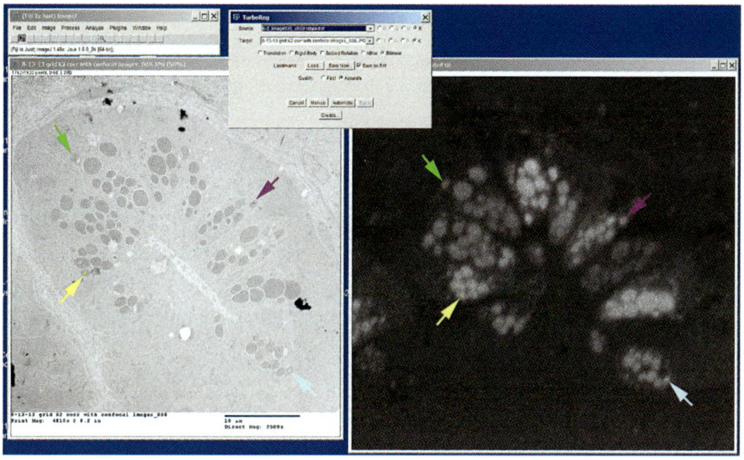

PLATE 50 (Fig. 18.11 on page 411 of this volume).

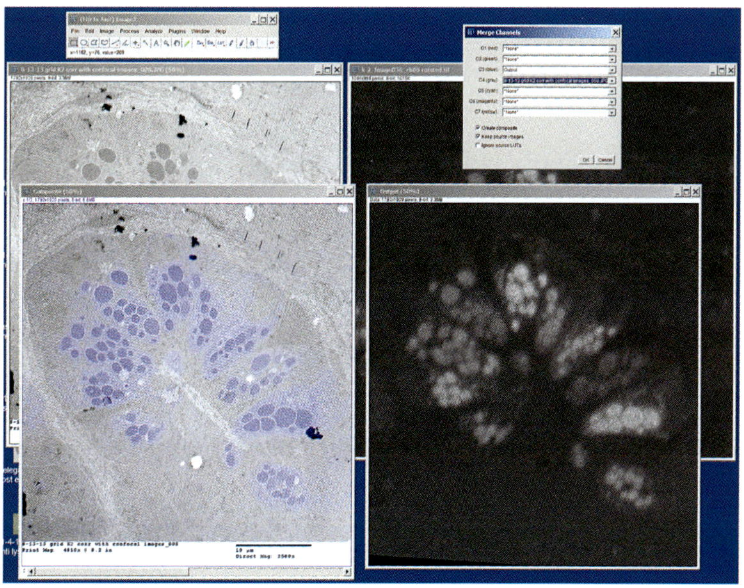

PLATE 51 (Fig. 18.12 on page 412 of this volume).

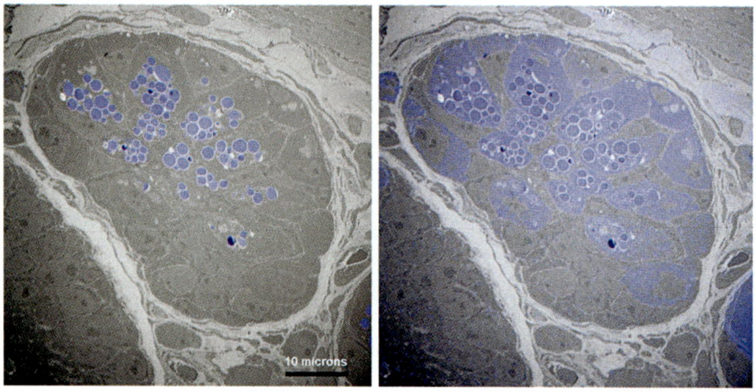

PLATE 52 (Fig. 18.13 on page 412 of this volume).

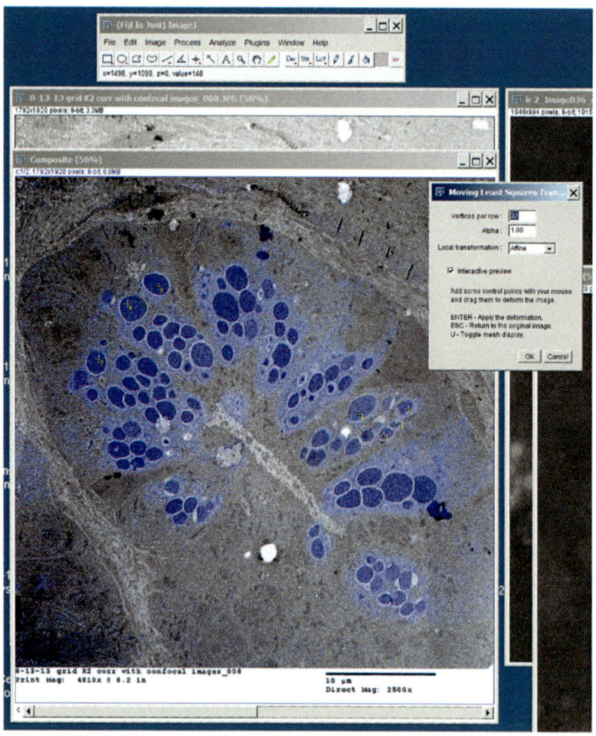

PLATE 53 (Fig. 18.14 on page 413 of this volume).

Edwards Brothers, Inc.
Thorofare, NJ USA
October 15, 2014